Molecular Scattering of Light

by
Immanuel L. Fabelinskii
P.N. Lebedev Physics Institute
Academy of Sciences of the USSR
Moscow, USSR

Translated from Russian by
Robert T. Beyer
Department of Physics
Brown University
Providence, Rhode Island

PLENUM PRESS • NEW YORK • 1968

Immanuil L. Fabelinskii, of the P. N. Lebedev Physics Institute of the Academy of Sciences of the USSR, was born in 1911. He completed the physics course at Moscow State University with a specialty in physical optics in 1936, and was named to the Physics Faculty, where he taught physics and carried on research in the Optics Laboratory until 1941. At various times, in addition to work at the university, he taught physics at the Moscow Energetics Institute, the Moscow Engineering Physics Institute, and the Moscow Physics-Technical Institute. For two years, he was involved in research at the Institute of Theoretical Geophysics of the Academy of Sciences of the USSR (1941-1943), and in 1942 was awarded the degree of Candidate of Physico-Mathematical Sciences. He has been at the Lebedev Physics Institute since 1943, where he directs a group of scientists engaged in various problems in physical optics, hypersound, and ultrasound. He attained the degree of Doctor of Physico-Mathematical Sciences in 1955. The present volume is his major work, but he has also published more than sixty technical papers throughout his career.

The original Russian text, published by Nauka Press, Moscow, in 1965, has been extensively revised and updated for the American edition.

Иммануил Лазаревич Фабелинский

Молекулярное рассеяние света

MOLEKULYARNOE RASSEYANIE SVETA

MOLECULAR SCATTERING OF LIGHT

Library of Congress Catalog Card Number 67-10534

ISBN 978-1-4684-1742-5 *ISBN 978-1-4684-1740-1 (eBook)*
DOI 10.1007/978-1-4684-1740-1

Softcover reprint of the hardcover 1st edition 1968
A Division of Plenum Publishing Corporation
227 West 17 Street, New York, N. Y. 10011

Molecular Scattering of Light

Translator's Preface

The development of the laser and the subsequent expansion of research in the field of molecular scattering of light tend to obscure the vast literature that accumulated on this subject during the last fifty years. The appearance of the Russian edition of Dr. Fabelinskii's book, just as this laser-induced explosion in light-scattering research took place, served to put the earlier work in its proper perspective. However, the book lacked any appreciable coverage of the laser work.

Fortunately, Dr. Fabelinskii has taken advantage of the time between the appearance of the Russian text and its translation into English to expand greatly the sections devoted to areas in which laser research has made such interesting and vital additions. At the same time, revisions and insertions have been made throughout the text, so that the English translation is virtually a second edition of this useful work.

The translator wishes to express his thanks here to Dr. Fabelinskii for making the revisions, corrections, and additions available for the English language work. He is also grateful to his graduate student, Mark B. Moffett, who prepared the index and who, during the course of its preparation, made a large number of critical comments and corrections that have enhanced the quality of the final product.

October 11, 1967 Robert T. Beyer

Preface to the American Edition

The news of a forthcoming English translation not only made me very happy, but also very anxious. This book was written before lasers became available as sources for the excitation of scattered light. The application of lasers to the investigation of the molecular scattering of light opened up entirely new possibilities in this region of physical optics, and some of these possibilities were quickly realized.

In the introduction to the Russian edition of this book, it was pointed out that I had managed to add in press the few new accomplishments then connected with laser use. Since that time, the flow of new researches has continued to increase and the region of investigation has grown ever broader. It appeared quite impossible for me to agree to this American edition of the book without including in it these important new researches.

I am very grateful to the publishers, who agreed with me and patiently waited for my additions. The amount of new publications and preprints was so great that it was not always possible to treat current research in detail; not to mention this work at all, however, would have been even worse. As a result of the additions made, more than 60 new references appear in the bibliography of the American edition, but even here it was quite impossible to include everything. Only the thought that the fewer the new publications included in the book, the more slowly the book will become out of date, consoled me.

The new results included in this book, it appears to me, will become permanent achievements in physical optics and acoustics.

For this edition of the book, a number of misprints and other errors have been corrected, and many additions have been made in the text. Section 34 has been almost completely reworked and significantly expanded, and a new Sec. 35 has been written. These two sections comprise the new Chapter X, which is devoted to the stimulated molecular scattering of light.

Immanuil L. Fabelinskii

Preface

The study of the molecular scattering of light, covering the broad sweep of molecular optics, has, in the last two decades, grown several fold in the volume of its investigations, and has attained significant new achievements.

The experimental and theoretical investigations of the molecular scattering of light have established a strong bond between this field and molecular acoustics, as well as with the different divisions of molecular physics, and the physics of the amorphous and crystalline states of matter.

The establishment of a close bond between light scattering and elastic thermal waves in the medium has led to the creation of a relaxation theory of sound propagation in liquids and to the discovery of a significant dispersion of the sound velocity in low-viscosity and viscous liquids, and to a number of other important results.

The study of molecular scattering in vapors and gases played, and continues to play, a significant role in the interpretation of molecular structure, and, together with the study of the Kerr effect and molecular refraction, gives quantitative data for finding the principal polarizabilities of molecules.

Molecular scattering of light is one of the most effective methods of investigating the kinetics of various fluctuations and intermolecular interactions and has been repeatedly used with success for these purposes.

The application of the methods of molecular light scattering to the study of solutions of polymers, proteins, and electrolytes has also made it possible to give valuable information in this area on the molecular weight of macromolecules, their dimensions, and shape.

A great deal of data have been accumulated over the years from investigations of molecular light scattering. These data are necessary for the development of modern ideas on the different aggregate states of matter.

In this developing region of molecular optics, while a number of fundamental problems have been solved, many serious problems have been formulated but still await solution. Extensive reviews devoted to molecular scattering of light exist in the book by Cabannes (1929), part of the book by Bhagavantam (1942), and in several papers by Stuart (1938-1952). None of these works were available in Russian.

Back in 1956, G. S. Landsberg and the author thought of writing a book on the molecular scattering of light. At this time, a significant amount of material had been collected in our laboratory concerning the spectral investigation of the Rayleigh scattering of light, and this material required detailed discussion. But we had only begun the discussion of the plan of the proposed book when Landsberg's illness and death in February 1957, prevented our writing the book together. It occurred to me to write a monograph, relying on the help of my colleagues. In this monograph, accounts are given of investigations carried out principally in the optical laboratories directed by G. S. Landsberg, and especially those in which the author had taken part. In addition, the author has attempted to present a picture of the modern state of the problem of molecular scattering of light in general, and to set forth the basic factual data. With this aim, sections devoted to the exposition of the theory of this problem, with some results of the investigation of molecular scattering of light in gases and vapors and in solutions of polymers and proteins, and with some other questions are included. Extensive tables and numerous graphs are given in the basic part of the book and in Appendix III.

This book does not pretend to be a complete exposition even of those problems which are discussed in it. In a number of cases in which the problem is considered only superficially, references are given to the literature in which the problem is discussed in detail. The bibliography given in the book is not exhaustive.

The first two chapters of the monograph contain the general theory of light scattering in isotropic media and crystals. The conclusions of the general theory are later combined with a series of calculations in order to discuss various problems in the molecular scattering of light.

After the manuscript of the book had been prepared for publication, researches appeared which made use of gas and ruby lasers for the study of scattered light. The author is grateful to the publisher for allowing him to include the fundamental results of these new researches in the already completed text of the book, and to supplement it in a short section on the stimulated Mandel'shtam–Brillouin (MB) scattering.

The author expresses his thanks to G. P. Motulevich, who wrote Sec. 9 and part of Sec. 31, and is grateful for the useful remarks of T. S. Velichkina and I. A. Yakovlev, who read the entire book in manuscript, to V. L. Ginzburg, S. M. Rytov, and V. S. Starunov, who read the manuscript of the theoretical part of the book, and to S. A. Akhmanov and R. V. Khokhlov, who read Sec. 34, as well as to E. V. Tiganov for help in reading the corrections. The author thanks T. K. Breus and N. A. Raiskui for numerous useful remarks and V. P. Zaitsev and G. A. Ivanov for help in shaping the book.

I. L. Fabelinskii

Contents

Notation

A, in Sec. 4, the refraction; in Secs. 6, 7, the coefficient of proportionality between the Maxwell constant and the relaxation time of the anisotropy; in Appendix II, the resolving power

α, the total amplitude coefficient of sound absorption

α_η, the amplitude coefficient of sound absorption due to shear viscosity

$\alpha_{\eta'}$, the amplitude coefficient of sound absorption due to bulk viscosity

B, magnetic induction; in Secs. 4 and 30, the Kerr constant ($K = B\lambda / n$)

B', the virial coefficient

b, the mean polarizability of the molecule; in Sec. 11, the brightness

b_1, b_2, and b_3, principal polarizabilities of a molecule

$\beta_S = -[(1/V)(\partial V/\partial p)]_S$, the adiabatic compressibility

$\beta_T = -[(1/V)(\partial V/\partial p)]_T$, the isothermal compressibility

c, concentration; the velocity of light

Δc, fluctuation of the concentration

c_{ik}, elastic constants

c_p, specific heat at constant pressure

c_V, specific heat at constant volume

c_i, specific heat of the vibrational degrees of freedom

D, diffusion coefficient

$\mathbf{D} = \varepsilon \mathbf{E} = \mathbf{E} + 4\pi \mathbf{P}$, electric displacement of the sound wave

$\mathbf{D}'$, electric displacement of the scattered sound wave

D_l, linear dispersion

---

*In the list of notation, we have not included abbreviations, the designation of geometric quantities, and some rarely encountered physical quantities.

d, lattice constant in Sec. 5

Δ, depolarization factor of scattered light

Δ_v, depolarization factor of scattered light in excitation by light polarized in the vertical plane

Δ_h, depolarization factor of scattered light in excitation by light polarized in the horizontal plane

Δ_u, depolarization factor of scattered light in excitation by natural light

Δ_0, depolarization factor

$\Delta\lambda$, path difference of the rays in a Kerr cell

δ, temporal coefficient of the attenuation

δ_λ, optical path length

$\mathbf{E}$, electric field intensity of the light wave

$\mathbf{E}_0$, amplitude of the electric field intensity of the light wave

$\mathbf{E}'$, field intensity of the scattered light wave

$\mathbf{E}'_1$, field intensity of the Stokes component of the scattered light wave in Sec. 34

$\mathbf{E}'_2$, field intensity of the anti-Stokes component of the scattered light wave in Sec. 34.

$\mathscr{E}$, intensity of the variable radiofrequency field

$\mathscr{E}_0$, amplitude of the intensity of the variable radiofrequency field

e, in Sec. 2, the charge on the electron; in Sec. 1, the illumination of the photoplate

e_{ik}, deformation tensor

η, coefficient of shear viscosity; in Sec. 2, expansion parameter of the thermodynamic potential

ε, optical dielectric constant

ε_0, optical dielectric constant of a continuous homogeneous medium

ε^0, static dielectric constant

$\Delta\varepsilon$, fluctuation of the optical dielectric constant

Φ, light flux

f, frequency of elastic vibrations

f^*, constant characterizing the intermolecular interaction

f_a, atomic scattering factor

f_c, critical frequency

f_h, frequency of hypersound

g, acceleration due to gravity

g_i, multiplicity of frequency degeneracy

$\gamma = c_p/c_V$, ratio of specific heats

γ_0, ratio of specific heats under static conditions

γ_∞, ratio of specific heats for infinitely high frequency

γ^*, coefficient of surface tension
H, magnetic field intensity
h, extinction coefficient,
$\hbar$, Planck's constant divided by 2π
I, intensity of scattered light
$\mathfrak{I}$, total intensity of the light from molecular scattering
I_0, intensity of the incident light
I_{90}, intensity of the light scattered at an angle of 90° to the direction of the incident light
I_{ad}, intensity of the light scattered by adiabatic density fluctuations
I_{is}, intensity of the light scattered by isothermal density fluctuations
I_c, intensity of the central component of the fine structure
I_{MB}, intensity of the Mandel'shtam–Brillouin (MB) component
I, moment of inertia
i, intensity of the light scattered by anisotropy fluctuations
K, Kerr constant, bulk modulus in Sec. 7
K_1, part of the Kerr constant which does not depend on the constant dipole moment
K_2, part of the Kerr constant which depends on the constant dipole moment
K_0, K_∞, limiting values of the bulk modulus
k, Boltzmann's constant
k, wave vector of the incident (primary) light
k', wave vector of the scattered light
$\mathbf{k}'_1$, $\mathbf{k}'_2$, wave vectors corresponding to the Stokes and anti-Stokes MB components
$\varkappa$, coefficient of thermal conductivity
L, distance between sçattering volume and observation point
$L(\omega) = 1/I$, the reciprocal of the intensity of the scattered light
l, mean free path length or mean distance bẹtween particles
Λ, wavelength of an elastic vibration
λ, wavelength of light in a vacuum
M, molecular weight; in Secs. 6 and 7, the Maxwell constant
M_w, M_n, mean by weight and mean by number molecular weights, respectively
m, mass of the electron; in the Appendix, order of interference
MB, Mandel'shtam–Brillouin
μ, constant electric moment of a molecule; shear modulus; in Sec. 34, small parameter of Van der Pol
μ', magnetic permeability of a substance

N, number of particles in the volume of fluctuations
ΔN, fluctuations of the number of particles
N_L, Loschmidt's number
N_A, Avogadro's number
N_{eff}, effective number of interfering rays
$n \equiv \sqrt{\varepsilon}$, index of refraction of a medium
n_p, index of refraction of a medium for light polarized parallel to the direction of the external electric field (Kerr effect)
n_s, index of refraction of a medium for light polarized perpendicular to the direction of the external electric field
$\nu = \omega/(2\pi c) = 1/\lambda$, spectroscopic frequency of a light wave, expressed in cm^{-1}
$\Delta\nu$, shift in the MB components relative to the frequency in the exciting line, cm^{-1}
$\delta\nu_{MB}$, true halfwidth of the MB component, cm^{-1}
$\delta\nu_c$, true halfwidth of the central component of the fine structure, cm^{-1}
$\delta\nu_{conc}$, true halfwidth of the central line of the fine structure, produced by scattering from concentration fluctuations
$\delta\nu_a$, halfwidth of the contour of the apparatus function
$\delta\nu_B$, halfwidth of the actual contour of the exciting line
$\delta\nu_D$, halfwidth of the line produced by the Doppler effect
$\delta\nu_w$, halfwidth of the visible spectral contour
$\delta\nu_A$, halfwidth of the visible contour of the exciting line
Ω, angular frequency of an elastic wave
Ω_L, angular frequency of a longitudinal hypersonic wave producing the MB components
Ω_T, angular frequency of the transverse hypersonic wave producing the MB components
ω, angular frequency of a light wave
ω_M, angular frequency of modulation
$\mathbf{\Pi}$, Hertzian vector
$\mathbf{P}$, polarization per unit volume
$\Delta\mathbf{P}$, fluctuations of the polarization
p, pressure; in Sec. 10, the coefficient of polarization of the light
Δp, fluctuation of the pressure
p_{ik}, elasto-optical constants
$\Delta\Psi$, change in the free energy, enthalpy, thermodynamic or chemical potential
Q, cross section of light scattering
$\mathbf{q}$, wave vector of an elastic wave
R, scattering coefficient

R_{90}, scattering coefficient at an angle of 90° to the direction of the incident light
R_g, gas constant
R_L, in Sec. 19, the coefficient of scattering in liquid helium
ρ, density of a substance
$\Delta\rho$, fluctuation of the density
$\Delta\rho_q$, spectral component of the density fluctuation
ρ_u, depolarization factor of light scattered by anisotropy fluctuations
$\rho(\omega)$, depolarization factor in the spectrum of scattered light
$\mathbf{S} = (c/4\pi)[\mathbf{E} \times \mathbf{H}]$, Poynting vector of the flux of electromagnetic energy
$\mathbf{S}_g$, vector of the flux of elastic energy in Sec. 34
S, entropy
ΔS, fluctuation of the entropy
S_{ik}, stress tensor
$e_{\alpha\alpha} = \sigma$, coefficient of volume expansion
T, absolute temperature
T_c, absolute temperature at the critical point
ΔT, fluctuation of the temperature
t, time; temperature in degrees Celsius
τ, relaxation time of the anisotropy
τ_D, relaxation time of the dipole moment
τ_V, relaxation time of the bulk viscosity
τ_ζ, in Sec. 8, relaxation time of internal friction
τ_S, relaxation time of the shear viscosity
τ_2 and τ_1, in Secs. 7 and 19, relaxation times of the anisotropy, associated with the narrow and broad parts of the wing
U, energy of elastic waves in Sec. 33; potential energy of molecules in an electric field
V, scattering volume
V^*, volume of fluctuations
v, velocity of propagation of elastic waves
v_0, velocity of sound at zero frequency
v_∞, velocity of sound at infinitely high frequency
ξ, parameter characterizing the internal state of the medium
ξ_{ik}, anisotropy tensor
$z = I_\theta / I_{\pi-\theta}$, asymmetry of scattering
ζ, coefficient of internal friction
$\zeta(x,y)$, fluctuating roughness of the interface between two media
$\chi = \varkappa / c_p \rho$, coefficient of temperature conductivity

Introduction

The molecular scattering of light or the scattering of light generally arises only as the result of optical inhomogeneities in the scattering medium or on the scattering surface of the substance.

The physical reasons for the generation of optical inhomogeneities are manifold and various. An optically inhomogeneous medium can be the result of gross inclusions of one substance in another, for example, as the result of foreign inclusions in crystals, droplets of water or solid particles in the atmosphere, in colloidal suspensions, etc. On the other hand, the scattering of light can be observed even in substances which are completely free from any sort of foreign contaminants, as the result of the statistical character of the thermal motion of the particles of the material.

In this case, the reason for the optical inhomogeneities has been established as the fluctuation of the optical dielectric constant, brought about in turn by the fluctuations of the density of the substance and fluctuations in the orientation of the molecules. In homogeneous solutions, an additional and very important reason for molecular light scattering is the fluctuation of the concentration of the components of the solution.

The scattering of light brought about by such different random fluctuations has the name of molecular scattering, in contrast with the scattering from colloidal particles and other impurities and from combination (Raman) light scattering.

In this book, we shall consider only the molecular scattering of light; a brief review of the development of the representation of molecular light scattering will be given in the Introduction; fundamental results obtained in this region will also be pointed out.

Each physical phenomenon contributing to the generation of scattered light makes its characteristic "imprint" on the intensity, polarization, and spectral composition of the scattered light.

The characteristic marks of the different physical phenomena in scattered light enable us to study these phenomena and the various problems connected with them. With the help of light scattering, the structure of molecules is studied, the molecular weights of solutions and proteins are determined, the dimensions of ions are found, Avogadro's number is determined, the kinetics of fluctuations and the propagation of hypersound in condensed media are investigated, as well as phase transitions and many other questions.

The first success in the investigation of scattered light occurred in 1869 when Tyndall [1] set up a laboratory experiment in which he observed light scattered by very fine particles.

A small quantity of particles, whose size was small in comparison with the wavelength of visible light, was put in a glass tube of about 1 m in length and 7 cm in diameter. An intense beam of white light was directed along the axis of the tube. The scattered light was observed at different angles to the direction of the incident light. As a result of this experiment, Tyndall found that the scattered light revealed a bluish hue and that complete or nearly complete linear polarization of the scattered light was observed at an angle of 90° to the direction of the incident light.

On the basis of these observations, Tyndall came to the idea that the blue color and the polarization of the light in the sky were determined by the scattering of sunlight by fine particles of dust which are always present in the earth's atmosphere in sufficient amounts.

The conclusions and experimental data of Tyndall clearly served as the starting point for the first theoretical researches of Rayleigh on light scattering.

At first, Rayleigh considered the problem of the passage of light through a continuous medium in which particles of spherical shape are randomly distributed on the basis of the elastic theory of light [2], and then from the viewpoint of electromagnetic theory [3,4]. These particles were assumed to be small in comparison with the wavelength of light, and to have a dielectric constant ε differing from the dielectric constant ε_0 of the continuous medium.

If the radii of the spherical particles are much smaller than the light wavelength λ, then the light field inside the particle can be considered to be homogeneous, and the induced electric moment of the particle, in accord with the conclusions of electrostatics, is equal to

$$P = \frac{\varepsilon - \varepsilon_0}{4\pi} \frac{3\varepsilon_0}{\varepsilon + \varepsilon_0} V_1 E \sin\vartheta. \tag{1}$$

Here V_1 is the volume of the particle and ϑ is the angle between the dipole and the direction of **E**, the intensity of the electric field of the light wave. The field of the scattered light wave in the wave zone can be found from the relation

$$E' = \frac{\ddot{P}\sin\vartheta}{Lc^2}, \tag{2}$$

where L is the distance from the radiating dipole to the point of observation.

Here and in the remainder of this book, the absolute value of the flux vector of electromagnetic energy will be called the intensity of the scattered light:

$$I = |S| = \frac{cn}{4\pi}|E'|^2. \tag{3}$$

It has already been taken into account that $\sqrt{\varepsilon}\,|\mathbf{E}| = \sqrt{\mu'}\,|\mathbf{H}|$ and the magnetic permeability $\mu' = 1$, so that the index of refraction $n \equiv \sqrt{\varepsilon}$.

It follows from Eq. (3) that the intensity has the dimensions erg-cm^{-2}-sec^{-1}, or watt-cm^{-2}, or watt-sterad^{-1}.

It is assumed that the scattering particles are located at great distances from one another, and that each of them radiates light independent of whether there is another particle present or not. In this case, the intensity (and not the amplitude) of the light scattered by each particle is summed. Taking into account (1), (2), and (3), we get for the illumination of the scattering volume by natural light:

$$I = I_0 \frac{9\pi^2\varepsilon_0^2 N_L V_1^2}{L^2\lambda^4}\left(\frac{\varepsilon - \varepsilon_0}{\varepsilon + \varepsilon_0}\right)^2 (1 + \cos^2\theta). \tag{4}$$

Here N_L is Loschmidt's number and θ is the scattering angle. This is the same conclusion which Rayleigh reached by a somewhat more complicated method [2, 3].

The value of $(\varepsilon - \varepsilon_0)^2$ here serves as a measure of the optical inhomogeneity. When the optical dielectric constants of the particles and the surrounding medium are the same, $\varepsilon - \varepsilon_0 = 0$, the optical inhomogeneity disappears and there is no light scattering.

It follows from (4) that the intensity of the scattered light is proportional to the square of the volume or to the sixth power of the radius of the particle; consequently, it depends strongly on the dimensions of the scattering particles.

An important conclusion which follows from Eq. (4) is the proportionality of the intensity of the light scattered by small particles to λ^{-4}. Such a dependence of the scattered light intensity on the wavelength is known as Rayleigh's law. It follows from Rayleigh's law that sunlight, scattered by the upper layers of the atmosphere (the light of the sky) must contain a relatively larger shortwave radiation than the direct sunlight. The blue color of the sky is explained in this natural way. But these representations lead to the conclusion that the higher the altitude above the surface of the earth, the fewer particles will be encountered by which light scattering takes place and, consequently, one must expect a decrease in the saturation of the blue light of the sky with increase in altitude above sea level.

However, observations have shown that just the opposite occurs. To be specific, the purer the air (mountain observatories) the fewer impurity particles are in it and the greater is the saturation of the blue and the more is the light of the sky polarized.

From this experimental fact, Rayleigh drew the following conclusion: it is possible that light is scattered from the gas molecules which comprise the air, and not from dust particles.

However, before it was possible to construct a statistical theory of the molecular scattering of light, it was necessary to overcome a serious difficulty. The difficulty was that in a medium free from any sort of impurities and consisting only of gas molecules, the mean free path in which is less than the wavelength of light, one must find a physical reason for the appearance of optical inhomogeneities.

Let us consider a homogeneous medium, i.e., such a medium for which identical numbers of molecules of a single type are found

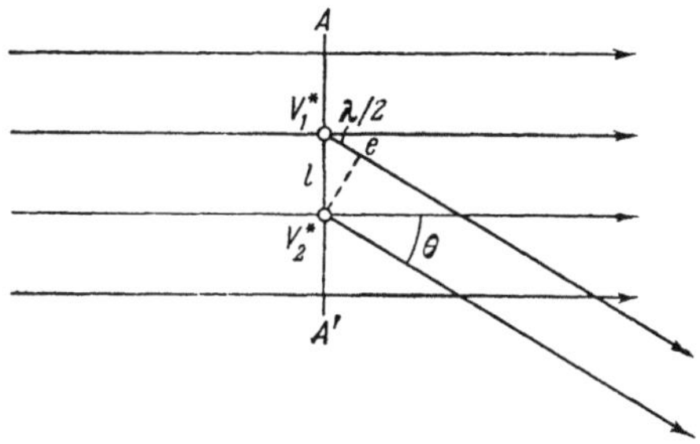

Fig. 1. Light scattering in homogeneous and inhomogeneous media.

in identical parts of the total volume. We first assume that all the molecules are fixed. A light wave, incident on such a medium, produces secondary radiation of the molecules in all possible directions.

For the calculation of the intensity at some point located outside of the beam of incident light, it is necessary to sum the amplitudes and take the phases of all waves entering this point into account.

We sum the amplitudes of the waves and not their intensities because the molecules of the medium (the secondary radiators) oscillate relative to one another with fixed phase differences. It is easy to see that as the result of such a summation, we obtain the result that the intensity of the secondary radiation in all directions except the direction of propagation of the exciting light will be equal to zero, and, consequently, the scattering of light does not take place in a homogeneous medium [5, 6].

Actually, let a plane wave be drawn in Fig. 1 with its front AA' propagating in a homogeneous medium. From the volume V_1^*, the linear dimensions of which are small in comparison with λ, a secondary wave with definite amplitude and phase is radiated in a direction making an angle θ with the direction of propagation of the incident light.

If the medium is really homogeneous, then one can always find such a volume V_2^* which radiates a wave in the same direction, with the same amplitude, but with opposing phase.

It is clear that such waves completely cancel one another and scattered light will be absent in the direction θ. It is not difficult to see that such a complete cancellation will take place for an arbitrary location of the front AA' in a homogeneous medium, and for any direction θ except $\theta = 0$, i.e., the direction of propagation of the exciting light.

We denote the distance from V_1^* to V_2^* by l, and the difference in path length by $\lambda / 2$ (waves in phase opposition). It is clear from Fig. 1 that

$$l = \frac{\lambda}{2 \sin \theta} \tag{5}$$

and, consequently, for any θ except $\theta = 0$, one can find V_1^* and V_2^* on AA' which radiate waves in opposite phases and which are located a finite distance l apart. In the latter case, it is impossible to find two points on AA' at a finite distance from one another that radiate waves of opposing phase. Therefore, in a homogeneous medium, light will be propagated only in the direction $\theta = 0$.

The effect of some change in the direction can be due to diffraction from the edges of the scattering body or the corresponding aperture, but since the dimension D of the cross section of the body or aperture is usually very large in comparison with λ, the angle of diffraction $\varphi \sim \lambda / D$ is a vanishingly small quantity, having no significance in the present case.

Consequently, it is sufficient that the medium be homogeneous and the secondary waves which are radiated from the molecules be coherent in order that the scattered wave be absent. Rayleigh, in explaining the blue sky by the molecular scattering of light in the air, understood this. But if the molecules execute disordered thermal motions, and if the mean free path length is much greater than the optical wavelength, then, as Rayleigh showed [2, 4], the thermal motion of the molecules changes the phase of the secondary radiation of the molecules in a random way and, consequently, this radiation becomes noncoherent. In this case, one must sum the intensities of the scattered waves rather than their amplitudes, just as has already been done for the calculation of the intensity of scattering by small particles.

The incident light of frequency ω with amplitude of the electric field intensity $\mathbf{E}_0$ creates an induced moment in each molecule of the scattering volume:

$$P = bE_0 \cos \omega t.$$

The intensity of the scattered light is easily computed by again using Eqs. (2) and (3).

By assuming the mean polarizability of the molecule

$$b = \frac{\varepsilon - 1}{4\pi N_L}$$

and taking into account the fact that the resulting intensity created by all the noncoherent radiators is equal to the radiation of a single molecule amplified VN_L times, we get the well-known Rayleigh formula

$$I = I_0 \frac{\pi^2 V}{2\lambda^4 L^2} \frac{(\varepsilon - 1)^2}{N_L}. \tag{6}$$

For a gas, ε is close to unity; therefore, without serious error, we can set $(\varepsilon - 1)^2 \approx 4(n - 1)^2$, and, instead of (6), we have

$$I = I_0 \frac{2\pi^2 V}{\lambda^4 L^2} \frac{(n - 1)^2}{N_L}. \tag{7}$$

Equations (6) and (7), as well as (5), indicate that the intensity of light scattering is the greater the shorter the wavelength of the incident light. The Rayleigh law ($I \sim 1/\lambda^4$), as has been said, gives a natural explanation for the blue of the sky and the reddish color of the sun at sunrise and sunset, but now it is from the viewpoint of molecular scattering. The extension of the Rayleigh theory to the scattering of light in dense media has been subjected to criticism [5].

From Eqs. (6) and (7), it is easy to calculate the extinction coefficient h in the law $I = I_0 e^{-hx}$, where x is the path traversed by the light in the scattering medium (see Sec. 1). Calculation gives

$$h = \frac{32\pi^3}{3} \frac{(n - 1)^2}{N_L}. \tag{8}$$

Having available sufficient reliable data for N_L, Abbo (1906) compared his experimental data on the transparency of a clean atmosphere, obtained by him on Mount Wilson (altitude 1780 m above sea level) in California with the results of calculations from Eq. (8). The comparison showed that the experimental data in this case can be explained only by molecular scattering of the light in the clean atmosphere [7].

Such measurements were repeated much later by Fowler (Mount Wilson) and by Dember [9] (Teneriffe peak) and excellent agreement was again found between theory and experiment.

A very satisfactory proof for the existence of molecular light scattering in air and in other gases has been obtained in laboratory experiments by Cabannes [10] and Strutt [11] (the younger Rayleigh), while Wood [12] compared the brightness of the scattering in a laboratory sample of dust-free air with the brightness of the sky, and confirmed that the blue of the sky can be explained entirely by the molecular scattering of light in the air.

Further careful experiments only increased the certainty that the Rayleigh formulas (6)-(8) correctly reproduce the phenomenon and are in need only of some refinement (see Sec. 4).

An understanding of the physical nature of the light scattering in condensed media came later. In the study of liquids and gases at high pressures and temperatures, it was discovered long ago that a completely transparent medium becomes milky near the critical point (critical opalescence). Critical opalescence is not described by thermodynamic equilibrium processes and therefore it remained unexplained for a long time.

Smoluchowski [13] explained the phenomenon of critical opalescence by fluctuations in the density of the substance, fluctuations which became large at the critical point.

The fluctuations of the density $\Delta\rho$ (the departure of the density in a small volume* from its average value) depend on the isothermal compressibility so that, as calculation shows [see Appendix I, Eq. (I.13)],

$$\overline{\Delta\rho^2} = \frac{\rho^2}{V} kT \left(-\frac{1}{V}\frac{\partial V}{\partial p}\right)_T .$$

* The linear dimensions of the volume are small in comparison with the wavelength of the light, but it contains a sufficient number of molecules that it can be subjected to statistical analysis.

At the critical point, $\partial V/\partial p$ increases strongly and $\overline{\Delta\rho^2}$ becomes a large quantity. But even far from the critical point $\overline{\Delta\rho^2} \neq 0$, as a consequence of the random motion of the molecules, although, in this case, $\Delta\rho^2$ is much less than at the critical point.

The density fluctuations in turn produce a fluctuation in the optical dielectric constant. These random fluctuations of the dielectric constant are in fact the sole physical reason for the optical inhomogeneities of the medium and the scattering of light in optically pure media.

The fruitful idea of Smoluchowski (1908) on fluctuations as the reason for light scattering lay at the base of the statistical theory of light scattering developed by Einstein [14]. In addition to fluctuations in the density, Einstein also took into account fluctuations of concentration which take place in solutions.

Fluctuations not only destroy the optical homogeneity within the substance, but also lead to the destruction of the mirror smoothness of the surface of a liquid or boundary of two immiscible liquids. Thermal fluctuations oppose the forces of surface tension that create the surface, thus making the surface "rough."

As a result of the molecular "roughness" of the surface of a liquid, molecular light scattering takes place in directions different from the direction of specular reflection of the primary light beam.

This phenomenon had already been pointed out in the work of Smoluchowski [13]. Mandel'shtam [15] gave the theory of the phenomenon and experimentally discovered the molecular scattering by the surface of a liquid.

As a result of these investigations of Rayleigh, Smoluchowski, Einstein, and Mandel'shtam, a new direction was opened up for molecular physics and molecular optics. Its development has been characterized by prominent physical investigations and discoveries.

From the viewpoint of the modern statistical theory of the scattering of light in gases, the reason for the disruption of optical homogeneity lies in the thermal fluctuations. For the present, we shall speak only of fluctuations of density or fluctuations in the number of particles in volumes whose linear dimensions are small in comparison with the light wavelength. We return to Fig. 1, which illustrates the propagation of light in an ideally homogeneous medium, and now take into account the fact that the medium under study has become optically inhomogeneous as a consequence of the fluctuations.

Without limiting the generality of the discussion, we shall assume that a number $\overline{N} + \Delta N$ particles is randomly distributed in the volume V_1^*, while the number in the volume V_2^* remains $\overline{N}$ as before. Therefore, the volume V_1^* radiates a wave of amplitude $\mathbf{E}_0 + \mathbf{E}'$, while the volume V_2^* radiates a wave of amplitude $\mathbf{E}_0$. Let these waves again be of opposite phase. However, the mean square of the algebraic sum of the amplitudes will now not be equal to zero for any one of the directions. The field $\mathbf{E}'$ is easily calculated if we assume that the fluctuations of the density generate an additional polarization under the action of the light wave which, for the fluctuation volume V^*, is equal to*

$$\Delta \boldsymbol{P} = \frac{\Delta\varepsilon V^*}{4\pi} E_0 \cos \omega t. \tag{9}$$

The connection between ΔP and E' is the same as in (2); therefore, we get for the intensity of the light scattered by the volume V^*

$$I_{V^*} = I_0 \frac{\pi^2 (V^*)^2}{2L^2\lambda^4} \overline{\Delta\varepsilon^2}. \tag{10}$$

Far from the critical point, the scattering by the separate volumes V_i^* can be regarded as independent if the linear dimensions of these volumes are large in comparison with the radius of the intermolecular interaction ($\sim 10^{-7}$ cm) and small in comparison with the wavelength of the incident light (10^{-5} cm), i.e., 10^{-7} cm $< (V^*)^{1/3} < 10^{-5}$ cm. In this simple and often-realized situation, the calculation of the scattering in the entire volume V reduces to a multiplication of (10) by V/V^*, and then, instead of (10), we get

$$I = I_0 \frac{\pi^2 V^* V}{2L^2\lambda^4} \overline{\Delta\varepsilon^2}. \tag{11}$$

The solution of the problem of scattering by optical inhomogeneities, much smaller than the wavelength of the incident light, has the form (11) in general. Calculation of the intensity of the scattered light

*We shall neglect quantities of second order of smallness $\sim \Delta\varepsilon E'$.

reduces to expressing $\overline{\Delta\varepsilon^2}$ in terms of quantities measured by experiment in the different cases. In the general case, such a calculation presents appreciable difficulties which have not been completely overcome at the present time. Some of them will be considered in this book.

We are now interested in the very simple case of light scattering in a gas as the result of density fluctuations. For a gas at low pressures, the following relation is easily satisfied:

$$\varepsilon - 1 = CN_L = CN\frac{1}{V^*}. \tag{12}$$

Here, N is the number of molecules in the volume V*. It follows from (12) that

$$\overline{\Delta\varepsilon^2} = \frac{(\varepsilon-1)^2}{N_L^2}\frac{\overline{\Delta N^2}}{(V^*)^2}.$$

Substituting the latter expression in (11), we obtain

$$I = I_0\frac{\pi^2 V(\varepsilon-1)^2}{2L^2\lambda^4 V^* N_L^2}\overline{\Delta N^2}. \tag{13}$$

Equation (13) is identical with the Rayleigh formula (6) only in the special case of an ideal gas, for which the statistical analysis gives [8, 16]:

$$\overline{\Delta N^2} = N = V^* N_L. \tag{14}$$

Substituting (14) in (13), we get exactly the same relation as was obtained by Rayleigh [Eq. (6)]. Actually, Eq. (6) is identical with (13) only when (14) is satisfied, i.e., when the molecules do not interact with one another.

A complete molecular theory of light scattering has been constructed for rarefied gases (Sec. 4). Numerous experimental investigations of molecular light scattering in gases have confirmed the correctness of the theory and permit us to obtain a great deal of valuable information on the structure of molecules. Moreover, they allow us to find Avogadro's number with great accuracy (see Sec. 18).

The dependence of the intensity of the scattered light on the wavelength is determined by the factor λ^{-4}, as is clear from (13). The weak dependence of ε on λ outside the absorption region can be neglected. But Landsberg and Mandel'shtam directed their attention to the fact that the factor should have a profound effect on the dependence of the scattered light intensity on the wavelength near the intense atomic absorption lines.

In their very difficult experiment such a selective scattering was also found for the radiation of a zinc arc with $\lambda_1 = 2502$ Å and $\lambda_2 = 2558$ Å in mercury vapor close to the resonance line 2537 Å (see Sec. 16).

Perhaps the largest amount of experimental and theoretical studies have been devoted to liquids, although it is exactly in this case that there remains the greatest number of unclear and unresolved problems.

Of course, this is closely connected with the inadequate development of the theory of the liquid state generally. But even in the study of scattering in a liquid, there are significant achievements. The formula obtained by Einstein [14] even now maintains its full force (Sec. 1). The theory of Einstein took into account the fluctuations of the density and concentration, but in liquids and solutions with strongly anisotropic molecules, the intensity of the light scattered by the anisotropy fluctuations exceeds the light intensity scattered by the density fluctuations. To compute the light intensity scattered by the fluctuations of the anisotropy in the liquid has not yet been possible. But Cabannes has proposed a successful method which allows us to study the light intensity scattered by the anisotropy fluctuations (the Cabannes correction). In this calculation, it is necessary to know only the depolarization factor of the total scattering and the depolarization factor of the light scattered by the anisotropy fluctuations (see Sec. 1).

For measurements of the absolute intensity of the scattered light, experimental difficulties are encountered which lead to confusion (Sec. 18). However, the agreement of the experimental data with the results of the calculation by means of the Einstein formula with the Cabannes correction now gives rise to no doubts whatsoever.

The Einstein theory of light scattering in solutions found a new region of application when Debye [42] and his coworkers found a practical application of the theory to the determination of the dimensions, shape, and molecular weight of solutions of polymers, proteins, and electrolytes (Sec. 21). The light-scattering method is now applied also to the study of the distribution of the scattering particles according to their dimensions and to a number of other problems. The material for this area of study is so great that entire books are devoted to its review [43].

Optical inhomogeneities, produced by the fluctuations, must clearly give rise to scattering even in an ideal crystal. But in real crystals, as a result of impurities and other imperfections, such a large flux of parasitic light appears that it has not been possible to establish whether or not molecular light scattering exists in the crystal without quantitative experiments.

The light of molecular scattering in a quartz crystal was first separated by Landsberg [17] after he had proposed a method which made it possible to separate the light of molecular scattering from the parasitic light (see Sec. 31).

After these first investigations, attempts were made to construct a theory of molecular light scattering in an ideal crystal. The first theoretical studies were carried out by Leontovich and S. Mandel'shtam [18], and by Matossi [19]. The general theory of molecular light scattering in crystals was given by Motulevich [20] (see Sec. 9). An anisotropy in the intensity of the light scattered in crystals was first observed by Vul'fson and Lombert [21] and was later studied in detail by Motulevich [20].

The theory of molecular light scattering in second-order phase transitions, which was developed by Ginzburg [23], predicted that the intensity of the scattered light near the transition temperature should become very large (see Sec. 2).

Careful investigations of the temperature dependence of the scattered light intensity in crystalline quartz, performed independently of the previously developed theory, allowed Yakovlev, Velichkina, and Mikheeva [22] to discover a new phenomenon that appears close to the transition temperature of quartz from the α to the β modification, which is similar to critical opalescence. In a temperature interval $\sim 0.1°$, the intensity of the scattered light changes by a factor of 14,000, while the character of the change in the intensity is represented by a λ curve (see Sec. 32).

The experimental results are excellently described in quantitative fashion by the theory of Ginzburg (see Sec. 32). The study of scattering in crystals has continued its successful development and has a great value in the study of dislocations and other imperfections of crystals (see Sec. 31).

Of great scientific interest is the spectroscopic study of scattered light. This book is devoted chiefly to the exposition of the results of the spectral investigations of molecular light scattering. Such investigations are one of the most effective means of study of the kinetics of different fluctuations.

The fluctuations of pressure, temperature or entropy, orientation of anisotropic molecules, or the fluctuations of concentrations arise and "dissipate" in time. The character of the process of the generation and "dissipation" of the fluctuations is different for the various fluctuations.

Actually, the incipient pressure fluctuations cannot remain unchanged but will dissipate with the speed of propagation of the elastic perturbation. Fluctuations of orientation and concentration equalize themselves because of diffusion, while temperature fluctuations will change with time because of the temperature conductivity. All these temporal changes of the molecular inhomogeneities naturally lead to temporal changes in the scattered light intensity. In other words, the temporal changes of the optical inhomogeneities produce a modulation of the scattered light, while the character of the modulation is closely connected with the physical process determining the particular form of the modulating function (see Secs. 5-9, 30).

The first theoretical investigation of the modulation of scattered light was made by Mandel'shtam [24] as early as 1918, although the corresponding publication appeared only in 1926. At this time, Brillouin [26] had independently found and published part of the results contained in [24].

The investigations of Mandel'shtam and Brillouin made it possible to say that if monochromatic light of frequency ν is incident on the scattering volume, then shifted frequencies will be observed in the light scattered by adiabatic fluctuations (the fine structure of the Rayleigh line).

The fine structure of the Rayleigh line was discovered by Landsberg and Mandel'shtam [6], and by Gross [27] in a quartz crystal, and by Gross [28] in a liquid. The work of Landsberg and Mandel'shtam on the fine structure of the scattering line led them to the discovery of combination light scattering in a quartz crystal. At the same time, Raman discovered this phenomenon in liquids.

Combination (Raman) scattering of light is now a powerful method of investigation of intermolecular and intramolecular interactions and of the structure of molecules while stimulated combination scattering of light serves for frequency conversion in other problems. An account of problems connected with combination scattering of light goes beyond the scope of this book.

A detailed investigation of the fine structure of the Rayleigh line has allowed the author and his coworkers to discover a notable dispersion of the sound velocity in low-viscosity liquids, brought about by the relaxation of the bulk viscosity coefficient [29-33]. These investigations made it possible to find the characteristics of the propagation of hypersound (sound of frequencies above 10^9 Hz) at frequencies $\sim 10^{10}$ Hz in liquids (see Sec. 24).

Later, Pesin and the author [34] succeeded in finding the fine structure of the Rayleigh line in viscous liquids over a great range of change of viscosity, up to the glassy state. Here, dispersion of the sound velocity was observed which exceeded 70%, and which was due to the relaxation of the shear viscosity (Ch. VII).

Application of the narrow spectral line $\lambda = 6328$ Å in stimulated emission of a Ne–He gas laser as a light source in liquids made it possible for Mash, Starunov, Tiganov, and the author to find the absorption coefficient of hypersound (frequency $\sim 10^{10}$ Hz) in several liquids by the halfwidth of the displaced components of the fine structure.

Thus, a close connection has been established between molecular light scattering and molecular acoustics.

The study of the fine structure at the present time is the only means of investigation of the propagation of hypersound in liquids (Ch. VI). But hypersound in liquids can in principle also be studied by acoustic means.

The interesting problem of the fine structure of the Rayleigh line in gases at pressures above atmospheric remains completely unsolved. The classical theory of the effect shows that for pressures of the order of 30 atm in hydrogen, for example, the same picture of the fine structure should be observed as in liquid benzene

(Sec. 17). But in the only current experiment, the fine structure has not been observed at high pressures in hydrogen and other gases.

As a result of the rapid temporal changes of the fluctuations of the anisotropy, modulation of the scattered light takes place which appears in the spectrum in the form of a depolarized broad continuous spectrum, which extends out 150 cm^{-1} from the unshifted line, and even further. This broad spectrum, which is usually called the wing of the Rayleigh line, was discovered by Cabannes and Daure [37], and by Raman and Krishnan [38], and has been frequently studied both theoretically and experimentally.

Using his experimental data and the formulas from the Leontovich [39] theory of the intensity distribution in the wing, the author found the relaxation time of the anisotropy (see Ch. VIII). A further development of the theory of the intensity distribution in the wing was obtained in the researches of Rytov [40] and Starunov [41]. In this region, the correct understanding of the nature of the phenomenon has only recently been achieved, and a rational form of investigation developed (Ch. VIII).

With the introduction into laboratory practice of such light sources as the gas and ruby laser, a number of entirely new investigations have been completed which could not have been dreamt of earlier.

It was pointed out above that the narrow exciting line of the gas laser made it possible to find the damping of the Debye elastic wave. This same light source also made it possible, by the width of the central component of the fine structure in a polymer solution, to proceed to the determination of the dimensions of the solute molecules [593].

A powerful giant pulse of a ruby laser made it possible for Chiao, Townes, and Stoicheff [595] to find a new phenomenon – stimulated Mandel'shtam–Brillouin scattering (see Sec. 34).

In the focused light of a gigantic pulse of the ruby laser, with a power of about 100 MW, Mash, Morozov, Starunov, and the author [631] observed another new phenomenon – stimulated light scattering of the wing of the Rayleigh line. Both these phenomena reveal new possibilities for the study of nonlinear optics and acoustical properties of the medium.

In this brief introduction, we could only sketch the various stages of development of the representations of molecular light scattering and point out only a few of the fundamental problems which still need to be solved.

Chapter I

Theory of Molecular Light Scattering in Condensed Isotropic Media and Gases

1. Thermodynamic Theory

The application of classical statistical thermodynamics to the calculation of the light intensity scattered in condensed isotropic media leads to fundamental results which will be described below. The results of such a calculation are applicable if the mean free path of the molecules of the medium l is much shorter than the light wavelength λ / n in the medium.

In addition, the classical thermodynamic calculation is applicable when the parameters characterizing the medium and entering into the formulas of the theory do not depend on the frequency.

If the condition $\bar{l} \ll \lambda/n$ is satisfied, the medium can be regarded as continuous and is characterized by some optical dielectric constant ε_0. When such a medium is homogeneous, i.e., when its properties do not depend on the coordinates (there are no optical inhomogeneities), then, as was pointed out in the Introduction, there will also be no light scattering.

Thermal motion of the molecules of the medium leads to the appearance of fluctuations of the density and of the orientation of anisotropic molecules, which in turn produce fluctuations of the optical dielectric constant. We shall assume that the dielectric constant of the medium is composed of the ε_0 of the undisturbed medium and a small correction $\Delta\varepsilon$ brought about by the fluctuations. If a medium that is isotropic as a whole is composed of anisotropic molecules, then an anisotropy appears in the volume of the fluctuations as a consequence of the fluctuation of their relative orientations; therefore, $\Delta\varepsilon$ becomes a tensor quantity, and, consequently,

$$\varepsilon_{ik} = \varepsilon_0 \delta_{ik} + \Delta\varepsilon_{ik}, \tag{1.1}$$

where δ_{ik} is the Kronecker delta. The first term of (1) determines the value of ε in the homogeneous medium in which the scattering of light is absent; therefore, this quantity will be considered no further.

The effect of this light scattering is determined only by the value of $\Delta\varepsilon_{ik}$, which is conveniently represented in the form of two parts [45]:

$$\Delta\varepsilon_{ik} = \Delta\varepsilon\delta_{ik} + \Delta\varepsilon_{ik}, \tag{1.2}$$

where $\sum_{i=1}^{3}\Delta\varepsilon'_{ii}=0$. The fluctuations $\Delta\varepsilon$ are isotropic, and are determined only by fluctuations of the pressure Δp and the entropy ΔS, or the density $\Delta\rho$ and temperature ΔT, or any other pair of independent thermodynamic variables. If one considers a solution or a mixture of gases, then $\Delta\varepsilon$ will also be a function of the concentration c of the components of the solution or mixture.

A certain part of $\Delta\varepsilon$, equal to $\Delta\varepsilon'$, as it is given in [45], cannot generally be reduced to fluctuations of thermodynamic quantities and the concentration, but the calculation of $\Delta\varepsilon'$ is somewhat difficult and we shall for the time being set $\Delta\varepsilon' = 0$.

The fluctuations $\Delta\varepsilon$ do not destroy the isotropic character of the medium, and therefore the light scattered by the fluctuations $\Delta\varepsilon$ must be completely polarized.

The fluctuations $\Delta\varepsilon'_{ik}$ determine the anisotropy of the medium that results from the thermal motion; therefore, the light scattered by $\Delta\varepsilon'_{ik}$ must be depolarized. The total intensity of the scattered light $\mathcal{I}$ is proportional to $\overline{|\Delta\varepsilon_{ik}|^2}$ (see Introduction):

$$\mathcal{I} \sim \overline{|\Delta\varepsilon_{ik}|^2} \sim \overline{|\Delta\varepsilon|^2} + \overline{|\Delta\varepsilon_{ik}^{(S)}|^2} + \overline{|\Delta\varepsilon_{ik}^{(a)}|^2}, \tag{1.3}$$

where $\Delta\varepsilon'_{ik}$ is represented in the form of two terms – the symmetric

$$\Delta\varepsilon_{ik}^{(S)} = \Delta\varepsilon_{ki}^{(S)} \tag{1.4}$$

and the antisymmetric

$$\Delta\varepsilon_{ik}^{(a)} = -\Delta\varepsilon_{ki}^{(a)}, \tag{1.5}$$

where it is assumed that the fluctuations $\Delta\varepsilon$, $\Delta\varepsilon_{ik}^{(S)}$, and $\Delta\varepsilon_{ik}^{(a)}$ are statistically independent.

It is seen from (1.3) that the intensity of the scattered light consists of the intensity of the light polarized by scattering from the fluctuations $\Delta\varepsilon$, and that depolarized by scattering from $\Delta\varepsilon_{ik}^{!}$.

Calculation of the intensity of the light scattered as a consequence of the fluctuations of $\Delta\varepsilon_{ik}^{!}$ is a very difficult problem, of which we shall take notice later. The light intensity scattered by the fluctuations $\Delta\varepsilon$ is computed comparatively simply and we shall begin with this case. For numerical calculation of the intensity of light scattered by a continuous medium, we make use of the macroscopic equations of Maxwell [16]:

$$\left.\begin{aligned} c\,\mathrm{rot}\,\boldsymbol{H}_a &= \frac{\partial \boldsymbol{D}}{\partial t}, \quad \mathrm{div}\,\boldsymbol{D}=0, \\ -c\,\mathrm{rot}\,\boldsymbol{E}_a &= \frac{\partial \boldsymbol{H}_a}{\partial t}, \quad \mathrm{div}\,\boldsymbol{H}_a=0. \end{aligned}\right\} \tag{1.6}$$

Here,

$$\boldsymbol{B}=\boldsymbol{H},\ \boldsymbol{D}=\varepsilon \boldsymbol{E}_a=\boldsymbol{E}_a+4\pi \boldsymbol{P},$$
$$\boldsymbol{P}=\frac{1}{4\pi}(\varepsilon-1)\boldsymbol{E}_a=\frac{1}{4\pi}(\varepsilon_0+\Delta\varepsilon-1)\boldsymbol{E}_a,$$

P is the electric moment per unit volume.

The electric field $\mathbf{E}_a$ and magnetic field $\mathbf{H}_a$ of the light wave in the scattered medium can be represented in the form of the sum of fields of the exciting wave **E** and **H**, and the fields of the scattered wave **E**' and **H**':

$$\boldsymbol{E}_a=\boldsymbol{E}+\boldsymbol{E}' \quad \text{and} \quad \boldsymbol{H}_a=\boldsymbol{H}+\boldsymbol{H}'. \tag{1.7}$$

It is clear from (1.6) that the fluctuations $\Delta\varepsilon$ produce an additional polarization per unit volume:

$$\Delta\boldsymbol{P}=\frac{\Delta\varepsilon}{4\pi}\boldsymbol{E}. \tag{1.8}$$

Assuming that $\Delta\varepsilon$ is small in comparison with ε_0, and that **E**' and **H**' are small in comparison with **E** and **H**, we neglect quantities of second order of smallness; further, taking (1.6), (1.7), and (1.8) into consideration, we obtain Maxwell's equations for the field of the scattered light wave:

$$\left.\begin{aligned} -c \operatorname{rot} \boldsymbol{E}' &= \frac{\partial \boldsymbol{H}'}{\partial t}, \quad \operatorname{div} \boldsymbol{D}' = 0, \\ c \operatorname{rot} \boldsymbol{H}' &= \frac{\partial \boldsymbol{D}'}{\partial t}, \quad \operatorname{div} \boldsymbol{H}' = 0, \end{aligned}\right\} \tag{1.9}$$

$$\boldsymbol{D}' = \varepsilon_0 \boldsymbol{E}' + \Delta\varepsilon \boldsymbol{E} = \varepsilon_0 \boldsymbol{E}' + 4\pi \Delta \boldsymbol{P}.$$

Elimination of **H'** from the set of equations (1.9) yields

$$\Delta \boldsymbol{D}' + \frac{\varepsilon_0}{c^2} \frac{\partial^2 \boldsymbol{D}'}{\partial t^2} = -\operatorname{rot} \operatorname{rot} (\Delta\varepsilon \boldsymbol{E}). \tag{1.10}$$

The solution of Eq. (1.10) will be [46-49]

$$\boldsymbol{D}' = \operatorname{rot} \operatorname{rot} \boldsymbol{\Pi}, \tag{1.11}$$

where **Π** is the Hertzian vector, which can be described at large distances from the scattering volume V (in the wave zone) in the following fashion:

$$\boldsymbol{\Pi} = \frac{e^{i k' L}}{4\pi L} \int_V \Delta\varepsilon \boldsymbol{E} e^{-i k' r} \, dV. \tag{1.12}$$

Here **k'** is the wave vector of the scattered light, **L** is the vector distance from an arbitrary but fixed point, inside the scattering volume, to the point of observation, **r** is the radius vector drawn from this fixed point to any element inside the scattering medium.

If the dependence of the vector **E** of the exciting light wave on **r** is represented in the form

$$\boldsymbol{E} = \boldsymbol{E}_0 e^{-i k r} \tag{1.13}$$

(**k** is the wave vector of the exciting light), then, by substituting (1.13) in (1.12), we get

$$\boldsymbol{\Pi} = \frac{E_0 e^{i k' L}}{4\pi L} \int_V \Delta\varepsilon e^{-i (k' - k) r} \, dV. \tag{1.14}$$

By assuming that $\mathbf{E}' = (1/\varepsilon_0)\mathbf{D}'$ at the point of observation, we calculate the density of the field of the scattered wave **E'** from (1.11), (1.14) if $\Delta\varepsilon$ can be expressed in terms of measured quantities.

Calculation of the Intensity of Scattered Light According to Einstein. For the calculation of the scattered light intensity in liquids and solutions, the method used by Einstein [14] consisted of the following: the fluctuation $\Delta\varepsilon$, which can be a function of any pair of independent thermodynamic variables, is taken to be equal to*

$$\Delta\varepsilon = \left(\frac{\partial\varepsilon}{\partial\rho}\right)_T \Delta\rho. \tag{1.15}$$

We represent the density fluctuations $\Delta\rho$ in the form of a spatial three-dimensional Fourier series

$$\Delta\rho = \sum_q \Delta\rho_q e^{iqr}, \tag{1.16}$$

where **q** is the expansion parameter with components q_x, q_y, q_z, taking on values equal to $2\pi m/2l$ (where $m = 0, \pm 1, \pm 2, \ldots$). It is assumed that the scattering volume V has the shape of a cube with $l = (V)^{1/3} \gg \lambda$. Expansion coefficients (1.16) are calculated from the formula [51]:

$$\Delta\rho_q = \frac{1}{V}\int \Delta\rho e^{-iqr}\, dV \tag{1.17}$$

and because $\Delta\rho$ is real, we have the relation

$$\Delta\rho_{-q} = \Delta\rho_q^*.$$

Substituting (1.15) and (1.16) in (1.14), we get

$$\Pi = \frac{E_0 e^{ik'L}}{4\pi L}\left(\frac{\partial\varepsilon}{\partial\rho}\right)_T \int \sum_q \Delta\rho_q e^{-i(k'-k-q)r}\, dV. \tag{1.18}$$

The integral in (1.18) differs from zero for large volumes of V and is equal to the product $\Delta\rho_q V$ only under the condition

$$k' - k - q = 0. \tag{1.19}$$

Thus, on satisfaction of the condition (1.19), the Hertzian vector (1.18) transforms into

$$\Pi = \frac{E_0 V e^{ik'L}}{4\pi L}\left(\frac{\partial\varepsilon}{\partial\rho}\right)_T \Delta\rho_q.$$

From the last expression and (1.11), we find the field of the scattered wave:

* Einstein neglected the quantity $(\partial\varepsilon/\partial T)_\rho \Delta T$.

$$E' = \frac{(k')^2 E_0 V e^{ik'L}}{4\pi L \varepsilon_0} \left(\frac{\partial \varepsilon}{\partial \rho}\right)_T \Delta\rho_q \sin \vartheta. \tag{1.20}$$

Here ϑ is the angle between the direction of the vector $\mathbf{k}'$ (direction of the scattered wave) and the direction of oscillation of the vector of the exciting wave $\mathbf{E}_0$. For a radiation by polarized light with vector $\mathbf{E}_0 \| z$,

$$E'_z = \frac{|k'|^2 V E_{0z} e^{ik'L}}{4\pi\varepsilon_0 L} \left(\frac{\partial \varepsilon}{\partial \rho}\right)_T \Delta\rho_q \sin \vartheta_z. \tag{1.21}$$

For radiation by polarized light with vector $\mathbf{E}_0 \| y$,

$$E'_y = \frac{|k'|^2 V E_{0y} V e^{ik'L}}{4\pi\varepsilon_0 L} \left(\frac{\partial \varepsilon}{\partial \rho}\right)_T \Delta\rho_q \sin \vartheta_y. \tag{1.22}$$

For natural exciting light,

$$\frac{cn}{4\pi}(E_{0y})^2 = \frac{cn}{4\pi}(E_{0z})^2 = \frac{1}{2} I_0,$$

where I_0 is the total intensity of the natural light. The light intensity is expressed in erg-cm^{-2}-sec^{-1} or watt-cm^{-2} or watt-sterad^{-1} [67, 68]. The intensity of the scattered light in excitation by natural light is the sum of the squares of (1.21) and (1.22):

$$I = I_0 \frac{(k')^4 V^2 \left(\frac{\partial \varepsilon}{\partial \rho}\right)_T^2 \overline{\Delta\rho_q^2}}{2(4\pi\varepsilon_0 L)^2} (\sin^2\vartheta_z + \sin^2\vartheta_y). \tag{1.23}$$

Taking into account that

$$\cos^2\vartheta_x + \cos^2\vartheta_y + \cos^2\vartheta_z = 1,$$

we obtain

$$\sin^2\vartheta_z + \sin^2\vartheta_y = 1 + \cos^2\vartheta_x,$$

where $\vartheta_x \equiv \theta$ is the angle between the direction of propagation of the exciting light $\mathbf{k}$ and the direction of observation of the scattered light $\mathbf{k}'$. Statistical calculation [Appendix I, Eq. (I.13)] of $\Delta\rho_q^2$ gives

$$\overline{\Delta\rho_q^2} = \frac{1}{V} \beta_T k T \rho^2. \tag{1.24}$$

Taking into account that $|\mathbf{k}| = (2\pi\sqrt{\varepsilon})/\lambda$, we get, by substituting (1.24) into (1.23), the well-known Einstein formula

$$I = I_0 \frac{\pi^2 V}{2\lambda^4 L^2} \left(\rho \frac{\partial \varepsilon}{\partial \rho} \right)_T^2 \beta_T kT (1 + \cos^2 \theta). \tag{1.25}$$

The fundamental formula (1.25) will be discussed below. Here it suffices to note that Einstein represented the density fluctuations in the form of a spatially periodic function. These Einstein "formal waves" bear a statistical character. The deep physical meaning and the dynamic character of these waves will be made clear only much later. We shall discuss this in Sec. 5. It now remains to turn our attention to the fact that the relation (1.19) is the well-known Bragg condition written in vector form.

Neglecting the small change in the wavelength in coherent scattering (see Sec. 5) and setting $|\mathbf{k}'| = |\mathbf{k}|$, we get

$$|q| = \frac{4\pi n}{\lambda} \sin \frac{\theta}{2}, \tag{1.26}$$

where $q = 2\pi/\Lambda$, and Λ is the length of the elastic wave. In place of (1.26), we can write

$$2n\Lambda \sin \frac{\theta}{2} = \lambda. \tag{1.27}$$

The maximum density of scattered light will be observed at the scattering angle θ for given λ and Λ in accord with the condition (1.27). For the same λ at a different angle θ_2, the scattered light will be determined by another Fourier component of the expansion (1.16), and consequently by another wavelength of the elastic wave, Λ_2. Thus, to each angle of scattering (for given λ), there corresponds its own Fourier component $\Delta\rho_q$, which determines the light intensity scattered in that direction.

Calculation of the Light Intensity Scattered by Spatially Independent Fluctuations. Another method of calculating the scattered light intensity is based on the summation of the light scattered by the different fluctuations. Such a calculation is especially simple if the fluctuations can be regarded as spatially independent [16].

Actually, the additional polarization in the volume of fluctuations V^*, according to (1.8), is equal to

$$\Delta P' = \frac{\Delta \varepsilon V^*}{4\pi} E. \tag{1.28}$$

Therefore, by using (1.11), (1.14), and (1.28), we find for the field of the scattered wave in the wave zone

$$\left.\begin{aligned} E_x' &= E_{0x}\frac{k^2\Delta\varepsilon V^*}{4\pi\varepsilon L}e^{ikL}\sin\vartheta_x, \\ E_z' &= E_{0z}\frac{k^2\Delta\varepsilon V^*}{4\pi\varepsilon L}e^{ikL}\sin\vartheta_z. \end{aligned}\right\} \tag{1.29}$$

From (1.29), and also from (1.21) and (1.22), we find for the intensity of the scattered light

$$I = I_0\frac{\pi^2 (V^*)^2}{2\lambda^4 L^2}\overline{\Delta\varepsilon^2}(1+\cos^2\theta). \tag{1.30}$$

Equation (1.30) refers to the scattering by very small volumes, the linear dimensions of which $\sim 0.05\lambda$ of the scattered light. Actually, the experimental investigation of the scattered light is always carried out in the volumes of matter V, the linear dimensions of which are tens and hundreds of thousands of times greater than the optical wavelength.

Therefore, strictly speaking, one ought to sum up the field irradiated by the small volumes V^*, with account of the phase shifts, and compute the intensity of the scattered light from such a sum of the intensities of the electric field. In the practically important cases of light scattering in gases and liquids, under conditions not very close to critical, the calculation of the intensity of the scattered light can be carried out much more simply. Actually, for ideal and real gases and for liquids, density fluctuation $\Delta\rho$ arising in two neighboring volumes V_1^* and V_2^* are statistically independent and therefore $\overline{\Delta\rho_1\Delta\rho_2} = 0$ [16]. This is satisfied under the condition already formulated in the Introduction. Such a condition is realized in liquids and real gases outside of the critical state region. For ideal gases, the dimensions of the volume V^* are generally not limited. From the statistical independence of these density fluctuations in two neighboring volumes V_1^*, V_2^*, it follows that the light scattered by these volumes will be noncoherent and, therefore, we can simply sum the intensities of the light scattered by the separate volumes V^* within the irradiated volume V. Such a summation reduces to multiplication of the expression (1.30) by V/V^*. Consequently, we get from (1.30)

$$I = I_0 \frac{\pi^2 V^* V}{2\lambda^4 L^2} \overline{\Delta\varepsilon^2} (1 + \cos^2\theta). \tag{1.31}$$

Assuming, after Einstein, that $\Delta\varepsilon$ is expressed by Eq. (1.15), we get

$$I = I_0 \frac{\pi^2 V^* V}{2\lambda^4 L^2} \left(\rho \frac{\partial\varepsilon}{\partial\rho}\right)_T^2 \overline{\Delta\rho^2} (1 + \cos^2\theta). \tag{1.32}$$

And, taking into account that, similarly to (1.24),

$$\overline{\Delta\rho^2} = kT\beta_T \frac{\rho^2}{V^*}, \tag{1.33}$$

we get an equation in place of (1.32) which is identical with (1.25).

In both the cases considered above, it was assumed that the scattering takes place from the scalar part of $\Delta\varepsilon_{ik}$. Here it would seem that the angular distribution of the scattered light intensity is expressed by (1.25) or (1.32) for illumination by natural light.

Placzek [50] and Landau and Lifshitz [47] considered the general scattering tensor $\Delta\varepsilon_{ik}$ and obtained, in addition to indicatrices of the scalar of scattering by the fluctuation $\Delta\varepsilon$, the indicatrix of scattering by the fluctuations $\Delta\varepsilon_{ik}^{(S)}$ and $\Delta\varepsilon_{ik}^{(a)}$ [see (1.3)].

Table 1 shows the scattering indicatrices by all three parts of $\Delta\varepsilon_{ik}$ [see (1.3)] for illumination of the scattering volume by polarized light $\mathbf{E}_{0z}$ (the vector $\mathbf{E}$ is perpendicular to the scattering plane) and by unpolarized light. Figure 2 shows the scattering indicatrices for the three forms of scattering (Table 1) for illumination by natural light.

Let us now return to the final formulas (1.25) and (1.31), (1.15) and (1.32) for the intensity of light scattered by the isotropic fluctuations $\Delta\varepsilon$ and emphasize again that these formulas are valid under the following assumptions:

1. $(\partial\varepsilon/\partial T)_\rho^2 \overline{\Delta T^2}/(\partial\varepsilon/\partial\rho)_T^2 \overline{\Delta\rho^2}$ is so small that it can be neglected in the calculation (see note on p. 23).
2. $\Delta\varepsilon'_{ik} = 0$.
3. $\Delta\varepsilon\,\delta_{ik}$ depends only on $\Delta\rho$ and ΔT.

Table 1. Angular Dependence of the Intensity of Scattered Light

Form of scattering	I_{0z}	I
$I \sim \overline{\vert\Delta\varepsilon\vert^2}$	$\sin^2\vartheta$	$1+\cos^2\theta$
$I \sim \overline{\vert\Delta\varepsilon_{ik}^{(S)}\vert^2}$	$1+\frac{\sin^2\vartheta}{6}$	$1-\frac{\sin^2\theta}{14}$
$I \sim \overline{\vert\Delta\varepsilon_{ik}^{(a)}\vert^2}$	$1+\cos^2\vartheta$	$1+\frac{\sin^2\theta}{2}$

ϑ: angle between E_{0z} and the direction of observation of the scattered light.
θ: scattering angle.

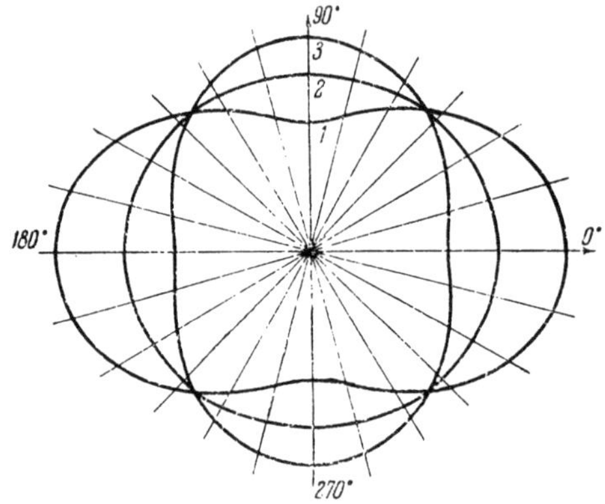

Fig. 2. Angular dependence of the light intensity of molecular scattering. 1) $I \sim \overline{|\Delta\varepsilon|^2}$; 2) $I \sim \overline{|\Delta\varepsilon_{ik}^{(S)}|^2}$; 3) $I \sim \overline{|\Delta\varepsilon_{ik}^{(a)}|^2}$.

We assume that $\Delta\varepsilon$ is a function of two independent thermodynamic variables, for example, ρ and T, and if we are concerned with a solution, then it is also a function of the concentration c of its components:

$$\Delta\varepsilon(\rho, T, c) = \left(\frac{\partial\varepsilon}{\partial\rho}\right)_{T,c}\Delta\rho + \left(\frac{\partial\varepsilon}{\partial T}\right)_{\rho,c}\Delta T + \left(\frac{\partial\varepsilon}{\partial c}\right)_{T,\rho}\Delta c.$$

The fluctuations of density, temperature, and concentration arise independently of one another and are statistically independent (see Appendix I) and therefore $\overline{\Delta\rho\Delta T} = 0$, $\overline{\Delta\rho\Delta c} = 0$, and $\overline{\Delta c\Delta T} = 0$. Consequently, we get from the expression written above for $\Delta\varepsilon(\rho, T, c)$

$$\overline{|\Delta\varepsilon(\rho, T, c)|^2} = \left(\frac{\partial\varepsilon}{\partial\rho}\right)^2_{T,c} \overline{\Delta\rho^2} + \left(\frac{\partial\varepsilon}{\partial T}\right)^2_{\rho,c} \overline{\Delta T^2} + \left(\frac{\partial\varepsilon}{\partial c}\right)^2_{T,\rho} \overline{\Delta c^2}. \quad (1.34)$$

The first two components in (1.34) characterize the isotropic scattering in a pure liquid or a pure gas, and the last term is characteristic of a solution or gas mixture.

Pure Liquids. We shall now make a detailed consideration of the general formula for the intensity of light scattered by adiabatic [first component of (1.34)] and isobaric [second component of (1.34)] fluctuations within the framework of statistical thermodynamics [29, 52, 53]. The detailed consideration is necessary because: (1) A large number of experimental determinations of the absolute intensity of scattered light in liquids have been completed. Here we note a definite divergence between theory and experiment (see Sec. 18); and (2) the spectroscopic study of the light scattered as a consequence of adiabatic and isobaric density fluctuations (fine structure of the Rayleigh line) allows us to find the relation of the integrated light intensity scattered by isobaric and adiabatic density fluctuations and to compare the experimental results with the theory. Here also there is a lack of agreement of theory [140] with the experimental data.

On the other hand, in following up the divergence between theory and experiment, there have been attempts to prove that the Einstein formula improperly takes into account the intensity of the light scattered by adiabatic and isobaric density fluctuations. It was asserted in [35] that the Einstein formulas (1.25) and (1.32) ought to contain the adiabatic compressibility β_S in place of the isothermal compressibility β_T, and it was assumed in [55, 56, 57] that slow fluctuations are not taken into account in the Einstein theory. The erroneousness of this information has already been pointed out [45, 52]. It can easily be shown that for the given assumptions (see p. 23) the Einstein formula takes all the adiabatic and isobaric density fluctuations into account.

We shall in fact carry out a thermodynamic calculation of $\Delta\rho$ by assuming that $\Delta\rho$ is a function of two independent thermodynamic variables – the pressure p and the entropy S. Then,

$$\Delta\rho(p, S) = \left(\frac{\partial\rho}{\partial p}\right)_S \Delta p + \left(\frac{\partial\rho}{\partial S}\right)_p \Delta S,$$

and because of the statistical independence of p and S, $\overline{\Delta p \Delta S} = 0$ [Appendix I, Eq. (I.10)], and, therefore,

$$\overline{\Delta\rho^2} = \left(\frac{\partial\rho}{\partial p}\right)_S^2 \overline{\Delta p^2} + \left(\frac{\partial\rho}{\partial S}\right)_p^2 \overline{\Delta S^2}. \tag{1.35}$$

The first term of the expression (1.35) characterizes the adiabatic (isentropic) and the second the isobaric density fluctuations.

It is clear from (1.35) and (1.32) that the expression for the intensity of scattered light is divided into two parts:

$$I = V^* G \left[\left(\frac{\partial\varepsilon}{\partial\rho}\right)_T^2 \left(\frac{\partial\rho}{\partial p}\right)_S^2 \overline{\Delta p^2} + \left(\frac{\partial\varepsilon}{\partial\rho}\right)_T^2 \left(\frac{\partial\rho}{\partial S}\right)_p^2 \overline{\Delta S^2}\right], \tag{1.36}$$

where the notation

$$G = I_0 \frac{\pi^2 V}{2\lambda^4 L^2}(1 + \cos^2\theta) \tag{1.37}$$

has been introduced. We shall calculate each component of (1.36) separately.

The intensity of the light scattered by adiabatic fluctuations of the density is

$$I_{ad} = V^* G \left(\rho \frac{\partial\varepsilon}{\partial\rho}\right)_T^2 \left(\frac{1}{\rho}\frac{\partial\rho}{\partial p}\right)_S^2 \overline{\Delta p^2}. \tag{1.38}$$

From the statistical calculation [Appendix II, Eq. (II.8)],

$$\overline{\Delta p^2} = -kT\left(\frac{\partial p}{\partial V}\right)_S = \frac{kT}{\beta_S V^*}, \tag{1.39}$$

and, in accord with the definition

$$\left(\frac{1}{\rho}\frac{\partial\rho}{\partial p}\right)_S = -\frac{1}{V}\left(\frac{\partial V}{\partial p}\right)_S = \beta_S. \tag{1.40}$$

We then have from (1.40), (1.39), and (1.38),

$$I_{ad} = G\left(\rho\frac{\partial\varepsilon}{\partial\rho}\right)_T^2 \beta_S kT. \tag{1.41}$$

Furthermore, it is possible to compute the intensity of the light scattered by isobaric density fluctuations:

$$I_{is} = V^* G \left(\rho \frac{\partial\varepsilon}{\partial\rho}\right)_T^2 \left(\frac{1}{\rho}\frac{\partial\rho}{\partial S}\right)_p^2 \overline{\Delta S^2}. \tag{1.42}$$

Calculation gives [see Appendix I, Eq. (I.9)]

$$\overline{\Delta S^2} = kc_p \rho V^*, \tag{1.43}$$

and taking into account that

$$\left(\frac{1}{\rho}\frac{\partial\rho}{\partial S}\right)_p = \frac{1}{V}\left(\frac{\partial V}{\partial S}\right)_p = -\frac{T}{c_p V^*}\left(\frac{1}{V}\frac{\partial V}{\partial T}\right), \tag{1.44}$$

we find the intensity from (1.42)-(1.44):

$$I_{is} = G\left(\rho\frac{\partial\varepsilon}{\partial\rho}\right)_T^2 \frac{kT^2\sigma^2}{c_p\rho}, \tag{1.45}$$

where σ is the coefficient of volume expansion

$$\sigma = \left(\frac{1}{V}\frac{\partial V}{\partial T}\right)_p.$$

Summing (1.41) and (1.45), we get

$$I = I_{ad} + I_{is} = GkT\left(\rho\frac{\partial\varepsilon}{\partial\rho}\right)_T^2 \left(\beta_S + \frac{T\sigma^2}{\rho c_p}\right). \tag{1.46}$$

Using the well-known thermodynamic relation [58],

$$\beta_T = \beta_S + \frac{T\sigma^2}{\rho c_p}, \tag{1.47}$$

we come to the Einstein equations (1.25) and (1.32) mentioned above.

Thus it is shown that the Einstein formulas for the intensity of scattered light include both the adiabatic and isothermal density fluctuations. Further, it follows from the derivation given here that the Einstein formulas (1.25) and (1.32) should actually contain the isothermal compressibility of the liquid.

However, the relations (1.25) or (1.32) which have been obtained, as has already been shown, are valid only under the condition that

$$\left(\frac{\partial\varepsilon}{\partial T}\right)_\rho^2 \overline{\Delta T^2} = 0.$$

Therefore, the error that arises in this case must be accounted for. For this purpose, we must first not neglect $(\partial\varepsilon/\partial T)_\rho^2\overline{\Delta T^2}$, and then carry out successive calculations. In this situation, it is better to assume the fluctuation of the dielectric constant $\Delta\varepsilon$ to be a function of the pressure p and the entropy S, which corresponds to the natural conditions of the problem of the separation into adiabatic and isobaric fluctuations of $\Delta\varepsilon$. Then

$$\Delta\varepsilon(p, S) = \left(\frac{\partial\varepsilon}{\partial p}\right)_S \Delta p + \left(\frac{\partial\varepsilon}{\partial S}\right)_p \Delta S. \qquad (1.48)$$

By virtue of the independence of p and S (see Appendix I) and for the case of statistical averaging, we obtain the result:

$$\overline{|\Delta\varepsilon(p, S)|^2} = \left(\frac{\partial\varepsilon}{\partial p}\right)_S^2 \overline{\Delta p^2} + \left(\frac{\partial\varepsilon}{\partial S}\right)_p^2 \overline{\Delta S^2}. \qquad (1.49)$$

Thus, $\overline{|\Delta\varepsilon(p, S)|^2}$ immediately breaks up into two parts. The first, $(\partial\varepsilon/\partial p)_S^2\overline{\Delta p^2}$, characterizes the intensity of the light scattered by the adiabatic density fluctuations, while the second, $(\partial\varepsilon/\partial S)_p^2\overline{\Delta S^2}$, gives the intensity of the light scattered by the isobaric density fluctuations.

The calculation of the intensity of the light scattered by the adiabatic and isobaric density fluctuations can be carried out in the following way. Since

$$\left(\frac{\partial\varepsilon}{\partial S}\right)_p = \left(\frac{\partial\varepsilon}{\partial T}\right)_p \left(\frac{\partial T}{\partial S}\right)_p, \quad \text{while} \quad \frac{\partial S}{\partial T} = \frac{c_p\rho V^*}{T},$$

then Eq. (1.49) is now rewritten as

$$\overline{|\Delta\varepsilon(p, S)|^2} = \left(\frac{\partial\varepsilon}{\partial p}\right)_S^2 \overline{\Delta p^2} + \left(\frac{\partial\varepsilon}{\partial T}\right)_p^2 \frac{T}{c_p\rho V^*} \overline{\Delta S^2}. \qquad (1.50)$$

Taking (1.40), (1.39), and (1.43) into account, and assuming that

$$\left(\frac{\partial\varepsilon}{\partial p}\right)_S = \left(\rho\frac{\partial\varepsilon}{\partial\rho}\right)_S \beta_S,$$

we obtain for the statistical average

$$\overline{|\Delta\varepsilon(p, S)|^2} = \left(\rho\frac{\partial\varepsilon}{\partial\rho}\right)_S^2 \beta_S kT\frac{1}{V^*} + \left(\frac{1}{\sigma}\frac{\partial\varepsilon}{\partial T}\right)_p^2 \frac{\sigma^2kT^2}{c_p\rho V^*}. \qquad (1.51)$$

Substituting (1.51) in Eq. (1.31) (assuming $\Delta \varepsilon'_{ik} = 0$), we get

$$I = G\left[\left(\rho \frac{\partial \varepsilon}{\partial \rho}\right)_S^2 kT\beta_S + \left(\frac{1}{\sigma}\frac{\partial \varepsilon}{\partial T}\right)_p^2 \frac{\sigma^2 T^2 k}{\rho c_p}\right]. \tag{1.52}$$

The intensity of the light scattered by the adiabatic fluctuations of the density is determined by the first component of (1.52), and is equal to

$$I'_{ad} = G\left(\rho \frac{\partial \varepsilon}{\partial \rho}\right)_S^2 \beta_S kT, \tag{1.53}$$

while the intensity of the light scattered by the isobaric density fluctuations is determined by the second component of (1.52) and is equal to

$$I'_{is} = G\left(\frac{1}{\sigma}\frac{\partial \varepsilon}{\partial T}\right)_p^2 \frac{\sigma^2 kT^2}{\rho c_p}. \tag{1.54}$$

Comparing (1.41) and (1.45), (1.53) and (1.54), respectively, we find

$$\frac{I'_{ad}}{I_{ad}} = \frac{\left(\rho \frac{\partial \varepsilon}{\partial \rho}\right)_S^2}{\left(\rho \frac{\partial \varepsilon}{\partial \rho}\right)_T^2} \text{ and } \frac{I'_{is}}{I_{is}} = \frac{\left(\frac{1}{\sigma}\frac{\partial \varepsilon}{\partial T}\right)_p^2}{\left(\rho \frac{\partial \varepsilon}{\partial \rho}\right)_T^2}. \tag{1.55}$$

These latter relations differ from unity; therefore, the formulas (1.52) that we have obtained differ from the Einstein formulas (1.36) or (1.25).

However, it must be made clear how large this quantitative difference is. In the derivation of the Einstein formulas, the quantity $(\partial\varepsilon / \partial T)_\rho$ was neglected. If this quantity is actually equal to zero, then Eqs. (1.36) and (1.52) must be identical to one another; consequently, the ratio (1.55) reduces to unity and we get

$$\left(\rho \frac{\partial \varepsilon}{\partial \rho}\right)_S^2 = \left(\rho \frac{\partial \varepsilon}{\partial \rho}\right)_T^2 \text{ and } \left(\frac{1}{\sigma}\frac{\partial \varepsilon}{\partial T}\right)_p^2 = \left(\rho \frac{\partial \varepsilon}{\partial \rho}\right)_T^2. \tag{1.56}$$

In order to find out in which approximation Eq. (1.56) is satisfied, we proceed in the following fashion. Let us consider $d\varepsilon$ as a function of the temperature T and volume V. Then

$$d\varepsilon = \left(\frac{\partial \varepsilon}{\partial V}\right)_T dV + \left(\frac{\partial \varepsilon}{\partial T}\right)_V dT.$$

Dividing both parts of the latter equation by dV, and considering the derivative for constant entropy, we find

$$\left(\frac{\partial \varepsilon}{\partial V}\right)_S = \left(\frac{\partial \varepsilon}{\partial V}\right)_T + \left(\frac{\partial \varepsilon}{\partial T}\right)_V \left(\frac{\partial T}{\partial V}\right)_S . \tag{1.57}$$

From the well-known thermodynamic expressions [58], we get

$$TdS = \left(\frac{\partial E}{\partial V} + p\right)_T dV + \left(\frac{\partial E}{\partial T}\right)_V dT. \tag{1.58}$$

(Here, E is the internal energy of the system.) Assuming the entropy to be a constant, we obtain

$$T\left(\frac{\partial p}{\partial T}\right)_V dV + \rho V c_V\, dT = 0,$$

where c_V is the specific heat at constant volume, whence

$$-\left(\frac{\partial T}{\partial V}\right)_S = \frac{T}{\rho c_V V}\left(\frac{\partial p}{\partial T}\right)_V = -\frac{T}{\rho V c_V}\left(\frac{\partial V}{\partial T}\right)_p \left(\frac{\partial p}{\partial V}\right)_T = \frac{T\sigma}{\rho V c_V \beta_T} . \tag{1.59}$$

Substituting (1.59) in (1.57), we get

$$V\left(\frac{\partial \varepsilon}{\partial V}\right)_S = V\left(\frac{\partial \varepsilon}{\partial V}\right)_T - \left(\frac{\partial \varepsilon}{\partial T}\right)_V \frac{T\sigma}{\rho c_V \beta_T} . \tag{1.60}$$

Using the identity

$$-\rho\frac{\partial \varepsilon}{\partial \rho} \equiv V\frac{\partial \varepsilon}{\partial V} ,$$

in place of (1.60), we find

$$\left(\rho\frac{\partial \varepsilon}{\partial \rho}\right)_T - \left(\rho\frac{\partial \varepsilon}{\partial \rho}\right)_S = -\left(\frac{\partial \varepsilon}{\partial T}\right)_\rho \frac{T\sigma}{c_v \beta_T \rho} . \tag{1.61}$$

It follows from Eq. (1.61) that for $(\partial \varepsilon / \partial T)_\rho = 0$, the isothermal and adiabatic values of $(\rho \partial \varepsilon / \partial \rho)$ are actually equal. Inasmuch as $(\partial \varepsilon / \partial T)_\rho$ is not measured directly by experiment, this quantity must be expressed in terms of parameters which are measurable experimentally. This is most simply done by considering ε as a function of the pressure p and the temperature T. In this case,

$$d\varepsilon = \left(\frac{\partial \varepsilon}{\partial T}\right)_p dT + \left(\frac{\partial \varepsilon}{\partial p}\right)_T dp. \tag{1.62}$$

Dividing (1.62) by dT and taking the derivative at constant ρ, we find

$$\left(\frac{\partial \varepsilon}{\partial T}\right)_\rho = \left(\frac{\partial \varepsilon}{\partial T}\right)_p + \left(\frac{\partial \varepsilon}{\partial p}\right)_T \left(\frac{\partial p}{\partial T}\right)_\rho = \left(\frac{\partial \varepsilon}{\partial T}\right)_p + \left(\frac{\partial \varepsilon}{\partial p}\right)_T \left(\frac{1}{V}\frac{\partial V}{\partial T}\right)_p \frac{1}{\beta_T},$$

whence

$$\left(\frac{\partial \varepsilon}{\partial T}\right)_\rho = \sigma \left[\left(\frac{1}{\sigma}\frac{\partial \varepsilon}{\partial T}\right)_p + \frac{1}{\beta_T}\left(\frac{\partial \varepsilon}{\partial p}\right)_T\right]. \tag{1.63}$$

Taking into account that

$$\frac{1}{\beta_T}\left(\frac{\partial \varepsilon}{\partial p}\right)_T = \left(\rho\frac{\partial \varepsilon}{\partial \rho}\right)_T$$

and that

$$\frac{\beta_T}{\beta_S} = \frac{c_p}{c_V} = \gamma, \tag{1.64}$$

and putting (1.63) in place of $(\partial\varepsilon/\partial T)_\rho$ in Eq. (1.61), we get, finally,

$$\left(\rho\frac{\partial \varepsilon}{\partial \rho}\right)_T - \left(\rho\frac{\partial \varepsilon}{\partial \rho}\right)_S = -\frac{\sigma^2 T}{\rho c_p \beta_S}\left[\left(\frac{1}{\sigma}\frac{\partial \varepsilon}{\partial T}\right)_p + \left(\rho\frac{\partial \varepsilon}{\partial \rho}\right)_T\right]. \tag{1.65}$$

In Eq. (1.65), the right side is expressed in terms of quantities measured experimentally.

A criterion follows from the relations (1.61) and (1.65), satisfaction of which makes possible the use of the Einstein formula. This criterion is the following: the Einstein formulas are applicable if

Table 2. Estimate of the Value of the Left-Hand Side of Eq. (1.67)

Liquid	$\frac{T\sigma^2}{c_p\beta_S\rho}\left[\left(\frac{1}{\sigma}\frac{\partial\varepsilon}{\partial T}\right)_p + \left(\rho\frac{\partial\varepsilon}{\partial\rho}\right)_T\right]\cdot 10^2$
Water	-0.4
Methyl alcohol	-1.6
Ethyl alcohol	-1.2
Ether	-1.7
Benzene	-1.6
Toluene	-3.1

the inequality

$$\left(\frac{\partial \varepsilon}{\partial T}\right)_{\rho} \frac{T\sigma}{c_V \beta_T \rho} \ll \left(\rho \frac{\partial \varepsilon}{\partial \rho}\right)_T \tag{1.66}$$

is satisfied, or the equivalent inequality following from (1.65):

$$\frac{T\sigma^2}{c_p \beta_S \rho} \left[\left(\frac{1}{\sigma} \frac{\partial \varepsilon}{\partial T}\right)_p + \left(\rho \frac{\partial \varepsilon}{\partial \rho}\right)_T\right] \ll \left(\rho \frac{\partial \varepsilon}{\partial \rho}\right)_T . \tag{1.67}$$

All the quantities entering into the criterion (1.67) can be determined by experiment and thus the decision as to the applicability of the Einstein formulas can be made for any case [14].

As an example, estimates are given in Table 2 of the values of the left side of (1.67) for certain liquids.*

The right-hand side of the inequality (1.67) $(\rho\, \partial\varepsilon/\partial\rho)_T$ is close to unity; therefore, on the basis of the estimates that have been made (Table 2), one can conclude that the Einstein formulas (1.25), (1.32), (1.36), and (1.52) give results that are consistent with one another with an accuracy to within about 2%. In those cases in which the requirements for accuracy exceed 2%, it is appropriate to use the relation (1.52).

The difference in the results of the calculation by Eqs. (1.46) and (1.57) becomes more marked if the ratio of the light intensity scattered by the adiabatic and isobaric density fluctuations is compared for the two cases. Here, the ratio, computed from (1.53) and (1.54) is larger than that calculated from the relations (1.41) and (1.42).

As has already been pointed out, the quantity $(\partial\varepsilon/\partial T)_\rho$ cannot be found directly from experiment [59, 60]. But it is calculated, for example, from (1.61). Such a calculation gives only the order of magnitude of $(\partial\varepsilon/\partial T)_\rho$, inasmuch as the values of $(\rho\, \partial\varepsilon/\partial\rho)_T$ and $(\rho\, \partial\varepsilon/\partial\rho)_S$ are very close to one another and, consequently, small errors in the values of either of them lead to large errors in the difference. However, a rough estimate of $(\partial\varepsilon/\partial T)_\rho$ in order of magnitude, made with the use of (1.61), is given in Table 3.

* The dynamic values of $(\rho\, \partial\varepsilon/\partial\rho)_T$ from Table 4 have been used. Another symbol is sometimes used for the static value. This refers to Tables 2 and 3.

Table 3. Estimate of Values of $(\partial \varepsilon / \partial T)_\rho$

Liquid	$(\partial\varepsilon/\partial T)_\rho$ 10^4, °C^{-1}
Water	-1.2
Methyl alcohol	-1.0
Ethyl alcohol	-0.7
Ether	-0.8
Benzene	-0.36
Toluene	-1.0

It is necessary to keep in mind that when we are dealing with calculations of the total light intensity scattered by the density fluctuations, we neglect in comparison with unity not the factor $(\partial\varepsilon / \partial T)_\rho$ but the considerably larger quantity

$$\left(\frac{\partial \varepsilon}{\partial T}\right)_\rho \frac{T\sigma}{cv\beta_T \rho} \sim 10^{-2}.$$

The Scattering Coefficient, the Extinction Coefficient, and the Transverse Scattering Cross Section. The intensity of the scattered light, as is seen from the formulas obtained earlier, depends not only on the physical characteristics of the scattering medium – $(\partial \varepsilon/ \partial \rho)$, $(\partial \varepsilon/ \partial T)$, β_T, β_S, σ – and on its own state, but also on the external conditions of the experiment – I_0, V, and L. Therefore, in place of the absolute intensity of the scattered light, it is convenient to introduce a quantity which does not depend on the conditions of the experiment.

As such a quantity, we introduce the scattering coefficient R, by definition equal to

$$R = \frac{I}{I_0} \frac{L^2}{V} \text{ cm}^{-1}. \tag{1.68}$$

From (1.25) and (1.68), we get for the scattering coefficient

$$R = \frac{\pi^2}{2\lambda^4} \left(\rho \frac{\partial \varepsilon}{\partial \rho}\right)_T^2 \beta_T k T (1 + \cos^2\theta). \tag{1.69}$$

For observation at a scattering angle $\theta = 90°$, Eq. (1.69) transforms to

$$R_{90} = \frac{\pi^2}{2\lambda^4} \left(\rho \frac{\partial \varepsilon}{\partial \rho}\right)_T^2 \beta_T \, kT.$$

It is not difficult to write the scattering coefficient also for all the other expressions for the absolute intensity of scattered light, making use of the definition (1.68). The scattering coefficient depends only on the physical properties of the medium. Instead of the scattering coefficient R, one often uses the extinction coefficient h, a quantity which characterizes the damping of a parallel beam of incident light, propagating in a homogeneous scattering, but not absorbing, optical medium. The extinction coefficient is simply related to the scattering coefficient R.

The connection of the extinction coefficient and the intensity of the incident light is given in the Introduction (see p. 7). In order to find the total attenuation of the light as a consequence of the scattering, it is necessary to know the dependence of the scattering coefficient on the angle and to carry out the corresponding integration over the complete sphere.

If we take the scattering coefficient and its dependence on angle from (1.69) and (1.70), then we can obviously write

$$h = \int_0^{2\pi} d\varphi \int_0^{\pi} R_{90} (1 + \cos^2 \theta) \sin \theta \, d\theta, \tag{1.71}$$

whence we find that

$$h = \frac{16}{3} \pi R_{90}. \tag{1.72}$$

The losses of light in the medium arise not only from molecular scattering, but also as a consequence of light absorption. If the attenuation of a parallel beam of light as the result of absorption is determined by the quantity a', then the total attenuation of the light will be [67]

$$I = I_0 e^{-mx}, \tag{1.73}$$

where

$$m = h + a'. \tag{1.74}$$

In a nonabsorbing medium, $a' = 0$ and the attenuation of the light results only from the scattering. Thus, it is evident that there are two methods of determining the scattering coefficient: (1) by a measurement of the absolute intensity of the scattered light (direct measurement of the scattering coefficient); and (2) by the measurement of the attenuation of a parallel beam of light passing through the scattering medium (direct measurement of h and calculation of R).

The second method can be effective when R is large and, consequently, the attenuation of the incident light is appreciable even at short distances, or when the light path length x is very long, for example, the attenuation of light in the atmosphere.

In the case of molecular light scattering in gases, crystals, and pure liquids, the determination of R from the laboratory measurements of extinction is accompanied by large errors and is therefore unsatisfactory.

On the other hand, for media which are characterized by large extinction coefficients, such as solutions of polymers and proteins, for example, the measurement of the extinction and the determination of the scattering coefficient from these measurements can give excellent results [43].

The term "transverse scattering cross section for light" is also used [50]. By definition, the transverse cross section Q is the coefficient of proportionality between the intensity of scattered light I and the intensity of the incident light:

$$Q = L^2 \frac{I}{I_0} = RV. \tag{1.75}$$

Comparing (1.75) with (1.25) or with (1.69), (1.70), we get

$$Q = \frac{\pi^2 V}{2\lambda^4} \left(\rho \frac{\partial \varepsilon}{\partial \rho}\right)_T^2 \beta_T kT (1 + \cos^2 \theta). \tag{1.76}$$

The transverse scattering cross section is especially useful when one is speaking of the scattering of light by light or by particles [62].

Solutions. The calculation of the intensity of the light scattered by fluctuations of concentration in binary solutions was first performed by Einstein [14].

For the computation of the intensity of light scattered by fluctuations in the concentration, we proceed in exactly the same fashion as in the case of pure liquids. To be precise, we substitute in Eq. (1.32) only that part of $\Delta\varepsilon$ from (1.34) which is determined by the concentration fluctuations. On the basis of (1.34) and (1.31), we obtain

$$I_{\text{conc}} = I_0 \frac{V^* \pi^2 V}{2\lambda^4 L^2} \left(\frac{\partial \varepsilon}{\partial c}\right)^2_{\rho, T} \overline{\Delta c^2} (1 + \cos^2 \theta), \tag{1.77}$$

where

$$\overline{\Delta c^2} = ckT \Big/ \frac{\partial p}{\partial c} V^* \tag{1.78}$$

[see Appendix I, Eq. (I.18a)]. Substituting (1.78) in (1.77), we get

$$I_{\text{conc}} = I_0 \frac{\pi^2 V}{2\lambda^4 L^2} \left(\frac{\partial \varepsilon}{\partial c}\right)^2_{\rho, T} \frac{ckT}{\frac{\partial p}{\partial c}} (1 + \cos^2 \theta).* \tag{1.79}$$

This formula can easily be transformed by means of the classical equation of Van't Hoff to a form which is more convenient for practical use.

The Van't Hoff equation for the osmotic pressure can be written in the following form:

$$p = \frac{ckT}{V_0} = \frac{n_1 kT}{V}, \tag{1.80}$$

where V_0 is the molecular volume of the solvent and n_1 is the number of dissolved molecules in the entire volume V occupied by the solvent. Here, it can be assumed without great error that the total volume occupied by the solution for highly dilute solutions is equal to the volume of the solvent.

In calculations based on the gram mole, the Van't Hoff equation takes the form

$$p = \frac{cR_g T}{M}, \tag{1.81}$$

* There is an unimportant difference in (1.79) from the formula of Einstein [14], in that (1.79) was obtained for the incidence of natural light, while Eq. (17) of reference [14] is for linearly polarized light.

where R_g is the gas constant and M is the molecular weight of the dissolved substance. Combining (1.81) and (1.79), we get

$$I_{\text{conc}} = I_0 \frac{\pi^2 V}{2\lambda^4 L^2} \left(\frac{\partial \varepsilon}{\partial c}\right)^2_{\rho, T} \frac{cM}{N_A}(1+\cos^2\theta),$$

where N_A is Avogadro's number.

On the basis of (1.68), we obtain an expression for the scattering coefficient as a consequence of the concentration fluctuations

$$R_{\text{conc}\,(90)} = \frac{\pi^2}{2\lambda^4} \left(\frac{\partial \varepsilon}{\partial c}\right)^2_{\rho, T} \frac{cM}{N_A}. \tag{1.82}$$

Equation (1.81), and consequently (1.82) also, are suitable only for ideal solutions.

For most solutions of practical interest, the Van't Hoff equation (1.81) must be replaced by a more complicated expression [61]

$$p = \frac{R_g T c}{M} + B'c^2 + D'c^3, \tag{1.83}$$

where B' and D' are the first and second virial coefficients. The first term of the series (1.83) is identical with (1.81).

On the basis of (1.79), (1.83), and (1.68), we get [in place of (1.82)]

$$K \frac{c}{R_{\text{conc}\,(90)}} = \frac{1}{M} + 2B'c + 3D'c^2 + \ldots, \tag{1.84}$$

where

$$K = \frac{\pi^2 \left(\frac{\partial \varepsilon}{\partial c}\right)^2_{\rho, T}}{2\lambda^4 N_A},$$

or, by means of the extinction coefficient,

$$H \frac{c}{h} = \frac{1}{M} + 2B'c + 3D'c^2 + \ldots, \tag{1.85}$$

where

$$H = \frac{16\pi}{3} K.$$

The expressions (1.84) and (1.85) show that the experimental determination of R_{90} or h in the solution for different concentrations, and the measurement of $(\partial \varepsilon / \partial c)_{\rho,T}$, suffice to find the molecular weight M. Although Eq. (1.79) was obtained by Einstein as early as 1910, it found practical application in 1944 when Debye [42] noted that the study of light scattered by solutions can give a reliable method for the determination of molecular weights in, and the sizes of the molecules of, the dissolved substance – polymers, proteins, or ions. The first successful investigations of Debye and his coworkers prompted the appearance of many researches devoted to the study of light scattering by solutions, and a great deal of interesting material has now been accumulated on the investigation of the light scattered by solutions of polymers, proteins, and polyelectrolytes. A most complete review of this material has been made by Stacey [43].

Study of the Intensity of Light Scattered by Fluctuations in Anisotropy. As the result of the calculations that have been made, relations have been obtained that express the intensity of the light scattered by adiabatic and isobaric fluctuations in the density and concentration, as a function of parameters accessible to experiment. However, it is still impossible to subject the theoretical relations to an experimental test, inasmuch as the intensity of the light scattered by fluctuations in the anisotropy have not been taken into account. Experiment shows that the intensity of light scattered by fluctuations in the anisotropy in most cases makes a substantial contribution to the total flux of scattered light. Sometimes, the amount of light involved in fluctuations of the anisotropy exceeds by several fold the light intensity scattered by the adiabatic and isobaric density fluctuations. Therefore, before comparing the theoretical calculations of the intensity of the scattered light in a liquid with experiment, it is highly imperative to take into account the scattering of light by fluctuations in the anisotropy.

It is not possible to carry out a calculation of $\overline{|\Delta\varepsilon_{ik}'|^2}$, as has been done for $\overline{|\Delta\varepsilon(p, S)|^2}$ and $\overline{|\Delta\varepsilon(c)|^2}$, inasmuch as the parameters characterizing $\Delta\varepsilon_{ik}'$ for the nonequilibrium state of the liquid are not known.

A number of attempts at the calculation of $\overline{|\Delta\varepsilon_{ik}'|^2}$ for liquids are known from the literature. Some of them are discussed in [63-66], but not one of them leads to satisfactory results.

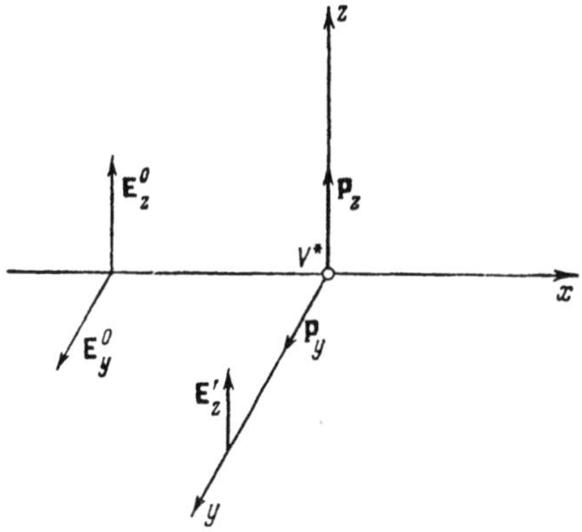

Fig. 3. The polarization of light scattered by isotropic fluctuations.

Considerable success has been attached to the theories of Leontovich [39], Rytov [40], and Ansel'm [69]; however, even these theories are incomplete, and cannot guarantee sufficient accuracy for a comparison with experiment of the computed intensity of scattering by $\overline{|\Delta\varepsilon'_{ik}|^2}$.

However, the theories mentioned have a value in principle for an understanding of the mechanism of scattering by fluctuations in the anisotropy, and we shall return to them; here, an account of the intensity of the depolarized scattering will be made in a way that is not connected with any specific theory.

It was already mentioned above that the fluctuations in the pressure and entropy or density, temperature and concentration in the approximation considered cannot produce an anisotropy in the volume of fluctuations, while fluctuations of orientation (fluctuations of anisotropy) produce such an anisotropy. It is not difficult to see that the light scattered by the fluctuations, which does not possess anisotropy, must be completely polarized in observation at an angle $\theta = 90°$, independent of whether the scattering volume is illuminated by linearly polarized or natural light, while scattering by fluctuations that produce anisotropy leads to a depolarization of the scattered light in all cases.

In confirmation of what has been said above, we consider a parallel beam of natural light, propagating along the x axis and in-

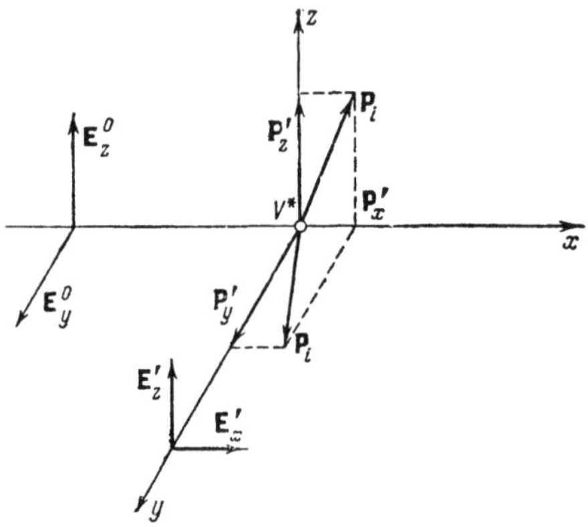

Fig. 4. The depolarization of light scattered by fluctuations in the anisotropy.

cident on an isotropic volume of fluctuations V*, located at the origin of the coordinate system (Fig. 3). In the incident natural light, $\mathbf{E}_z^0 = \mathbf{E}_y^0$. If the scattering element is isotropic, then the direction of the induced light wave of (electric) polarization $\mathbf{P}_z$ and $\mathbf{P}_y$ is identical with the direction of $\mathbf{E}_z^0$ and $\mathbf{E}_y^0$, respectively. The field of the scattered light wave $\mathbf{E}' \sim \Delta\varepsilon\ddot{\mathbf{P}}$, where $\Delta\varepsilon$ is a scalar, and therefore the electric vector $\mathbf{E}_z' \| \mathbf{P}_z$ and the vector $\mathbf{E}_y' \| \mathbf{P}_y$.

For observation along the y axis, only the $\mathbf{E}_z'$ component of the field of the scattered wave will be present, inasmuch as the $\mathbf{E}_y'$ component cannot be propagated, as a consequence of the transverse character of the light oscillations in the direction of the y axis.

In the case in which the scattering element V* has optical anisotropies, the induced polarizations $\mathbf{P}_i$ are not parallel with $\mathbf{E}_i^0$, since $\mathbf{P}_i \sim \Delta\varepsilon_{ik}\mathbf{E}_i^0$, where $\Delta\varepsilon_{ik}$ is a tensor [49, 70]. Therefore, $\mathbf{E}_z'$ and $\mathbf{E}_x'$ components appear in the electric field of the scattered light wave (Fig. 4). The depolarization factor for scattering by the fluctuations $\Delta\varepsilon_{ik}'$ will be

$$\rho_u = \frac{|E_x'|^2}{|E_z'|^2} = \frac{i_x}{i_z}. \tag{1.86}$$

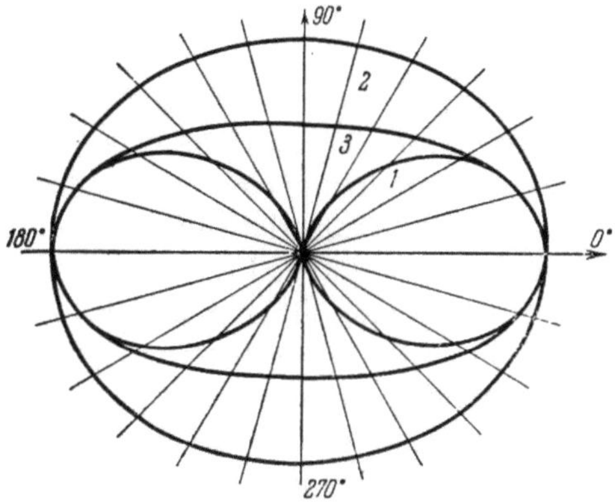

Fig. 5. Angular dependence of the depolarization factor of optical molecular scattering. 1) $\rho_u(\Delta\varepsilon)$; 2) $\rho_u(\Delta\varepsilon_{ik}^{(S)})$; 3) $\rho_u(\Delta\varepsilon_{ik}^{(a)})$.

For account of the intensity of the light scattered by fluctuations of the anisotropy, we proceed in the following fashion: we denote the intensity of the light scattered by the fluctuations of density and anisotropy by I and i, respectively, and the total intensity by $\mathcal{I} = I + i$.

If we now assume that the incident natural light is propagated along the x axis, the observation is made along the y axis, while the z axis is perpendicular to the plane of scattering; then, obviously, $I = I_z$ and $i = i_z + i_x$. Thus, the total scattered light intensity will be

$$\mathcal{I} = I_z + i_x + i_z. \tag{1.87}$$

On the other hand, the depolarization of the total scattered light is

$$\Delta_u = \frac{i_x}{I_z + i_z}. \tag{1.86a}$$

We express $\mathcal{I}$ in terms of the experimentally measured Δ_u and ρ_u and the value $I = \mathcal{I}_z$ computed earlier. Using the relations (1.87), (1.86a), and (1.86), this is not difficult to do. Eliminating i_x and i_z from the two relations, we get

$$\mathcal{J} = I\left\{1 + \frac{\Delta_u(1+\rho_u)}{\rho_u - \Delta_u}\right\}. \tag{1.88}$$

Thus, (1.88) gives an expression for the total intensity of light scattered by thermodynamic fluctuations of pressure and entropy and takes into account scattering from the fluctuations $\Delta\varepsilon_{ik}$ [see (1.3)]. The light scattering from the fluctuations $\Delta\varepsilon_{ik}^{(a)}$ far from the absorption band can be regarded as small in comparison with the other sources of scattering [45]. Placzek [50] and Landau and Lifshitz [47] made theoretical investigations of the depolarization of light scattered by all three components of the tensor $\Delta\varepsilon_{ik}$ in (1.3) and found that for scattering by $\Delta\varepsilon_{ik}^{(S)}$ in observation of the scattered light at the angle $\theta = 90°$, $\rho_u = 6/7$. In the theories of Leontovich [39], Rytov [40], and Ansel'm [69], it was also found that $\rho_u = 6/7$ for $\theta = 90°$. This result of the theory is found to be in excellent agreement with experiment [70-73].

Values of the depolarization of scattered light for the three components of $\Delta\varepsilon_{ik}$ in (1.3) are given in Table 4 for illumination of the scattering volume by polarized and natural light, as a function of the angles ϑ and θ (see also Fig. 5).

We now take into account only the depolarization of the scattered light in the calculation, arising as a consequence of the scattering by the fluctuations $\Delta\varepsilon_{ik}^{(S)}$. Consequently, it is necessary to set $\rho_u = 6/7$ in (1.88) and then

$$\mathcal{J}_{90} = I_{90}\frac{6+6\Delta_u}{6-7\Delta_u}. \tag{1.89}$$

Account of scattering by fluctuations of the anisotropy was first made by Cabannes [74, 75] and the factor

$$f(\Delta) = \frac{6+6\Delta_u}{6-7\Delta_u} \tag{1.90}$$

bears the name of the Cabannes factor.

It also follows from (1.89) and (1.87) that the intensity of light scattered by the anisotropy fluctuations is equal to

$$i = I\frac{13\Delta_u}{6-7\Delta_u}. \tag{1.91}$$

Thus the formula for the intensity of light scattered by the fluctuations of density and anisotropy, which is subject to experi-

Table 4. Depolarization Factor of the Scattered Light for Different Types of Scattering as a Function of the Scattering Angle

Depolarization / Type of scattering	Excitation by polarized light E_{0z} ρ_v	Excitation by natural light ρ_u	ρ_u for $\theta=\pi/2$
$I\sim\overline{\lvert\Delta\varepsilon\rvert^2}$	$\sin^2\vartheta$	$\cos^2\theta$	0
$I\sim\overline{\lvert\Delta\varepsilon_{ik}^{(S)}\rvert^2}$	$\frac{3}{3+\sin^2\vartheta}$	$1-\frac{1}{7}\sin^2\theta$	$\frac{6}{7}$
$I\sim\overline{\lvert\Delta\varepsilon_{ik}^{(a)}\rvert^2}$	$\cos^2\vartheta$	$\frac{1}{1+\sin^2\theta}$	$\frac{1}{2}$

mental test, has the form

$$\mathcal{J}_{90}=\frac{I_0V\pi^2}{2L^2\lambda^4}\left[\left(\rho\frac{\partial\varepsilon}{\partial\rho}\right)_S^2\beta_S kT+\left(\frac{1}{\sigma}\frac{\partial\varepsilon}{\partial T}\right)_p^2\frac{\sigma^2kT^2}{c_p}\right]\frac{6+6\Delta_u}{6-7\Delta_u}. \qquad (1.92)$$

For the scattering coefficient,

$$R_{90}=\frac{\pi^2}{2\lambda^4}\left[\left(\rho\frac{\partial\varepsilon}{\partial\rho}\right)_S^2\beta_S kT+\left(\frac{1}{\sigma}\frac{\partial\varepsilon}{\partial T}\right)_p^2\frac{\sigma^2kT^2}{c_p}\right]\frac{6+6\Delta_u}{6-7\Delta_u}. \qquad (1.93)$$

If the condition (1.67) is satisfied with sufficient accuracy, then Eq. (1.93) goes over into the Einstein–Cabannes formula

$$R_{90}=\frac{\pi^2}{2\lambda^4}\left(\rho\frac{\partial\varepsilon}{\partial\rho}\right)_T^2\beta_T kT\frac{6+6\Delta_u}{6-7\Delta_u}. \qquad (1.94)$$

In Eqs. (1.92) and (1.94), the scattering by adiabatic and isobaric fluctuations of density are taken into account, as well as those by anisotropy fluctuations. These formulas can now be compared with the results of experiment.

2. Critical Opalescence and Light Scattering in Second-Order Phase Transitions

Critical Opalescence. Near the critical point of individual substances or near the critical temperature of a layer or mixture of binary solutions, a great increase is observed in the intensity of scattered light [78]. The phenomenon bears the name of critical opalescence. Smoluchowski explained this phenomenon by the strong growth of local density fluctuations in the approach to the critical state.

The fact is that the isothermal compressibility β_T increases strongly in the approach to the critical state (it goes to infinity at the critical point if one uses the Van der Waals equation). Moreover, as follows from Eqs. (1.25) and (1.69), the intensity of the scattered light I and the scattering coefficient R increase without limit, going to infinity at the critical temperature. Equation (1.92) allows us to find what component is responsible for the tremendous increase in the intensity of the scattered light.

By taking (1.47) into account, we rewrite (1.93) in the following form:

$$R_{90}=\frac{\pi^2 kT}{2\lambda^4}\left[\left(\rho\frac{\partial\varepsilon}{\partial\rho}\right)_S^2\beta_S+\left(\frac{1}{\sigma}\frac{\partial\varepsilon}{\partial T}\right)_p^2(\beta_T-\beta_S)\frac{1}{\rho}\right]f(\Delta). \tag{2.1}$$

The intensity of the light scattered by the adiabatic density fluctuations close to the critical point and at this very point does not represent any catastrophic change; β_S changes slightly, remaining finite throughout the entire range of temperature change, while the second component, which determines the light intensity scattered by the isobaric fluctuations, goes to infinity together with β_T at the critical point.

This result also means that the density fluctuations in the immediate vicinity of the critical temperature cannot be calculated as was done earlier. The approach of β_T toward infinity as the critical temperature is neared means that in a power series expansion of the change of the free energy, the terms with the second derivative of the free energy with respect to the density go to zero. Smoluchowski [13], attempting to get over this difficulty, kept the fourth derivatives of the free energy with respect to the density in the series just mentioned.

In this case, the intensity of the scattered light at the critical point remains finite, but here another absurdity arises. To be precise, the intensity of the scattered light is now shown to be proportional not to the volume, as it is natural to expect, but to $V^{3/2}$.

This is explained by the fact that the assumption made by Smoluchowski lies beyond the limits of applicability of the Boltzmann principle, which establishes the connection between the free energy of the nonequilibrium state and the probability of this state [16].

An escape from these difficulties was found by Ornstein and Zernike [80, 82]. The physical idea of Ornstein and Zernike, on which their theory is based, is that a strong increase in the compressibility close to the critical point leads to a great change in the density of the medium even under the action of very small forces. Under such conditions, the resultant density fluctuation can have an effect on the state of the medium at distances that are much greater than the radius of molecular interaction under normal conditions. The free energy now depends not only on $\Delta\rho/\rho$, but also on the gradient of this quantity. Thus, two density fluctuations, $\Delta\rho_1$ and $\Delta\rho_2$, arising in two neighboring volumes V_1^* and V_2^*, are no longer statistically independent, as was earlier assumed, although even in this case, it is assumed that the interaction decreases with increase in the distance between the two neighboring fluctuations and, finally, the two fluctuations can again be regarded as independent at large distances.

The calculation of the fluctuations in such a case should be carried out for the condition that the change in the free energy depends on a term which takes the spatial inhomogeneity into account. The density fluctuations $\Delta\rho$ under the circumstances mentioned is seen to be equal to (see Appendix I):

$$\overline{\Delta\rho^2} = \frac{kT}{V\left[\frac{1}{\beta_T} + 8\pi^2 \frac{f^*}{\lambda^2}(1-\cos\theta)\right]}. \tag{2.2}$$

Here f^* is a constant which characterizes the intermolecular interaction; it is larger the greater the correlation radius between the fluctuations in the neighboring volumes.

Substituting (2.2) in (1.23), we get, without taking $f(\Delta)$ into account,

$$R = \frac{\pi^2}{2\lambda^4}\left(\rho\frac{\partial\varepsilon}{\partial\rho}\right)_T^2 \frac{kT(1+\cos^2\theta)}{\left[\frac{1}{\beta_T} + 8\pi^2\frac{f^*}{\lambda^2}(1-\cos\theta)\right]}. \tag{2.3}$$

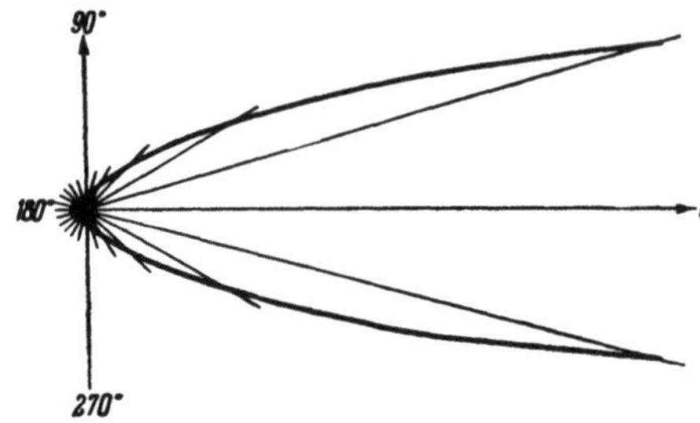

Fig. 6. The angular dependence of the intensity of light scattered at the critical temperature.

Equation (2.3) was first obtained by Ornstein and Zernike [80] and is a generalization of the Einstein formula to an arbitrary range of temperatures, including the critical.

At the critical temperature, part of the intensity of the scattered light is associated with fluctuations of the anisotropy; therefore, (2.3) must be multiplied by the Cabannes factor $f(\Delta)$ [see (1.90)].

One can see how the depolarization Δ changes in the approach to the critical temperature from the examples given in Sec. 19.

At the critical point itself, $\beta_T \to \infty$ and (2.3) goes over into

$$R = \frac{kT}{16\lambda^2}\left(\rho\frac{\partial\varepsilon}{\partial\rho}\right)_T^2 \frac{1+\cos^2\theta}{f^*(1-\cos\theta)}. \qquad (2.4)$$

It is important to note that the intensity of the scattered light at the critical temperature is proportional to λ^{-2} and not to λ^{-4} (as it is far from the critical temperature), while the scattering indicatrix is markedly different from the Rayleigh value.

As is clear from Fig. 6, the principal part of the light scattered in a material at its critical point is directed forward.

For the calculation of the extinction coefficient h in the case of critical opalescence, one must write, in accord with (1.71),

$$h = \int_0^{2\pi} d\varphi \int_0^{\pi} R \sin\theta \, d\theta, \quad (2.5)$$

where R is determined by the expression (2.4).

In this case, the integral (2.5) diverges for small angles θ. However, this apparent difficulty is explained by the fact that the integration must be carried out over angles not less than the diffraction angle $\theta \sim \lambda/d$, where d is the linear dimension of the scattering volume [47, 85]. Consequently, the global scattering remains finite and is a function of the dimensions of the scattering body.

As one approaches the critical temperature of mixing or separation of two liquids, the change in the osmotic pressure with concentration, $\partial p/\partial c$, tends to zero and the intensity of the scattered light (1.79) increases without limit.

Here, as in the previous case, account of the correlation of fluctuations in two neighboring volumes of the liquid is necessary; however, in the case of critical opalescence in a solution, it is necessary to take into account that $\overline{\Delta c_1 \Delta c_2} \neq 0$, where Δc_1 and Δc_2 are the fluctuations of the concentration in the neighboring volumes V_1^* and V_2^*. A calculation that is entirely analogous to that given above also leads to a solution of the problem. Questions of the theory of light scattering at the critical temperature have been widely studied since Ornstein and Zernicke [80-82] first gave the solution of this problem. The most general phenomenological theory is that of Leontovich [16, 83]; this theory makes it possible to find the fluctuations of the different parameters under various conditions, including the case of the critical state of the substance. Vladimirskii [46] introduced generalized thermodynamic functions which take into account the intermolecular interaction and permit a formula to be found for the intensity of scattered light that is identical with (2.3). Improvement and generalization of the theory of Ornstein and Zernike from the viewpoint of molecular theory was made by Klein and Tisza [92] and by Münster [84]. Interesting attempts at carrying out a molecular calculation of the intensity of scattered light are due to Yvon [93], Davydov [88], Zimm [94], Fixman [90], and others. There is no systematic molecular theory in these investigations, inasmuch as to date there is no molecular theory of liquids and, therefore, it is not possible to find, for example, the internal field acting on the molecule. One must introduce a whole series of simplifying assumptions; for example, it is assumed that the field acting on the molecules of the liquid is equal to the Lorentz effective field. In considering interacting, optically isotropic molecules, Davydov [88] did not find the value of the mean-square fluctuations of the density of the medium as was done in all the phenomenological theories, but instead summed the radiation of separate molecules with account of the amplitude and phase of their oscillation.

As a result of such a calculation, a formula was obtained which is valid for a liquid (isotropic molecules) over the whole temperature interval, including the critical temperature.

The formula obtained by Davydov [88] is identical with the formula of Ornstein and Zernike [82]. If we write the Davydov formula in our notation, then we have for the scattering coefficient

$$R = \frac{\frac{\pi^2 V}{2\lambda^4}\left(\rho\frac{\partial\varepsilon}{\partial\rho}\right)_T^2 kT\,(1+\cos^2\theta)}{\frac{1}{\beta_T}+\frac{4}{3}\pi^2\frac{N_A kT}{V}\frac{r^2}{\lambda^2}\,(1-\cos\theta)}, \tag{2.6}$$

where r has the meaning of the correlation radius and V the molar volume. Comparing (2.6) and (2.2), we find that

$$f^* = r^2\frac{N_A kT}{6V} = \frac{1}{6}r^2 N_L\, kT. \tag{2.7}$$

In [88], a connection was also obtained between r and the Van der Waals constants under the assumption that the molecules are solid spheres, the energy of interaction between which $\sim A/r^6$. Under these assumptions,

$$r^2 = 2\rho\left(\frac{3b}{2\pi}\right)^{2/3}\frac{\left(\frac{a}{R_g T}-\frac{b}{5}\right)}{N_A^{5/3}}.$$

By using the well-known connection of the Van der Waals constants a and b with the critical pressure and temperature, we obtain

$$r^2 = 4.42 \cdot 10^{-28} \left(\frac{T_{cr}}{p_{cr}}\right)^{6/5} \rho_{cr}, \tag{2.8}$$

where p_{cr} is expressed in atmospheres and ρ_{cr} in cm^{-3}. This simplified model gives for r in ether (T_{cr} = 467°K, p_{cr} = 35.6 atm, ρ_{cr} = 0.2625 cm^{-3}) the value $r \approx 0.8 \cdot 10^{-7}$ cm, while the experimental value [91] gives $r \approx 26 \cdot 10^{-7}$ cm. In the work of Fixman [90], the calculation is carried out for temperatures up to the critical, but the anisotropy of the molecules ($\Delta \varepsilon_{ik}^{(S)}$) is taken into account. A formula is then obtained which is identical with the Einstein–Cabannes formula (see Sec. 1). Although the molecular calculations are by necessity limited and do not give new results, they have a value in principle and sometimes can explain the physical significance of the parameters of phenomenological theory.

The formula which expresses the scattering coefficient or the intensity of scattered light can be written in a form equivalent to (2.3) by means of the radial distribution function for the density g(r), in place of the factor f^* which characterizes the intermolecular interaction.

Such a formula can be obtained by starting either from an expression for the intensity of scattering of x rays [84, 87, 95, 97, 98, 101, 103], or from an expression for $\overline{\Delta\rho^2}$ written in terms of the density radial distribution function or the correlation function [16, 48, 80–84, 86, 87, 103, 104]. For irradiation of a liquid consisting of monatomic molecules by a polarized beam of x rays, the intensity of the scattered light is equal to

$$I = \frac{A^2}{L^2} \sum_i \sum_j \exp(iqr_{ij}), \tag{2.9}$$

where $A^2 = I_0 a^2 f_a^2$ is the square of the scattering amplitude, $a = e^2/mc^2$ is the Thomson radius of the electron, f_a is the atomic scattering factor, and m and e are the mass and charge of the electron, respectively.

Averaging over all configurations gives, for irradiation by an unpolarized beam of x rays, an intensity

$$I(\theta) = I_0 \frac{a^2 f_a^2}{L^2} \frac{(1+\cos^2\theta)}{2} \int \rho^{(1)}(q)\, dq + \int \rho^{(2)}(qq') \exp(iqr)\, dq\, dq' \Big\} \tag{2.10}$$

After integration of (2.10) over dq and dq', we get

$$I(\theta) = I_0 \frac{a^2 f_a^2}{L^2} N \frac{(1+\cos^2\theta)}{2} \left\{ 1 + 4\pi\rho \int_0^\infty g(r) \frac{\sin(qr)}{qr} r^2 dr \right\}. \tag{2.10a}$$

Introducing the new function

$$I(q) = R' \frac{1}{(1+\cos^2\theta)} - 1 \text{ where } R' = \frac{I(\theta)\, 2L^2}{I_0 a^2 f^2 N},$$

we find from the two previous expressions

$$I(q) = 4\pi\rho \int_0^\infty g(r) \frac{\sin(qr)}{qr} r^2 dr$$

Carrying out the inverse Fourier transform, we obtain

$$rg(r) = \frac{1}{2\pi^2\rho} \int_0^\infty qI(q) \sin(qr)\, dq. \tag{2.10b}$$

Thus, as is clear from the last formula, the study of the angular distribution of diffuse scattering of x rays makes it possible to find the radial distribution function for the density or the correlation function. For visible light, we get from (2.10a) for the scattering coefficient [87, 103]:

$$R = \frac{\pi^2}{2\lambda^4} \left(\frac{1+\cos^2\theta}{2} \right) \left(\rho \frac{\partial\varepsilon}{\partial\rho} \right)_T^2 \left\{ 1 + 4\pi\rho \int_0^\infty g(r) \frac{\sin qr}{qr} r^2 dr \right\}, \tag{2.11}$$

where, instead of the compressibility β_T [see (1.69)], we have the quantity

$$\beta_T = \frac{1}{kT\rho} \left\{ 1 + 4\pi\rho \int_0^\infty g(r) \frac{\sin(qr)}{qr} r^2 dr \right\}.$$

Equation (2.11) is equivalent to (2.3).

Equation (2.11) can be transformed to a form corresponding to the expression for I(q), and g(r) can be found from the angular dependence of R close to the critical temperature [see (2.6)] by the inverse Fourier transform.

Measurements of the angular distribution of the intensity of scattered light near the critical point for a pure substance are of special interest in connection with the new theoretical investigations of Giterman and Kontorovich [606], in which the effect of spatial dispersion on the propagation of sound waves and light scattering is taken into account (see also Sec. 5). The function g(r) is not given by rigorous theoretical calculation even in the case of comparatively simple organic liquids. However, it can be obtained from experiment by the study of the angular dependence of the intensity of the scattered x rays.

The correlation function, obtained by such a method, has been used successfully for the calculation of the depolarization factor of scattered light in a condensed medium (see Sec. 19).

The determination of g(r) from the angular dependence of the intensity of scattered light near the critical temperature is undoubtedly of interest, but, so far as we know, such a determination has not yet been made.

Attempts have already been made to find the correlation function from the angular dependence of the intensity of scattered light close to the critical temperature of mixing of liquids [98] on the basis of formulas of the type (2.10b). It was later shown that such a determination is possible even in the case of mixtures, but the calculation must of necessity be complicated [99].

Komarov and Fisher [103] have shown that the space–time correlation function $G(|\mathbf{r}|, t)$ can be obtained from the intensity distribution over the frequencies in the components of the fine structure of the Rayleigh line. But for this purpose it is necessary to know the true intensity distribution. Pecora [604] has put forward an investigation which leads to the same conclusion. However, the practical determination of the spatial part of the correlation function from the scattering of visible liquids far from the critical point is hardly possible, inasmuch as $rg \ll 1$ even for scattering angles of about 1°.

Scattering of Light in Second-Order Phase Transitions. The intensity of scattered light near the temperature of a second-order phase transition increases so strongly that this phenomenon recalls that of critical opalescence.

The theory of light scattering for second-order phase transitions in the vicinity of the critical Curie point has been given by Ginzburg [23, 105], and later developed by him in collaboration with Levanyuk [106].

Independently of theoretical researches, Yakovlev, Velichkina, and Mikheeva [22] discovered that in a narrow range of temperatures ($\Delta T \sim 0.1°$) for the $\alpha \rightleftharpoons \beta$ transition in quartz, the scattered light intensity increases approximately by a factor of 10^4 relative to the intensity of the light scattered on both sides of the transition temperature. This singular phenomenon is excellently explained and described quantitatively by the Ginzburg theory.

The physical reason which lies behind such a significant increase in the light scattering near the critical Curie point is that at this temperature of the region, great thermal fluctuations are generated of some characteristic quantity η which figures in the theory of second-order phase transitions developed by Landau, and serves there as a quantitative measure of the "degree of ordering" [107, 51]. The fluctuations of the quantity η lead to the result that in a crystal, for example, the regions whose properties differ from the properties of the crystal at the equilibrium temperature T can appear and disappear. As a parameter η one can select various characteristics of the medium. To make clear the physical meaning of η, we point out several examples. Thus, in ferroelectrics, one can usually use the component of the spontaneous polarization $\mathbf{P}_S$ as the characteristic parameter η. The fluctuations of η in this case bring about the appearance and disappearance of small regions of spontaneous polarization at a temperature higher than the Curie temperature, i.e., where such an appearance is impossible in the equilibrium state.

In the transition from He I to He II, the characteristic parameter η is associated with the density of the superconducting phase ($\rho_S = \eta^2$). As another illustration we may take the case where the characteristic parameter η describes the $\alpha \rightleftharpoons \beta$ transition in quartz. Here we choose as η a quantity which is proportional to the shift in the equilibrium position of the atoms in the crystalline lattice. The problem of the selection of η in [23, 105, 106] has not yet been worked out in detail, inasmuch as it is not important for further developments of the theory, and the quantity η itself does not enter into the final result. The temperature of the $\alpha \rightleftharpoons \beta$ transition is close to the temperature of the critical Curie point T_c for quartz.*

Near the transition temperature in crystalline quartz, the fluctuations of η indicate that above the temperature of transformation of α-phase into β-phase, regions with symmetry of the α-phase appear and disappear. Inasmuch as the α and β modifications of quartz are characterized by different dielectric constants, it is clear that the fluctuations of η produce a fluctuation of ε and, consequently, an additional reason for light scattering.

It was shown earlier that the fluctuations of η ought to generate an anomalous scattering of x rays [108] and lead to an increase in the thermal noise [109]. In [92], the possibility was also pointed out of an increase in light scattering at the λ point in a solid. The difference in the dielectric constants above and below the transition temperature, $\varepsilon_2 - \varepsilon_1$, cannot depend on the sign of η and, therefore, it is necessary to assume [106] that $\varepsilon_2 - \varepsilon_1 = a\eta_0^2$, where η_0 is the equilibrium value of η. Therefore, the fluctuation of the dielectric constant near the transition temperature must be regarded (as before) as a function of thermodynamic parameters, for example, ρ and T, and, furthermore, as a function of η^2. The case of an isotropic medium will be considered here. Near the transition temperature, the neglect of the anisotropies leads to an insignificant error [106]. For an isotropic medium, one can write, similar to (1.34),

$$\Delta\varepsilon(\rho, T, \eta^2) = \left(\frac{\partial\varepsilon}{\partial\rho}\right)_{T,\eta^2}\Delta\rho + \left(\frac{\partial\varepsilon}{\partial T}\right)_{\rho,\eta^2}\Delta T + \left(\frac{\partial\varepsilon}{\partial(\eta^2)}\right)_{T,\rho}\Delta(\eta^2), \quad (2.12)$$

where $\Delta(\eta^2) = \Delta(\eta^2 - \overline{\eta^2})$ and $\eta = \eta_0 + \Delta\eta$, whence

$$\Delta(\eta^2) = (\Delta\eta)^2 + 2\eta_0\Delta\eta - \overline{(\Delta\eta)^2}. \quad (2.13)$$

The intensity of the scattered light is proportional to $\overline{\Delta\varepsilon^2}(\rho, T, \eta^2)$ [see (1.30)]. Consequently, as is clear from (2.12) and (2.13), terms appear with a product of the fluctuations, the average statistical value of which does not vanish, inasmuch as η is a param-

*In the p, T diagram, T_c corresponds to a temperature at which the phase lines of first-order transitions go over to the line of second-order phase transitions.

eter dependent on ρ and T. However, an estimate of the value of the terms with a product of fluctuations, made in [106], shows that they remain two orders of magnitude smaller than $[\partial \varepsilon / \partial(\eta^2)]_{\rho, T} \cdot \Delta(\eta^2)$ near the phase transition temperature. Therefore, we neglect all terms containing a product of fluctuations and return to the fundamental component

$$\Delta\varepsilon(\eta^2) = \left(\frac{\partial\varepsilon}{\partial(\eta^2)}\right)_{\rho, T}\left[2\eta_0\Delta\eta + (\Delta\eta)^2 - \overline{\Delta\eta^2}\right].$$

From the latter formula, we get for the mean square

$$\overline{\Delta\varepsilon^2}(\eta^2) = \left(\frac{\partial\varepsilon}{\partial(\eta^2)}\right)^2_{\rho, T}\left\{4\eta_0^2\overline{(\Delta\eta)^2} + \overline{(\Delta\eta)^4} - [\overline{(\Delta\eta)^2}]^2\right\} \equiv \overline{\Delta\varepsilon_1^2} + \overline{\Delta\varepsilon_2^2}, \quad (2.14)$$

where it is taken into account that $\overline{\Delta\eta} = 0$.

Calculation of the mean square of the fluctuations of the quantity η has been carried out by Ginzburg and Levanyuk [106].

An expression suggested by Landau [51, 107] in his theory of second-order phase transitions was used as a thermodynamic potential. It is written for the spatially inhomogeneous case:

$$\Phi = F_0 + A\eta^2 + \frac{1}{2}C\eta^4 + \frac{1}{6}D\eta^6 + \ldots + f^*(\nabla\eta)^2, \quad (2.15)$$

where A, C, and D depend on the temperature and pressure.

In the calculation of $\overline{\Delta\eta^2}$ in (2.14) in [106], the procedure was the same as in the calculation of the density fluctuations near the critical temperature [see Appendix I, Eqs. (I.20)-(I.24)]. As a result of this calculation, an equation was obtained which we write for the scattering coefficient from $\overline{\Delta\varepsilon_1^2}(4\eta_0\overline{\Delta\eta^2})$ in the following way:

$$R_1 = \frac{\pi^2 kT}{2\lambda^4}\left(\frac{\partial\varepsilon}{\partial(\eta^2)}\right)^2_{\rho, T}\frac{(1+\cos^2\theta)\dfrac{4D}{}}{(C^2-2AD)^{1/2} + \dfrac{\pi^2 n^2}{\lambda^2}f^*\dfrac{}{[(C^2-2AD^{1/2}-C)]}(1-\cos\theta)}. \quad (2.16)$$

In structure, the formula is very similar to (2.3).

At the critical Curie point, A = C = 0 and $R_1 = 0$. But in the immediate vicinity of the temperature T_C, R_1 has a sharp maximum, located below the transition temperature. Above the Curie point, $\eta_0^2 = 0$ and R_1 also vanishes [see (2.14)].

For the intensity and its temperature dependence above the transition temperature, respectively, $R_2 \sim \overline{(\Delta\varepsilon_2^2)} \sim \overline{(\Delta\eta)^4}$ in (2.14). In the research of [106], $\overline{\Delta\eta^4}$ was computed in the same way as $\overline{\Delta\eta^2}$ was computed, and a formula was obtained for the scattering resulting from the fluctuations $\overline{\Delta\varepsilon_2^2}$. Although the formula for $R_2 \sim \overline{(\Delta\eta)^4}$ [see (2.14)] obtained in [106] and its interpretation are reasonable, the derivation of the formula is obviously less well founded than in the case of Eq. (2.16). For the application of the theory of light scattering in second-order phase transitions, see Sec. 32.

3. Molecular Scattering of Light by the Interface of Two Media

In the previous sections, we have briefly considered the results of the theory of light scattering developed by Einstein for the volume of an isotropic medium, based on an idea of Smoluchowski on the statistical fluctuations of the density.

As has already been pointed out in the Introduction, Mandel'shtam [15] constructed a theory of light scattering by the surface of a liquid or by the interface of two liquids for the case in which the molecular roughness is much smaller than the wavelength of the light.

The formulas found by him express the dependence of the intensity of the light scattered by the surface of the liquid on the properties of the liquid, the wavelength of the primary light, and the angles that determine the direction of the incident and scattered light. For simplicity of calculation, Mandel'shtam obtained formulas for the case in which the scattered ray lay in the plane of incidence and made a small angle with the specularly reflected ray.

The theory of light scattering from the molecularly inhomogeneous surface of a liquid was developed for the general case by Andronov and Leontovich [111]. Using the ideas of Mandel'shtam, Gans [112] analyzed the case of molecular light scattering by a surface of metallic mercury and also, independently of Andronov and Leontovich, found general formulas which describe the intensity of the light scattered by the interface between two transparent media [113].

Even in the first work of Mandel'shtam [15], there was an experimental part devoted to the experimental test of his theory. The experiments had a qualitative character, but they made it possible to verify that the theory was in qualitative agreement with experiment.

Subsequently, other experiments were carried out which confirmed the theory; we shall discuss them in Sec. 20. Here, a brief mention will be made of the course of the discussions of Mandel'shtam, Andronov, and Leontovich and, following [111], we shall obtain the final formulas necessary for the analysis of the experimental material. The reader can find detailed calculations in the original papers [15, 111, 112].

In the calculations of the light intensity scattered by the surface between two transparent media, carried out by Mandel'shtam, and later by Andronov and Leontovich, a method was used which was employed by Rayleigh [114] for the solution of the problem of the diffraction of a plane light wave on a rough boundary separating two transparent media. However, Rayleigh considered a one-dimensional problem, while a two-dimensional problem is solved in the present case.

We shall assume that the plane $z = 0$ corresponds to the surface of separation of two transparent media when this surface is ideally smooth. We select an area on this surface S that is large but bounded, in the shape of a square with side a, and place the origin of the coordinate system at the center of the square.

As a result of thermal motion inside the chosen square, a deformation of the interface arises – the roughness ζ (x,y), which represents the departure from the plane $z = 0$.

We represent this roughness or inhomogeneity of the surface in the form of a two-dimensional Fourier series

$$\zeta(x, y) = \sum_{l,m=-\infty}^{+\infty} \zeta_{l,m} e^{i\Omega(lx+my)}, \tag{3.1}$$

where l and m are integers and $\Omega = 2\pi/a$.

We restrict the problem to the consideration only of such cases in which ζ (x,y) is much less than the optical wavelength.

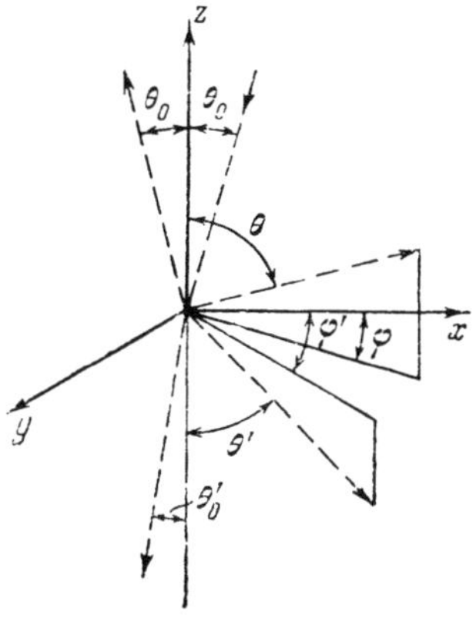

Fig. 7. For the theory of light scattering by an interface between two media.

This limitation, however, allows us to consider real phenomena over a wide range of temperatures, with the exception of a narrow temperature interval in the vicinity of the critical temperature of mixing of the liquids.

The problem consists of the calculation (under the assumptions made) of the intensity of the light reflected and refracted in different directions from the deformed interface (of area S), the equation for which is given by (3.1).

It follows from the solution of this problem [15, 111] that there will be reflected and refracted waves, the intensity and phase of which (with the accuracy assumed) are the same as in the corresponding reflected and refracted waves calculated according to the formulas of Fresnel for an ideally smooth interface. In addition to these waves, there will also exist diffraction (scattered) waves, the intensity of which is calculated in the following way. Each term of (3.1) describes a reflecting (and transmitting) diffraction lattice with its own period.

The condition $|\zeta_{l,m}| \ll \lambda$ given above simplifies the problem by allowing us to neglect the terms $|\zeta_{l,m}/\lambda|^2$ and those of higher order, and to keep the principle of superposition for the light diffracted by infinite lattices. This means that it is sufficient to find the solution of the problem for a single diffraction grating:

$$\zeta = \zeta_{l,m} e^{i\Omega\,(lx+my)},$$

in order to obtain a solution for all possible infinite diffraction gratings [see (3.1)].

If a plane light wave is incident on an infinite plane sinusoidal grating, then such a grating transmits a single diffracted plane wave, whose direction is determined by the conditions [111] (Fig.7):

$$\left.\begin{aligned} -\sin\theta\cos\varphi &= \sin\theta_0 + \frac{\Omega l\lambda}{2\pi}, \\ -\sin\theta\sin\varphi &= \frac{\Omega m\lambda}{2\pi}, \\ -\sin\theta'\cos\varphi' &= \sin\theta_0' + \frac{\Omega l\lambda'}{2\pi}, \\ -\sin\theta'\sin\varphi' &= \frac{\Omega m\lambda'}{2\pi} \end{aligned}\right\} \qquad (3.2)$$

[The primed quantities refer to the lower halfspace, the unprimed to the upper (Fig. 7).] A real grating with a large but finite area S sends out a set of waves differing somewhat in direction when a plane wave is incident on it.

One must compute what energy is sent out by the surface S in the direction θ, φ in the solid angle $d\Omega$ in which is contained a number of waves, traveling out from the individual sinusoidal gratings. It is assumed here that all the rays inside $d\Omega$ are of the same intensity. Moreover, it is assumed, as in the calculation of the ordinary diffraction grating, that the field of the incident wave on the surface of the finite grating is the same as in the case of the corresponding infinite grating.

For the assumptions that have been made, the total energy of the light diffracted in all directions from a finite diffraction grating is equal to the energy diffracted in the direction of the polar angles θ and φ from the part S of an infinite grating. Consequently,

$$I d\Omega = I'_{m,l} S \cos\theta \Delta l \Delta m, \qquad (3.3)$$

where $\Delta l \Delta m$ is the number of rays in $d\Omega$. The intensity of the plane wave diffracted by a single infinite sinusoidal lattice is

$$I'_{l,m} = I_0 \overline{|\zeta_{l,m}|^2}\, \Phi(\theta,\varphi). \qquad (3.4)$$

From (3.1), it can be found that

$$\Delta l \Delta m = \left|\frac{2\pi}{\lambda\Omega}\right|^2 \cos\theta d\Omega.$$

Combining this last equation with (3.3) and (3.4), we get

$$I d\Omega = I_0 \left|\frac{2\pi}{\lambda\Omega}\right|^2 \overline{|\zeta_{l,m}|^2}\, S\Phi(\theta,\varphi)\cos^2\theta\, d\Omega. \qquad (3.5)$$

Thus, the problem reduces to the calculation of the mean square of the fluctuations $\overline{|\zeta_{l,m}|^2}$ and the function $\Phi(\theta, \varphi)$.

The calculation of $\overline{|\zeta_{l,m}|^2}$ was made by Mandel'shtam [15]. In a later work, Leontovich [16, 83] developed a general method for the calculation of the fluctuations considered as a function of the location in space.

The calculations of [16] (see Appendix I) give for our case,

$$\overline{|\zeta_{l,m}|^2} = \frac{c_1\left(\frac{\Omega}{2\pi}\lambda\right)^2}{c_2 + \sin^2\theta_0 + \sin^2\theta + 2\sin\theta_0\sin\theta\cos\varphi}, \tag{3.6}$$

where $c_1 = kT/4\pi\gamma^*$, $c_2 = (g\delta/4\pi\gamma^*)\lambda^2$. Here, γ^* is the surface tension, g is the acceleration due to gravity, and δ is the difference in the density of the media making up the interface. The function $\Phi(\theta, \varphi)$ was computed by Andronov and Leontovich [111] for all possible cases.

If the interface of the two media is irradiated by natural light, and the angles θ and θ_0 are less than the angle of total internal reflection, then

$$\Phi(\theta_i, \varphi) = \frac{\mathrm{tg}^2(\theta_0 - \theta_0')\,\mathrm{ctg}^2\theta_0\sin^2\theta\cos^2\theta}{\sin^2(\theta + \theta')}\{\sin^2\varphi\cos^2\theta_0' + \cos^2\varphi\cos^2(\theta_0 - \theta_0') + \sec^2(\theta - \theta')[(\sin\theta\sin\theta_0 + \cos\theta'\cos\theta_0'\cos\varphi)^2 + \cos^2\theta'\cos^2(\theta_0 - \theta_0')\sin^2\varphi]\}. \tag{3.7}$$

In the case in which the surface is irradiated by natural light as before, but θ_0 is less and θ is greater than the angle of total internal reflection, we get

$$\Phi(\theta_i, \varphi) = \frac{\mathrm{tg}^2(\theta_0 - \theta_0')\,\mathrm{ctg}^2\theta_0\cos^2\theta}{(n^2 - 1)(n^2\sin^2\theta - \cos^2\theta)}\{\sin^2\theta_0\sin^2\theta + \cos^2(\theta_0 - \theta_0')(n^2\sin^2\theta + \sin^2\theta\cos^2\varphi - 1) + \cos^2\theta_0(n^2\sin^2\theta + \sin^2\theta\sin^2\varphi - 1)\}. \tag{3.8}$$

If the surface is irradiated by linearly polarized light, with its electric vector vibrating in the plane of incidence, then

$$\Phi(\theta_i, \varphi) = \frac{\operatorname{tg}^2(\theta_0 - \theta')\operatorname{ctg}^2\theta_0 \sin^2\theta\cos^2\theta}{\sin^2(\theta+\theta')}[\sin^2\varphi\cos^2\theta_0' + \sec^2(\theta-\theta')(\sin\theta\sin\theta_0 + \cos\theta'\cos\theta_0'\cos\varphi)^2]. \quad (3.9)$$

The direction of the vibrations of the electric vector in the scattered light [111] in the case of (3.9) is determined by the angle μ, where

$$\operatorname{tg}\mu = -\frac{\sin\theta\sin\theta_0 + \cos\theta'\cos\theta_0'\cos\varphi}{\cos(\theta-\theta')\sin\varphi\cos\theta_0'}.$$

In the case when the scattering by the surface is excited by linearly polarized light with the electric vector vibrating in a plane perpendicular to the plane of incidence, then

$$\Phi(\theta_i, \varphi) = \frac{\sin^2(\theta_0 - \theta')\operatorname{ctg}^2\theta_0 \sin^2\theta\cos^2\theta}{\sin^2(\theta+\theta')}[\cos^2\varphi + \sec^2(\theta-\theta')\cos^2\theta'\sin^2\varphi],$$

$$\operatorname{tg}\mu = \frac{\cos\theta'\operatorname{tg}\varphi}{\cos(\theta-\theta')}. \quad (3.10)$$

Similar formulas are obtained for the lower halfspace (Fig. 7): for irradiation of the surface by natural light

$$\Phi(\theta_i, \varphi) = \cos^2\theta_0\cos^2\theta + (\sin\theta_0\sin\theta\cos\varphi + 1)^2; \quad (3.11)$$

for plane polarized incident light, with the electric vector directed parallel to the plane of incidence,

$$\left.\begin{aligned} \Phi(\theta_i, \varphi) &= \cos^2\theta\sin^2\varphi + (\sin\theta_0\sin\theta + \cos\varphi)^2, \\ \operatorname{tg}\mu &= -\frac{\sin\theta\sin\theta_0 + \cos\varphi}{\cos\theta\sin\varphi}; \end{aligned}\right\} \quad (3.12)$$

and for a plane polarized incident light with the electric vector directed perpendicular to the plane of incidence,

$$\left.\begin{aligned} \Phi(\theta_i, \varphi) &= [1 - \sin^2\theta\cos^2\varphi]\cos^2\theta_0, \\ \operatorname{tg}\mu &= \frac{\operatorname{tg}\varphi}{\cos\theta}. \end{aligned}\right\} \quad (3.13)$$

The expression for the intensity of the scattered light can be obtained if the value of (3.6) is substituted in (3.5) and, in the corresponding specific case, one of the values of $\Phi(\theta, \varphi)$ described above [(3.7)-(3.13)] is also substituted in Eq. (3.5).

From the resulting formulas, we can draw the conclusion that the intensity of scattered light is approximately inversely proportional to λ^2. Actually, one can almost always neglect the term $\lambda^2 c_{2\gamma^*} = g\delta\lambda^4/4\pi$ in the formulas for the intensity in comparison with the remaining terms in the denominator.

In this case, the intensity of the scattered light is inversely proportional to the surface tension γ^*. Consequently, the intensity of the light scattering must increase rapidly as one approaches the critical temperature of mixing of the two media that form the interface. This effect was first observed by Mandel'shtam [15].

4. Calculation of the Intensity and the Depolarization Factor of Light Scattered in Gases and Vapors

In rarefied gases, one can neglect the forces of intermolecular interaction; therefore, the theoretical consideration of the phenomenon of molecular scattering of light is greatly simplified and can be carried through to the end both for the intensity and for the depolarization of the molecularly scattered light.

Such a consideration has been undertaken repeatedly; therefore, we shall only make a few remarks here about this simple, but in principle very important case of molecular light scattering. It has already been noted earlier that the Einstein formula (1.25) or its equivalent (1.94) transforms into the Rayleigh formula for gases.

The condition (1.67) is well satisfied for gases in the optical region of the spectrum and therefore there is practically no difference between the relations (1.93) and (1.94)

The parameters defining the quantity R_{90} in (1.94) are found from the very well-known relations which connect the optical dielectric constant ε with the density ρ and from the equation of state of an ideal gas.

The connection between ε and ρ for gases has a clear physical meaning, and is expressed by the relation

$$\varepsilon - 1 = C\rho, \tag{4.1}$$

where C is a constant characterizing the mean polarizability of the gas molecules.

From (4.1) we find

$$\left(\rho \frac{\partial \varepsilon}{\partial \rho}\right)_T = \varepsilon - 1. \tag{4.2}$$

On the other hand, from the equation of state of an ideal gas pV = $N_L kT$, we find

$$\beta_T = -\frac{1}{V}\left(\frac{\partial V}{\partial p}\right)_T = \frac{1}{N_L kT}. \tag{4.3}$$

Substituting (4.2) and (4.3) in (1.94), we find the scattering coefficient for gases

$$R_{90} = \frac{\pi^2}{2\lambda^4}(n^2-1)^2 \frac{1}{N_L} \frac{6+6\Delta_u}{6-7\Delta_u}. \tag{4.4}$$

Equation (4.4) differs from the Rayleigh formula by the Cabannes factor, which takes into account the light scattered by the fluctuations of the anisotropy.

The correction of Cabannes is important in the study of scattering in vapors and gases that consist of molecules with large optical anisotropies, and makes a small contribution to R for optically symmetric molecules. The depolarization factor for hydrogen, for example, $\Delta = 0.009$ and the Cabannes factor $f(\Delta) = 1.009$, need not be taken into account in the calculation of R_{90}, while for nitrous oxide (N_2O), $\Delta_u = 0.125$ and $f(\Delta) = 1.45$. Consequently, in nitrous oxide, one-third of the total flux of scattered light is included in the depolarized scattering and, therefore, it should not be neglected. In other vapors and gases, Δ is not so large as in nitrous oxide or carbon bisulfide but even for them the depolarization factor varies from 0.015 to 0.11, and the corresponding Cabannes correction factor $f(\Delta)$ differs from unity.

The dependence of the scattering coefficient in (4.4) on the wavelength of the incident light is determined by Rayleigh's λ^{-4} law and by some dependence of the index of refraction n on the wavelength. Far from the absorption line, the dependence of n on λ is unimportant and it can be considered that the entire dependence of R on λ is determined by the Rayleigh law. In the immediate vicinity of the absorption band. on the other hand, the factor $(n^2 - 1)^2$ plays the decisive role, and not λ^{-4}.

Loschmidt's number N_L (or, if the numerator and denominator are multiplied by the volume of a gram molecule V, then Avogadro's number N_A appears in place of N_L) enters into Eq. (4.4) and therefore a new possibility is opened up by the measurement of R for the experimental determination of Avogadro's number. Such a determination of N_A has been made many times, and the results have been shown to be in excellent agreement with the values of this quantity obtained by other methods (see Sec. 15).

The Depolarization Factor. By using the already well-known fact that the intensity of the light scattered in a gas is proportional to the number of molecules of the gas per unit volume, one can compute the intensity of the light scattered from an individual molecule and sum over the number of molecules. We proceeded in essentially this way in obtaining Eq. (4.4) or, more precisely, that part of it which includes the scattering by density fluctuations. The scattering by fluctuations of the anisotropy was taken into account by the introduction of the Cabannes factor $f(\Delta)$ [75]. For gases, the intensity of the light scattered by the fluctuations of the anisotropy and the depolarization factor can be computed comparatively simply; the general outlines of this calculation will be given here.

By considering the light scattered by an individual molecule as the radiation of a dipole vibrating under the action of the field **E** of the incident light wave, we can write that its intensity (see the Introduction) $I \sim \overline{\ddot{P}^2}$. Here, **P** is the moment induced in the molecule, equal to b**E**.

If the field $\mathbf{E} = \mathbf{E}_0 e^{i\omega t}$, then $|\ddot{\mathbf{P}}| \sim \mathbf{P}\omega^2$; therefore, it is obvious that it suffices to compute $I_z \sim P_z^2$ and $I_x \sim P_x^2$ in order to find the total intensity of the scattered light and its depolarization factor Δ. If the optical polarizability of the molecule is the same in any direction, i.e., if the polarizability b is a scalar, then the directions of the moment **P** and the field **E** coincide. However, in real gases, it is almost never thus. Molecules are more or less anisotropic. The optical anisotropy can be characterized by the three axes of an ellipsoid and only in the simplest cases does the molecule have an axis of symmetry. In that case, polarizability is characterized by an ellipsoid of revolution.

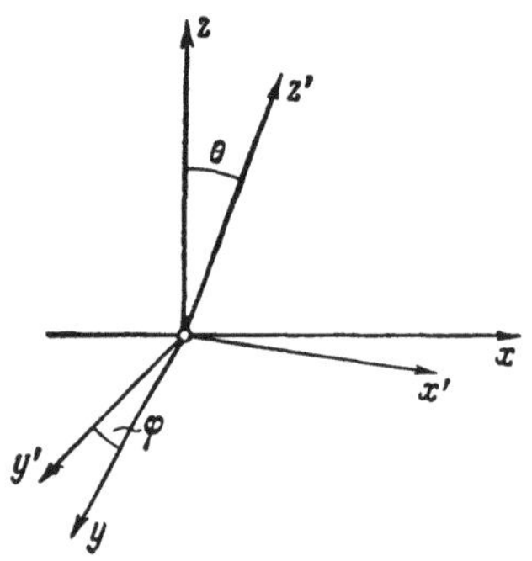

Fig. 8. Laboratory system of coordinates (x,y,z) and the system of coordinates attached to the molecule.

In the general case, the polarizability b_{ik} is a tensor of second rank. If we exclude optically active molecules and the region of absorption bands from consideration, then the tensor b_{ik} is a symmetric tensor of second rank which can be transformed to the principal axes so that only the three diagonal elements are left in the tensor, and all the nondiagonal elements vanish [115, 116].

We choose a coordinate system x', y', z' (Fig. 8), attached to the molecule, so that it is initially located at the center of the molecule, and the remaining three diagonal elements – the principal polarizabilities $b_1 = b_{x'x'}$, $b_2 = b_{y'y'}$, and $b_3 = b_{z'z'}$ – are directed along the axes of the coordinate system.

If the direction of the electric field vector of the incident plane-polarized light is the same as the direction of one of the axes of the set of coordinates x', y', z', then the corresponding induced moment is in the same direction as the field. The light scattered by such a molecule is also linearly polarized (see the Introduction). But if the direction of **E** in the incident wave does not coincide with any of the axes of the molecular system of coordinates, then a finite depolarization will be observed in the scattered light (Fig. 4). Of course, the observer cannot adjust his observations to the set of coordinates attached to the molecule, which is continuously changing its location. Therefore, it is appropriate to choose the fixed laboratory set of coordinates x, y, z (Fig. 8) and transform from the system attached to the molecule to the fixed system, attached to the laboratory. Since the molecules occupy a very different position relative to the laboratory system of coordinates, the resultant quantities must be averaged over the angles θ and φ (Fig. 8).

Without limiting the generality of the problem, we give the coordinate systems x, y, z and x', y', z' a common origin and assume that y' lies in the xy plane. Let the exciting light be incident in the direction of the x axis, and let the observation be carried out along the y axis; we consider the case in which the molecule has an axis of symmetry, i.e., $b_2 = b_3 \neq b_1$.

The transformation of the polarizability tensor from one system of coordinates to the other can be made by means of the appropriate transformation formulas [102, 115, 116].

If the incident light is linearly polarized, with the direction of vibration of the electric field **E** along the z axis, then we get for the components of the induced moment

$$\left.\begin{aligned} \mathbf{P}_z &= (b_2\cos^2\theta + b_1\sin^2\theta)\,E_z, \\ \mathbf{P}_x &= (b_2 - b_1)\sin\theta\cos\theta\sin\theta E_z. \end{aligned}\right\} \quad (4.5)$$

It then follows that the component $\mathbf{P}_x$ arises only as a consequence of the anisotropy of the molecule. If the molecule is isotropic, then $b_1 = b_2 = b_3$ and $\mathbf{P}_x$ vanishes. The components of $\mathbf{P}_y$, as was noted above, cannot radiate in the direction of observation – along the axis 0y. The average intensity of the scattered light for one molecule is equal to the total mean intensity divided by the total number of molecules.

We shall assume that the molecules take on arbitrary orientations relative to the laboratory system of coordinates. Then the average over the Eulerian angles φ and θ in Eq. (4.5) gives

$$\left.\begin{aligned} \overline{P_z^2} &= \frac{\int_0^{2\pi} d\varphi \int_0^{\pi} P_z^2 \sin\theta d\theta}{\int_0^{2\pi} d\varphi \int_0^{\pi} \sin\theta d\theta} = \frac{1}{15}(3b_1^2 + 8b_2^2 + 4b_1b_2)(E_z)^2, \\ \overline{P_x^2} &= \frac{\int_0^{2\pi} d\varphi \int_0^{\pi} P_x^2 \sin\theta d\theta}{\int_0^{2\pi} d\varphi \int_0^{\pi} \sin\theta d\theta} = \frac{1}{15}(b_2 - b_1)^2 (E_z)^2. \end{aligned}\right\} \quad (4.6)$$

By means of data for P_x^2 and P_z^2, we can find the total light intensity scattered in the gas. A simple calculation leads to the Einstein–Cabannes formulas that we obtained earlier for gases (4.4) or to the Rayleigh formula. One must expect, of course, that the intensities of the scattered light for the case of gases and vapors, computed from the molecular and thermodynamic theories, are in complete agreement. Even a quantum-mechanical calculation of the intensity yields nothing new [50].

For the depolarization factor Δ_V in the irradiation by polarized light with the vertical (along the z axis) direction of the electric vector of the light wave, we get from (4.6)

$$\Delta_v = \frac{I'_x}{I'_z} = \frac{\overline{P_x^2}}{\overline{P_z^2}} = \frac{(b_2 - b_1)^2}{3b_1^2 + 8b_2^2 + 4b_1 b_2} \,. \tag{4.7}$$

For irradiation of the scattering volume of the gas by natural light, an $\mathbf{E}_y$ appears in the primary light, in addition to $\mathbf{E}_z$, wherein $|\mathbf{E}_z| = |\mathbf{E}_y|$.

The component $\mathbf{E}_y$ produces an additional induced moment, the projections of which on the x and z axes will be identical and the same as the component $\mathbf{P}_x$ in the irradiation by polarized light, with $\mathbf{E}_z$ directed along the z axis.

Therefore, the depolarization factor for irradiation by natural light is

$$\Delta_u = \frac{2(b_2 - b_1)^2}{4b_2^2 + 9b_1^2 + 2b_1 b_2} \,. \tag{4.8}$$

One can calculate the corresponding values for a molecule without an axis of symmetry ($b_1 \neq b_2 \neq b_3$) in the same way as for a molecule having a symmetry axis. Without repeating the calculations, because of their cumbersome nature, we write down the final result:

For linearly polarized incident light, the depolarization factor of the scattered light is equal to

$$\Delta_v = \frac{I'_x}{I'_z} = \frac{b_1^2 + b_2^2 + b_3^2 - b_1 b_2 - b_2 b_3 - b_3 b_1}{3(b_1^2 + b_2^2 + b_3^2) + 2(b_1 b_2 + b_2 b_3 + b_3 b_1)} \tag{4.9}$$

and for excitation by unpolarized light,

$$\Delta_u = \frac{I_x}{I_z} = \frac{2I'_x}{I'_z + I'_x} = \frac{2(b_1^2 + b_2^2 + b_3^2 - b_1 b_2 - b_2 b_3 - b_3 b_1)}{4(b_1^2 + b_2^2 + b_3^2) + b_1 b_2 + b_2 b_3 + b_3 b_1} \,. \tag{4.10}$$

It follows from (4.7)-(4.10) that for the extreme anisotropic case, in which the ellipsoid of polarizability of the molecule becomes a straight line $b_2 = b_3 = 0$, and only $b_1 \neq 0$. The limiting values of the depolarization are equal to $\frac{1}{3}$ for irradiation by linearly polarized light, and $\frac{2}{3}$ for irradiation by natural light.

For convenience in the following discussion of Eqs. (4.9) and (4.10), we introduce the notation $\gamma_1^2 = b_1^2 + b_2^2 + b_3^2 - b_1b_2 - b_2b_3 - b_3b_1$, and assume that $b = \frac{1}{3}(b_1 + b_2 + b_3)$. Then Eqs. (4.9) and (4.10) are rewritten in the form

$$\Delta_v = \frac{\frac{3}{45}\gamma_1^2}{b^2 + \frac{4}{45}\gamma_1^2}, \qquad (4.11)$$

$$\Delta_u = \frac{\frac{6}{45}\gamma_1^2}{b^2 + \frac{7}{45}\gamma_1^2}. \qquad (4.12)$$

The intensity of completely polarized scattered light expressed by the Rayleigh formula is proportional to $b^2 \sim (n^2 - 1)^2$. The depolarization factors Δ_V and Δ_u are inversely proportional to b^2; this is to be expected, because the larger the fraction of polarized scattered light in the total flux of scattered light, the smaller the depolarization factor.

If we eliminate the completely polarized part of the scattered light (scattering by the density fluctuations) from consideration, i.e., set $b^2 = 0$ in Eqs. (4.11) and (4.12), which is the same as equating the trace of the scattering tensor to zero, we obtain the depolarization factors $\Delta_V = \frac{3}{4}$ for irradiation by polarized light, and $\Delta_u = \frac{6}{7}$ for irradiation by natural light. Thus, in the example considered, we reach an important conclusion: if $\Delta\varepsilon_{ik}$ [see (3.1)] is a symmetric tensor of second rank with a trace equal to zero, then the depolarization factor in the scattering of natural light by a molecule, for observation at right angles to the direction of the incident light, will be equal to

$$\rho_u = \frac{6}{7}.$$

The same is valid for scattering by a volume that is small in comparison with the wavelength.

The general study of the scattering tensor [50, 47] leads to the same result, which agrees excellently with the experimental data.

Connection between the Kerr Constant and the Depolarization Factor in Gases. The formulas (4.9) and (4.10) can be used for the determination of the principal polarizabilities of the molecule. Equation (4.10) can be written in the following form:

$$\frac{10\Delta_u}{6-7\Delta_u}=\frac{(b_2-b_1)^2+(b_2-b_3)^2+(b_3-b_1)^2}{(b_1+b_2+b_3)^2}=\delta^2, \tag{4.13}$$

where δ^2 is regarded as a measure of the optical anisotropy [77], changing from zero (complete isotropy) to unity (maximum anisotropy). It is clear from (4.13) that it is not possible to determine all three fundamental polarizabilities from a single measurement of Δ_u. Perhaps the simplest, and at the same time real, case is the one in which the molecule has an axis of symmetry. In this case, the principal polarizabilities can be found if, along with (4.13), we take into account the connection of the index of refraction with the average polarizability. The well-known relation for the index of refraction in gases gives the second equation

$$n-1=2\pi N_L\frac{1}{3}(b_1+b_2+b_3). \tag{4.14}$$

Actually, in the case $b_2 = b_3$, we get a set of two equations from (4.13) and (4.14) for the determination of the two unknowns b_1 and b_2:

$$\frac{45\Delta_u}{6-7\Delta_u}b^2=(b_1-b_2)^2,$$

$$\frac{n-1}{2\pi N_L}=\frac{1}{3}(b_1+2b_2).$$

However, this system contains an equation of second degree and, therefore, two solutions are obtained: (1) $b_1 > b_2$ and, (2) $b_1 < b_2$. The correct choice between the two solutions makes possible the formulation of the qualitative theory of induced dipoles of Sil'berstein [117].

For a molecule which does not have an axis of symmetry, the two equations (4.13) and (4.14) do not suffice to determine the three principal polarizabilities and, therefore, it is necessary to find another phenomenon, the theory of which would add the needed equation. Gans [118, 126] chose the Kerr phenomenon [119] as this necessary link.

The Kerr effect consists of the fact that a medium that is purely isotropic as a whole takes on the properties of an optically uniaxial crystal when placed in an electric field with its optic axis directed along the external electric field.

A path difference arises between the extraordinary and ordinary rays, which are propagated perpendicularly to the optic axis (it is expressed in wavelengths):

$$\Delta_\lambda = \frac{nl}{\lambda}(n_p - n_s), \tag{4.15}$$

where l is the optical path length in the specimen, n is the index of refraction in the medium without the field, n_p and n_s are the indices of refraction for rays with polarization parallel and perpendicular to the external field, respectiyely.

Kerr found that for glasses [119] and liquids [120] in an electric field

$$\Delta_\lambda = lBE^2, \tag{4.16}$$

where B is the Kerr constant which characterizes the medium. The same law has been found for gases [121].

It is convenient to introduce the constant

$$K = \frac{\lambda B}{n}, \tag{4.17}$$

then, we get from (4.15)-(4.17)

$$\Delta n = n_p - n_s = KE^2. \tag{4.18}$$

The relation (4.18) is valid for an arbitrary aggregate state, but in this section we shall speak in what follows only of gases and vapors. The Kerr effect is connected with the presence of an optical anisotropy of the molecules and, therefore, the theory of the effect must connect the Kerr constant B or K with the anisotropies of the optical polarizability of the molecule.

Two completely different points of view were initially advanced for an explanation of the Kerr effect. One of them was due to Voigt [122] and consisted of the following: in the electric field of an atom and a molecule, there is a different polarizability in different directions, and thus the external constant electric field changes the polarizability and the tensor of polarizability of the molecule. The Voigt theory predicts the existence of the Kerr effect independently of whether the medium consisted originally of isotropic or anisotropic molecules. The Kerr constant, computed from the Voigt theory, for example for carbon bisulfide, was three orders of magnitude smaller than that obtained experimentally. Moreover, the Voigt theory predicts the temperature dependence of the effect incorrectly, although in principle the effect pointed out by Voigt takes place, and it is shown to be appreciable close to an absorption band.

The other point of view belongs to Langevin [123]. He concluded that the anisotropic molecules which, in the absence of an external field are randomly oriented, so arrange themselves in an electric field that they take the position of minimum potential energy. Such a position corresponds to the direction of maximum polarizability along the electric field. Consequently, the molecules in the electric field tend to be extended, with the axes of largest polarizability along the field; this thus creates an anisotropy of the medium. At the same time, the thermal motion tends to disorganize this distribution of the molecules. As a result of the orienting action of the external electric field and the disorienting action of the thermal motion, there is a certain preferential equilibrium orientation, which depends on the temperature. The energy of the molecule in the external electric field is small in comparison with the energy of the thermal motion, even for intensities of the electric field corresponding to an electric breakdown.

If this is so, then the Boltzmann distribution function shows that for $U \ll kT$, the number of oriented molecules will be small or, in other words, the average total component of the maximum polarizability in the direction of the external field is very small. But the entire effect is accounted for by just this weak ordering of the anisotropies of the molecules. The Langevin theory gives the correct quantitative description of the Kerr effect and establishes the experimentally observed temperature dependence.

The theory of the Kerr effect for the case in which the molecule has a constant dipole moment has been worked out by Born [124, 125], and later extended by Gans [126], while the theory for the general case of a molecule lacking an axis of symmetry, and for optically active molecules, has been constructed by Mallemann [127].

For the calculation of the Kerr constant, we proceed in the same way as was done earlier in the calculation of the depolarization factor [77, 102, 115, 116, 128]. We again choose a set of coordinates rigidly attached to the molecule, x', y', z', and a fixed laboratory set, x, y, z. We give the two systems the same origin and then the relative positions of their axes will be given by the three Eulerian angles θ, φ, ψ. If the axes of the molecules in the absence of a field are randomly oriented, then, when the electric field is applied, dN molecules will have a definite orientation [77]:

$$dN = Ce^{-\frac{U}{kT}} \sin\theta \, d\theta \, d\varphi \, d\psi, \quad (4.19)$$

where U is the potential energy of the molecule in the electric field.

We again assume that the principal polarizabilities b_1, b_2, and b_3 are directed along the x', y', z' axes, respectively, and, further, that the external field is directed along the z axis and induces a moment in the molecule corresponding to the static polarizabilities a_1, a_2, and a_3. The potential energy of the molecule U can be written in the following form:

$$U = -A_1 E_{0z} + \frac{1}{2} A_2 E_{0z}^2, \quad (4.20)$$

where

$$A_1 = -\mu_1 \sin\theta \cos\psi + \mu_2 \sin\theta \sin\psi + \mu_3 \cos\theta,$$
$$A_2 = a_1 \sin^2\theta \cos^2\psi + a_2 \sin^2\theta \sin^2\psi + a_3 \cos^2\theta.$$

Here, μ_i is the component of the constant electric dipole moment in the individual molecule.

Now let a linearly polarized light wave be incident on a sample placed in the electric field, with the direction of the vector of the light field $\mathbf{E}_z$ coinciding with the direction of the external field $\mathbf{E}_{0z}$. The component of the electric moment $\mathbf{P}_z$ induced by the light wave is equal to

$$P_z = (b_1 \sin^2\theta \cos^2\psi + b_2 \sin^2\theta \sin^2\psi + b_3 \cos^2\theta) E_{0z}. \tag{4.21}$$

The mean value of the moment is

$$\overline{P_z} = \frac{\int\limits_0^{2\pi} d\varphi \int\limits_0^{2\pi} d\psi \int\limits_0^{\pi} P_z e^{-\frac{-A_1E_{0z}+\frac{1}{2}A_2E_{0z}^2}{kT}} \sin\theta\, d\theta}{\int\limits_0^{2\pi} d\varphi \int\limits_0^{2\pi} d\psi \int\limits_0^{\pi} e^{-\frac{-A_1E_{0z}+\frac{1}{2}A_2E_{0z}^2}{kT}} \sin\theta\, d\theta}. \tag{4.22}$$

As has been pointed out, the inequality $U \ll kT$ is always satisfied; therefore, there is no need of computing (4.22) exactly. Sufficient accuracy is obtained if we expand the exponent under the integral in powers of U/kT and keep terms up to second order in $\mathbf{E}_{0z}$, inclusively.

In place of (4.21), we introduce

$$P_z = b_z E_{0z}. \tag{4.23}$$

Then (4.20), (4.21), (4.22), and (4.23) permit us to find b_z for the assumptions made. The value of b_x can be computed in the same way.

Without going through the elementary but cumbersome calculations, we write the final result:

$$b_z = \frac{1}{3}(b_1 + b_2 + b_3) + 2(\theta_1 + \theta_2)\frac{1}{2}(E_{0z})^2, \tag{4.24}$$

$$b_x = \frac{1}{3}(b_1 + b_2 + b_3) + (\theta_1 + \theta_2)\frac{1}{2}(E_{0z})^2. \tag{4.25}$$

Here the following abbreviations have been used:

$$\theta_1 = \frac{1}{45kT}[(a_1 - a_2)(b_1 - b_2) + (a_2 - a_3)(b_2 - b_3) + (a_3 - a_1)(b_3 - b_1)], \tag{4.26}$$

$$\theta_2 = \frac{1}{45k^2T^2}[(\mu_1^2 - \mu_2^2)(b_1 - b_2) + (\mu_2^2 - \mu_3^2)(b_2 - b_3) + (\mu_3^2 - \mu_1^2)(b_3 - b_1)]. \tag{4.27}$$

For gases in an electric field, similar to (4.14), we have the relations

$$n_p - 1 = 2\pi N' b_z,$$
$$n_s - 1 = 2\pi N' b_x, \tag{4.28}$$

where N' is the number of particles per cm^3 in the presence of the field. As a consequence of electrostriction, $N' \neq N_L$, but the effect of electrostriction will not play an important role in the determination of the difference $n_p - n_s$ which we need to find for (4.18).

From (4.28), the desired difference is equal to

$$n_p - n_s = 2\pi N_L (b_z - b_x).$$

Substituting (4.24) and (4.25) in the last equation, we get

$$n_p - n_s = 3\pi N_L (\theta_1 + \theta_2) E_{0z}^2. \tag{4.29}$$

Strictly speaking, $\mathbf{E}_{0Z}$ has the sense of a field acting on the molecule, and it is therefore different from the external field **E**. If we compute the external field according to Lorentz [89], then $\mathbf{E}_{0Z} = [(\varepsilon + 2)/3]\mathbf{E}$, but for gases we can set $[(\varepsilon + 2)/3] \sim 1$ with great accuracy, while the calculation by the Lorentz scheme for condensed media generally does not correspond to actual conditions. Thus, in the case of gases and vapors, we can take $\mathbf{E}_{0Z}$ in Eq. (4.29) as the external field. Comparing (4.18) with (4.29), we find that the Kerr constant is

$$K = 3\pi N_L (\theta_1 + \theta_2) = K_1 + K_2. \tag{4.30}$$

It follows from (4.30), (4.26), and (4.27) that if the molecule is optically isotropic, then K = 0 and the Kerr effect does not appear.* If the molecule does not have a constant dipole moment, then $\theta_2 = 0$, and

$$K_1 = 3\pi N_L \theta_1. \tag{4.31}$$

---

*More precisely, even in a medium consisting of optically isotropic molecules, double refraction does arise, due to the mechanism pointed out by Voigt [122], but this effect is extremely small.

Following Gans [118] we make use of the approximate equality

$$\frac{a_1}{b_1} = \frac{a_2}{b_2} = \frac{a_3}{b_3} = \frac{\varepsilon^0 - 1}{\varepsilon - 1}, \tag{4.32}$$

where ε^0 is the static dielectric constant. Using (4.32), we write (4.26) in the following form:

$$\theta_1 = \frac{1}{45kT} \frac{\varepsilon^0 - 1}{\varepsilon - 1}[(b_1 - b_2)^2 + (b_2 - b_3)^2 + (b_3 - b_1)^2],$$

while for the Kerr constant

$$K_1 = \frac{3\pi N_L}{45kT} \frac{\varepsilon^0 - 1}{\varepsilon - 1} [(b_1 - b_2)^2 + (b_2 - b_3)^2 + (b_3 - b_1)^2]. \tag{4.33}$$

Adding the equality (4.32) to the relations (4.13) and (4.14), and taking into account that the molecule does not have an axis of symmetry, we find it possible to determine all three principal polarizabilities from the measurement of the index of refraction, the depolarization factor of the scattered light, and the Kerr constant K_1 by applying the considerations of Sil'berstein [117]. If the molecule has a dipole moment then, as a rule, $K_2 \gg K_1$, and we can neglect the value K_1 in a number of cases. In such a case, when $K_1 \sim K_2$, one can separate these two quantities by their different temperature dependences.

The expression for θ_2 is greatly simplified if the moment for a plane molecule lies in the plane formed by two principal axes and makes an angle α with b_3, i.e., $\mu_1 = \mu \sin\alpha$ and $\mu_2 = \mu\cos\alpha$, while $\mu_3 = 0$. In this special case,

$$\theta_2 = \frac{1}{45k^2T^2} \mu^2 [3\cos^2\alpha\,(b_3 - b_1) + 2b_1 - b_2 - b_3] \tag{4.34}$$

or for the very simple case in which the moment is directed along one of the principal axes, we get, in place of (4.34),

$$\theta_2 = \frac{1}{45k^2T^2} \mu^2 [2b_3 - b_2 - b_1], \tag{4.35}$$

$$K_2 = 3\pi N_L \theta_2 = \frac{3\pi N_L}{45k^2T^2} \mu^2 [2b_3 - b_2 - b_1]. \tag{4.36}$$

Thus, it is now shown to be possible to compute all three principal polarizabilities of the molecule by using expressions for the refraction, Kerr constant, and the depolarization factor.

Actually, we get from (4.14)

$$b_1+b_2+b_3=\frac{3\,(n-1)}{2\pi N_L}=A \qquad \text{(refraction)}$$

from (4.36),

$$2b_3-b_2-b_1=\frac{45\,k^2T^2K_2}{3\pi N_L\mu^2}=B \qquad \text{(Kerr effect)}$$

and from (4.13)

$$(b_1-b_2)^2+(b_2-b_3)^2+(b_3-b_1)^2=\frac{90\Delta_u}{6-7\Delta_u}\left(\frac{n-1}{2\pi N_L}\right)^2=C \quad \text{(scattered light)}$$

Solving the last three equations relative to the three polarizabilities, we get

$$b_3=\frac{A+B}{3},$$

$$b_2=\frac{A}{3}-\frac{B}{6}\pm\frac{1}{6}(6C-3B^2)^{1/2}, \qquad (4.37)$$

$$b_1=\frac{A}{3}-\frac{B}{6}\mp\frac{1}{6}(6C-3B^2)^{1/2}. \qquad (4.38)$$

The results of the measurements of Δ_u, K, K_1, and K_2, along with the principal polarizabilities of a wide range of molecules, are given in Table 1.

Chapter II

Some Theoretical Studies of the Spectral Composition of Molecularly Scattered Light

5. The Spectral Composition of Light Scattered by Density Fluctuations in Liquids and Gases

In Chapter I, we considered the scattering of light produced by inhomogeneities in the index of refraction, which in turn arose because of various thermal fluctuations. In the derivation of the formulas of Chapter I, the values of the corresponding fluctuations were sought in each case, but their change with time was not taken into account. Such an approach is fully justified if we are seeking the total light intensity, without touching on the problem of the spectral composition of the scattered light.

Actually, it was assumed that a definite amount of local optical inhomogeneities (fluctuations in the index of refraction) exist in a certain volume at a given temperature, an amount which does not depend on the time; they were regarded as "frozen." In this case, the dependence of the total intensity of the scattered light on the experimentally measured parameters of the medium and on the wavelength of the incident light was found.

The scattered light had the same wavelength as the incident light in this case if the incident light were monochromatic. If the excitation of the scattering takes place in a part of the continuous spectrum, then the intensity distribution of the scattered light will be changed according to the law $I \sim 1/\lambda^4$.

Actually, the statistical fluctuations do not remain "frozen," but change continuously in time. The fluctuations characterize the departure from the equilibrium state; they are therefore unstable.

Different fluctuations change differently with time and, therefore, they can modulate the scattered light in different ways. The fact that each form of fluctuation imposes its own characteristic "imprint" on the spectrum of the scattered light, allows us to use the spectrum of scattered light for the study of the kinetics of fluctuations. Above we investigated the intensity of the light scattered by fluctuations in density and concentration and reference was made to the scattering by the fluctuations of the orientations of anisotropic molecules.

In the present section we shall consider briefly those changes which are produced in the spectrum of scattered light by temporal changes in the fluctuations of density and concentration. As has already been shown, the density fluctuations are composed of adiabatic fluctuations (pressure fluctuations) and isobaric fluctuations (entropy fluctuations). The pressure fluctuations represent random local compressions or rarefactions which, as a consequence of the elastic properties of the medium, do not remain fixed in position but "travel" throughout the volume of the sample. The numerous random compressions and rarefactions arising from the thermal pressure fluctuations can be considered as elastic waves of different frequencies, which are propagated in all possible directions inside the volume under study.

Such a representation was not made immediately. We recall that, in the calculation of the total intensity of the scattered light, Einstein expanded the density fluctuations in a spatial Fourier series and found the intensity of the light scattering from the statistical amplitudes of the expansion. No problems of the dynamics of the fluctuations were considered in this research of Einstein, and no mention was made of the problems relating to the specific heat [14].

On the other hand, the study of problems of the specific heat of a solid led Debye [25] to the fruitful idea that the energy which one must have in all 3N degrees of freedom associated with the atomic oscillators of the solid can be considered as the energy of 3N normal elastic waves. Thus, Debye considered the energy of thermal motion of a solid as the energy of elastic waves. From this viewpoint, the fluctuation is the result of the interference of the Debye waves. But Debye did not consider and did not mention the problem of scattered light in this research.

A great success in the development of the problems of light scattering was achieved when Mandel'shtam [24, 129] established the fact that the Einstein "formal waves" of density (the components of the spatial Fourier series) and the Debye thermal elastic waves were one and the same thing.

The scattered light in such a case is considered to be the result of the diffraction of light by elastic Debye waves and its spectral composition is determined by the character of the modulation. If the damping of thermal waves, which are important in light scattering, is small per wavelength in such media as a liquid and a dense gas (the quantitative criterion will be given below), then the Debye representation of the energy of thermal motion can also be extended to those media, at least for problems associated with light scattering.

On the basis of the analysis just described, Mandel'shtam [24] predicted the phenomenon of the fine structure of the Rayleigh line. Brillouin arrived at this same conclusion [26]. Mandel'shtam and Landsberg [130, 131] and Gross [27] experimentally found the predicted effect in a quartz crystal and later Gross discovered the fine structure of the Rayleigh line in a liquid [28].

The fundamental theories of the phenomenon were developed by Mandel'shtam, Landsberg, and Leontovich [24, 132-135], Brillouin [26], and others.

If, following Debye, one considers the crystal as a continuum, then its normal vibrations are determined by the equations of elasticity theory with the corresponding boundary conditions. The total number of normal vibrations with frequencies in the interval from Ω to $\Omega + d\Omega$ is equal to [16]:

$$dz(\Omega)=\frac{3\Omega^2\, d\Omega}{2\pi^2 v^3}\, V, \qquad (5.1)$$

where v is the mean velocity of elastic waves in a solid amorphous body, $(3/v^3) = (2/v_t^3) + (1/v_l^3)$; here, v_t and v_l are the velocities of transverse and longitudinal elastic waves, respectively. In a crystal, anisotropy of the velocity must be taken into account (see Sec. 9). The maximum frequency of the elastic vibrations is determined approximately from the condition

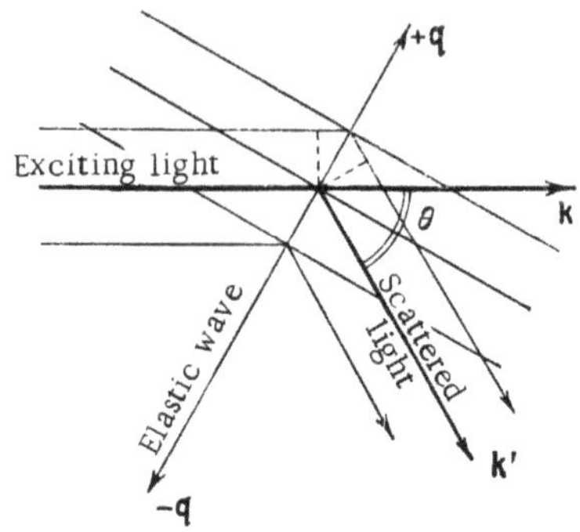

Fig. 9. Directions of the wave vectors of the incident light (**k**), the scattered light (**k'**), and the elastic thermal wave (**q**), corresponding to the Bragg condition.

$$z(\Omega_{max}) = \int_0^{\Omega_{max}} dz(\Omega) = \frac{V\Omega^3_{max}}{2\pi^2 v^3} = 3N, \tag{5.2}$$

where N is the number of particles in the volume V. Recalling that $d^3 = V/N$ (d is the lattice constant in the crystal or the average distance between particles in the liquid), we find

$$\begin{aligned} \Omega_{max} &= \frac{2\pi v}{d}\left(\frac{3}{4\pi}\right)^{1/3}, \\ \Lambda_{min} &= \left(\frac{4\pi}{3}\right)^{1/3} d. \end{aligned} \tag{5.3}$$

The relations (5.3) give the order of magnitude of the maximum value of the frequency of the waves $\Omega_{max} \sim 10^{14}$ Hz and their minimum wavelength $\Lambda_{min} \sim 1.5$ Å. It will be made clear below that only a certain part of the spectrum of elastic waves is important for the light scattering.

All 3N Debye waves in the body are propagated in all possible directions, thus forming a complicated spatial "net" of optical inhomogeneities. However, if such a body is irradiated by a parallel light beam with a definite wave vector **k**, and the scattered light is observed in the direction of the wave vector of the scattered light **k'** (Fig. 9), then the maximum intensity of the scattered (diffracted) light in this direction will occur when the wave vectors of the elastic wave **q**, together with the wave vectors **k** and **k'**, satisfy the Bragg condition (1.19). Thus, the furnishing of a parallel beam of direction **k** and the furnishing of an observation direction **k'** (or angle of observation θ) selects two waves from the great number of waves with identical frequencies Ω and with wave vectors **q** equal in magnitude but opposite in direction. Here, the absolute value of $|\mathbf{q}|$ is determined by the relation (1.26).

As has already been shown, the time-varying optical inhomogeneity more or less modulates the scattered light. Let the modulating function be $\Phi(t)$ and the monochromatic light wave undergoing modulation $\mathbf{E}'(t) = \mathbf{E}_0 e^{i\omega_0 t}$. The resulting vibration will have the form

$$E(t) = E'(t)\,\Phi(t). \tag{5.4}$$

Let us first consider the modulation of the scattered light by adiabatic fluctuations of density or of pressure Δp. The change of Δp in time satisfies the wave equation. By starting from the assumption that $\Delta p \sim \Delta n \sim \Phi(t)$, where Δn is the change in the index of refraction, and neglecting the absorption of the elastic waves, we can write the wave equation in the form

$$\ddot{\Phi} - v^2 \nabla^2 \Phi = 0. \tag{5.5}$$

The solution of Eq. (5.5) will be

$$\Phi(t) = \Phi_0 \cos[\Omega_0 t - (qr)], \tag{5.6}$$

where

$$\Omega_0^2 = v^2 q^2.$$

Using (5.4), we find that the vibration in the light wave diffracted by the elastic wave with wave vector **q** in the direction **k'** will be

$$E'(t) = \frac{\Phi_0 E_0}{2}\left\{e^{i[(\omega_0+\Omega_0)t-(qr)]} + e^{i[(\omega_0-\Omega_0)t+(qr)]}\right\}. \tag{5.7}$$

It is then immediately evident that there are two satellites in the spectrum of the scattered light with frequencies $\omega_0 + \Omega_0$ and

$\omega_0 - \Omega_0$, symmetrically located on the two sides of the initial frequency ω_0. The frequency difference between the two satellites will be $2\Omega_0$. Inasmuch as we have neglected the absorption of the elastic wave, the satellites are monochromatic.

The intensity of the diffracted (scattered) light changes in time according to the law [24]

$$E(t)\,E^*(t) = I = I_0\{1 + \cos[2\Omega_0 t - 2(\boldsymbol{qr})]\},$$

which shows that this change does not depend on the form of the perturbation, but depends on the ratio v/c, the frequency ω_0 of the exciting light, and the angle of observation θ. Actually,

$$\Omega_0 = vq = v\left(\frac{2\pi}{\Lambda}\right) = \frac{4\pi n v}{\lambda}\sin\frac{\theta}{2} = 2\omega_0 n\frac{v}{c}\sin\frac{\theta}{2}. \tag{5.8}$$

The relative frequency change of the satellites is equal to

$$\pm\frac{\Delta\omega}{\omega_0} = \pm\frac{\Omega_0}{\omega_0} = \pm\frac{\Delta\nu}{\nu} = 2n\frac{v}{c}\sin\frac{\theta}{2}. \tag{5.9}$$

Equation (5.9) was first obtained by Mandel'shtam [24] and Brillouin [26]. It follows from a consideration of (5.9) that the relative change in the frequency depends only on the ratio v/c, the angle of observation, and the index of refraction n. Equation (5.9) was confirmed at least qualitatively in the earliest experiments [27].

In the first researches of Mandel'shtam and Brillouin, attention was turned to the fact that here we are dealing with an unusual Doppler effect. However, we are speaking now of changes in the light frequency as a result of the reflection from a traveling elastic wave as from a moving mirror, and not as the result of the motion of the individual molecules. Such a consideration leads to the same Eq. (5.9) [134].

Up to this point, it has been assumed that there is no absorption of the elastic wave, and, therefore, the Mandel'shtam–Brillouin (MB) components are infinitely sharp, if they are excited by a monochromatic line of frequency ω_0. In reality, there is always some damping of the elastic wave in the medium, and, therefore, the MB components will always have a line width different from zero.

The effect of absorption of the elastic wave on the width of the MB component was first considered by Leontovich [135, 138], and later in a number of other researches [29, 30, 45, 47]. The problem for the absorbing medium differs from the idealized case just considered only in this respect, that one must use the Stokes equations for the propagation of sound in a viscous medium in place of Eq. (5.5). These equations can be written in the following form [136, 137]:

$$\ddot{\Phi} - v^2\nabla^2\Phi - \Gamma\nabla^2\dot{\Phi} = 0, \tag{5.10}$$

where

$$\Gamma = \frac{1}{\rho}\left\{\frac{4}{3}\eta + \eta' + \frac{\varkappa}{c_p}(\gamma - 1)\right\}, \tag{5.11}$$

here η and η' are the shear and bulk viscosity coefficients, respectively, $\varkappa$ is the coefficient of thermal conductivity, and $\gamma = c_p/c_V$. We shall seek the solution of Eq. (5.10) in the form

$$\Phi(t) = \Phi_0 e^{i[mt - (qr)]}. \tag{5.12}$$

Substituting (5.12) in Eq. (5.10), we find

$$m = i\frac{\Gamma q^2}{2} \pm \left(q^2 v^2 - \frac{\Gamma^2 q^4}{4}\right)^{1/2}.$$

Taking (5.6) into account, we rewrite the last relation in the following form:

$$m = i\frac{\Gamma q^2}{2} \pm \left(\Omega_0^2 - \frac{\Gamma^2 q^4}{4}\right)^{1/2}. \tag{5.13}$$

Substituting (5.13) in (5.12), we find, keeping the real part:

$$\Phi(t) = \Phi_0 e^{-\delta t} \cos\left[\Omega t - (\boldsymbol{qr})\right], \tag{5.14}$$

where

$$\delta = \frac{\Gamma q^2}{2}, \quad \Omega = (\Omega_0^2 - \delta^2)^{1/2}. \tag{5.15}$$

The resultant light perturbation, in accord with (5.4) and (5.14), will be

$$E(t) = \Phi(t)\,E'(t) = \frac{E_0\Phi_0}{2} e^{-\delta t}\left\{e^{i[(\omega_0+\Omega)t - (qr)]} + e^{i[(\omega_0-\Omega)t - (qr)]}\right\}. \tag{5.16}$$

It follows from (5.16) that the dependence of the intensity I(t) on time is determined by the expression

$$I(t) = I_0 e^{-2\delta t}\{1 + \cos 2[\Omega t - (\mathbf{qr})]\}.$$

The distribution of the intensity over the frequencies in the MB components can be determined by the expansion of $\mathbf{E}(t)$ in a Fourier integral

$$E(t) = \int_{-\infty}^{\infty} E(\omega') e^{i\omega' t} d\omega', \tag{5.17}$$

where the desired coefficients of the expansion are

$$E(\omega') = \frac{1}{\pi}\int_{0}^{\infty} E(t) e^{-i\omega' t} dt.$$

Completing the integration, we find that

$$I(\omega') = \overline{E(\omega')\ E^*(\omega')} = \frac{\text{const}}{(\omega^2 + \Omega_0^2)^2 + 4\delta^2\omega^2}. \tag{5.18}$$

Here we have introduced a new variable $\omega = \omega' - \omega_0$, which is measured from the unshifted frequency ω_0. The integrated intensity of the fine structure components, $2I_{MB}$, is

$$\int_{-\infty}^{\infty} I(\omega)\, d\omega = 2I_{MB}. \tag{5.18a}$$

In a liquid, where there are two MB components, the total intensity of the fine structure $2I_{MB} = I_{ad}$, where I_{ad} is the intensity of the light scattered by the adiabatic pressure fluctuations computed earlier; therefore, we can write (5.18) in the form

$$I(\omega_0 \pm \omega) = \frac{\Omega_0^2 \frac{2\delta}{\pi} I_{ad}}{(\omega^2 - \Omega_0^2)^2 + 4\delta^2\omega^2}. \tag{5.19}$$

The location of the maximum of the MB components (when there is damping of the sound wave) is determined by $\omega_{max} = \sqrt{\Omega_0^2 - 2\delta^2}$.

The halfwidth of the fine structure line, as is easily seen from (5.19), is equal to

$$\delta\omega = 2\delta\left(1 + \frac{1}{2}\frac{\delta^2}{\Omega^2}\right) \approx 2\delta = 2\alpha v, \tag{5.20}$$

inasmuch as the temporal absorption coefficient $\delta = \alpha v$, where α is the amplitude coefficient of the damping of an elastic wave over 1 cm of path.

The hydrodynamic theory of the absorption of elastic waves in a viscous medium leads to the well-known expression

$$\alpha = \alpha_\eta + \alpha_{\eta'} + \alpha_\varkappa = \frac{\Omega^2}{2v^3}\Gamma = \frac{q^2}{2v}\Gamma = \frac{q^2}{2v\rho}\left\{\frac{4}{3}\eta + \eta' + \frac{\varkappa}{c_p}(\gamma - 1)\right\}. \tag{5.21}$$

Therefore, substituting (5.21) in (5.20), we get

$$\delta\omega_{MB} = q^2\Gamma. \tag{5.22}$$

Putting the value (1.26) in (5.22), we find

$$\delta\omega_{MB} = 2n^2\Gamma\,|\boldsymbol{k}|^2(1 - \cos\theta), \tag{5.23}$$

(**k** is the wave vector of the incident light). As is clear from Eq. (5.23), the halfwidth $\delta\omega_{MB}$ for unchanged n, **k**, and Γ depends materially on the scattering angle θ. Thus, in observation at the angle $\theta = \pi/2$, $\delta\omega_{MB}$ is half as large as for observation at the angle $\theta = \pi$. On the other hand, as follows from (5.9), the distance between the unshifted line ω_0 and the maximum of the MB component $\Delta\omega$ also depends on θ. The ratio of $\Delta\omega$ to $\delta\omega_{MB}$ has the form

$$\frac{\Delta\omega}{\delta\omega_{MB}} = c\,\frac{1}{\sin\theta/2}, \tag{5.24}$$

where $c = v/2nk\Gamma$.

In the observation of scattered light at angles θ equal to 180°, 90°, and 30°, the ratio $\Delta\omega/\delta\omega_{MB}$ (5.24) is equal, respectively to 1c, 1.41c, and 4c; therefore, taking into account only the characteristic width of the MB component, one can draw the conclusion that the fine structure will be seen more clearly at smaller angles of observation.

However, until the advent of the laser, the situation was just the opposite; this is principally because the halfwidth of the exciting line, together with the apparatus halfwidth (see Sec. 12) greatly exceeded the halfwidth $\delta\omega_{MB}$, so that at small θ (where $\Delta\omega$ is small) one could say that the fine structure is not visible, while for θ = 90° or 180°, it is completely accessible to investigation.

In those cases in which such sources as lamps operating on a single mercury isotope, or lasers with a limitingly sharp spectral line are used as sources for the excitation of scattered light, the dependence of the characteristic width of the fine structure line on the scattering angle takes on considerable importance.

We turn our attention also to the fact that, with increase in the absorption coefficient α (5.20) or with increase in the viscosity (5.21) (the term depending on the thermal conductivity does not play a significant role for most liquids and solids) the halfwidth increases in proportion with α or $(\eta + \eta')$ and the fine structure components broaden. In addition, the location of the maximum of the fine structure shifts, upon increase in α, in the direction of the unshifted line ω_0. Actually, on the basis of (5.15), (5.20), and (5.21), one can write that

$$\Omega_{\max} = (\Omega_0^2 - 2\alpha^2 v^2)^{1/2} \approx \Omega_0 \left(1 - \frac{\alpha^2 v^2}{\Omega_0^2}\right), \qquad (5.25)$$

and thus it is clear that for increase in α, $\Omega_{\max} \to \omega_0$. Consequently, for increase in the viscosity or, what amounts to the same thing, increase in the absorption coefficient of hypersound,* the MB components broaden and their maximum is displaced toward ω_0.

Upon significant increase in the viscosity, one must expect a diminution and diffusion of the discrete components of the fine structure. But, speaking of a significant increase in the viscosity, one must not lose sight of the fact that Eq. (5.10) is the approximate equation of Stokes, the conclusions from which are valid only in the case when

$$\frac{\Gamma\Omega}{v^2} \ll 1. \qquad (5.26)$$

Taking (5.21) into account, we can put (5.26) in the form of the condition

$$\alpha\Lambda \ll \pi. \qquad (5.27)$$

*According to generally accepted terminology, elastic waves with frequency above 10^9 Hz are known as hypersound.

The condition (5.27) means that the absorption of an elastic wave in a distance equal to its wavelength is small [138, 139]. It is quite evident that one can study the fine structure experimentally, when the condition (5.27) is satisfied. This is clearly so, both qualitatively and quantitatively. Actually, if the halfwidth $\delta\omega$ is equal to the interval of frequencies between the unshifted line and the maximum of the MB component $\Delta\omega$, then it is clear that the fine structure will not be observed. In order that the fine structure be seen, it is necessary that $\Delta\omega \gg \delta\omega_{MB}$. By using Eqs. (5.9) and (5.20), the condition for the existence of the fine structure can be represented by the inequality $2n\omega_0(v/c)\sin(\theta/2) \gg 2\alpha v$, which, by taking (5.8) into consideration, can be rewritten in the form $\alpha\Lambda \ll \pi$, which is identical with (5.27). If we assume that the sound wave is weakly damped when at a distance $x = \Lambda$, the amplitude falls off by less than a factor e, we can then require that

$$\alpha\Lambda \ll 1, \tag{5.28}$$

since

$$A = A_0 e^{-\alpha x}.$$

Equation (5.28) is essentially no different than (5.27). At the beginning of this section, it was pointed out that not all elastic waves of the spectrum of eigenfrequencies of the body produce scattered light. This is easily seen on the basis of Eqs. (5.8) and (1.26), which can be transformed to the expression

$$f = \frac{\Omega}{2\pi} = \frac{v}{\Lambda} = \frac{2nv\sin\theta/2}{\lambda}. \tag{5.29}$$

For irradiation of the scattering volume of the green line of the mercury spectrum (λ = 4358 Å) and for observation at an angle $\theta = 90°$, $\Lambda = 2.3 \cdot 10^{-5}$ cm ($n \sim 1.5$) and frequency $f_{90} = 0.75 \cdot 10^{10}$ Hz for $v = 1.5 \cdot 10^5$ cm/sec [the velocity of sound in liquids at low viscosities varies approximately in the range $(0.8\text{–}1.9) \cdot 10^5$ cm per sec].

In observation at an angle $\theta = 0$, $f_0 = 0$, and at $\theta = 180°$, $f_{180} = 1.05 \cdot 10^{10}$ Hz. In crystals, the velocity of elastic waves is 5-6 times greater than in a liquid. If we use ultraviolet light λ = 2000 Å for excitation, then the frequencies of the elastic waves will be $f_{180} \sim 10^{11}$ Hz. All the very high frequencies do not play a role in the process of light scattering.

In a real experiment, observations at $\theta = 180°$ and $\theta = 0°$ are virtually impossible because of the large amount of parasitic light at these scattering angles. Moreover, for small θ, $\Delta\omega$ is very small, which makes the study of the fine structure difficult or even impossible. In practice, it is convenient to study the scattering at angles not smaller than 20-30°, and not larger than 175-160°, with respect to the direction of propagation of the exciting light.* In this case, the range of frequency shift for liquids lies within the limits from $1 \cdot 10^{10}$ Hz to $0.25 \cdot 10^{10}$ Hz, i.e., the frequency changes at most by a factor of 4 or even less. Therefore, the tempting view of the study, for example, of the hypersonic velocity v over the frequency range of f from zero to 10^{10} Hz by means of light scattering (5.9) cannot be realized at the present time. However, measurement of the hypersonic velocity (frequency $f \sim 10^{10}$ Hz) and a comparison of this quantity with the corresponding ultrasonic measurements furnishes information on the dispersion of the sound velocity and the relaxation of the bulk viscosity coefficient η' [23, 30]. As a consequence of the finiteness of the line width of the exciting light and the finiteness of the aperture of the light incident on the scattering volume, a certain selection of elastic waves $\Delta\Lambda$ is used. However, for a strongly limited angle of scattering (small $\Delta\theta$), $\Delta\Lambda$ is very small, so that one can speak in practice of the study of a monochromatic elastic wave. In a liquid, for $\theta = 90°$ and $\lambda = 4358$ Å, $f_{90} \sim 10^{10}$ Hz, and all the other frequencies play no role.

Spectral Composition of Light Scattered by Isobaric Fluctuations of Density and by Fluctuations in the Concentration. We have considered thus far only the light scattered by density fluctuations, which is determined by the fluctuations of Δp and is expressed by the first component of the right-hand side of (1.52) or (1.53).

This part of the scattered light appears as the result of its diffraction by elastic thermal waves. However, there is still a part of the light which is scattered by the isobaric fluctuations in the density, a scattering associated with the temperature fluctua-

*These discussions do not apply to stimulated MB scattering.

tions ΔT or the entropy fluctuations ΔS. These fluctuations, which lead to light scattering, and which are of special importance in liquids, are not taken into account in the framework of the exposition of the MB theory, which is based on the ideas of Debye.

Therefore, we shall consider the scattering of light by the isobaric fluctuations of the density separately, but in exactly the same way as was done above for the adiabatic fluctuations of the density. We shall begin from (5.4) and, as before, base our arguments on the natural supposition that the optical inhomogeneity $\Delta n \sim \Delta S \sim \Phi(t)$ or $\Delta n \sim \Delta T \sim \Phi(t)$. The rate of change of ΔS and ΔT with time will be determined by the coefficient of temperature conductivity χ, while $\Phi(t)$ is determined by the equation of thermal conductivity

$$\dot{\Phi} - \chi \nabla^2 \Phi = 0, \tag{5.30}$$

the solution of which is written in the form

$$\Phi(t) = \Phi_0 e^{i[mt-(qr)]}. \tag{5.31}$$

Substituting (5.31) in (5.30), we find that

$$\Phi(t) = \Phi_0 e^{-\delta t} e^{-iqr}, \tag{5.32}$$

where

$$\delta_c = \chi q^2. \tag{5.33}$$

Equation (5.32) shows that $\Phi(t)$ is damped out exponentially with time, and that the exciting optical inhomogeneities remain fixed in space, in contrast with the elastic waves, which are propagated in accord with (5.14).

We can reach exactly the same conclusions by considering the time change in the concentration fluctuations. If we again assume that $\Delta c \sim \Delta n \sim \Phi(t)$, then $\Phi(t)$ obeys the diffusion equation, which is formally completely identical with (5.30) but with the replacement of the temperature conductivity χ by the diffusion coefficient D.

The resulting light perturbation, determined by (5.4), will be

$$E(t) = E_0 \Phi_0 e^{-\delta t} e^{i\omega_0 t}, \tag{5.34}$$

whence the dependence of the intensity of the scattered light on time is

$$I(t)=I_0 e^{-2\delta t}. \tag{5.35}$$

Even in his first work, Mandel'shtam [24] considered the problem of scattering as the result of fluctuations of temperature and concentration, and obtained Eq. (5.35). Thus, the scattering considered by him leads to the appearance of an unshifted line with frequency ω_0 at the maximum just as is obtained from (5.34). We now find the intensity distribution of the scattered light with respect to the frequency for the unshifted line. Here, as before, we expand (5.34) in a Fourier series. Entirely similar calculations as in the previous case lead to the formula

$$I(\omega)=\frac{\frac{\delta_c}{\pi} I_{is}}{\omega^2+\delta_c^2}. \tag{5.36}$$

Here the total intensity of the central component is

$$\int_{-\infty}^{\infty} I(\omega)\,d\omega = I_{is},$$

where I_{is} is the light intensity scattered by the isobaric density fluctuations, determined by (1.45). As follows from (5.36) and (5.33), the halfwidth of the central line of the fine structure is

$$\delta\omega_c=\delta_c=\chi q^2=2n^2\chi k^2(1-\cos\theta). \tag{5.37}$$

In the same way, we obtain the distribution of the intensity in the contour of the line scattered by the concentration fluctuations

$$I(\omega)=\frac{\frac{\delta_{\text{conc}}}{\pi} I_{\text{conc}}}{\omega^2+\delta_{\text{conc}}^2}.$$

Here,

$$\int_{-\infty}^{\infty} I(\omega)\,d\omega = I_{\text{conc}},$$

where I_{conc} is determined by the relation (1.79). The corresponding halfwidth of the scattering line due to the concentration fluctuations is

$$\delta\omega_{\text{conc}} = Dq^2 = 2n^2Dk^2(1-\cos\theta). \quad (5.38)$$

All the expressions (5.23), (5.37), and (5.38) give the same dependence of the line width of the fine structure on the scattering angle θ. A rough estimate of the halfwidth for $\theta = 90°$ and $\lambda = 4358$ Å illustrates the order of the quantity for a liquid ($n \sim 1.5$, $v = 1.5 \cdot 10^5$ cm/sec):

$$\delta\omega_{\text{MB}} \sim 7\cdot 10^9 \text{ sec}^{-1}, \quad \delta\nu_{\text{MB}} \approx 4\cdot 10^{-2} \text{ cm}^{-1} \quad (\alpha\Lambda = 0.5),$$
$$\delta\omega_c \sim 10^8 \text{ sec}^{-1}, \quad \delta\nu_c \sim 5\cdot 10^{-4} \text{ cm}^{-1} \quad (\chi \sim 10^{-3} \text{ cm}^2/\text{sec}),$$
$$\delta\omega_{\text{conc}} \sim 10^5 \text{ sec}^{-1}, \quad \delta\nu_{\text{conc}} \sim 5\cdot 10^{-7} \text{ cm}^{-1} \quad (D \sim 10^{-5} \text{cm}^2/\text{sec}).$$

Ratio of Intensities of the Fine Structure Components. The ratio of the integrated intensities in the fine structure components was first calculated by Landau and Placzek [140]; they showed that

$$\frac{I_c}{2I_{\text{MB}}} = \frac{c_p - c_V}{c_V} = \gamma - 1, \quad (5.39)$$

where c_p and c_V are the specific heats at constant pressure and constant volume, respectively.

Equation (5.39) was given in [140] without a derivation. Later, Eq. (5.39) was derived by Vladimirskii [46] with account of the intermolecular interaction, by Ginzburg [45, 42], and later by Gross [143]. In the derivation of Eq. (5.39), which is known as the Landau–Placzek formula, the quantity $(\partial\varepsilon/\partial T)_\rho$ is neglected; actually, this quantity cannot be neglected in a number of cases.

We shall carry out an exact thermodynamic calculation of $I_c/2I_{\text{MB}}$ [29, 53, 144, 145]. The ratio of the integrated intensities of the light scattered by the adiabatic and isobaric density fluctuations is obtained from (1.53) and (1.54):

$$\frac{I'_{is}}{I'_{ad}} = \frac{I_c}{2I_{\text{MB}}} = L\,\frac{\sigma^2 T}{\rho c_p \beta_S}, \quad (5.40)$$

where

$$L = \frac{\left(\frac{1}{\sigma}\frac{\partial\varepsilon}{\partial T}\right)_p^2}{\left(\rho\frac{\partial\varepsilon}{\partial\rho}\right)_S^2}.$$

Formula (5.40) gives the exact expression for the ratio of the integral intensities of the fine structure components in terms of the thermodynamic calculation. On the basis of (1.47), we can write that

$$\frac{\sigma^2 T}{\rho c_p \beta_S} = \frac{\beta_T}{\beta_S} - 1 = \gamma - 1. \tag{5.41}$$

If, moreover, we set L = 1, then (5.40) goes over into the Landau–Placzek formula (5.39). One cannot always take L = 1, inasmuch as for water, for example, L = 1.7, and for other liquids its value ranges from 1.5 to 1. There is an especially large difference between (5.39) and (5.40) if there is dispersion in the sound velocity and β_S differs from its static value.

In the case in which the "thermodynamic parameters" such as β_S reveal dispersion, a very general question arises – the applicability of the conclusions based on classical thermodynamics.

It is clear that in such a case the thermodynamic approach is not suitable; however, even in this case, Eq. (5.40) can be preserved. The basis for this statement will be given below. Here we shall show only that the calculation for I_{ad} can be made without using the methods of classical thermodynamics by considering the intensity of the light diffracted by a thermal elastic wave of frequency f.

The isobaric density fluctuations change with time much more slowly than the adiabatic fluctuations and, therefore, up to frequencies $\sim 10\delta\omega_c$, measured from the maximum of the unshifted line, one can use the thermodynamic expression for I_{is} with excellent approximation.

Therefore, we can compare Eq. (5.40) with experiment even in the case in which β_S undergoes appreciable dispersion ($\beta_S = 1/v^2\rho$).

Upon approach to the critical point, Eqs. (5.39) and (5.40) become inapplicable. In this case, the intensity of the central line

$I_c = I'_{is}$ [the second term of (2.1)] increases without limit. In order to avoid this, Vladimirskii [46] made a calculation of the ratio of the intensity of the fine structure components by taking into account (in the expression for the free energy) the gradient of the density fluctuations and obtained the formula

$$\frac{I_c}{2I_{\text{MB}}} = (\gamma - 1)\frac{\left(\frac{\partial p}{\partial \rho}\right)_T}{\left(\frac{\partial p}{\partial \rho}\right)_T + f^* q^2}, \tag{5.42}$$

which remains finite even at the critical point. If we neglect the constant of interaction ($f^* = 0$), then (5.42) transforms into the Landau–Placzek formula (5.39). On the basis of (5.40) and (5.42), we can write the general expression

$$\frac{I_c}{2I_{\text{MB}}} = L\frac{\sigma^2 T}{\rho c_p \beta_s}\frac{\left(\frac{\partial p}{\partial \rho}\right)_T}{\left(\frac{\partial p}{\partial \rho}\right)_T + f^* q^2}. \tag{5.43}$$

Equation (5.43) in principle allows us to find the constant characterizing the intermolecular interaction at the critical point and beyond. Actually, it is easy to get from (5.43) the result that

$$f^* = \frac{1}{q^2}\left(\frac{\partial p}{\partial \rho}\right)\left(L\frac{\sigma^2 T}{\rho c_p \beta_S}\frac{2I_{\text{MB}}}{I_c} - 1\right). \tag{5.43a}$$

The determination of f^* from the study of the fine structure near the critical temperature has not been made experimentally to date. Far from the critical point, f^* is virtually equal to zero. Giterman and Kontorovich [606] obtained an expression for the ratio of the integrated intensities of the doublet and the intensity of all three components of the fine structure with account of spatial dispersion. It follows from their calculation that

$$\frac{I_c}{2I_{\text{MB}}} = \frac{\rho Q(q) + \frac{T}{c_V \rho}\left(\frac{\partial p}{\partial T}\right)_\rho^2}{\rho Q(q)} - 1 = \frac{v_T^2(q)}{v_S^2(q)} - 1. \tag{5.42a}$$

The function Q(q) was determined in [606]. Here, we shall illustrate its meaning by means of the well-known expression for the Fourier component of the density fluctuations: $\overline{|\Delta\rho_q|^2} = VkT/Q(q)$ [Appendix I, Eq. (I.24)].

For a weak spatial dispersion, $Q(q) = [(1/\rho)(\partial p/\partial\rho)]_T + f^* q^2$ and Eq. (5.42a) transforms into (5.42). Near the critical point, as a result of the fact that $c_V \to \infty$ (see [606]), it follows from (5.42a) that, because of the decrease in the intensity of I_c and the growth of $2I_{MB}$, the ratio $I_c/2I_{MB} \to 0$. When there is no spatial dispersion, then, in the approach to the critical point, I_c increases strongly ($\gamma \to 0$) and $I_c/2I_{MB} \to \infty$.

This alternative derivation of the theory is of great interest and should be subjected to experimental verification. The experimental studies of the fine structure are frequently limited to the measurement of the maximum values of the intensity of the components.

Here it must be recalled that the maximum intensities I(0) are connected with the total intensities [see (5.19) and (5.36)] by the relation [142, 145]:

$$I_{MB}(0) = \frac{1}{2\pi}\frac{I_{ad}}{\delta\omega_{MB}}, \quad I_c(0) = \frac{1}{\pi}\frac{I_{is}}{\delta\omega_c}. \tag{5.44}$$

From Eq. (5.44), we get

$$\frac{I_c(0)}{2I_{MB}(0)} = \frac{I_{is}}{I_{ad}}\frac{\delta\omega_{MB}}{\delta\omega_c}, \tag{5.45}$$

where $\delta\omega_{MB}$ and $\delta\omega_c$ are given by the relations (5.37), (5.22), and (5.23).

It follows from the calculation made above that the spectrum of the light scattered by adiabatic and isobaric density fluctuations is a triplet, consisting of a central line and the two MB components, located on the opposite sides of the central line and at equal distances. This conclusion refers to liquids and gasses, for which the equations of hydrodynamics are applicable.

Ginzburg [147] has pointed out He II as an interesting exception to the theoretical relation, and has given a calculation for this case (see below).

At very low temperatures, the classical theory of light scattering, which we have used throughout, ceases to be valid. Therefore, it is necessary to call on the quantum theory of the phenomenon.

The quantum theory of light scattering was constructed by Tamm [148]. It follows from the Tamm theory that for the scattering coefficient of the Stokes component $(\omega_0 - \Omega)$ one can obtain the expression

$$R_{(\omega_0 - \Omega)} = \frac{\pi^2}{4\lambda^4}\left(\rho \frac{\partial \varepsilon}{\partial \rho}\right)_T^2 \frac{\hbar\Omega}{\rho v^2}\left\{\frac{1}{e^{\frac{\hbar\Omega}{kT}}-1}+1\right\}. \quad (5.46a)$$

For the anti-Stokes line,

$$R_{(\omega_0 + \Omega)} = \frac{\pi^2}{4\lambda^4}\left(\rho \frac{\partial \varepsilon}{\partial \rho}\right)_T^2 \frac{\hbar\Omega}{\rho v^2}\left\{\frac{1}{e^{\frac{\hbar\Omega}{kT}}-1}-1\right\}. \quad (5.46b)$$

In obtaining Eq. (5.46), it was assumed that $\omega_0 \pm \Omega \sim \omega_0$ and that dispersion was lacking, i.e.,

$$\beta_T \sim \beta_S = \frac{1}{\rho v^2}.$$

If the energy of the phonon is small, i.e., if the following condition is satisfied:

$$\hbar\Omega \ll kT, \quad (5.47)$$

then $R_{(\omega_0 - \Omega)} + R_{(\omega_0 + \Omega)}$ is equal to the scattering coefficient R_{90}, which is expressed by Eq. (1.70). Taking into account that in a liquid $\Omega \sim 10^{10}$ sec^{-1}, we find that the classical theory of light scattering is applicable down to $T \sim (\hbar\,\Omega/k) \sim 0.1$°K, and only for much lower temperatures need one use Eqs. (5.46).

Thus, the Einstein formula can be used even for He II. In this relation, He II does not differ in its anomalous properties. The values of the scattering intensity in He II, measured and computed from (1.70), are in qualitative agreement with each other. One of the unusual properties of He II is that, in addition to ordinary sound, which is propagated with the speed v_1, "second sound" can propagate in it with a speed v_2, as has been considered in detail by Landau [149] and found by Peshkov [150], where $v_2/v_1 \sim 0.1$. Therefore, [147], two doublets will be observed in He II, one corresponding to ordinary sound $(\Delta\omega/\omega_0) \sim (v_1/c) \sim 10^{-6}$, and the second corresponding to "second sound," $(\Delta\omega/\omega) \sim (v_2/c) \sim 10^{-7}$.

However, the displacement of the components of the second doublet, called "second sound," is so small, and its intensity so negligible that there is no hope of finding these components experimentally.

The integral intensity of each of the doublets was calculated by Ginzburg [147] and expressed by the formula

$$I=\frac{\pi^2 V}{\lambda^4 L^2}\left(\rho\frac{\partial\varepsilon}{\partial\rho}\right)_T^2\frac{\beta_T kT}{\dfrac{[(\partial\rho/\partial p)_T v^2-1]^2}{\gamma-1}+1}. \qquad (5.48)$$

The velocities of the waves of ordinary sound and "second sound" are determined by the Landau equation [149]:

$$v^4-v^2\left[\left(\frac{\partial p}{\partial\rho}\right)_S+\frac{TS^2\rho_s}{c_V\rho_n}\right]+\frac{S^2\rho_s\left(\dfrac{\partial p}{\partial\rho}\right)_T}{c_V\rho_n}=0, \qquad (5.49)$$

where ρ_n and ρ_s are thedensity of normal and superfluid phases, respectively. For ordinary sound, $v = v_1$ in (5.49) and for "second sound," $v = v_2$. Finding v_1 and substituting in (5.48), we determine the integrated intensity of the ordinary doublet I_1. Substituting $v = v_2$, we find integrated intensity of the "second doublet" I_2. The ratio of the intensity of these two doublets I_2/I_1 is equal to

$$\frac{I_2}{I_1}=\frac{\left[\left(\dfrac{\partial\rho}{\partial p}\right)_T v_1^2-1\right]^2+\gamma-1}{\left[\left(\dfrac{\partial\rho}{\partial p}\right)_T v_2^2-1\right]^2+\gamma-1}. \qquad (5.50)$$

For normal liquids and, in this case, for He II also,

$$v_1^2\equiv\left(\frac{\partial p}{\partial\rho}\right)_S\equiv\gamma\left(\frac{\partial p}{\partial\rho}\right)_T,\quad v_2^2=0,$$

and Eq. (5.50) goes over into Eq. (5.39), but even in He II, (5.50) is close to (5.39), inasmuch as even at the λ point, $\gamma \sim 1.008$ and, consequently, I_2 is negligibly small.

6. Relaxation Theory of the Spectral Composition of Light Scattered in Liquids

The spectral composition of light scattered by adiabatic and isobaric density fluctuations in low-viscosity liquids was considered in Sec. 5. In such liquids, the transverse elastic waves are so strongly damped [136, 151] that the inner doublet, which is brought about by the diffraction and modulation of light by the transverse wave, is absent. Moreover, the spectral composition of the light scattered by the fluctuations $\Delta\varepsilon^{\prime}_{ik}$, which determines the significant part of the intensity of scattered light and which on the whole is responsible for its depolarization, is not considered.

A continuous spectrum is observed in the experiment (in addition to the fine structure); the maximum intensity of this spectrum is the same as the maximum of the unshifted line, and falls off monotonically on both sides of it, extending out to 100-150 cm^{-1}. It is strongly depolarized. This continuous spectrum is known as the "wing" of the Rayleigh line [37, 38]. Although 35 years have passed since its discovery, the nature of the wing is still not completely explained, both in its theoretical and experimental aspects. Of the different opinions on the nature of the wing that have been set forth by various investigators, we shall say more later in Sec. 28. Here, we shall merely note that in our judgment, the most rational viewpoint on the nature of the wing belongs to Landau and Placzek [140], who proposed that the wing is determined by relaxational processes in the liquid, similar to that relaxation process which governs the dispersion of electromagnetic waves in polar liquids [77, 152].

Starting from the Maxwell concept of viscosity [153], Leontovich [39, 154] developed a quantitative theory of the scattering of light by pressure fluctuations (the MB components) and by that part of the fluctuations of the orientation of the molecules which are brought about by fluctuations in the deformations; consequently, he considered only the scattering from part of $\Delta\varepsilon^{\prime}_{ik}$. It is possible that in a number of cases this part plays the principal role.

The Leontovich theory played a large role in the development of spectral investigations of scattered light, and made it possible to conduct meaningful experiments and to obtain valuable information on the relaxation times of anisotropies in liquids.

Later, a correlation theory of light scattered was developed by Rytov [40, 155, 156), from which the results of the Leontovich theory follow as a special case. In this section, we shall set forth the basic assumptions and conclusions of the Leontovich theory, without giving his computations in detail but insofar as possible keeping his notation to make easier the acquaintance with the details of the calculation carried out in the original work [39].

Basic Assumptions and the Initial Equations of the Theory. It is assumed in the Leontovich theory [39] that the state of the liquid at any point can be described by the deformation tensor e_{ik}, the time derivative of which is connected with the components of the velocity u_i by the well-known relation [157]:

$$\dot{e}_{ik}=\frac{1}{2}\left(\frac{\partial u_i}{\partial x_k}+\frac{\partial u_k}{\partial x_i}\right),$$

where the trace of the deformation tensor $e_{\alpha\alpha} = \sigma$ is the coefficient of volume expansion. Moreover, the state of the liquid at the given point is described in the Leontovich theory by the temperature T and the anisotropy tensor ξ_{ik} which characterizes the departure of the axes of anisotropic molecules from an isotropic distribution. Inasmuch as the tensor ξ_{ik} is a measure of the deviation from isotropy, one can take its trace $\xi_{\alpha\alpha} = \xi = 0$ although the latter assumption is not necessary.

Thus, the fundamental assumption is that the state of a liquid at a given point and for a given temperature can be completely described by two tensors e_{ik} and ξ_{ik}. This limits the region of applicability of the theory, as was pointed out in [39].

The free energy per unit volume Ψ described by two tensors e_{ik} and ξ_{ik} will, at small departures from equilibrium, be a linear combination of nonvanishing invariants of these tensors; for simplicity, we do not take into account the dependence of Ψ on the temperature, i.e., we have eliminated from consideration the isobaric fluctuations of the density. For these cases,

$$2\Psi=L\sigma^2+2Me_{\alpha\beta}e_{\beta\alpha}+4Ne_{\alpha\beta}\xi_{\beta\alpha}+2\mu\xi_{\alpha\beta}\xi_{\beta\alpha}, \tag{6.1}$$

where L, M, N, and μ are constants.

The derivatives of Ψ are connected with the components of the stress tensor S_{ik} in the following way:

$$\frac{\partial\Psi}{\partial e_{ii}}=S_{ii},\quad \frac{\partial\Psi}{\partial e_{ik}}=2S_{ik}. \tag{6.2}$$

In correspondence with (6.1) and (6.2), we find

$$S_{ik}=L\sigma\delta_{ik}+2Me_{ik}+2N\xi_{ik}. \tag{6.3}$$

Using the conditions of minimum free energy for the state of equilibrium, and introducing a new variable

$$\zeta_{ik} = e_{ik} - \frac{1}{3}\sigma\delta_{ik} - \xi_{ik}, \tag{6.4}$$

for which $\zeta = \zeta_{\alpha\alpha} = 0$; after appropriate calculations, we obtain

$$2\Psi = K\sigma^2 + 2\mu\zeta_{\alpha\beta}\zeta_{\beta\alpha}, \tag{6.5}$$

and in place of (6.3), we then have

$$S_{ik} = K\sigma\delta_{ik} + 2\mu\zeta_{ik}. \tag{6.6}$$

In the state of equilibrium, $\zeta_{ik} = 0$ and $S_{ik} = K\sigma\delta_{ik}$. It is then clear that K has the meaning of the isothermal bulk modulus. The equation of motion of the liquid, for small departures from equilibrium, is written in the following form:

$$\rho\dot{u}_i = \frac{\partial S_{ik}}{\partial x_k}, \tag{6.7}$$

where ρ is the equilibrium value of the density. The time dependence of the anisotropy tensor follows from the reaction equation

$$\dot{\xi}_{ik} = \frac{1}{\tau}\zeta_{ik}, \tag{6.8}$$

or, taking (6.4) into account,

$$\dot{\zeta}_{ik} - \dot{e}_{ik} + \frac{1}{3}\dot{\sigma}\delta_{ik} = -\frac{1}{\tau}\zeta_{ik}. \tag{6.9}$$

Here, τ has the meaning of the relaxation time of the anisotropy and is the same for ξ_{ik} and S_{ik}. In (6.8) and (6.9), the second time derivatives of ξ_{ik} were not included. These take into account the inertial terms of the equation which, as a rough estimate shows [39], suffices for frequencies below 10^{12} $\sec^{-1}$.

Equation (6.6), together with (6.8), is the Maxwellian formulation of the relaxation theory of a viscous liquid. Actually, for tangential stresses, we get from (6.6) and (6.8)

$$S_{ik} + \frac{1}{\tau}S_{ik} = 2\mu e_{ik}, \tag{6.10}$$

where the constants τ and μ are connected with the viscosity η by the relation

$$\eta = \mu\tau, \tag{6.11}$$

where μ is the shear modulus.

In all the foregoing discussions, no assumptions were made as to a definite molecular model of the liquid. This phenomenological character of the theory is also preserved in the following for calculation of the spectral composition of the scattered light. However, it is of undoubted interest to explain what is obtained for the quantity ξ_{ik} introduced here when we introduce a definite model of the liquid into the calculation. If we choose the old Debye model as such a model [152], and the corresponding distribution function [158], we then get from (6.9)

$$\zeta_{ik} = \frac{5}{2}\left(\overline{s_i s_k} - \frac{1}{3}\delta_{ik}\right).$$

Together with (6.8), this gives $\tau = {}^1/_3\tau_D$ where τ_D is the relaxation time of the dipole moment in a polar liquid. Then, setting

$$\tau = \frac{4\pi a^3\eta}{3kT} \tag{6.12}$$

we find, by comparing (6.12) with (6.11),

$$\mu = \frac{3kT}{4\pi a^3}.$$

For finding the spectrum of the scattered light, it is necessary to find the time dependence of the fluctuations which have the form of spatial sinusoidal waves, with wave vector $\mathbf{q}$ satisfying the Bragg condition.

The time dependence follows from the equations of motion and the kinetics for the case under consideration. On the basis of (6.6), (6.7), and (6.9), these equations will have the following form:

$$\begin{aligned} \rho\dot{u}_i &= K\frac{\partial\sigma}{\partial x_i} + 2\mu\frac{\partial\zeta_{ik}}{\partial x_k}, \\ \dot{\zeta}_{ik} + \frac{1}{\tau}\zeta_{ik} &= \dot{e}_{ik} - \frac{1}{3}\dot{\sigma}\delta_{ik}, \end{aligned} \tag{6.13}$$

where it is assumed that

$$u_i = v_i e^{i(qr)}, \qquad e_{ik} = \mathcal{E}_{ik} e^{i(qr)}, \qquad \zeta_{ik} = Z_{ik} e^{i(qr)}$$

We introduce the rectangular set of coordinates **n**, **t**, **s**, and set it in such a way that the vector **n** coincides with the normal to the wave front, and **t** and **s** lie in the plane of the wave front. We further introduce the notation

$$Z_{nn} = \Phi, \quad Z_{tt} = -\frac{1}{2}\Phi + X, \quad Z_{ss} = -\frac{1}{2}\Phi - X, \quad Z = Z_{\alpha\alpha} = 0.$$

Here

$$\Phi = -Z_{tt} - Z_{ss}, \quad X = \frac{1}{2}(Z_{tt} - Z_{ss}).$$

Then the equations of motion are decomposed into independent groups

$$\rho\dot{u}_n - iqK\mathcal{E} = 2\mu iq\Phi, \quad \dot{\Phi} + \frac{1}{\tau}\Phi = \frac{2}{3} iqv_n, \tag{6.14}$$

$$\left.\begin{aligned} \rho\dot{u}_t = 2\mu iqZ_{tn}, \quad \dot{Z}_{tn} + \frac{1}{\tau}Z_{tn} = \frac{1}{2} iqv_t, \\ \rho\dot{u}_s = 2\mu iqZ_{sn}, \quad \dot{Z}_{sn} + \frac{1}{\tau}Z_{sn} = \frac{1}{2} iqv_s, \end{aligned}\right\} \tag{6.15}$$

$$\dot{X} + \frac{1}{\tau}X = 0, \qquad \dot{Z}_{st} + \frac{1}{\tau}Z_{st} = 0. \tag{6.16}$$

The group of equations (6.14) describe the longitudinal waves which correspond to the formation of the MB components, considered in the previous section. Equations (6.14) correspond to Eq. (5.10). Inasmuch as isobaric fluctuations of density are not considered, K in Eqs. (6.14) is the adiabatic bulk modulus.

It also follows from Eqs. (6.14) that the propagating longitudinal wave accompanies the change of anisotropy, inasmuch as $\Phi = Z_{nn} \neq 0$ in this case.

The group of equations (6.15) describe the propagation of transverse waves which, in the case of slow motions, represent the ordinary "viscous" transverse waves, while for fast motions they go over into shear waves in a solid. The propagation of these waves is connected with the changes of the anisotropy in the liquid, inasmuch as the components of the anisotropy tensors Z_{tn} and Z_{sn} do not vanish. Finally, the group of equations (6.16) describes the perturbation of the anisotropy, which is not accompanied by motion of the liquid. In these equations, only those components of the tensor Z_{ts} appear which refer to the directions lying in the plane of the wave front. It is therefore possible to speak of transverse anisotropy waves in this case, although one cannot speak about any real propagation of such waves. The components of the tensor Z_{ts} and $X = \frac{1}{2}(Z_{tt} - Z_{ss})$ fall off with time at any point of the wave independently of what happens at the other points.

The Intensity Distribution in the Spectrum of Scattered Light. For the calculation of the intensity distribution in the spectrum of scattered light one must, as before, find the field of the scattered wave and then seek the coefficients of the corresponding Fourier transform of the light field. Let the exciting light be propagated in the scattering medium in the x direction. We shall further assume that this light is linearly polarized, and that its electric vector is directed along the l axis, which can be either the z or y axis of the cartesian coordinate system. The total component of the field of the light wave scattered by the entire scattering volume in the irradiation of it by polarized light is equal to [39]:

$$E_p' = e^{i\omega_0 t} P_i \int_V \Delta\varepsilon_{ik}' e^{i(qr)} dV, \tag{6.17}$$

where ω_0 is the frequency of the incident light, $\mathbf{q}$ is determined by the relation (1.26), and $\Delta\varepsilon_{ik}$ is determined by Eq. (1.2). Taking into account the smallness of the quantity $\Delta\varepsilon_{ik}'$, we can assume that it is proportional to the anisotropy tensor ζ_{ik} and, consequently,

$$\Delta\varepsilon_{ik}' = A\zeta_{ik}. \tag{6.18}$$

The relation (6.18) allows us to conclude that the relaxation time of the anisotropy τ will also characterize the duration of the existence of double refraction, which is determined by the tensor $\Delta\varepsilon_{ik}'$. Therefore, it can be thought that the time τ is also the time of appearance or disappearance of the Kerr effect or of double refraction in the flux (the Maxwell effect). For slow motions, when the quantity $\dot{\zeta}_{ik}$ can be neglected, $\Delta\varepsilon_{ik}'$, with account of (6.9), is written in the form

$$\Delta\varepsilon_{ik}' = M\left(e_{ik} - \frac{1}{3}\sigma\delta_{ik}\right), \tag{6.19}$$

where M = $A\tau$ is the Maxwell constant or the constant of double refraction in the flux. If the flux is directed along the x axis and the gradient along the y axis, then the only nonvanishing component of e_{ik} will be e_{xy}, and, consequently,

$$\Delta\varepsilon'_{ik} = \Delta\varepsilon'_{xy} = \frac{1}{2} M \frac{\partial u_x}{\partial y}. \quad (6.20)$$

The scalar part of $\Delta\varepsilon_{ik}$ can be represented in the form

$$\Delta\varepsilon\delta_{ik} = \left(\frac{\partial\varepsilon}{\partial\rho}\right) \Delta\rho\delta_{ik}. \quad (6.21)$$

Using (6.19) and (6.21), we write the general expression for $\Delta\varepsilon_{ik}$ in the following form:

$$\Delta\varepsilon_{ik} = \left(\frac{\partial\varepsilon}{\partial\rho}\right) \Delta\rho\delta_{ik} + M\zeta_{ik}. \quad (6.22)$$

The electric field of the light wave scattered in the direction **k'** can be computed from (6.17) and (6.22). By introducing the cartesian coordinates **n, t, s** (**n** is directed along **q** and **t** and **s** are perpendicular to **n**) and the notation $\varphi = \zeta_{nn}$ and $\chi = \frac{1}{2}(\zeta_{tt} - \zeta_{ss})$, we get

$$E'_p = e^{i\omega_0 t} \Big\{ p_e \left(-\rho \frac{\partial\varepsilon}{\partial\rho}\right) \mathscr{E} + A \left[\Phi \left(\frac{3}{2} n_e (pn)\right) - \frac{1}{2} p_e\right] +$$
$$+ Z_{nt} [(pn) t_e + (pt) n_e] + Z_{ns} [(pn) s_e + (ps) n_e] +$$
$$+ X [(pt) s_e + (ps) t_e] + X [(pt) - (ps) s_e] \Big\}, \quad (6.23)$$

where

$$Z_{jh} = \frac{1}{V} \int \zeta_{jh} e^{i(qr)} dV, \quad X = \frac{1}{V} \int \chi e^{i(qr)} dV,$$

$$\Phi = \frac{1}{V} \int \varphi e^{i(qr)} dV, \quad \mathscr{E} = \frac{1}{V} \int e e^{i(qr)} dV. \quad (6.24)$$

The intensity in the spectrum $I(\omega)$ is proportional to the mean square of the coefficients $|E(\omega)|^2$ of the Fourier transform (6.17). For the calculation of the mean squares of the pairs of coefficients of any random quantity, it is necessary to know the equation which determines the change of this quantity with time (the phenomenological kinetic equation) and the mean square of this quantity in the equilibrium state. As kinetic equations, we use here the Eqs. (6.14), (6.15), and (6.16) for the quantities $\mathscr{E}$, Φ, X, Z_{nt}, Z_{ns}, and

Z_{ts}. The mean squares of these quantities are computed in [16, 39, 51] by a thermodynamic method with the help of Eq. (6.5) for the free energy.

The path of the rather complicated calculations can be traced by turning to the original research; here we shall write only the final results of the calculation.

For the case in which the scattering volume is illuminated by monochromatic light with the vector **E** directed along the z axis, the intensities $I_{zz}(\omega_0 + \omega)$ and $I_{zx}(\omega_0 + \omega)$ are obtained as follows:*

$$I_{zz}(\omega_0+\omega)=\beta\left(\rho\frac{\partial\varepsilon}{\partial\rho}\right)^2 S_{\mathscr{E}}(\omega)+\left(\rho\frac{\partial\varepsilon}{\partial\rho}\right)^2 A\left\{\frac{1}{2}\beta S_{\Phi\mathscr{E}}(\omega)+\right.$$
$$\left.+\frac{1}{6\mu}S_{\mathscr{E}\Phi}(\omega)\right\}+A^2\left\{\frac{1}{2\mu}S_{\Phi}(\omega)+\frac{1}{4\mu}S_A(\omega)\right\};$$
$$I_{zx}(\omega_0+\omega)=\frac{A^2}{8\mu}\{S_T(\omega)+S_A(\omega)\}. \tag{6.25}$$

If the vector **E** in the incident, linearly polarized light, is directed along the y axis, then

$$\left.\begin{aligned} I_{yz}(\omega_0+\omega)&=\frac{A^2}{8\mu}\{S_T(\omega)+S_A(\omega)\},\\ I_{yx}(\omega_0+\omega)&=\frac{A^2}{8\mu}\left\{\frac{3}{2}S_{\Phi}(\omega)+S_A(\omega)\right\}.\end{aligned}\right\} \tag{6.26}$$

In the case in which the scattered volume is irradiated by monochromatic natural light, the intensities are as follows:

$$I_z(\omega_0+\omega)=\left(\rho\frac{\partial\varepsilon}{\partial\rho}\right)^2\beta S_{\varepsilon}(\omega)+\frac{1}{2}\left(\rho\frac{\partial\varepsilon}{\partial\rho}\right)^2 A\left\{\rho S_{\Phi\varepsilon}(\omega)+\right.$$
$$\left.+\frac{1}{3\mu}S_{\varepsilon\Phi}(\omega)\right\}+\frac{A^2}{4\mu}\left\{\frac{1}{3}S_{\Phi}(\omega)+\frac{1}{2}S_T(\omega)+\frac{3}{2}S_A(\omega)\right\}, \tag{6.27}$$
$$I_x(\omega_0+\omega)=\frac{A^2}{8\mu}\left\{\frac{3}{2}S_{\Phi}(\omega)+S_T(\omega)+\frac{3}{2}S_A(\omega)\right\}.$$

*When I has two indices, then the first of them denotes the direction of the electric vector of the light wave in the incident light, and the second in the scattered light. One index on I denotes the direction of **E** in the scattered light for excitation by natural light.

In this investigation of Leontovich [39], an expression was also found for the depolarization of the spectrally undecomposed Rayleigh scattering line Δ_u, namely,

$$\Delta_u = \frac{A^2/2\mu}{\beta\left(\rho\frac{\partial\varepsilon}{\partial\rho}\right)^2 + \frac{7}{12}\frac{A^2}{\mu}} = \frac{M^2/2\eta\tau}{\beta\left(\rho\frac{\partial\varepsilon}{\partial\rho}\right)^2 + \frac{7}{12}\frac{M^2}{\eta\tau}}. \tag{6.28}$$

By replacing A^2/μ by Δ_u [(6.28)] in Eq. (6.27), we get

$$I_z(\omega_0+\omega) = kT\beta\left(\rho\frac{\partial}{\partial\rho}\right)\left\{S_g(\omega)\left[1 \pm \frac{\omega^2-\Omega_L}{2\Omega_L\Omega_T}\left(\frac{\Delta_u}{2\left(1-\frac{7}{6}\Delta_u\right)}\right)^{1/2}\right] + \right.$$
$$\left. + \frac{\Delta_u}{4\left(1-\frac{7}{6}\Delta_u\right)}\left[S_T(\omega)+3S_A(\omega)\right]\right\}, \tag{6.29}$$

$$I_x(\omega_0+\omega) = \frac{kT}{4}\beta\left(\rho\frac{\partial\varepsilon}{\partial\rho}\right)^2 \frac{\Delta_u}{1-\frac{7}{6}\Delta_u}\left\{\frac{3}{2}S_\Phi(\omega)+S_T(\omega)+\frac{3}{2}S_A(\omega)\right\}.$$

In Eqs. (6.25)-(6.27) and (6.29), the following contractions are introduced:

$$\left.\begin{aligned}
&S_g(\omega) = \frac{\frac{8}{3}\Omega_L^2\Omega_T^2\tau}{(\omega^2-\Omega_L^2)^2+\omega^2\tau^2(\omega^2-\Omega_S^2)^2}, \quad S_\Phi = \frac{3(\omega^2-\Omega_L^2)}{4\Omega_T^2\Omega_L^2}S_g(\omega),\\
&S_T(\omega) = \frac{2\omega^2\tau}{\omega^2+\tau^2(\omega^2-\Omega_T^2)^2}, \quad S_A(\omega) = \frac{2\tau}{1+\omega^2\tau^2},\\
&S_{\Phi g}(\omega) = \frac{\frac{4}{3}\Omega_L^2(\omega^2-\Omega_L^2)\tau}{(\omega^2-\Omega_L^2)^2+\omega^2\tau^2(\omega^2-\Omega_S^2)^2},\\
&S_{g\Phi} = \frac{4\Omega_T^2(\omega^2-\Omega_L^2)\tau}{(\omega^2-\Omega_L^2)^2+\omega^2\tau^2(\omega^2-\Omega_S^2)^2},
\end{aligned}\right\} \tag{6.30}$$

where

$$\left.\begin{aligned}
&\Omega_L^2 = \frac{q^2}{\rho\beta}, \quad \Omega_L = q\sqrt{\frac{1}{\beta\rho}} = \frac{4\pi v_L}{\lambda}\sin\frac{\theta}{2}, \quad \Omega_T^2 = \frac{\mu}{\rho}q^2,\\
&\Omega_T = q\sqrt{\frac{\mu}{\rho}} = \sqrt{\frac{\eta}{\rho\tau}}\frac{4\pi}{\lambda}\sin\frac{\theta}{2}, \quad \Omega_S^2 = \frac{\frac{1}{\beta}+\frac{4}{3}\mu}{\rho}q^2 = \Omega_L^2+\frac{4}{3}\Omega_T^2.
\end{aligned}\right\} \tag{6.31}$$

The intensity distribution depends strongly on the frequencies of the longitudinal (Ω_L) and transverse (Ω_T) Mandel'shtam–Brillouin doublets, on the relaxation time τ and on the value and sign of the ratio $A/(\rho\partial\varepsilon/\partial\rho)$. Far from the fine structure lines, Eqs. (6.27) and (6.29) describe the distribution of intensity in the wing of the Rayleigh line. Assuming $\Omega_L \ll \omega$ and $\Omega_T \ll \omega$ in this case, we find that $S_{\mathscr{E}}(\omega)$, $S_{\Phi\mathscr{E}}(\omega)$, and $S_{\mathscr{E}\Phi}(\omega)$ are negligibly small, and that

$$S_{\Phi}(\omega)=S_T(\omega)=S_A(\omega)=\frac{2\tau}{1+\omega^2\tau^2}.$$

Formulas (6.27) go over in this case to the following:

$$I_x(\omega_0+\omega)=\frac{A^2kT}{2\mu}\frac{2\tau}{1+\omega^2\tau^2},\quad I_z(\omega_0+\omega)=\frac{7A^2kT}{12\mu}\frac{2\tau}{1+\omega^2\tau^2}. \qquad (6.32)$$

It is clear from Eqs. (6.32) for I_x and I_z that the theory predicts the constancy of the depolarization factors over the entire wing of the Rayleigh line, while

$$\frac{I_x}{I_z}=\rho(\omega)=\frac{6}{7}. \qquad (6.33)$$

The intensity of the polarization of the undecomposed light of the wing is described by a formula which follows from (6.32):

$$I(\omega_0+\omega)=\frac{13}{12}\frac{A^2kT}{\mu}\frac{2\tau}{1+\omega^2\tau^2}. \qquad (6.34)$$

For the intensity at the maximum of the wing ($\omega = 0$), we get from (6.34)

$$I(\omega_0)=\frac{13}{6}\frac{A^2}{\mu}\tau kT.$$

For a comparison of the theory with experiment, it is convenient to write (6.34) or (6.32) in the following form:

$$L(\omega)=\frac{1}{I(\omega)}=\frac{1}{I(\omega_0)}+\frac{1}{I(\omega_0)}\omega^2\tau^2. \qquad (6.35)$$

By plotting $L(\omega)$ along the ordinate and ω^2 along the abscissa, we get the straight line of Fig. 10. The intersection of this line with the L axis has the value $1/I(\omega_0)$, while the slope of the line determines the anisotropy relaxation time

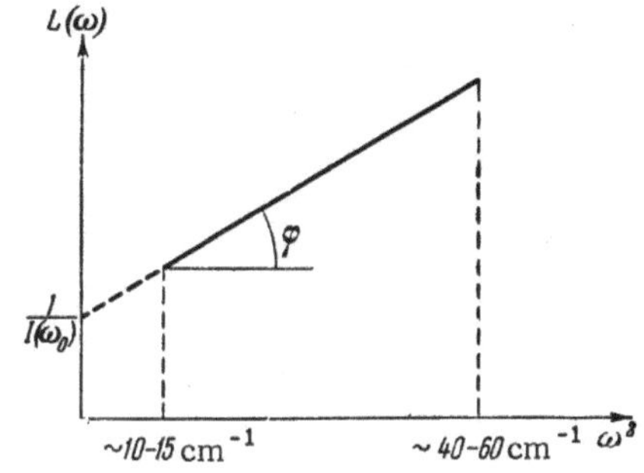

Fig. 10. Dependence of $1/I(\omega) = L(\omega)$ on ω^2 [Eq. (6.35)].

$$\tau = \left(I(\omega_0) \frac{d\,[L(\omega)]}{d\,(\omega^2)} \right)^{1/2}. \tag{6.36}$$

In the region where the fine structure components are located, the character of the distribution of the intensity over the spectrum is strongly dependent on $\Omega_L\tau$ and $\Omega_T\tau$. One can construct I_x and I_z graphically from Eqs. (6.29) only by supplying the values of $\Omega_L\tau$, the coupling Ω_T and Ω_L, the quantity Δ_u, and the sign of the quantity $A/(\rho\,\partial\varepsilon/\partial\rho)$. It was assumed in [39] that $(\Omega_T/\Omega_L)^2 = 0.075$, $\Delta_u = 0.50$, and cases were considered for which $\Omega_L\tau = 0.1$, and $\Omega_L\tau = 10$. The corresponding graphs are plotted in Figs. 11a and 11b. Figure 12 shows $\rho(\omega)$ as a function of the frequency for different signs of $A/(\rho\,\partial\varepsilon/\partial\rho)$ and the two values of $\Omega_L\tau$ shown. In the graphs of Figs. 11a and 11b, the intensity is plotted in arbitrary units while ω/Ω_L is plotted along the abscissa.

From Eqs. (6.29) one can draw an interesting conclusion on the depolarization of the longitudinal components of the fine structure. Although the adiabatic fluctuations are isotropic, the propagation of the plane elastic wave connected with them generates an anisotropy wave, as has been pointed out above and, consequently, an additional depolarization of the light scattering appears at the frequency of the longitudinal doublet. This depolarization depends on the values of $\Omega_L(\tau)$, $\Omega_T\tau$, and the depolarization factor Δ_u. The coupling of the depolarization at the frequency of the longitudinal doublet with the quantities mentioned is easily obtained from Eqs.

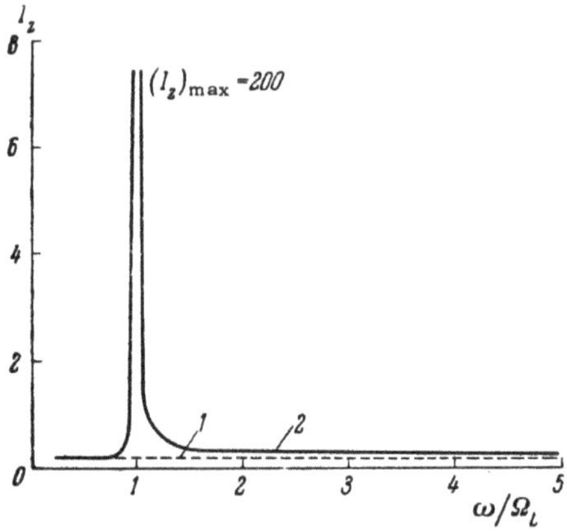

Fig. 11a. Intensity distribution in the spectrum of scattered light (Leontovich [39]). 1) I_x; 2) I_z; $A/(\rho\,\partial\varepsilon/\partial\rho) > 0$, $\Omega_L\tau = 0.1$, $\Omega_T^2 = 0.075\,\Omega_L^2$.

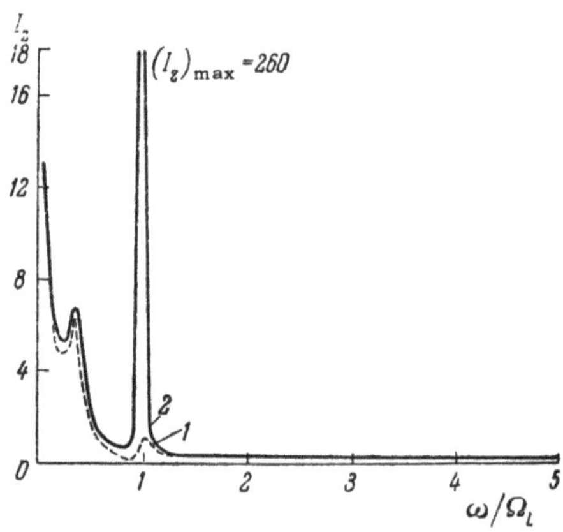

Fig. 11b. Intensity distribution in the spectrum of scattered light (Leontovich [39]). 1) I_x; 2) I_z; $A/(\rho\,\partial\varepsilon/\partial\rho) > 0$; $\Omega_L\tau = 10$, $\Omega_T^2 = 0.075\,\Omega_L^2$.

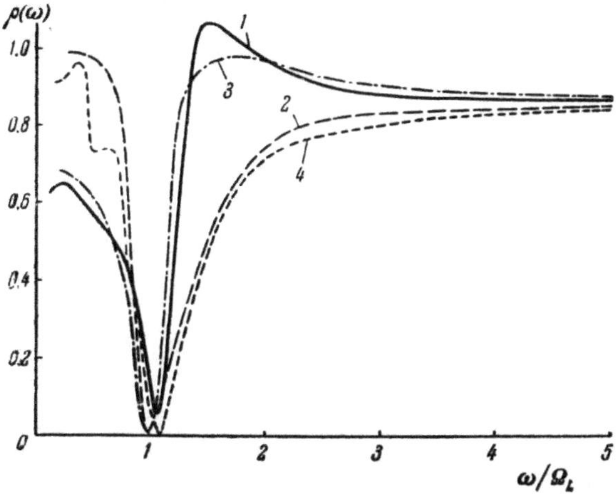

Fig. 12. Distribution of the depolarization factor $\rho(\omega)$ over the spectrum of scattered light (Leontovich [39]). 1) $A/(\rho\partial\varepsilon/\partial\rho) < 0$; 2) $A/(\rho\partial\varepsilon/\partial\rho) > 0$, $\Omega_L\tau = 0.1$; 3) $A/(\rho\partial\varepsilon/\partial\rho) < 0$; 4) $A/(\rho\partial\varepsilon/\partial\rho) > 0$, $\Omega_L\tau = 10$.

(6.29) if we set $\omega = \Omega_L$:

$$\rho(\Omega_L) = \frac{I_x}{I_z} = \frac{\Delta_u (2z+3)}{\left(\dfrac{1+\Omega_L^2\tau^2}{\Omega_T^2\tau^2}\right)(6-7\Delta_u) + 2(z+3)\Delta_u}, \tag{6.37}$$

where the following notation has been introduced:

$$z = \frac{1+\Omega_L^2\tau^2}{1+\Omega_L^2\tau^2\left[1-\left(\dfrac{\Omega_T}{\Omega_L}\right)^2\right]}. \tag{6.38}$$

If $\Omega_L\tau$ and $\Omega_T\tau$ tend to zero, then $z \to 1$ and $\rho(\Omega_L) \to 0$. If $\Omega_L\tau$ and $\Omega_T\tau$ go to infinity and $(\Omega_T/\Omega_L)^2$ is small (for example, in [39]: $(\Omega_T/\Omega_L)^2 = 0.075$), then z goes to unity as before, and the depolarization factor approaches a maximum value

$$\rho_{max}(\Omega_L) = \frac{5\Delta_u}{\left(\dfrac{\Omega_L}{\Omega_T}\right)^2(6-7\Delta_u)+8\Delta_u}. \tag{6.39}$$

In the case considered by Leontovich [39]: (1) $\Omega_L\tau = 0.1$, and (2) $\Omega_L\tau = 10$ for $(\Omega_T/\Omega_L)^2 = 0.075$. From (6.37) and (6.38), for $\Delta_u = 0.4$, it is easy to find that in the first case $\rho(\Omega_L) = 5 \cdot 10^{-4}$, while in the second case $\rho(\Omega_L) = 4 \cdot 10^{-2}$.

It must be noted that the large increase in τ due to the increase in viscosity is hardly possible because of the relaxation process. However, the combination of the effects of viscosity and the dimensions of the molecules possibly leads to $\Omega_L\tau \gg 1$.

Account of Inertial Terms in the Reaction Equation. The reaction equation, as was pointed out above, is obtained under the assumption that the inertial effects do not have a role and therefore the second derivative can be omitted in (6.8). Such a simplification is possible up to frequencies $\sim 10^{12}$ Hz, as has been shown here and in [39]. It then follows that Eqs. (6.32) are valid in the range of frequencies shown in Fig. 10.

Experimental studies of the wing of the Rayleigh line cover a range of frequencies extending above $\sim 10^{12}$ Hz. In the experimental results, $L(\omega)$ and ω^2 are plotted as the coordinates, and the absence of the linear dependence predicted by (6.32) is observed in all cases.

Therefore, the formulas for the intensity distribution in the wing, obtained by Leontovich [39], must be recalculated for the case in which one cannot neglect the inertial terms. Such a calculation has been made by Zhivlyuk [504], in which the reaction equation of the form (6.8) is replaced by the equation

$$a\frac{\partial^2\zeta}{\partial t^2} + \tau\frac{\partial\zeta}{\partial t} + \zeta = 0. \tag{6.40}$$

Zhivlyuk carried out the calculation of the distribution of the intensity in the wing of the Rayleigh line according to the method of Leontovich [39]. The results of the calculation for the intensity distribution of the wing are the following:

$$I_x(\omega_0+\omega) = \frac{A^2\tau}{4\mu}\,\frac{4-3a\Omega_T^2}{\omega^2\tau^2+(1-a\omega^2)^2}, \quad I_z(\omega_0+\omega) = \frac{A^2\tau}{12\mu}\,\frac{14-\frac{17}{3}a\Omega_T^2}{\omega^2\tau^2+(1-a\omega^2)^2}. \tag{6.41}$$

It follows from these formulas that the depolarization factor in the spectrum is determined by the quantity

$$\rho(\omega) = \frac{12 - 9a\Omega_T^2}{14 - \frac{17}{3}a\Omega_T^2} . \tag{6.42}$$

In those cases in which the inertial terms do not play an important role ($a \to 0$), Eqs. (6.41) transform into (6.32), and (6.42) into (6.33). However, when such neglect is impossible, the experimental material is best described by Eqs. (6.41), beginning from 15-20 cm^{-1} and higher.

Thus, for example, the wing of the scattering line in carbon bisulfide is well described by (6.41) from 15 cm^{-1} up to the highest frequencies measured for $a = 6.2 \cdot 10^{-25}$ sec^2, obtained from the experimental curve. For such an a, the value of $\rho(\omega)$ according to (6.42) is practically unchanged from $^6/_7$. However, if it were shown for any liquid that $a\Omega_T^2$ was of the order of unity, then it would follow from (6.42) (for $a > 0$) that $\rho(\omega) > {}^6/_7$.

7. Nonthermodynamic Theory of Light Scattering in Liquids

Elementary Nonthermodynamic Calculation of the Light Intensity Scattered by Adiabatic Density Fluctuations. All the calculations given above of the intensity of the scattered light were based on the use of statistical thermodynamics.

Strictly speaking, the Einstein formula and the other expressions written above for the absolute intensity cannot be applied to the problems of light scattering, inasmuch as thermodynamics assumes very slow (reversible) changes in the parameter, while, in the scattering process the changes take place extraordinarily rapidly (in each case for the adiabatic fluctuations). Where it has not been established that certain parameters which determine the scattering are functions of frequency, the application of the thermodynamic relations can be regarded as legitimate although, in principle, it can be assumed that a dependence on the frequency, for example of the compressibility [45], must take place, inasmuch as the adiabatic fluctuations of the density "decay" with the rate of elastic waves of frequency about 10^{10} Hz.

After the dispersion of sound velocity had been discovered in a number of liquids [29-32, 34-36], the application of thermodynamic formulas was made clearly invalid. For example, in the derivation of Eq. (1.52), thermodynamics is applied for the determination of the quantities $\overline{(\Delta p)^2}$ and $\overline{(\Delta S)^2}$. The expressions for these quantities are obtained under the supposition that by furnishing the values of p and S (which differ from $\bar{p}$ and $\bar{S}$), one can characterize the state of incomplete equilibrium. In other words, it is implicitly assumed that the relaxation time of these quantities is large in comparison with the relaxation time of the other processes of approach to equilibrium which are possible in the system under consideration [51]. With this assumption, we obtain expressions for the total fluctuation of p or S, or other thermodynamic quantities.

For the theoretical calculation of the intensity of the scattered light in a liquid, the expression for the spectral density of the corresponding fluctuations $\overline{(\Delta \rho_\omega)^2}$ and $\overline{(\Delta S_\omega)^2}$ is of importance, and these quantities are expressed in terms of parameters which can be functions of the frequency.

The general solution of the problem of the calculation of $\overline{(\Delta \varepsilon_\omega)^2}$ as a function of thermodynamic and nonthermodynamic parameters, which completely determine the light scattering, is still lacking. But significant steps have already been made along this path, at least in principle.

An attempt to calculate the intensity of the light scattering and its spectral distribution, without having recourse to thermodynamics, was made by Rytov [40, 155, 156] who developed the correlation theory of Rayleigh light scattering, taking into account the fluctuations of the pressure and entropy, and the anisotropy fluctuations which are called the deformation fluctuations.

Before introducing the basic results of this theory, we shall give some elementary discussions [29, 52, 53, 142] which make it possible simply, without recourse to thermodynamics, to find the intensity of the light scattered by the adiabatic density fluctuations, and which show that the formulas obtained above – for example for the ratio of the intensity of the fine structure components and for the total scattering, can be preserved even in the presence of a significant sound velocity dispersion. Of course, the values of the parameters measured at the corresponding frequency must be used here.

The calculation of the intensity of the light scattered by the adiabatic density fluctuations, in correspondence with the representations developed in Sec. 5, can be reduced to the calculation of the intensity of the light diffracted by the Debye thermal wave in the direction corresponding to the Bragg condition (1.26).

We express the incident light wave in the form $E = E_0 \cdot e^{i(\omega t - \mathbf{kr})}$ and assume that the elastic wave produced periodic departures of the dielectric constant from its average value, according to the law

$$\Delta\varepsilon = \frac{\Delta\varepsilon_0}{2}\left\{e^{i[\Omega t-(qr)]} + e^{-i[\Omega t-(qr)]}\right\}, \qquad (7.1)$$

where $\Delta\varepsilon_0$ is the amplitude of the oscillations of $\Delta\varepsilon$.

By computing the intensity of the scattered light or the scattering coefficient of the light by means of Eqs. (1.11)-(1.14), and neglecting the difference in the wavelengths of both MB components, we get

$$2I_{MB} = G\Delta\varepsilon_0^2. \qquad (7.2)$$

The amplitude of the change in the dielectric constant can be written in the form

$$\Delta\varepsilon_0 = \left(\rho\frac{\partial\varepsilon}{\partial\rho}\right)_S \frac{\Delta\rho_0}{\rho}. \qquad (7.3)$$

The adiabatic value $(\rho\partial\varepsilon/\partial\rho)_S$ is taken because the process of sound propagation at a frequency $\sim 10^{10}$ Hz is an adiabatic process. The propagation of sound in a liquid becomes isothermal only for frequencies $f = v^2/4\pi\chi$ (χ is the coefficient of temperature conductivity, v is the sound velocity), i.e., for frequencies $> 10^{13}$ Hz.

According to the hydrodynamic theory of sound [157],

$$\left|\frac{\Delta\rho}{\rho}\right|^2 = \frac{u^2}{v^2} = \frac{V\rho u^2}{V\rho v^2},$$

where u is the velocity of vibration of the particles of the liquid. Equating the average energy of the normal vibrations of the medium to kT, and taking into account that $1/\rho v^2 = \beta_S$, where β_S is the adiabatic compressibility, we get the final expression for the intensity of the components

$$I_{MB} = G\left(\rho\frac{\partial\varepsilon}{\partial\rho}\right)_S^2 kT\beta_S. \tag{7.4}$$

From a comparison of Eqs. (7.4) and (1.53), obtained by the thermodynamic method, it is seen that these expressions coincide only formally, but they are essentially different.

It is evident from the very character of the derivation of (7.4) that $(\rho\partial\varepsilon/\partial\rho)_S$ and β_S are quantities corresponding to the frequency of sound waves on which the diffraction of light is considered. Thus, a formula is obtained for the intensity of the scattered light that is free from thermodynamic limitations. For the calculation of the intensity of Eq. (7.4), it is necessary to use the values of $(\rho\partial\varepsilon/\partial\rho)_S$ and β_S at the corresponding frequency of the process. Strictly speaking, this refers to all liquids, but the difference between (7.4) and the corresponding thermodynamic formula comes out very strongly when an appreciable dispersion of the sound velocity is found in the liquid. Those liquids such as benzene, carbon bisulfide, carbon tetrachloride, etc., in which dispersions of the order of 10-20% were observed by us can serve as examples (see Chapter VI). As a consequence of the dispersion of the sound velocity, the integrated intensity in these liquids should be about 20-40% lower than that calculated from the thermodynamic value of the parameters. The intensity of the central component, just as for I_{MB}, should be calculated with consideration of the rate of "decay" of the isobaric fluctuations. But if the fluctuations of pressure change rapidly, which is expressed in a shift of the MB components by a value $\sim 10^{10}$ Hz, then the isobaric density fluctuations change much more slowly. In scattered light they are in the form of the unshifted line, the maximum of which corresponds to $\omega = \omega_0$. The halfwidth of this line for the different liquids is somewhat different but, in the medium, differs only slightly from $\delta\omega \sim 10^7$ Hz. Thus, processes take part in the formation of the central component that are slower by three orders of magnitude than the processes which bring about a shift in the line.

Generally speaking, the quantities σ, $(\partial\varepsilon/\partial T)_p$, and c_p which enter into Eq. (1.54) can depend on the frequency. In accord with relaxation theory, the change in c_p will be determined by the same relaxation time as the sound velocity. Then, in each case, for benzene, carbon bisulfide, and carbon tetrachloride, one can show (see Chapter VI) what is the value of the relaxation frequency f_c. It is shown that for these liquids f_c is higher by one or two orders of magnitude than the values of the halfwidth of the central component. Consequently, within the limits of the halfwidth of the central component, where the principal part of the intensity is contained, one can use the static value of c_p for calculation according to (1.54). Only at very large distances from the maximum of the central component (~10 halfwidths) can the static value c_p be shown to be significantly larger than the dynamic, and the corresponding small part of the total intensity is shown to be minimized.

The relaxation coefficient of the expansion of σ is connected with the relaxation of the elasticity. Therefore, the corresponding f_c here is the same as for c_p. This identity also holds for the frequency dependence of $(\partial \varepsilon / \partial T)_p$. To what degree I_c, computed from (1.54) for static values of σ, c_p, and $(\partial \varepsilon / \partial T)_p$ differs from the true value it is difficult now to say. Only the comparison of the computed values with the experiment can clear up this question.

From Eqs. (7.4) and (1.54) for the ratio of the integrated intensities I_c and $2I_{MB}$, we get Eq. (5.40); however, it is suitable also in the case when there is dispersion of the parameters entering into Eq. (5.40).

Correlation Theory of Light Scattering in an Isotropic Medium. In the correlation theory of the molecular scattering of light developed by Rytov [40], as in the previous "thermodynamic" theories, the fundamental problem was the calculation of the mean square $\Delta\varepsilon_{ik}$. However, the difference is that in the correlation theory the mean squares of the spectral components of the tensor $\Delta\varepsilon_{ik}$ are computed.

Taking into consideration that the fluctuations are small departures from the equilibrium state, we assume that there is a linear coupling between the spectral amplitudes $\Delta \varepsilon_{ik}$ and the spectral amplitudes of the quantities describing the departure from the equilibrium state, for example, the deformations and the temperature. Consequently, in order that the problem be solved, one must know the spectral correlation functions for the thermal fluctuations of temperature and deformation. The correlation theory of thermal fluctuations was developed by Landau and Lifshitz [47, 159] for a viscous liquid not exhibiting dispersion, and by Rytov [156] for the case of an isotropic but otherwise arbitrary medium. The correlation theory of thermal fluctuations allows us to solve the problem of the intensity and spectral composition of scattered light in sufficiently general form for those cases of interest to us in which the parameters of the medium can depend on the frequency.

The phenomenological theory of Rytov does not contain any presuppositions as to the nature of the dispersion, and does not assume the form of the dispersion laws in advance. It is only required from these laws that they do not violate the dissipation behavior of the medium. To establish the specific form of the dispersion laws, we need other, additional, assumptions lying outside the theory being considered. A liquid with an arbitrary viscosity or a solid amorphous body is characterized in this theory by two complex moduli of elasticity, by scalar thermal coefficients, and scalar elasto-optical and thermo-optical parameters.

Further, the theory considers only those internal processes (which are developed in the scattering medium) which are connected with fluctuations of the deformations and of the temperature. In this framework, the theory describes the isotropic and partially anisotropic fluctuations $\Delta\varepsilon_{ik}$ and the spectral and total intensity of the scattered light associated with them.

Left out of consideration are the fluctuations in the orientation of the molecules, which are not connected with deformation fluctuations [313], the isotropic fluctuations measured in Sec. 1, which do not reduce to temperature and density fluctuations, and the scattering by the antisymmetric part of the tensor $\Delta\varepsilon'_{ik}$. It is most probable that these last two unconsidered theories of the process make an extremely small contribution to the total flux of scattered light. The part of the scattered light connected with $\Delta\varepsilon'^{(a)}_{ik}$ cannot generally be very far from the absorption line.

Correlation theory, even in its present form, gives a complete, nonthermodynamic derivation of the formulas for the intensity distribution in the spectra and for the integrated intensities of the light scattered by the adiabatic and isobaric density fluctuations, and probably for a large part of the light scattered by the anisotropy fluctuations.

The approach to the solution of the problem of the molecular light scattering that has been described is of undoubted interest, and, therefore, we shall report here some of the fundamental results of the theory without detailed calculations, which latter can be found in the original literature [40, 155, 156].

The solution of the Maxwell equations for the field of the scattered wave (1.9) is sought in spectral form, while the tensor character of the fluctuations of the dielectric constant is taken into account; it is represented as follows:

$$\Delta\varepsilon_{ik}(t, \boldsymbol{r}) = \int_{-\infty}^{\infty} \Delta\varepsilon_{ik}(\omega, \boldsymbol{r})\, e^{i\omega t} d\omega,$$

while the field of the scattered light wave is represented in the form

$$\boldsymbol{E}_i(t, \boldsymbol{r}) = \int_{-\infty}^{\infty} \boldsymbol{E}_i(\omega, \boldsymbol{r})\, e^{i(\omega_0+\omega)t}\, d\omega.$$

Then, in the Fraunhofer zone [compare with (1.14)]:

$$\boldsymbol{E}_i(\omega, \boldsymbol{r}) = \frac{(k')^2 e^{-ik'L}}{4\pi\varepsilon_0 L} \int_V e^{-iqr} \Delta \boldsymbol{P}_i(\omega, \boldsymbol{r})\, dr, \tag{7.5}$$

where $\Delta\mathbf{P}_i$ is the component of the fluctuations of the polarization perpendicular to the direction of observation:

$$\Delta \boldsymbol{P} = 4\pi \{\Delta \boldsymbol{P} - \boldsymbol{k}'(\boldsymbol{k}'\Delta \boldsymbol{P})/k'^2\}; \quad \Delta \boldsymbol{P}_i = \frac{\Delta\varepsilon_{ik}}{4\pi} \boldsymbol{E}_k \tag{7.6}$$

($\mathbf{E}$ is the amplitude of the incident wave).

The intensity of the scattered light is equal to

$$I_{\omega_0+\omega} = \frac{1}{VA^2} \int_V \int_V \overline{\Delta \boldsymbol{P}_\alpha(\omega, \boldsymbol{r})\, \Delta \boldsymbol{P}_\alpha^*(\omega \boldsymbol{r}')\, e^{-iq\Delta r}}\, dr dr, \tag{7.7}$$

where $\Delta\mathbf{r} = \mathbf{r} - \mathbf{r}'$.

According to (7.7), the intensity at the frequency $\omega_0 + \omega$ is determined by the spatial correlation function of the spectral amplitude of the vector $\Delta\mathbf{P}$, while the correlation function depends only on $\Delta\mathbf{r}$ by virtue of the statistical homogeneity of the field $\Delta\mathbf{P}$.

Under experimental conditions, the linear dimensions of the volume of the scattering medium are always large in comparison with the correlation radius of the optical inhomogeneities of the medium; therefore, the integration over $\mathbf{r}$ is replaced by integration over $\Delta\mathbf{r}$ within infinite limits.

Let the incident light be propagated along the x axis, and the scattered light be observed along the y axis; then $q^2 = 2k^2$. Further, we shall consider the case in which the exciting light is polarized along the z axis and along the y axis, while we shall denote the components of the scattered light in the same way as in Sec. 6 (see the note on p. 108).

It follows from Eqs. (7.6) and (7.7) that, under the assumptions given,

$$\left.\begin{aligned} I_{zx} &= (2\pi)^3\,\overline{|\Delta\varepsilon_{13}(\omega,\ \mathbf{q})|^2}, & I_{zz} &= (2\pi)^3\,\overline{|\Delta\varepsilon_{33}(\omega,\ \mathbf{q})|^2}, \\ I_{yx} &= (2\pi)^3\,\overline{|\Delta\varepsilon_{12}(\omega,\ \mathbf{q})|^2}, & I_{yz} &= (2\pi)^3\,\overline{|\Delta\varepsilon_{32}(\omega,\ \mathbf{q})|^2}. \end{aligned}\right\} \tag{7.8}$$

Calculation of the four components of the tensor $\Delta\varepsilon_{ik}$ which determine the value of the intensities (7.8) can be completed by establishing the connection of $\Delta\varepsilon_{ik}$ with the quantities which determine the fluctuations in the scattering medium. As such quantities, we take the amplitudes of the deformation tensor e_{ik} and the temperature fluctuations ΔT. As in Sec. 6, $e_{\alpha\alpha} \equiv \sigma = -\Delta\rho/\rho$, and the purely shear deformation is $e'_{ik} = e_{ik} - {}^1/_3 e\delta_{ik}$. The smallness of the fluctuations allows us to restrict ourselves to linear coupling between e'_{ik} and $\Delta T/T = \vartheta$ on the one hand, and the tensor $\Delta\varepsilon_{ik}$ on the other. For the spectral amplitudes, we can write the coupling in the following fashion:

$$\Delta\varepsilon_{ik}(\omega,\ \mathbf{r}) = X(i\omega)\,e'_{ik}(\omega,\ \mathbf{r}) + Y(i\omega)\,e(\omega,\ \mathbf{r})\,\delta_{ik} + Z(i\omega)\,\vartheta(\omega,\ \mathbf{r})\,\delta_{ik}, \tag{7.9}$$

where X, Y, Z are the elasto-optical and thermo-optical characteristics of the medium.

In the absence of dispersion $Y = (\partial\varepsilon/\partial e)_\vartheta = -(\rho\partial\varepsilon/\partial\rho)_T$ and

$$Z = \left(\frac{\partial\varepsilon}{\partial\vartheta}\right)_e = \left(T\frac{\partial\varepsilon}{\partial T}\right)_\rho. \tag{7.10}$$

The same expression can be written for $\Delta\varepsilon_{ik}(\omega,\ \mathbf{r})$ as for $\Delta\varepsilon(\omega, \mathbf{q})$. Therefore, taking (7.8) into account, we obtain

$$I_{zx} = (2\pi^3)\,|X|^2\,\overline{|e'_{13}|^2}, \quad I_{zz} = (2\pi)^3\,\overline{|Xe'_{33} + Ye + Z\vartheta|^2}, \tag{7.11}$$

$$\left. I_{yx} = (2\pi)^3 |X|^2 \overline{|e'_{12}|^2}, \quad I_{y'z} = (2\pi)^3 |X|^2 \overline{|e'_{32}|^2}. \right\} \tag{7.11}$$

Consequently, for the calculation of the intensities, it is first necessary to compute the spectral amplitudes of the fluctuations of deformation and temperature. Such a computation has already been made by Rytov [40, 156] and general formulas were obtained for the intensity of the scattered light:

$$\left.\begin{aligned}
I_{yx} &= \frac{kT|X|^2}{8\pi i\omega}\left\{\frac{1}{\overline{\mu}^*}\left(1+\frac{d_1\overline{\mu}q^2}{\Delta} - \text{complex conjugate}\right)\right\},\\
I_{yz} = I_{zx} &= \frac{kT|X|^2}{16\pi i\omega}\left\{\frac{1}{\overline{\mu}^*}\left(1+\frac{\rho^2\omega^4}{|A_u|^2} - \text{complex conjugate}\right)\right\},\\
I_{zz} &= \frac{kT}{2\pi i\omega}\left\{\frac{|X|^2}{3\overline{\mu}^*}\left(1+\frac{d_1\overline{\mu}q^2}{3\Delta_1}\right) + \left(XY^*C^* + X^*YC\frac{\overline{\mu}}{\overline{\mu}^*}\right)\frac{KCq^2}{3\Delta_1} -\right.\\
&-\left(XY^*\frac{\overline{K}}{\overline{K}^*} - X^*Y\frac{\overline{\mu}}{\overline{\mu}^*}\right)\frac{dq^2}{3\Delta_1} + \left(XZ^* + X^*Z\frac{\overline{\mu}}{\overline{\mu}^*}\right)\frac{C\overline{K}q^2}{3\Delta_1} +\\
&+\frac{|Y|^2}{\Delta_1}\left(\frac{d_1A_3}{\overline{K}^*} - CC^*\overline{K}q^2\right) - \Big[YZ^*C\overline{K}q^2 -\\
&\left.-Y^*Z\left(C^*A_1 - C\frac{\overline{K}}{\overline{K}^*}A_3\right)\Big]\frac{1}{\Delta_1} + \frac{|Z|^2}{\Delta_1}A_1 - \text{complex conjugate}\right\},
\end{aligned}\right\} \tag{7.12}$$

where

$$d = D + \frac{\varkappa T q^2}{i\omega}, \quad d_1 = d - C^2\overline{K} = D_1 + \frac{\varkappa T q^2}{i\omega},$$

$$D_1 = D - C^2\overline{K}, \quad \Delta_1 = dA_1 - C^2\overline{K}A_3 = d_1A_3 - d\overline{K}q^2,$$

$$A_1 = \rho\omega^2 - \left(\overline{K} + \frac{4}{3}\overline{\mu}\right)q^2, \quad A_3 = \rho_0\omega^2 - \frac{4}{3}\overline{\mu}q^2, \quad A_u = \rho\omega^2 - \overline{\mu}q^2.$$

Here, $\overline{K}$ and $\overline{\mu}$ are the complex elastic moduli

$$\overline{K} = k(\omega^2) + i\omega^2\eta'(\omega^2)$$

and

$$\overline{\mu} = \mu(\omega^2) + i\omega\eta(\omega^2),$$

$\varkappa$ is the complex coefficient of thermal conductivity, and C and D_1 are complex coefficients of the equation connecting the spectral amplitudes of the fluctuations of the specific entropy S, the volume deformation e, and the temperature $\vartheta = \Delta T/T$:

$$\rho TS(\omega, r) = C\overline{K}e(\omega, r) + D_1/\vartheta(\omega, r).$$

In the case when there is no dispersion, the quantities $\overline{K}$, $\overline{\mu}$, $\varkappa$, C, D, and D_1 are real constants equal, respectively, to

$$K = \frac{1}{\beta_T}, \quad C = \sigma T, \quad D = \rho T c_p, \quad D_1 = \rho T c_V. \tag{7.13}$$

The formulas (7.12) are within the framework of the above assumptions, the general phenomenological expression for the intensities and spectral composition of the scattered light. However, in such a general form, which contains many currently unknown functions of the frequency, Eqs. (7.12) cannot be compared with experiment. But, by using specific dispersion laws in (7.12) and employing other assumptions, one can obtain a series of interesting conclusions.

<u>Isotropic (Polarized) Light Scattering.</u> If we first consider only polarized scattering, we need keep only those terms in Eqs. (7.12) which do not contain X; this yields

$$I_{zz} = \frac{kT}{2\pi i\omega}\left\{\frac{1}{\Delta_1}\left[|Y|^2\left(\frac{d_1 A_3}{\overline{K}^*} - CC^*\overline{K}q^2\right) - YZ^*C\overline{K}q^2 + \right.\right.$$
$$\left.\left. + Y^*Z\left(C^*A_1 - C\frac{\overline{K}}{\overline{K}^*}A_3\right) + |Z|^2 A_1\right] - \text{complex conjugate.}\right\} \tag{7.14}$$

Making use of a specially developed computational method, Rytov calculated the integrated intensity of the polarized scattering, which is equal to

$$I = \int_{-\infty}^{\infty} I_{zz}d\omega = kT\left\{Y^2\left(\frac{1}{K_\infty} - \frac{4\mu_0/3}{K_0\left(K_0 + \frac{4}{3}\mu_0\right)}\right) + \right.$$
$$\left. + \frac{Z^2}{D_{1,\infty}} + \frac{2(K_\infty - K_0)}{K_0 K_\infty \Delta_a} Y(Yd_1 - ZCK_a)\right\} \tag{7.15}$$

under the assumptions: (1) Y, Z, and C do not show dispersion and consequently are real thermodynamic functions [see (7.10) and (7.13)]; and, (2) the bulk modulus $\overline{K}$ obeys a dispersion law with a single relaxation time τ':

$$K = K_\infty \frac{z + a}{z + \frac{1}{\tau'}}, \quad \text{where} \quad z = i\omega, \quad a = \frac{K_0}{K_\infty \tau'}.$$

The indices 0 and ∞ denote the values of the parameters as $\omega \to 0$ and $\omega \to \infty$, respectively. The index a denotes the values (real) of quantities at $\omega = \alpha$.

Equation (7.15) is applicable to viscous liquids and solid amorphous media under the stated assumptions.

For liquids in the absence of dispersion in the bulk modulus ($K_\infty \equiv K_0$), and with neglect of any change in the dielectric constant at constant density ($Z = 0$), Eq. (7.15) transforms into the Einstein formula (1.32). For a low-viscosity liquid ($\mu = 0$) in which there is small dispersion of K, Eq. (7.15) goes over to

$$I = kT\left\{Y^2\left(\frac{1}{K_\infty} + \frac{2(K_\infty - K_0)S}{K_0^2(1+S)}\right) + \frac{Z^2}{D_{1,0}}\right\}, \qquad (7.16)$$

where

$$S = -\left(\frac{d_1 A_3}{dKq^2}\right)_a \approx \frac{1}{\Omega_L^2 \tau'^2}.$$

The value of S shows to what extent the modulus K is relaxed at the frequency Ω of the MB doublet. If $S \ll 1$, then (7.16) gives

$$I = kT\left(\frac{Y^2}{K_\infty} + \frac{Z^2}{D_{1,0}}\right) = kT\beta_T\left\{\left(\rho\frac{\partial\varepsilon}{\partial\rho}\right)_T^2 \frac{K_0}{K_\infty} + (\gamma - 1)\left(\frac{1}{\sigma}\frac{\partial\varepsilon}{\partial T}\right)_\rho^2\right\}. \qquad (7.17)$$

Introducing the additional assumptions that D and $\varkappa$ are frequency independent, and that the modulus μ obeys the dispersion law

$$\bar{\mu} = \mu_\infty \frac{z+b}{z+\frac{1}{\tau}}, \text{ where } b = \frac{\mu_0}{\mu_\infty \tau}. \qquad (7.18)$$

We can compute from (7.14) the value of all four parts into which the intensity of the isotropic scattering is divided [40, 155]; the integrated intensity of the central component will be

$$I_c = \frac{kTq^2(\gamma - 1)}{\Omega_L^2\rho\left(1 + \frac{4}{3}\beta_T\mu_0\right)}\left[Y + \frac{1 + \frac{4}{3}\beta_T\mu_0}{\sigma T} Z\right]^2. \qquad (7.19)$$

The integrated intensity of the MB component is

$$I_{MB}=\frac{kTq^2}{\rho}\left\{\frac{Y_1^2}{\Omega_L^2}-\frac{\gamma(\tau')^2 Y_1\Delta K}{K_0\left[1+\Omega_L^2(\tau')^2\right]}\left[\frac{\gamma q^2(\tau')^2 Y_1}{\beta_T\rho\left[1+\Omega_L^2(\tau')^2\right]}+\right.\right.$$

$$\left.\left.+\frac{2(\gamma-1)Z}{\sigma T}\right]-\frac{q^2\tau^4 Y_1^2\Delta M}{\rho\left[1+\Omega_L(\tau')^2\right]^2}\right\}. \quad (7.20)$$

Moreover, it follows from the theory that there must be two wings of weak intensity — the compressional wing, for which

$$I_{cw}=kT\beta_T\frac{\Delta K}{K_0}\left(\frac{1+\frac{4}{3}\frac{q^2}{\rho}\mu_0(\tau')^2}{1+\Omega_L^2(\tau')^2}\right)^2 Y_1^2 \quad (7.21)$$

and the shear wing, for which

$$I_{sw}=kT\frac{\Delta M q^2\tau^4}{\rho\left(1+\Omega_L^2\tau^2\right)^2}Y_1^2. \quad (7.22)$$

In Eqs. (7.19)-(7.22), the following notation has been used:

$$\Delta K=K_\infty-K_0,\quad \Delta M=M_\infty-M_0,\quad Y_1=-\left(\rho\frac{\partial\varepsilon}{\partial\rho}\right)_S.$$

Proceeding again to the liquid case, i.e., setting $\mu_0 = 0$, introducing $Z_1 = T(\partial\varepsilon/\partial T)_p$, and again using the parameter $S_1 = 1/[\Omega_L^2(\tau')^2]$ we get, from (7.19)-(7.21),

$$\left.\begin{aligned}
I_c&=kT\frac{Z_1^2}{\rho T c_p},\\
I_{MB}&=kT\beta_T Y_1^2\left\{\frac{1}{\gamma}-\frac{\Delta K}{K_0(1+S_1)}\left[\frac{1}{1+S_1}+\frac{2(\gamma-1)Z}{\sigma T Y_1}\right]\right\}-I_{sw},\\
I_{cw}&=kT\beta_T Y_1^2\frac{\Delta K}{K_0}\left(\frac{S_L}{1+S_L}\right)^2
\end{aligned}\right\} \quad (7.23)$$

It follows from Eqs. (7.21) and (7.22) that the integrated intensity of the compressional and shear wings is very small in comparison with the intensity of the central component. Therefore, they are entirely omitted from consideration in the calculation of the ratio I_c/I_{MB}; then,

$$\frac{I_c}{I_{MB}}=\left(\frac{Z_1}{\sigma T Y_1}\right)^2(\gamma-1)\left\{1+\frac{\gamma\Delta K}{K_0(1+S_1)}\left[\frac{2(\gamma-1)Z}{\sigma T Y_1}+\frac{1}{1+S_1}\right]\right\}. \quad (7.24)$$

Neglecting S_1 and Z in the dispersion correction in this formula, we again obtain the formula (5.40) which, under the assumptions made in Sec. 5, transforms into the Landau–Placzek formula (5.39).

Anisotropic (Depolarized) Light Scattering. The relaxation theory of light scattering due to Leontovich [39] was presented briefly in the previous section. In this theory, attention is chiefly paid to the spectral composition of the depolarized scattering. In the Leontovich theory [39], and in the more general theory of Rytov described in this section, other fluctuations of the dielectric tensor $\Delta \varepsilon'_{ik}$ are connected only with the tensor e_{ik}. It is of interest to make clear under what conditions the basic formula of the Leontovich theory follow from the general formula (7.12). Such an investigation has already been carried out in [40]. There it was shown that the formulas of [39] are obtained from (7.12) for the following conditions [155]:

1. The coefficient X is proportional to the complex shear modulus

$$\left.\begin{array}{c} X(i\omega) = B\bar{\mu}(i\omega) \\ B = \text{const} \end{array}\right\} \qquad (7.25)$$

2. The coefficient Y does not depend on the frequency and has real values

3. The temperature dependence of $\Delta\varepsilon_{ik}$ is neglected (for constant density), i.e., Z = 0*

4. The thermal expansion is not taken into consideration (C = 0)*

5. The dispersion of the bulk modulus is neglected*

$$\bar{K} = K_0 = \frac{1}{\beta_T}. \quad \eta' = 0$$

6. The shear modulus obeys the following dispersion law:

* These assumptions were made by Leontovich for simplification of his fundamental problem – the explanation of the wing of the Rayleigh line.

$$\bar{\mu}=\mu+i\omega\eta=\frac{\mu_\infty i\omega}{1+\frac{1}{\tau}} \tag{7.26}$$

Under these assumptions, and taking account of the fact that

$$B=\frac{A}{\mu_\infty},\quad \Omega_T^2=\frac{\mu_\infty}{\rho}q^2,\quad \Omega_L^2=\frac{K_0}{\rho}q^2,$$

equations are obtained from (7.12) which are identical with the formulas of the Leontovich theory [39], which can be written here in the following form:

$$\left.\begin{aligned}
I_{yx}&=\frac{kTA^2\tau}{16\pi\mu_\infty}\left\{\frac{3(\omega^2-\Omega_L^2)^2}{|\Delta_L|^2}+\frac{1}{1+\omega^2\tau^2}\right\},\\
I_{yz}&=I_{zy}=\frac{kTA^2\tau}{8\pi\mu_\infty}\left\{\frac{\omega^2}{|\Delta_T|^2}+\frac{1}{1+\omega^2\tau^2}\right\},\\
I_{zz}&=\frac{kT\tau}{2\pi\mu_\infty}\left\{\frac{A^2}{2}\left[\frac{(\omega^2-\Omega_L^2)^2}{3|\Delta_L|^2}+\frac{1}{1+\omega^2\tau^2}\right]-\right.\\
&\qquad\left.-\frac{4AY\Omega_T^2(\omega^2-\Omega_L^2)}{3|\Delta_L|^2}+\frac{8Y^2\Omega_T^4}{3|\Delta_L|^2}\right\},
\end{aligned}\right\} \tag{7.27}$$

where

$$\Delta_L=\omega^2-\Omega_L^2+i\omega\tau(\omega^2-\Omega_S^2),\quad \Delta_T=i\omega+\tau(\omega^2-\Omega_T^2),$$

and Ω_S has the same value as in Sec. 6. For the wing of the Rayleigh line, i.e., for $\omega \gg \Omega_L$, it follows from (7.27) that

$$I_{yx}=I_{yz}=I_{zx}=\frac{3}{4}I_{zz}=\frac{A^2kT}{4\pi\mu_\infty}\frac{\tau}{1+\omega^2\tau^2}, \tag{7.28}$$

which corresponds to Eqs. (6.32).

It follows from (7.28) and (6.32) that the reciprocal of the intensity, $L(\omega)$, depends linearly on ω^2 (see Fig. 10), while experiment shows a significant departure from a linear dependence in the region of small (up to 20-25 cm^{-1}), and large (above 60-80 cm^{-1}), values of ω.

Therefore, the investigation of the intensity distribution in the wing of the Rayleigh line, which follows from the general formulas of (7.12) without the limitations applied in the derivation of Eqs. (6.29) and (6.32), is of interest.

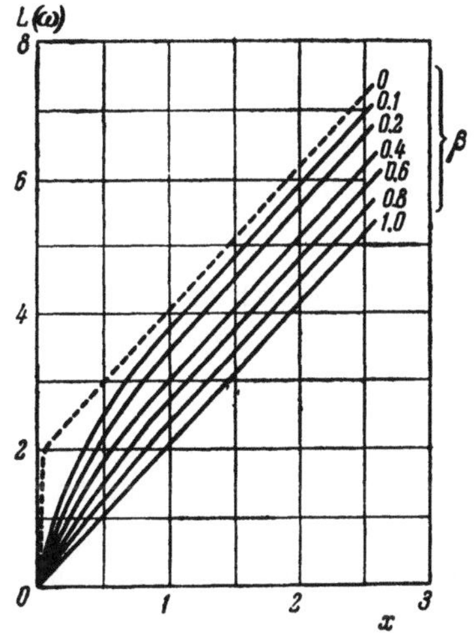

Fig. 13. Dependence of L(ω) on $x = \omega^2\tau^2$, corresponding to Eq. (7.31) (Rytov [40]).

By considering the distribution of the intensity over frequencies much larger than Ω_L and $1/\tau$, we get from (7.12)

$$I_{yx}=I_{zx}=I_{yz}=\frac{3}{4}I_{zz}=\frac{kT\,|X|^2}{8\pi i\omega}\left(\frac{1}{\bar{\mu}^*}-\frac{1}{\bar{\mu}}\right)=\frac{kT}{4\pi}\left|\frac{X}{\bar{\mu}}\right|^2\eta. \quad (7.29)$$

By keeping condition (1), according to which $X = B\bar{\mu}$, we get a linear dependence $1/I_{yx} \sim \text{const}/\eta$ on the function ω^2 from (7.29). Therefore, Rytov [40] gave up condition (1), assuming that

$$X(i\omega)=\frac{X_\infty i\omega+X_0}{i\omega+\dfrac{1}{\tau_1}}, \quad (7.30)$$

and keeping the dispersion law (7.18) for $\bar{\mu}$. In this case, for each inverse of the intensity (7.29), a frequency dependence is obtained of the form

$$L(\omega)\sim\left|\frac{\bar{\mu}}{X}\right|^2\frac{1}{\eta}\sim x\,\frac{x+1}{x+\beta}, \quad (7.31)$$

where

$$x = \omega^2\tau_1^2, \quad \beta = \frac{X_0^2}{X_\infty^2}.$$

The dependence of $L(\omega)$ on the function x for various values of $\beta \leq 1$ is shown in Fig. 13.

The curves of Fig. 13 recall the experimental curves for $\beta \neq 0$ and $\beta \neq 1$. Both the parameters τ_1 and β can be determined from the experimental curves.

Equation (7.31) has, however, the disadvantage that X differs from zero as $\omega \to 0$ ($X = X_0$), which means the presence of a photoelastic anisotropy in the liquid and gives $L(\omega) = 0$ as $x \to 0$. Therefore, in [40, 155], the presence of at least one relaxation time τ_2 is assumed in X and $X(i\omega)$ is written as

$$X(i\omega) = \frac{i\omega}{i\omega + \frac{1}{\tau_2}} \frac{X_\infty i\omega + \frac{X_0}{\tau_1}}{i\omega + \frac{1}{\tau_1}}. \tag{7.32}$$

For the inverse of the intensity, this gives the formula

$$L(\omega) \approx \left(\frac{\tau_1^2}{\tau_2^2} + x\right)\frac{x+1}{x+\beta}, \tag{7.33}$$

which is free from the disadvantages of (7.31) at the price of the introduction of a new parameter τ_2. This parameter determines the halfwidth of the narrow and most intense part of the wing.

From (7.33) for $x \gg (\tau_1/\tau_2)^2$, we get the same curve for $L(\omega)$ as is shown in Fig. 13, but now $L(\omega) \neq 0$ as $x \to 0$. The value of τ_2 can be estimated either by the halfwidth of the narrow and intense part of the wing, or by the value of $L(\omega) \neq 0$ as $x \to 0$. We note that at low frequencies ($\omega \ll 1/\tau_2$) in accord with (7.32), $X = i\omega\tau_2 X_0$ and, consequently, $\Delta\varepsilon_{ik}^{!} = i\omega\tau_2 X_0 e_{ik}^{!}$. On the other hand, the empirical formula describing the Maxwell effect gives $\Delta\varepsilon_{ik}^{!} = 2i\omega\varepsilon_0\eta M e_{ik}^{!}$. Comparing the last two expressions, we find that $X_0 = 2\varepsilon_0\eta(M/\tau_2)$, where M is the Maxwell constant. The possibility then follows of the determination of the frequency variation of X in the study of the diffraction of linearly polarized light by ultrasound up to frequencies of about 10^8 Hz, and hence a test of the validity of (7.33).

By considering that the assumptions made relative to the dispersion laws for $\overline{\mu}$ and X satisfactorily describe the frequency dependence of the intensity in the wing, after the elimination of the high-frequency region where the intensity of the wing is very low, there is meaning in calculating the total depolarization of the scattered light Δ. This calculation is carried out in [40, 155] for the case of a low-viscosity liquid, when $\Omega_T^2\tau^2 = (\eta q^2\tau)/\rho \sim 10^{-4}$; further simplifications are possible, namely, in (7.12), we neglect terms with Z; we assume that C, Y, and $\overline{K}$ do not depend on the frequency, i.e., they are equal to their own thermodynamic values, and for the intensity of the polarized scattering we use the expression corresponding to the Einstein formula (1.32). If we assume that $1/\tau_1 \sim 1/\tau \gg \Omega_L$, then we get for the total intensity in the wing

$$\left.\begin{aligned} I_{yx} = I_{zx} = I_{yz} &= \frac{kTX_0^2}{4\eta}\frac{\tau_2\left(\tau_2+\frac{\tau_1}{\beta}\right)}{\tau_1+\tau_2}, \\ I_{zz} = \frac{4}{3}I_{yx} &= \frac{kTX_0Y}{3}\beta_T\frac{1-\Omega_L^2\tau_2^2}{1+\Omega_L^2\tau_2^2}+kTY^2\beta_T. \end{aligned}\right\} \tag{7.34}$$

For low viscosities, $\Omega_L^2\tau_2^2 \ll 1$, which enables us to carry out a further simplification of (7.34) and finally obtain

$$\Delta_u = \frac{I_{zx}+I_{yx}}{I_{zz}+I_{yz}} = \frac{\frac{6}{7}}{1+\frac{12}{7}\beta_T\eta k''\frac{\tau_2+\tau_1}{\tau_2+\frac{\tau_1}{\beta}}\left(k''\tau_2+\frac{1}{3}\right)}, \tag{7.35}$$

where

$$k'' = \frac{|Y|}{X_0\tau_2} = \frac{\rho}{2\varepsilon\eta M}\left(\frac{\partial\varepsilon}{\partial\rho}\right)_T.$$

The calculation of Δ_u, carried out by Rytov according to (7.35), and on the basis of the experimental data of the author [144, 53] for τ_1 and τ_2, leads to satisfactory agreement with the experimental data for Δ_u (see Sec. 19). However, the low accuracy of determination of τ_1, τ_2, and M, and also the fact that one case has been considered thus far (benzene) still does not permit us to draw far-reaching conclusions, although the estimate given is encouraging. Broadening of the range of experimental data will enable us to make more definite conclusions in the future.

8. Calculation of the Spectral Distribution of the Intensity of Depolarized Scattered Light on the Basis of a Simplified Molecular Model

The phenomenological theory of the spectral composition of scattered light and, in particular, the theory which describes the broad continuous depolarization spectrum (the wing of the Rayleigh line) of light scattered by the fluctuations of the anisotropy apply to that part of the scattering which is associated with the fluctuations of the deformation tensor [313].

It should be noted that, although the phenomenological theories that have been discussed do not take into consideration all the reasons for scattering, account of the inertial terms in the reaction equation of Sec. 6 and the two relaxation terms in the correlation theory of Sec. 7 give a satisfactory description of the distribution of the intensity in the wing of the Rayleigh line. This description of the experimental data is accomplished with the help of Eqs. (6.41) and (7.27). Formulas (6.41) and (7.27) can obviously be combined into one equation, but it is still unknown what part of the total scattered light is described by the theory, which takes into account only the scattering from the deformation fluctuations. As was noted by Ginzburg [313], the scattering by fluctuations of the deformation tensor in low-viscosity liquids evidently involves only a small part of the total scattering from the anisotropy fluctuations.

Therefore, the attempt of Starunov [41, 505] to compute as completely as possible the intensity distribution in the spectrum of depolarized scattering brought about by fluctuations of the anisotropy, by starting out from a simplified molecular model, is of interest.

In the simplified picture of molecular motion, it is assumed that the molecule of the liquid experiences Brownian motion which can be divided somewhat artificially into two parts: (1) the molecule can undergo rotational "migrations" from one potential well to another (this process reduces to rotational diffusion); and (2) near the equilibrium position, the molecule can undergo rocking or rotational oscillations. If we consider the motion of an anisotropic molecule, then both these forms of periodic motion will modulate the light. The slower motion of the first type gives a much narrower continuous spectrum, while the second, more rapid type of motion leads to a much broader continuous spectrum. It is difficult to consider both forms of periodic motion at the same time; therefore, they will be considered separately.

Modulation of Scattered Light as the Result of Periodic (Rotational) Diffusion of Anisotropic Molecules. For the calculation of the scattered spectrum brought about by rotational diffusion, Starunov [41, 505] used the Einstein–Fokker equation. He considered the thermal motion of a molecule, represented in the form of a bar having only a single moment of inertia. The orientation of such a molecule is determined by the angular coordinates ϑ and φ.

A light wave polarized in the z direction induces a moment in the molecule, a component of which can be described in the given case (on the basis of [151]) in the form (the intensity of the electric field of the light wave is taken to be unity)

$$P_z = \cos^2\vartheta - \frac{1}{3}, \qquad P_x = \sin\vartheta\cos\vartheta\cos\varphi, \tag{8.1}$$

where the angles ϑ and φ are functions of time. We assume that the process is stationary; then the correlation functions $f_Z(t)$ and $f_X(t)$ depend only on the time delay:

$$f_z(t) = \overline{P_z(t)\,P_z(t'+t)}, \qquad f_x(t) = \overline{P_x(t)\,P_x(t'+t)}.$$

The spectral density of the intensity of the scattered light can easily be obtained from the latter formulas. Omitting the unessential factor $(\omega^4/L^2c^4)(b_1 - b_2)^2$ in what follows, we get

$$\left.\begin{aligned} I_z(\omega) &= f_z(\omega) = \frac{1}{\pi}\int_0^\infty f_z(t)\,\bar{e}^{i\omega t}\,dt, \\ I_x(\omega) &= f_x(\omega) = \frac{1}{\pi}\int_0^\infty f_x(t)\,e^{-i\omega t}\,dt. \end{aligned}\right\} \tag{8.2}$$

To find the correlation function contained in the integrands of (8.2), we must use the method proposed by Vladimirskii [506]. To find products of the form $\overline{A(t')B(t'+t)}$ by this method, it is sufficient to find the average value of A(t) in an ensemble with additional energy αB (α is an arbitrary parameter), take the derivative with respect to α of the resultant average, and set $\alpha = 0$ in it.

To find the mean $\overline{P_Z(t)}$ in a system with the additional energy $\alpha P_Z(t)$, it is necessary to know the probability density, which is determined [41, 505] from the Einstein–Fokker equations [507, 508]. The probability density $w(t, \vartheta, v_\vartheta, v_\varphi)$ for $P_Z(t)$ will be

$$\frac{\partial w}{\partial t}+v_\vartheta\frac{\partial w}{\partial \vartheta}+\operatorname{ctg}\vartheta\left(v_\varphi\frac{\partial w}{\partial v_\vartheta}-v_\vartheta\frac{\partial w}{\partial v_\varphi}\right)=$$

$$=\frac{\zeta}{I}\frac{\partial}{\partial v_\vartheta}\left(v_\vartheta w+\frac{kT}{I}\frac{\partial w}{\partial v_\vartheta}\right)+\frac{\zeta}{I}\frac{\partial}{\partial v_\varphi}\left(v_\varphi w+\frac{kT}{I}\frac{\partial w}{\partial v_\varphi}\right); \quad (8.3)$$

for $t = 0$,

$$w(0,\ \vartheta,\ v_\vartheta,\ v_\varphi)=C\exp\left[-\frac{\alpha P_z(\vartheta)+\frac{1}{2}I\left(v_\vartheta^2+v_\varphi^2\right)}{kT}\right]\approx$$

$$\approx\frac{1}{2}\left[1-\frac{\alpha}{kT}P_z(\vartheta)\right]\exp\left[-\frac{I\left(v_\vartheta^2+v_\varphi^2\right)}{2kT}\right], \quad (8.4)$$

where ζ is the internal friction, I is the moment of inertia, and v_ϑ and v_φ are the angular velocities of the molecule. For convenience of calculation, following Sack [508], we replace $w(t, \vartheta, v_\vartheta, v_\varphi)$ by another function:

$$\Psi(t,\ \vartheta,\ u_\vartheta,\ u_\varphi)=\exp\left[\frac{kT}{2I}\left(u_\vartheta^2+u_\varphi^2\right)\right]\times$$

$$\times\iint\limits_{\infty}^{\infty} w(t,\ \vartheta,\ v_\vartheta,\ v_\varphi)\exp\left(-iu_\vartheta v_\vartheta-iu_\varphi v_\varphi\right)dv_\vartheta dv_\varphi \quad (8.5)$$

and write the velocity in polar coordinates in the form

$$u_\vartheta=u\cos\chi,\ u_\varphi=u\sin\chi. \quad (8.6)$$

Averaging over the velocities v_ϑ and v_φ reduces to equating them to zero. Taking (8.5) and (8.6) into account, and rewriting Eq. (8.3) in terms of the function (8.5), we shall seek a solution of the resultant equation in the form

$$\Psi=\Psi_0+\Psi_1(t,\ u)\cos^2\vartheta-i\Psi_2(t,\ u)\sin\vartheta\cos\vartheta\cos\chi+$$
$$+\frac{1}{2}\Psi_3(t,\ u)\sin^2\vartheta\sin^2\chi. \quad (8.7)$$

Substituting (8.7) in (8.3) (transformed to the variable Ψ), we get three differential equations for Ψ_1, Ψ_2, and Ψ_3.

As a result of cumbersome calculations, we obtain the following expression for the correlation function:

$$f(t)=\frac{1}{15}\frac{I}{kT}\int_0^\infty \lambda R(\lambda,\ t)\exp\left(-\frac{I}{2kT}\lambda^2\right)d\lambda, \quad (8.8)$$

where the function $R(\lambda, t)$ satisfies the equation

$$\frac{d^3R}{dt^3}+3\beta\frac{d^2R}{dt^2}+2\beta^2\frac{dR}{dt}+4\lambda^2e^{-2\beta t}\frac{dR}{dt}+\frac{12\beta kT}{I}R=0. \quad (8.9)$$

Here, $\beta = \xi/I$; for $t = 0$,

$$R=\frac{4}{3},\quad \frac{dR}{dt}=0,\quad \frac{d^2R}{dt^2}=-4\lambda^2.$$

The spectral density of the correlation function $f(\omega)$ can be found from (8.9), (8.8), and (8.2), and, consequently, the distribution of the intensity in the wing of the Rayleigh line can be evaluated.

It has not been possible to find a general solution of the problem; therefore, equations have been obtained in [41, 505] for two cases:

$$\left.\begin{array}{ll} 1) & \frac{\zeta}{I}\to 0; \\ 2) & \left(\frac{\zeta}{I}\right)^2\gg\frac{4kT}{I},\quad \frac{2\zeta}{I}\gg\omega. \end{array}\right\} \quad (8.10)$$

In the first case, we get

$$I_z=\frac{4}{3}I_x=\frac{I}{120kT}\exp\left(-\frac{I\omega^2}{8kT}\right)+\frac{1}{45}\delta(\omega). \quad (8.11)$$

The equation gives the distribution over the frequencies of the intensity of the light scattered in gases. In the problem under study, other mechanisms of broadening have not been considered; therefore, the second component in (8.11) is a delta function. Equation (8.11) was obtained even earlier [509] by another method.

In the second case, which is the situation for ordinary liquids, the following final result was obtained in [41, 505]:

$$I_z = \frac{4}{3} I_x = \frac{4}{45\pi} \frac{\frac{\zeta}{6kT}}{\left(1 + \frac{I}{4kT}\omega^2\right)^2 + \omega^2 \left(\frac{\zeta}{6kT}\right)^2}, \tag{8.12}$$

where $\zeta / 6kT = \frac{1}{3}\tau_D$, and τ_D is determined by Eq. (6.12).

The exact solution of the problem should furnish the kinetics of the change in the shape of the wing in the transition from (8.11) to (8.12). Equation (8.12) can be said to describe the intensity distribution in the wing in the region adjoining the unshifted line of Rayleigh scattering.

Equation (8.12) can also be easily written for the case in which the excitation is carried out by means of natural light.

Modulation of the Scattered Light as the Result of Rotational Oscillations of Anisotropic Molecules. In addition to the rotational diffusion considered above, the molecule can vibrate about its equilibrium orientation or undergo rotational rocking. Such a rotational rocking of an anisotropic molecule produces modulation of the scattered light.

For small angles of departure from the equilibrium position, the motion of the molecule can be described by a stochastic equation of the following form:

$$I \frac{d^2 (\Delta\vartheta)}{dt^2} + \zeta \frac{d (\Delta\vartheta)}{dt} + \mu (\Delta\vartheta) = M (t). \tag{8.13}$$

The same equation is obtained for $\Delta\varphi$. In Eq. (8.13), μ is the shear modulus while, on the right-hand side, M(t) is a random force. The change in the moment brought about the oscillations of the molecule is equal to

$$\Delta P_z = -2 \sin\vartheta \cos\vartheta \Delta\vartheta. \tag{8.14}$$

The spectral densities corresponding to the correlation functions

$$\overline{\Delta P_z(t')\,\Delta P_z(t'+t)} \quad \text{and} \quad \overline{\Delta\vartheta(t')\,\Delta\vartheta(t'+t)}$$

are related to one another by the expression

$$\overline{(\Delta P_z)^2_\omega} = 4\sin^2\vartheta\cos^2\vartheta\overline{(\Delta\vartheta)^2_\omega}. \tag{8.15}$$

By following [505], one can find the values of $\overline{(\Delta\vartheta)^2_\omega}$ from Eq.(8.13) and, by carrying out averaging in the latter equation over the angles, we find the expression for the intensity

$$I_z = \frac{4}{3} I_x = \frac{4}{45\pi}\frac{6kT}{\mu}\frac{\frac{\zeta}{\mu}}{\left(1-\frac{I\omega^2}{\mu}\right)^2 + \omega^2\left(\frac{\zeta}{\mu}\right)^2}. \tag{8.16}$$

Equation (8.16) should describe the same region as the Leontovich theory [39], with account of the inertia [504]. In Eq. (8.13), it is assumed that the internal friction and the shear modulus do not depend on the time and, therefore, these quantities are independent of frequency in (8.16).

It has already been pointed out that the dependence of these parameters on frequency must be investigated [41, 53, 230]. Starunov took into account the frequency dependence of ζ and μ, writing Eq. (8.13) for large t ($t \gg \tau_\zeta$) in the form

$$I\frac{d^2(\Delta\vartheta)}{dt^2} + \zeta_0\frac{d(\Delta\vartheta)}{dt} + \mu_0\,\Delta\vartheta = M(t)$$

and for small time intervals $t \ll \tau_\zeta$,

$$I\frac{d^2(\Delta\vartheta)}{dt^2} + \mu_\infty\,\Delta\vartheta = M(t),$$

where τ_ζ is the relaxation time of internal friction (viscosity) or of the shear modulus. The same value is obtained from the last two equations for I_z as in (8.16), but with parameters which depend on the frequency in the following way:

$$\zeta = \frac{\zeta_0}{1+\omega^2\tau_\zeta^2} \quad \text{and} \quad \mu = \mu_0 + \tau_\zeta\zeta_0\frac{\omega^2}{1+\omega^2\tau^2}. \tag{8.17}$$

The index "zero" means that the value of the quantity is taken up to the region of relaxation, and the index ∞, beyond the relaxation region:

$$\mu_\infty = \mu_0 + \frac{\zeta_0}{\tau_r}. \tag{8.18}$$

For frequencies $\omega > 1/\tau_{re}$ the distribution of the intensity in the wing will be determined by the modulation of the scattered light as the result of vibrations of the anisotropic molecules, and is described by Eq. (8.16).

If we consider intervals of time greater than the time of reorientation of the molecule τ_{re}, then this means that the rotational diffusion is taken into account. Neglecting the vibrations of the molecules that take place at the same time, we obtain the distribution (8.12). The region where the distribution (8.12) transforms into the distribution (8.16) with parameters expressed by (8.17) should correspond to the point of inflection on the experimental curve, which in turn corresponds to the frequency $\omega_{re} = 1/\tau_{re}$. The point of inflection, which corresponds to ω_{re}, should tend toward the unshifted frequency ω_0 upon increase in the viscosity.

For a large value of the relaxation time of the viscosity, it follows from (8.16) and (8.17) that a depolarized line appears on each side of the unshifted line at the frequency

$$\omega_{max} \sim \left(\frac{\mu_\infty}{I}\right)^{1/2}. \tag{8.19}$$

The frequency ω_{max} characterizes the line of low frequency. Here, only a single pair of lines is obtained, inasmuch as only a single moment of inertia has been taken into account.

The experimental results will be compared with the theory in Chapter VII. The processes of rotational diffusion and the vibration of the molecule are different forms of Brownian motion; therefore, it is a natural wish to encompass the intensity distribution in the wing by a single relation, a single formula. However, this wish still remains unsatisfied and represents a difficult problem.

As is shown in [505], the continuous transition from (8.16) to (8.12) would be possible (with accuracy up to the inertia terms) if we considered it as a process of the relaxation of the parameters (8.17), and if it were possible to assume that $\mu_0 = 6kT$. But, as has been pointed out earlier, in such a case an incorrect temperature dependence is obtained for μ_0 and the transition from diffusion to oscillation is overlooked; otherwise one must assume that the time to bring the molecule to the state of equilibrium τ_{re} and the relaxation time of the viscosity τ_{ζ} are of the same order [505].

The results of the molecular theory of the wing, which have been presented briefly here, do not agree with the rotational theory and the quasicrystalline theories of this phenomenon.

9. Theory of Light Scattering in Crystals

General Calculation of the Intensity, Polarization, and Frequency of Light Scattered in Crystals. It was shown in Sec. 5 that light scattering can be regarded as the result of the diffraction of the incident light wave by the elastic Debye waves. Following this idea, let us carry out the calculation of the intensity, polarization, and frequency of the scattered light. Here, we shall give an account of the theory of molecular scattering in crystals in general form (as was done in [20]) with necessary refinements.

Let us consider the propagation of a single, plane ultrasonic wave. The displacement **u** is given by the expression

$$\mathbf{u} = \frac{\mathbf{A}}{2}\left[e^{i(\Omega t - \mathbf{qr})} + e^{-i(\Omega t - \mathbf{qr})}\right], \tag{9.1}$$

where $\mathbf{q} = q\boldsymbol{\varkappa}$ is the wave vector of the ultrasonic wave, $\mathbf{A} = A\boldsymbol{\gamma}$ is the amplitude of the displacement of the ultrasonic wave, $\boldsymbol{\varkappa}$ is a unit vector in the direction of propagation of the elastic wave, $\boldsymbol{\gamma}$ is the unit polarization vector of the elastic wave, and Ω is the frequency of the ultrasonic wave.

The deformation tensor, produced by this elastic wave, will be

$$e_{ik} = \frac{1}{2}\left(\frac{\partial u_i}{\partial x_k} + \frac{\partial u_k}{\partial x_i}\right) =$$
$$= -\frac{1}{4} iqA\left(\gamma_i \varkappa_k + \gamma_k \varkappa_i\right)\left[e^{i(\Omega t - qr)} - e^{-i(\Omega t - qr)}\right]. \tag{9.2}$$

The elastic wave produces a change in the optical dielectric tensor which, in accord with Pockels [160], will be equal to

$$\left.\begin{aligned} &\varepsilon_{xx} - \varepsilon_{0xx} = \\ &= -n_x^4\left(p_{11}e_{xx} + p_{12}e_{yy} + p_{13}e_{zz} + 2p_{14}e_{yz} + 2p_{15}e_{zx} + 2p_{16}e_{xy}\right), \\ &\ldots\ldots\ldots\ldots\ldots\ldots\ldots\ldots\ldots \end{aligned}\right\} \tag{9.3}$$

$$\left.\begin{aligned}&\varepsilon_{xy}-\varepsilon_{0xy}=\\&=-n_x^2 n_y^2(p_{61}e_{xx}+p_{62}e_{yy}+p_{63}e_{zz}+2p_{64}e_{yz}+2p_{65}e_{zx}+2p_{66}e_{xy}).\end{aligned}\right\} \quad (9.3)$$

Here, p_{ik} are the elasto-optical constants. The coordinate axes are chosen along the chief axes of the dielectric tensor (i.e., $\varepsilon_{0xy} = \varepsilon_{0zy} = \varepsilon_{0zx} = 0$).

This change in the dielectric tensor can be written in the form

$$\varepsilon_{ij}-\varepsilon_{0ij}=\Delta\varepsilon_{ij}=\frac{1}{2}\,iqA\varphi_{ij}\,[e^{i(\Omega t-qr)}-e^{-i(\Omega t-qr)}], \quad (9.4)$$

$$\varphi_{ij}=n_i^2 n_j^2 p_{ij\alpha\beta}\varkappa_\beta\gamma_\alpha. \quad (9.5)$$

The presence of the second Greek index in the product indicates summation over this index from 1 to 3; p_{ijlm} are the same elasto-optical constants. For convenience, we introduce four indices, running over values from 1 to 3, in place of the two indices running over the values from 1 to 6. Thus, $p_{1111} = p_{11}$; $p_{1122} = p_{12}$; $p_{1133} = p_{13}$; . . . ; $p_{1223} = p_{64}$; etc. Since $p_{ijlm} = p_{jilm} = p_{ijml} \neq p_{lmij}$, then $\varphi_{ij} = \varphi_{ji}$.

In most cases, the differences between n_x, n_y, and n_z are small, and we shall assume that

$$\varphi_{ij}=n^4\varphi'_{ij}, \quad \text{where} \quad \varphi'_{ij}=p_{ij\alpha\beta}\varkappa_\beta\gamma_\alpha. \quad (9.6)$$

Here n is the mean index of refraction.

Let us consider the problem of the propagation of light through a crystal whose dielectric constant changes according to the law given by Eq. (9.4).

The equations for electrodynamics for an anisotropic medium are

$$\begin{gathered}c\,\mathrm{rot}\,\boldsymbol{H}_a=\frac{\partial \boldsymbol{D}}{\partial t},\quad -c\,\mathrm{rot}\,\boldsymbol{E}_a=\frac{\partial \boldsymbol{H}}{\partial t},\\ \mathrm{div}\,\boldsymbol{D}_a=0,\qquad \mathrm{div}\,\boldsymbol{H}_a=0.\\ D_{ai}=\varepsilon_{i\alpha}E_{a\alpha}\end{gathered} \quad (9.7)$$

In the absence of an elastic wave, the solution is well known; we represent it by **E, H, D**, where $D_i = \varepsilon_{0i\alpha}E_\alpha$. Just as in Sec. 1, we seek a solution in the form

$$\left.\begin{aligned} &E_a = E + E', \quad H_a = H + H', \\ &D_{ai} = \varepsilon_{i\alpha} E_{a\alpha} = (\varepsilon_{0i\alpha} + \Delta\varepsilon_{i\alpha})(E_\alpha + E'_\alpha) = \\ &= \varepsilon_{0i\alpha} E_\alpha + \varepsilon_{0i\alpha} E'_\alpha + \Delta\varepsilon_{i\alpha} E_\alpha + \Delta\varepsilon_{i\alpha} E'_\alpha, \\ &\varepsilon_{0i\alpha} E_\alpha = D_i, \quad \varepsilon_{0i\alpha} E'_\alpha = D'_i. \end{aligned}\right\} \tag{9.8}$$

Here, $\Delta\varepsilon_{ji}$, and also E', are small quantities. Limiting ourselves to terms of first order of smallness, we get*

$$D_a = D + D' + \{\Delta\varepsilon_{ik}\} E.$$

Using the equations for **E**, **H**, and **D**, we get, similarly to (1.9),

$$\left.\begin{aligned} &c \operatorname{rot} H' = \frac{\partial D'}{\partial t} + 4\pi j, \qquad -c \operatorname{rot} E' = \frac{\partial H'}{\partial t}, \\ &\operatorname{div} D' = 4\pi\rho, \qquad \operatorname{div} H' = 0, \\ &D'_i = \varepsilon_{0i\alpha} E'_\alpha, \end{aligned}\right\} \tag{9.9}$$

where

$$\left.\begin{aligned} &j_m = \frac{1}{4\pi} \frac{\partial}{\partial t} (\Delta\varepsilon_{m\sigma} E_\sigma), \\ &\rho = -\frac{1}{4\pi} \operatorname{div} (\{\Delta\varepsilon_{ik}\} E). \end{aligned}\right\} \tag{9.10}$$

It is not difficult to see that j and ρ satisfy the equation of continuity:

$$\frac{\partial\rho}{\partial t} = -\operatorname{div} j. \tag{9.11}$$

The result can be interpreted in the following way. The problem of the propagation of light in a crystal excited by an elastic wave can be replaced by the problem of the propagation of light in an unperturbed crystal in which the various currents and charges determined by (9.10) are distributed.

Let the plane light wave

$$E = E_0 \beta_j e^{i(\omega t - kr)}, \tag{9.12}$$

---

*The expression $\{\Delta\varepsilon_{ik}\}$**E** denotes a vector equal to the product of the tensor $\{\Delta\varepsilon_{ik}\}$ by the vector **E**, i.e., $(\{\Delta\varepsilon_{ik}\}\mathbf{E})_m = \Delta\varepsilon_{m\gamma} E_\gamma$.

be incident on the crystal, where E_0 is the amplitude of the incident wave in the crystal, β_j is a unit vector characterizing the polarization of the incident wave with components β_{jx}, β_{jy}, and β_{jz}:

$$\left.\begin{aligned} j &= -\frac{qAE_0}{8\pi}\{\varphi_{ik}\}\,\boldsymbol{\beta}_j\{(\omega+\Omega)\,e^{i[(\omega+\Omega)t-(k+q)r]} - \\ &\qquad -(\omega-\Omega)\,e^{i[(\omega-\Omega)t-(k-q)r]}\}, \\ \rho &= -\frac{qAE_0}{8\pi}\varphi_{\sigma\tau}\beta_{j\tau}\{(k_\sigma+q_\sigma)\,e^{i[(\omega+\Omega)t-(k+q)r]} - \\ &\qquad -(k_\sigma-q_\sigma)e^{i[(\omega-\Omega)t-(k-q)r]}\}. \end{aligned}\right\} \tag{9.13}$$

The solution of the problem of the radiation of light of an anisotropic medium for given fluxes and charges was given in the research of Ginzburg [161]. In the anisotropic case, it is impossible to limit oneself to the introduction of a single Hertzian vector, as was done in Sec. 1 for the isotropic case. The application of the solution of Ginzburg for our problem is given in detail in [20]. Here, all the intermediate steps are omitted because of their cumbersome mathematics, and only the final result is written out.

Light with wave vector **k**, diffracted by an elastic wave with wave vector **q** will be propagated in the direction **k'** satisfying the Bragg condition (1.19).

Let $\mathbf{k}' - \mathbf{k} = +\mathbf{q}$. Then the frequency of the scattered wave will be $\omega + \Omega$. The normal component of the Poynting vector of the scattered wave S'_n will be equal to

$$S'_n = \frac{V}{L^2}\frac{(\omega+\Omega)^4}{64\pi^3c^3}\,\frac{n'_i}{\cos^2\delta'_i}\cdot\frac{q^2}{\rho\Omega^2}\,\frac{\rho\Omega^2A^2}{2}(E_0)^2(\varphi_{\sigma\tau}\alpha_{i\sigma}\beta_{j\tau})^2. \tag{9.14}$$

Here V is the scattering volume; n'_i is the index of refraction for the scattered wave, the polarization of which is characterized by a unit vector $\boldsymbol{\alpha}_i$ with components α_{ix}, α_{iy}, and α_{iz}; δ'_i is the angle between the vectors **E'** and **D'** for the scattered wave of given polarization; $q^2/\rho\Omega^2$ is a function only of the elastic constants; $\rho\Omega^2A^2/2$ is the kinetic energy of the elastic wave; and $(\varphi_{\sigma\tau}\alpha_{i\sigma}\beta_{j\tau})^2$ is a function of the elastic constants (p_{ijlm}), the direction of propagation ($\boldsymbol{\varkappa}$), and the direction of polarization ($\boldsymbol{\gamma}$) of the ultrasonic wave, and also of the directions of polarization of the incident wave ($\boldsymbol{\beta}_j$) and the scattered wave ($\boldsymbol{\alpha}_i$).

In the following, we assume that

$$\frac{\Omega}{q} = v, \tag{9.15}$$

where v is the velocity of the ultrasonic wave.

For the case of light scattering by thermal elastic waves,we have

$$\frac{\rho\Omega^2 A^2}{2} = \frac{kT}{2}. \tag{9.16}$$

The normal component of the Poynting vector S_n of the incident light wave will be equal to

$$S_n = \frac{c}{4\pi} n_j' \cos^2\delta_j' (E_0)^2. \tag{9.17}$$

Here, n_j' is the index of refraction of the incident light wave, the polarization of which is characterized by the unit vector $\boldsymbol{\beta}_j$; δ_j' is the angle between the vectors **E** and **D** for the incident wave of given polarization.

The scattering coefficient, analogous to (1.68), is equal to

$$R = \frac{S_n'}{S_n} \frac{L^2}{V}.$$

It follows from (9.14) and (9.17) that, for irradiation by linearly polarized light,

$$R = \frac{(\omega+\Omega)^4}{32\pi^2 c^4} \frac{n_i'}{n_j'} \frac{1}{\cos^2\delta_i' \cos^2\delta_j'} \frac{kT}{\rho v^2} (\varphi_{\sigma\tau}\alpha_{i\sigma}\beta_{j\tau})^2. \tag{9.18}$$

The elastic wave propagating in the direction opposite to that considered above ($\mathbf{q}' = -\mathbf{q}$), gives scattered light in this same direction $\mathbf{k}'$ with the same intensity and with frequency $(\omega - \Omega)$. Therefore, the light-scattering coefficient for both MB components will be determined by the formula

$$R = \frac{\pi^2}{\lambda^4} \frac{n_i'}{n_j'} \frac{1}{\cos^2\delta_i' \cos^2\delta_j'} \frac{kT}{\rho v^2} (\varphi_{\sigma\tau}\alpha_{i\sigma}\beta_{j\tau})^2. \tag{9.19}$$

Here it has been taken into account that $\omega \gg \Omega$.

Account of the fact that the radiating dipole is found in an anisotropic medium led to the appearance of the factor

$n_i'/n_j' \cos^2\delta_i' \cos^2\delta_j'$. This factor differs slightly from unity for crystals, the ellipsoid of the optical dielectric constant of which is close to a sphere. For example, for quartz, it differs from unity by 1-2%. But for crystals with a large optical anisotropy, the difference from unity will be much greater. Thus, for Iceland spar, this factor can differ from unity by 10-20%.

Further specific calculation of the intensity, frequency, and polarization of scattered light requires a knowledge of the elastic and elasto-optical constants. Data corresponding to calculations for several crystals are given in Sec. 31.

We note only that if the wave vectors of the incident and scattered waves are given, then the wave vector of the elastic wave $\pm\mathbf{q} = \mathbf{k}' - \mathbf{k}$ is uniquely given; then $|\mathbf{q}|$, $\varkappa_x$, $\varkappa_y$, and $\varkappa_z$ are known. The polarization of this elastic wave, i.e., γ_x, γ_y, and γ_z, and its frequency are determined by the set of equations

$$\left.\begin{aligned} (Q_{xx} - X)\gamma_x + Q_{xy}\gamma_y + Q_{xz}\gamma_z = 0, \\ Q_{yx}\gamma_x + (Q_{yy} - X)\gamma_y + Q_{yz}\gamma_z = 0, \\ Q_{zx}\gamma_x + Q_{zy}\gamma_y + (Q_{zz} - X)\gamma_z = 0. \end{aligned}\right\} \quad (9.20)$$

Here $X = \rho\Omega^2/q^2 = \rho v^2$. The values of Q_{ij} are taken from Table 5. Thus, for example, $Q_{xx} = c_{11}\varkappa_x^2 + c_{66}\varkappa_y^2 + c_{55}\varkappa_z^2 + 2c_{56}\varkappa_y\varkappa_z + 2c_{15}\varkappa_z\varkappa_x + 2c_{16}\varkappa_x\varkappa_y$, etc.; $Q_{ij} = Q_{ji}$; the c_{ik} are the usual elastic constants.

This system of equations has three sets of nontrivial solutions. In other words, for a given value of the wave vector in the crystal, three plane waves of different velocity, different polariza-

Table 5. Elastic Constants

	$\varkappa_x^2$	$\varkappa_y^2$	$\varkappa_z^2$	$\varkappa_y\varkappa_z$	$\varkappa_z\varkappa_x$	$\varkappa_x\varkappa_y$
Q_{xx}	c_{11}	c_{66}	c_{55}	$2c_{56}$	$2c_{15}$	$2c_{16}$
Q_{yy}	c_{66}	c_{22}	c_{44}	$2c_{24}$	$2c_{46}$	$2c_{26}$
Q_{zz}	c_{55}	c_{44}	c_{33}	$2c_{34}$	$2c_{35}$	$2c_{45}$
Q_{yz}	c_{56}	c_{24}	c_{34}	$c_{23}+c_{44}$	$c_{36}+c_{45}$	$c_{46}+c_{25}$
Q_{zx}	c_{15}	c_{46}	c_{35}	$c_{45}+c_{36}$	$c_{13}+c_{55}$	$c_{14}+c_{56}$
Q_{xy}	c_{16}	c_{26}	c_{45}	$c_{25}+c_{46}$	$c_{14}+c_{56}$	$c_{12}+c_{66}$

tion, and different frequency will be propagated in each given direction. One of the waves will be quasilongitudinal, the other two quasitransverse.

Since, in a crystal in the general case, for given directions of the incident light and the observed scattered light, four different values are possible for the wave vector of the elastic wave **q** (this is the case because of the possibility of different polarizations for the incident and scattered waves), twelve Stokes and twelve anti-Stokes satellites will be present in the scattered wave, a circumstance which was noted by Vladimirskii [162].

In the calculation carried out, the central unshifted component of the scattered light was not taken into account. In the crystal, its intensity is extraordinarily small because of the smallness of the coefficients of the expansion [see (1.54)].

Scattering of Light in Cubic Crystals. In order to make a more understandable general scheme for the calculation of the intensity, polarization, and frequency of the scattered light, a detailed calculation has been carried out for the case of a crystal of the cubic system.

We choose a system of coordinates directed along the edges of the cube. The matrices of the elastic and elasto-optical constants of the cubic crystal have the form

Matrix of the elastic constants

$$\begin{vmatrix} c_{11} & c_{12} & c_{12} & 0 & 0 & 0 \\ c_{12} & c_{11} & c_{12} & 0 & 0 & 0 \\ c_{12} & c_{12} & c_{11} & 0 & 0 & 0 \\ 0 & 0 & 0 & c_{44} & 0 & 0 \\ 0 & 0 & 0 & 0 & c_{44} & 0 \\ 0 & 0 & 0 & 0 & 0 & c_{44} \end{vmatrix}$$

Matrix of the elasto-optical constants

$$\begin{vmatrix} p_{11} & p_{12} & p_{12} & 0 & 0 & 0 \\ p_{12} & p_{11} & p_{12} & 0 & 0 & 0 \\ p_{12} & p_{12} & p_{11} & 0 & 0 & 0 \\ 0 & 0 & 0 & p_{44} & 0 & 0 \\ 0 & 0 & 0 & 0 & p_{44} & 0 \\ 0 & 0 & 0 & 0 & 0 & p_{44} \end{vmatrix}$$

The frequency and polarization of the elastic waves are determined by the coefficients Q_{ij}. For our case, they are equal to

$$Q_{xx} = c_{11}\varkappa_x^2 + c_{44}\varkappa_y^2 + c_{44}\varkappa_z^2, \quad Q_{yz} = (c_{12} + c_{44})\,\varkappa_y\varkappa_z,$$
$$Q_{yy} = c_{44}\varkappa_x^2 + c_{11}\varkappa_y^2 + c_{44}\varkappa_z^2, \quad Q_{zx} = (c_{12} + c_{44})\,\varkappa_z\varkappa_x,$$
$$Q_{zz} = c_{44}\varkappa_x^2 + c_{44}\varkappa_y^2 + c_{11}\varkappa_z^2, \quad Q_{xy} = (c_{12} + c_{44})\,\varkappa_x\varkappa_y.$$

The intensity and the polarization of the scattered light are determined by means of the coefficients φ_{ij}. For a cubic crystal, $\varphi_{ij} = n^4\varphi'_{ij}$, where the φ'_{ij} are determined by Eq. (9.6),

$$\varphi'_{xx} = p_{11}\gamma_x\varkappa_x + p_{12}\gamma_y\varkappa_y + p_{12}\gamma_z\varkappa_z, \quad \varphi'_{yz} = p_{44}(\gamma_y\varkappa_z + \gamma_z\varkappa_y),$$
$$\varphi'_{yy} = p_{12}\gamma_x\varkappa_x + p_{11}\gamma_y\varkappa_y + p_{12}\gamma_z\varkappa_z, \quad \varphi'_{zx} = p_{44}(\gamma_z\varkappa_x + \gamma_x\varkappa_z),$$
$$\varphi'_{zz} = p_{12}\gamma_x\varkappa_x + p_{12}\gamma_y\varkappa_y + p_{11}\gamma_z\varkappa_z, \quad \varphi'_{xy} = p_{44}(\gamma_x\varkappa_y + \gamma_y\varkappa_x).$$

Let the direction of observation of the scattered light be perpendicular to the direction of the incident light. We consider two different orientations of the crystal relative to the direction of observation and direction of incidence of the beam.

Case 1. Light is incident along the edge of the cube (an axis of symmetry of fourth order). The scattered light is observed at an angle of 90° along another edge of the cube (Fig. 14a). Let the light be incident along the x axis, and be scattered along the y axis. Thus, we have the wave vectors: for the incident wave $\mathbf{k}$ (k, 0, 0), for the scattered wave $\mathbf{k}'$ (0, k, 0); for the elastic wave, $\mathbf{q} = \mathbf{k}' - \mathbf{k}$, $\mathbf{q}(-k, k, 0)$, $q^2 = 2k^2$, $\varkappa_x = -\sqrt{2}/2$, $\varkappa_y = +\sqrt{2}/2$, $\varkappa_z = 0$.

The elastic wave is propagated along the diagonal to the boundary.

$$Q_{xx} = Q_{yy} = \frac{1}{2}(c_{11} + c_{44}), \quad Q_{zz} = c_{44}, \quad Q_{yz} = Q_{zx} = 0,$$
$$Q_{xy} = -\frac{1}{2}(c_{12} + c_{44}).$$

The set of equations for the determination of the frequency and polarization of the elastic wave has the form

$$(Q_{xx} - X)\gamma_x + Q_{xy}\gamma_y = 0,$$
$$Q_{xy}\gamma_x + (Q_{xx} - X)\gamma_y = 0,$$
$$(Q_{zz} - X)\gamma_z = 0,$$

where $X = \rho\Omega^2/q^2 = \rho v^2$. For a nontrivial solution, it is required that X satisfy the equation

$$\begin{vmatrix} Q_{xx} - X & Q_{xy} & 0 \\ Q_{xy} & Q_{xx} - X & 0 \\ 0 & 0 & Q_{zz} - X \end{vmatrix} = 0$$

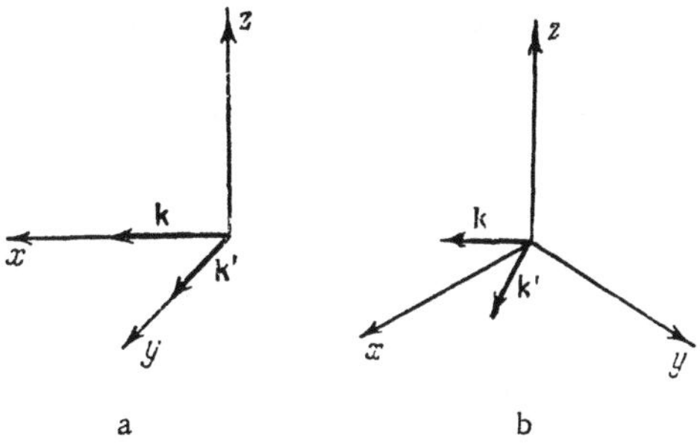

Fig. 14. For the calculation of the scattered light in a cubic crystal: (a) incident and scattered light are propagated along the corresponding edges of the cube (an axis of fourth order); (b) the incident light is directed along the diagonal of the boundary of the cube. The scattered light is observed at an angle of 90° along another diagonal to the boundary of the cube.

or

$$[(Q_{xx}-X)^2-Q_{xy}^2](Q_{zz}-X)=0.$$

We have three solutions:

$$X_{\text{I}}=Q_{zz}=c_{44},\ \gamma_x=\gamma_y=0,\ \gamma_z=1 \text{ -- transverse wave,}$$

$$X_{\text{II}}=Q_{xx}+Q_{xy}=\frac{1}{2}(c_{11}-c_{12}),\ \gamma_x=\gamma_y=\frac{\sqrt{2}}{2},\ \gamma_z=0 \text{ — transverse wave,}$$

$$X_{\text{III}}=Q_{xx}-Q_{xy}=\frac{1}{2}\left(c_{11}+c_{12}+2c_{44}\right),\ \gamma_x=-\gamma_y=\frac{\sqrt{2}}{2},$$
$$\gamma_z=0 \text{ — longitudinal wave.}$$

In a cubic crystal, the polarization of the incident and scattered light is indeterminate. We consider separately the cases of polarization of the incident light perpendicular and parallel to the plane of scattering, i.e., along the z axis and along the y axis. Then we do the same for the polarization of the scattered light. The polarization perpendicular to and parallel to the plane of scattering corresponds to polarization of the scattered light along the z axis and along the x axis, respectively.

(a) The incident light is polarized perpendicular to the plane of scattering, i.e., along the z axis; the scattered light is polarized perpendicular to the plane of scattering, i.e., along the z axis,

$$\left.\begin{array}{l}\beta_x=0,\ \beta_y=0,\ \beta_z=1;\\ \alpha_x=0,\ \alpha_y=0,\ \alpha_z=1;\end{array}\right\}\ \varphi'_{\sigma\tau}\alpha_\sigma\beta_\tau=\varphi'_{zz}\alpha_z\beta_z=-\frac{\sqrt{2}}{2}p_{12}(\gamma_x-\gamma_y).$$

For the I elastic wave, $\varphi'_{ZZ} = 0$. This means that the intensity of the light of the given polarization scattered by this wave is equal to zero.

For the II elastic wave, $\varphi'_{ZZ} = 0$. The intensity of the scattered light is equal to zero.

For the III elastic wave, $\varphi'_{ZZ} = -p_{12}$. As a result,

$$R_{zz}=\frac{\pi^2}{\lambda^4}n^8kT\frac{p_{12}^2}{\frac{1}{2}(c_{11}+c_{12}+2c_{44})}. \tag{9.21}$$

The first index on R denotes the polarization of the incident light; the second, the polarization of the scattered light.

(b) The incident light is polarized perpendicular to the plane of scattering, i.e., along the z axis; the scattered light is polarized parallel to the plane of the scattering, i.e., along the x axis,

$$\left.\begin{array}{l}\beta_x=0,\ \beta_y=0,\ \beta_z=1;\\ \alpha_x=1,\ \alpha_y=0,\ \alpha_z=0;\end{array}\right\}\ \varphi'_{\sigma\tau}\alpha_\sigma\beta_\tau=\varphi'_{xz}\alpha_x\beta_z=-\frac{\sqrt{2}}{2}p_{44}\gamma_z.$$

For the I elastic wave, $\varphi'_{XZ} = -(\sqrt{2}/2)p_{44}$,

$$R_{zx}=\frac{\pi^2}{\lambda^4}n^8kT\frac{0.5p_{44}^2}{c_{44}}. \tag{9.22}$$

For the II elastic wave, $\varphi'_{XZ} = 0$. The intensity of the scattered light is equal to zero.

For the III elastic wave, $\varphi'_{XZ} = 0$. The intensity of the scattered light is equal to zero.

(c) The incident light is polarized parallel to the plane of scattering, i.e., along the y axis; the scattered light is polarized perpendicular to the plane of scattering, i.e., along the z axis,

$$\left.\begin{array}{l}\beta_x=0,\ \beta_y=1,\ \beta_z=0;\\ \alpha_x=0,\ \alpha_y=0,\ \alpha_z=1;\end{array}\right\}\ \varphi'_{\sigma\tau}\alpha_\sigma\beta_\tau=\varphi'_{zy}\alpha_z\beta_y=\frac{\sqrt{2}}{2}p_{44}\gamma_z.$$

For the I elastic wave, $\varphi'_{yz} = (\sqrt{2}/2)p_{44}$,

$$R_{yz}=\frac{\pi^2}{\lambda^4}n^8kT\frac{0.5p_{44}^2}{c_{44}}. \tag{9.23}$$

For the II elastic wave, $\varphi'_{yz} = 0$. The intensity of the scattered light is equal to zero.

For the III elastic wave, $\varphi'_{yz} = 0$. The intensity of the scattered light is equal to zero.

(d) The incident light is polarized parallel to the plane of scattering, i.e., along the y axis; the scattered light is polarized parallel to the plane of scattering, i.e., along the x axis,

$$\left.\begin{array}{l}\beta_x=0,\ \beta_y=1,\ \beta_z=0;\\ \alpha_x=1,\ \alpha_y=0,\ \alpha_z=0;\end{array}\right\}\ \varphi'_{\sigma\tau}\alpha_\sigma\beta_\tau=\varphi'_{xy}\alpha_x\beta_y=\frac{\sqrt{2}}{2}p_{44}(\gamma_x-\gamma_y).$$

For the I elastic wave, $\varphi'_{xy} = 0$. The intensity of the scattered light is equal to zero.

For the II elastic wave, $\varphi'_{xy} = 0$. The intensity of the scattered light is equal to zero.

For the III elastic wave, $\varphi'_{xy} = p_{44}$,

$$R_{yx}=\frac{\pi^2}{\lambda^4}n^8kT\frac{p_{44}^2}{0.5(c_{11}+c_{12}+2c_{44})}. \tag{9.24}$$

The distribution of the intensity of the scattered light in frequency and in direction of polarization is given in Table 5a. For this case, the factor $(\pi^2/\lambda^4)n^8kT$, the value of which does not change for the different directions of polarization of the incident and scattered light, serves as unity for the scattering coefficient. The relative shift of the frequency of the scattered light is equal to $\Omega/\omega = \sqrt{2X/\rho}\,(n/c)$.

Table 5a. Fine Structure of Light Scattered in a Cubic Crystal, in the Case in which the Light Is Incident and Observed along an Edge of the Cube

Scattering coefficient / Frequency shift	R_{zz}	R_{zx}	R_{yz}	R_{yx}
$\frac{\Omega_I}{\omega}=\sqrt{\frac{c_{44}}{\rho}}\frac{n}{c}$ Transverse wave, polarized perpendicular to the plane of scattering	0	$\frac{1}{2}\frac{p_{44}^2}{c_{44}}$	$\frac{1}{2}\frac{p_{44}^2}{c_{44}}$	0
$\frac{\Omega_{II}}{\omega}=\sqrt{\frac{c_{11}-c_{12}}{\rho}}\frac{n}{c}$ Transverse wave, polarized parallel to the plane of scattering	0	0	0	0
$\frac{\Omega_{III}}{\omega}=\sqrt{\frac{c_{11}+c_{12}+2c_{44}}{\rho}}\frac{n}{c}$ Longitudinal wave	$\frac{2p_{12}^2}{c_{11}+c_{12}+2c_{44}}$	0	0	$\frac{2p_{44}^2}{c_{11}+c_{12}+2c_{44}}$

For irradiation of the crystal by natural light*

$$R_1=\frac{\pi^2}{2\lambda^4}n^8kT\left[\frac{2\,(p_{12}^2+p_{44}^2)}{c_{11}+c_{12}+2c_{44}}+\frac{p_{44}^2}{c_{44}}\right]. \tag{9.25}$$

The depolarization of the scattered light is given by the formula

$$\Delta_u=\frac{\frac{1}{2}\frac{p_{44}^2}{c_{44}}+\frac{2p_{44}^2}{c_{11}+c_{12}+2c_{44}}}{\frac{1}{2}\frac{p_{44}^2}{c_{44}}+\frac{2p_{12}^2}{c_{11}+c_{12}+2c_{44}}}. \tag{9.26}$$

The intensity of one of the components of the fine structure, corresponding to a transverse elastic wave polarized in the plane

* This formula is identical with the formula obtained by Leontovich and Mandel'shtam [18].

of scattering, is equal to zero. Therefore, in this case, there will be only two Stokes and two anti-Stokes satellites.

Case 2. The light is incident along the diagonal of the boundary of the cube. The scattered light is observed at an angle of 90° along another diagonal to the boundary of the cube (Fig. 14b). Let the z axis be perpendicular to the plane of scattering. Thus, we have the wave vectors: for the incident wave, $\mathbf{k}[(\sqrt{2}/2)k, -(\sqrt{2}/2)k, 0]$, for the scattered wave $\mathbf{k}'[(\sqrt{2}/2)k, (\sqrt{2}/2)k, 0]$, and for the elastic wave $\mathbf{q} = \mathbf{k}' - \mathbf{k}$,

$$q(0, \sqrt{2}k, 0), \; q^2 = 2k^2, \; \varkappa_x = 0, \; \varkappa_y = 1, \; \varkappa_z = 0.$$

The elastic wave is propagated along the edge of the cube. Proceeding along the lines of the scheme outlined above, we obtain

$$Q_{xx} = c_{44}, \; Q_{yy} = c_{11}, \; Q_{zz} = c_{44}, \; Q_{yz} = Q_{zx} = Q_{xy} = 0,$$

$$(c_{44} - X)\gamma_x = 0, \; (c_{11} - X)\gamma_y = 0, \; (c_{44} - X)\gamma_z = 0.$$

We have three solutions:

$$X_{\text{I}} = c_{44}, \; \gamma_x = 1, \; \gamma_y = 0, \; \gamma_z = 0 \text{ — transverse wave,}$$

$$X_{\text{II}} = c_{44}, \; \gamma_x = 0, \; \gamma_y = 0, \; \gamma_z = 1 \text{ — transverse wave,}$$

$$X_{\text{III}} = c_{11}, \; \gamma_x = 0, \; \gamma_y = 1, \; \gamma_z = 0 \text{ — longitudinal wave.}$$

Carrying out the subsequent calculation, which is completely analogous to the foregoing, we get the intensity distribution of the scattered light in frequency and in direction of polarization (Table 5b); the units for the scattering coefficient of the light are the same as in Table 5a.

For irradiation of the crystal by natural light

$$R_2 = \frac{\pi^2}{2\lambda^4} n^8 kT \left[\frac{p_{12}^2 + 0.25\,(p_{12} - p_{11})^2}{c_{11}} + \frac{p_{44}^2}{c_{44}} \right]. \tag{9.27}$$

Depolarization of the scattered light is given by the formula

$$\Delta_u = \frac{\dfrac{p_{44}^2}{2c_{44}} + \dfrac{(p_{12} - p_{11})^2}{4c_{11}}}{\dfrac{p_{44}^2}{2c_{44}} + \dfrac{p_{12}^2}{c_{11}}}. \tag{9.28}$$

Table 5b. Fine Structure of Light Scattered by a Cubic Crystal in the Case in which the Light Is Incident along a Diagonal to the Faces of the Cube, and Observed along Another Diagonal to the Faces of the Cube

Frequency shift \ Scattering coefficient	$R_{\perp\perp}$	$R_{\perp\parallel}$	$R_{\parallel\perp}$	$R_{\parallel\parallel}$
$\frac{\Omega_I}{\omega}=\sqrt{\frac{2c_{44}}{\rho}}\frac{n}{c}$ Transverse wave, polarized in the plane of scattering	0	0	0	0
$\frac{\Omega_{II}}{\omega}=\sqrt{\frac{2c_{44}}{\rho}}\frac{n}{c}$ Transverse wave, polarized perpendicular to the plane of scattering	0	$\frac{1}{2}\frac{p_{44}^2}{c_{44}}$	$\frac{1}{2}\frac{p_{44}^2}{c_{44}}$	0
$\frac{\Omega_{III}}{\omega}=\sqrt{\frac{2c_{11}}{\rho}}\frac{n}{c}$ Longitudinal wave	$\frac{p_{12}^2}{c_{11}}$	0	0	$\frac{(p_{12}-p_{11})^2}{4c_{11}}$

The indices ⊥ and ∥ denote the polarization of light perpendicular and parallel to the plane of scattering.

It is important to emphasize that for the MB components the cubic crystal is not isotropic. This refers to the intensities and frequencies of the satellites. Such a situation is produced by the fact that there are anisotropies of the elastic and elasto-optical properties in the cubic crystal.

In Tables 5a and 5b the components produced by the elastic longitudinal wave are different for both cases considered.

Molecular Light Scattering in an Isotropic Solid. Molecular light scattering in an isotropic solid can be obtained by using the formulas for the cubic crystal. Here we need to take it into account that

$$c_{44}=\frac{c_{11}-c_{12}}{2},\quad p_{44}=\frac{p_{11}-p_{12}}{2}, \tag{9.29}$$

Table 5c. Fine Structure of Light Scattered in an Isotropic Solid

Scattering coefficient / Frequency shift	R_{zz}	R_{zx}	R_{yz}	R_{yx}
$\frac{\Omega_{I}}{\omega}=\sqrt{\frac{c_{11}-c_{12}}{\rho}}\frac{n}{c}$ Transverse wave, polarized perpendicular to the plane of scattering	0	$\frac{1}{4}\frac{(p_{11}-p_{12})^2}{c_{11}-c_{12}}$	$\frac{1}{4}\frac{(p_{11}-p_{12})^2}{c_{11}-c_{12}}$	0
$\frac{\Omega_{II}}{\omega}=\sqrt{\frac{c_{11}-c_{12}}{\rho}}\frac{n}{c}$ Transverse wave, polarized parallel to the plane of scattering	0	0	0	0
$\frac{\Omega_{III}}{\omega}=\sqrt{\frac{2c_{11}}{\rho}}\frac{n}{c}$ Longitudinal wave	$\frac{p_{12}^2}{c_{11}}$	0	0	$\frac{1}{4}\frac{(p_{11}-p_{12})^2}{c_{11}}$

and then the intensity distribution of the scattered light over the frequencies and over the directions of polarization is (for scattering at 90°) given in Table 5c.

The total coefficient of light scattering in the case in which natural light is incident is equal to

$$R=\frac{\pi^2}{2\lambda^4}n^8kT\left[\frac{4p_{12}^2+(p_{11}-p_{12})^2}{4c_{11}}+\frac{(p_{11}-p_{12})^2}{2(c_{11}-c_{12})}\right]. \tag{9.30}$$

The depolarization is

$$\Delta_u=\frac{\frac{1}{4}\frac{(p_{11}-p_{12})^2}{c_{11}-c_{12}}+\frac{1}{4}\frac{(p_{11}-p_{12})^2}{c_{11}}}{\frac{1}{4}\frac{(p_{11}-p_{12})^2}{c_{11}-c_{12}}+\frac{p_{12}^2}{c_{11}}}. \tag{9.31}$$

Equation (9.30) coincides with the formula obtained by Landau and Lifshitz [47] for this same case.

By using the Neiman constants p and q, which are associated with the elasto-optical constants by the relations

$$p_{11}=\frac{q}{n}, \quad p_{44}=\frac{p}{n}, \quad p_{12}=\frac{q-p}{2n}, \tag{9.32}$$

we get

$$R=\frac{\pi^2}{2\lambda^4}n^6kT\left[\frac{4p^2}{c_{11}}+\frac{(q-p)^2}{c_{11}}+\frac{(q-p)^2}{c_{44}}\right]. \tag{9.33}$$

Application of the basic formulas obtained in this section to specific crystals will be given in Sec. 31.

Chapter III

Apparatus and Methods of Measurement of the Basic Characteristics of Scattered Light and Auxiliary Parameters

The experimental study of scattered light, aside from qualitative observations, reduces in its main outlines to the following measurements:

1. Visual, photographic, or photoelectric measurements of the intensity of the scattered light I or the scattering coefficient R, the depolarization factor Δ, and the distribution of the intensity over the various directions (scattering indicatrix) of the spectrally undecomposed scattered light.

2. The photographic or photoelectric measurements of the frequency distribution of the same quantities in the spectrum of the scattered light.

In case (1), the excitation of the light scattering is brought about by a narrow spectral line or by some range of wavelengths or, finally, by white light. In case (2), the scattered light is excited by the sharpest spectral lines possible.

The smallness of the scattering coefficient ($R \sim 10^{-7}$-10^{-5} cm^{-1}) requires the use of the most intense light sources. However, as is well known, the requirements of narrow width and simultaneous high intensity are almost always contradictory in the usual gas discharge light sources. Therefore, one settles on a reasonable compromise, conforming in each case with the specific problem. In addition to this difficulty, a host of common specific difficulties and sources of error arise, complicating the measurement of the depolarization factor, the scattering coefficient, and the distribution of these quantities over the spectrum and over the scattering angles. Here we shall mention only the most important sources of errors.

In order to establish to what degree the experiment is in agreement with theory, a quantitative comparison of the measured and computed values of the scattering coefficient is necessary. However, such a calculation cannot be carried out without measurement of a whole series of auxiliary quantities, such as the adiabatic (β_S) and isothermal (β_T) compressibilities, $(\rho\, \partial \varepsilon / \partial \rho)_S$, $(\rho\, \partial \varepsilon / \partial \rho)_T$, $(\partial \varepsilon / \partial T)_\rho$, and certain other factors.

The measurement of some auxiliary quantities, for example $(\partial \varepsilon / \partial \rho)_S$, is nontrivial and even requires the creation of special methods. Therefore, the methodology of the measurement of these quantities will be discussed at the end of the chapter.

10. Vessels for the Scattering Material and Apparatus for the Measurement of the Depolarization Factor in Spectrally Undecomposed Scattered Light

The experimental study of the phenomenon of molecular light scattering under laboratory conditions in dust-free gases began with the researches of Cabannes [10] and Strutt [11] (the younger Rayleigh). The vessels for the gases to be studied, used in the researches of these authors, had much in common and were cross-shaped jointed sections of metallic pipes. The end faces of the pipes through which the light comes (which produces the scattering), and through which the scattered light is observed, were covered with optical windows. The vessels described by Bhagavantam [163], Weiler [164], and Volkmann [165] et al. were intended for work with gases at different pressures and represent different variants of the cross-shaped vessels of Cabannes and Rayleigh.

For the study of scattering in gases and liquids at normal pressure, Cabannes [75] used a glass cross-shaped vessel which guaranteed much higher purity of the inner volume. Glass vessels were also used by Ananthakrishnan [166] and Rao [167].

The horn-shaped endings of the cross-shaped vessel of Cabannes were blackened on the inside or out (see vessel in Fig.17).

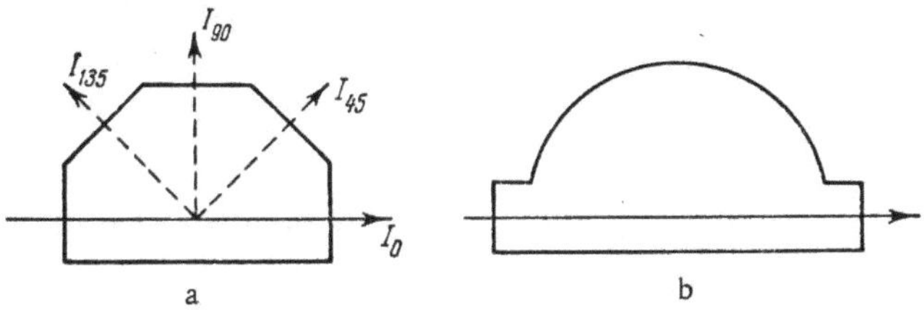

Fig. 15. Vessels for the measurement of the angular dependence of the intensity of scattered light in liquids and solutions. (a) Vessel of Brice et al. [168]; (b) vessel of Stacey [43].

In such a horn-shaped branch, absorption of practically the entire direct light beam takes place and a fine "dark background" is created for observation of the scattered beam.

For the study of the scattered light in liquids and solutions, vessels of another shape are necessary. In particular, vessels of spherical or cylindrical shape are used, in which it is convenient

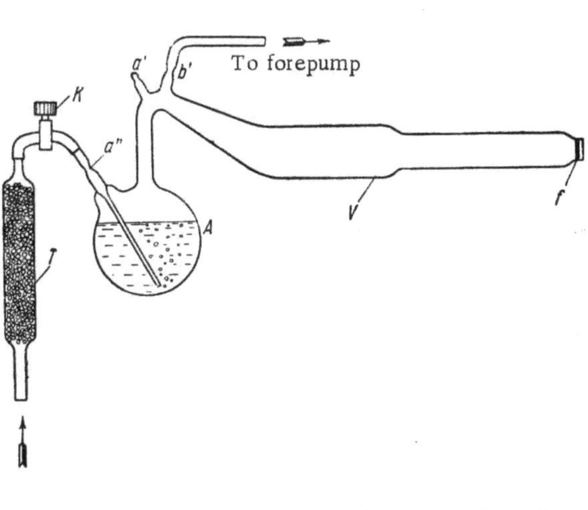

Fig. 16. Apparatus for the optical purification of a viscous liquid (Fabelinskii [53]). A – Flask for the liquid; T – tube with drying material; a'', a', and b' – constrictions; K – clamp regulating the entry of dust into the liquid; V – vessel with scattering liquid; f – plane exit window.

to seek the angular dependence of the intensity of the scattered light, but in such vessels, the "dark background" is much poorer and therefore one can expect a larger amount of parasitic light than in the horn-shaped vessels.

In the study of the intensity of the scattered light and its angular dependence in solutions of polymers and proteins, Stacey [43], Brice et al. [168], used the vessels shown schematically in Fig. 15. In such vessels it is convenient to measure both the angular dependence of the intensity of scattered light and the intensity of the directly transmitted light.

A vessel was suggested by Wood [169], different modifications of which have been used especially widely and successfully for the investigation of the spectral composition of the scattered light in gases, liquids, and solutions. But such a vessel can also be used with success in a number of cases [170] for the measurement of the total scattering. The illumination of such vessels is carried out from the side while observations of the scattered light are made along the axis of the vessel. The Wood vessel, somewhat modified by the author, is shown in Fig. 16.

The range of vessels used in laboratory practice is extraordinarily large, but they all, in one degree or another, are variants of what have been described above, and are chosen in correspondence with the character of the specific experimental problem.

Preparation and Choice of an Optically Pure Scattering Medium. In any sample of a scattering medium which has not been specially prepared, there are, as a rule, contaminants (dust, colloidal particles, defects and impurities in crystals) with an index of refraction differing from that of the surrounding medium. Naturally, such foreign contaminants give a much more intense light scattering than the statistical fluctuations of the index of refraction. Therefore, a serious problem arises in the procedures for cleaning the scattering medium of the contaminants — the preparation of an optically "empty," molecularly scattering medium.

The technique for cleaning gases of dust of a different origin reduces to the multiple filtration of the gas through a thick layer of wadding or, even better, through special filters of porous glass. Such gas cleaning gives excellent results [75, 66].

A very effective method of obtaining an optically "empty" liquid was proposed by Martin [171]. This method is widely used in all investigation with liquids and solutions. The principle of the Martin method consists of the following. A flask A is soldered to the vessel V in which the scattering will later be studied (Fig. 16) (for the purification of low-viscosity liquids, the branches a' and b', and the capillary a'' are not needed). Liquid that has already been cleansed of impurities and suspended matter is distilled in the flask A. The branch a'' is sealed off and the vessel is evacuated by means of the branch b', which is sealed off also after the desired vacuum has been obtained. Then the flask A is heated in a water or oil bath and the vessel V is cooled by flowing water or air.

After a certain amount of liquid has been slowly boiled off from the flask A into the vessel V, the apparatus is taken from its stand and the liquid is shaken energetically, thus washing the walls of the vessel V; then the liquid is poured back again into A. Such an operation is repeated several times, after which the vessel V is completely filled with liquid by slow boiling. It is better not to seal off the flask A from the vessel V so that one can repeat the cleaning if such a necessity arises later.

It is much more complicated to carry out the cleaning of liquids of high viscosity by such a method. Viscous liquids are easily overheated and boil up violently, transferring part of the liquid from the flask A to the vessel V. In order to avoid this, the capillary a'' is attached to the flask A. A small amount of air is admitted through this capillary, air which was first cleaned and dried in the tube T without blocking the pumping of the entire system. The air which passes through the capillary a'' mixes the liquid and removes the heat energy. Its evaporation takes place normally. Figure 16 shows the arrangement used by the author. It is similar to other setups used in such cases [172]. If the liquid decomposes with a small amount of heating, it is necessary to resort to a more complicated method of molecular distillation [173]. Of course, it is not possible to purify crystals artificially from foreign impurities and inclusions.

In the study of crystals, a proper criterion is needed to distinguish the light of molecular scattering from the light scattered by foreign impurities. Such a criterion was found by Landsberg

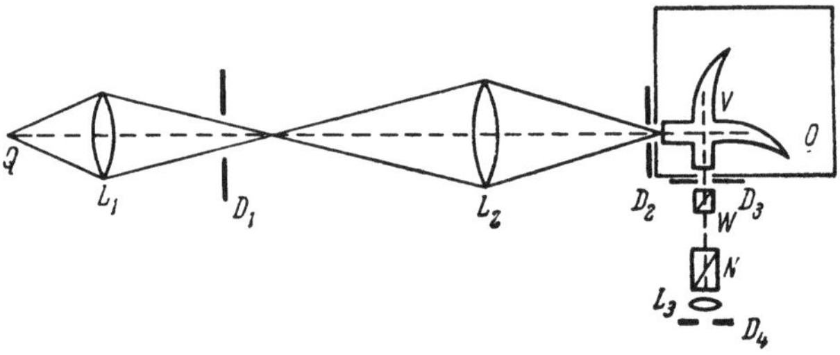

Fig. 17a. Arrangement for the visual measurement of the depolarization factor by a Cornu photometer. Q – light source; L_1, L_2 – objectives; D_1, D_2, D_3, D_4 – diaphragms; V – cross-shaped vessel for the liquid being measured; W – Wollaston prism; N – Nicol prism; L_3 – eyepiece.

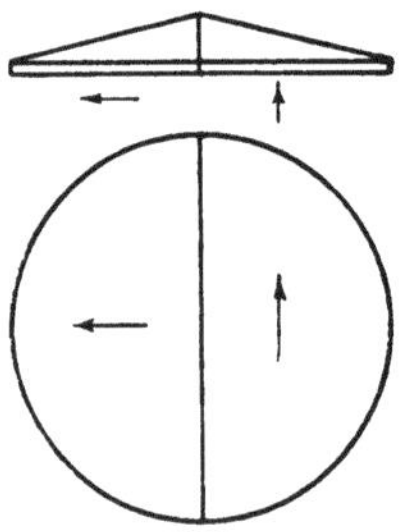

Fig. 17b. Biprism with polaroid (Fabelinskii).

[17] in his first work on the scattering of light in a solid. Landsberg analyzed the character of the temperature dependence of the intensity of the scattered light and found that the intensity of the molecular scattering in the solid is essentially proportional to the absolute temperature. If the intensity of the scattering in the crystal, measured at different temperatures, is plotted as a function of the absolute temperature, then one should obtain (and actually does get) a straight line [17]. By extrapolating the straight line to absolute zero, one can estimate the fraction of the light intensity scattered by the foreign impurities (which does not depend on the temperature). In the absence of impurity scattering, this straight line passes through the origin of the coordinates [20]. The Landsberg criterion permits us to select specimens of excellent crystals or to take into account the fraction of parasitic light scattered by each specimen.

Measurement of the Depolarization Factor of Scattered Light. The coefficient of depolarization Δ, in accord with Eqs. (1.88), (4.9), and (4.10), is the ratio of the light intensity polarized in the scattering plane, I_x, to the light intensity polarized in the plane perpendicular to the scattering, I_z, under the condition that the observation of the scattered light is made at right angles to the direction of the incident light and that the incident and scattered light are collimated beams.

The principle of the method of measurement of the depolarization of the scattered light follows from its definition, and amounts to the following: under the conditions mentioned, the I_X and I_Z components are separated in some way and are measured separately, or their ratio is measured.

It can quickly be noted that the production of beams of light which are well collimated and which carry appreciable energy became possible only when lasers were invented. Lasers are the best light sources [592, 594] for the measurement of the depolarization of scattered light. All the measurements of which we shall speak in what follows were made with finite apertures, and corrections have been carried out on the measured quantities (see below). In setting up the apparatus, it should be added that the aperture was as small as possible, and in making the corrections, it is necessary that the aperture angle be accurately known.

The scheme for one of the numerous experiments for measurement of the depolarization of scattered light is shown in Fig. 17a. Light from the source Q is shone into the vessel V by means of the lenses L_1 and L_2. Observation of the scattered light is carried out in a direction perpendicular to the direction of the incident light. A polarization Cornu photometer is placed in the path of the scattered light. Two fields are observed through the diaphragm D_4, irradiated by the polarized components of the scattered light I_X and I_Z. The scattered light is split into two components, which are polarized in mutually perpendicular directions, corresponding to the two principal planes of the Wollaston prism. The Wollaston prism is so located that its principal planes W_1 and W_2 are parallel to the corresponding z and x axes, while the scattering plane is perpendicular to z (Fig. 18). Then I_Z is equal to the sum of the total intensity of the linearly polarized scattered light and half the intensity of the natural light I_X. In addition, a Glan prism N or some other analyzer is placed so that its principal plane is parallel to the z axis. In order to establish the equality of illumination of the two fields by means of the Glan prism, as observed through the diaphragm D_4 (Fig. 17a), it is necessary to orient the Glan prism so that the components of the electric field vec-

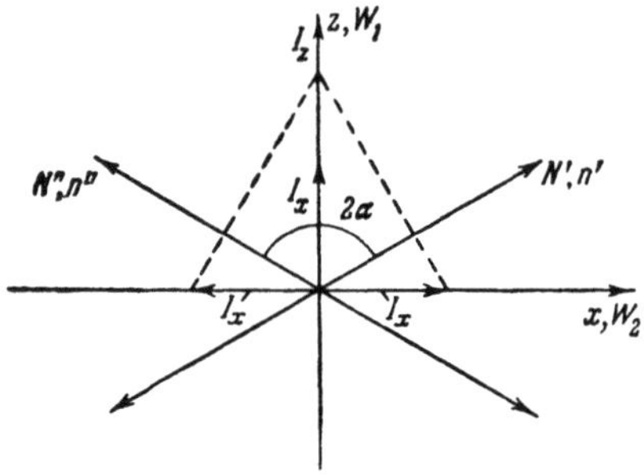

Fig. 18. Determination of the degree of depolarization by the Cornu photometer.

tor of the light wave in the N' direction were identical (first quadrant of Fig. 18). This equality of the fields is also obtained for the setting N". The same will also be the case for the other two quadrants. It is clear from Fig. 18 that $\alpha = (n' - n'' + 360)/2$, where n' and n" are measured from the limb of the analyzer, corresponding to the settings N' and N". The degree of depolarization is

$$\Delta = \frac{I_x}{I_z} = \mathrm{ctg}^2\,\alpha. \tag{10.1}$$

The construction of the Wollaston prism and other doubly refracting prisms is well known [110, 209].

Here we shall comment briefly on the construction of a prism suggested by the author for polarization observations and measurements. This is a biprism, prepared from ordinary optical glass. The edge of the biprism divides it into two halves (Fig. 17b). Polaroid films are attached on each of the halves of the biprism on the plane side; these polarize the light in two mutually perpendicular directions (in Fig. 17b, the directions are shown by the arrows). The film is covered by protective glass. Such a biprism spatially divides the incident light beam into two parts, polarized in mutually perpendicular directions. The angle of separation of the beams emanating from the prism is equal (for small α) to $\Psi = 2\alpha(n - 1)$, where α is the refracting angle of the prism and n is its index of refraction.

Of course, one can make a prism with three boundaries; then it will split the incident light beam into three parts, etc. The value of this prism lies in the fact that its dimensions and angle of separation of the light beams can be set at practically any value necessary in the research. Such a prism also has small chromatism and can be made achromatic if necessary.

The disadvantage of the prism is that for quantitative measurements it is necessary that a homogeneous beam of light be incident on the prism. Moreover, the polaroid polarizes the light less satisfactorily than the Glan prism.

The prism that has been described can be used with special effectiveness in astronomical investigations. The random error in measuring Δ is connected with the error in measuring α (in the arrangement of the analyzer for equality of the illumination) by the following relation: $d\Delta/\Delta = 4d\alpha/\sin 2\alpha$. For the measurement of the degree of depolarization by the photographic method, a photographic camera is placed behind the Wollaston prism (Fig. 17a) or the prism shown in Fig. 17b. Two images are obtained on the photographic film D_4, which is irradiated by the x and z polarized components of the scattered light.

In order to eliminate the inequality of the conditions of propagation of the two polarized beams through the apparatus, a quarter-wave plate is usually placed behind the doubly refracting prism, transforming the linearly polarized light into circularly polarized light.*

Under these conditions, when the degree of depolarization is very small, the blackening of the corresponding fields of the film are quite different. To reduce errors in the measurement, the illumination of the fields can be equalized either by the analyzer, as

*If the spectrum of the scattered light consists of an appreciable range of wavelengths, then a thin wedge (2-3 mm) of Iceland spar must be put in place of the quarter-wave plate. This wedge has its optic axis located at an angle of 45° to the principal planes of the Wollaston prism. The correctly selected angle of the wedge for the given spectral interval creates identical propagation conditions in the apparatus for the x and z components.

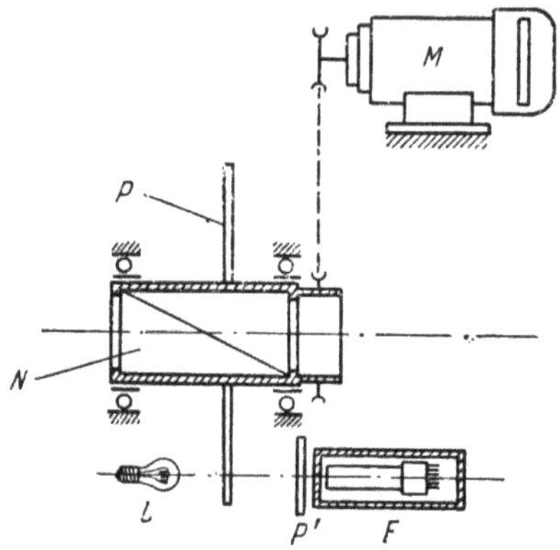

Fig. 19a. Photoelectric polarimeter. N – rotating Glan prism; P – polaroid put on N; L – auxiliary light source; P' – polaroid for changing phase; F – photoconductive cell; M – synchronous motor.

in the previous case of visual method of measurement of Δ, or by the introduction of a calibrated photometric wedge in the more intense light beam. In such a method, the photographic film plays the role of a null indicator and, if equal illumination of the fields is achieved with sufficient accuracy, one can then equate the blacknesses without transforming them into intensities.

We note that if the equalizing of the fields is carried out by means of the Glan prism, then the quarter-wave plate or the wedge would be located immediately after the analyzer.

Subjective or photographic methods of investigation are now being replaced by photoelectric methods. In such cases, the Wollaston prism must be replaced by a Nicol prism, a Glan prism, or another analyzer, while the photographic film is replaced by a photomultiplier, so located that the scattered light which passes through the analyzer is focused by the lens L_3 (see Fig. 17a) on the photocathode of the photomultiplier. The photocurrent of the photomultiplier can be measured directly by a galvanometer or preamplifier and then recorded by the galvanometer or some other indicator. If the polarizer is turned slowly, then the current in-

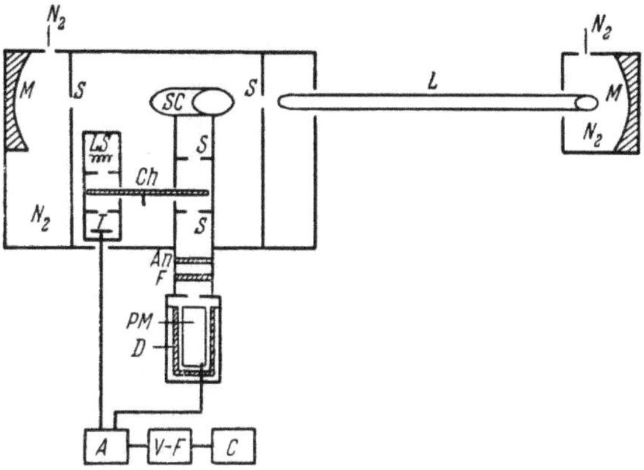

Fig. 19b. Scheme for the photoelectric determination of the depolarization factor of scattered light (Leite, Moore, and Porto [594]). The scattering is excited by the stimulated emission of a Ne–He laser λ = 6328 Å. M – spherical dielectric-coated mirror; L – Ne–He laser; N_2 – nitrogen atmosphere for protection of the mirror and window of the vessel SC and the laser L from dust; PM – photomultiplier; V–F – voltage to frequency converter; C – electronic counter; A – lock-in amplifier; SC – sample cell; T – solar cell; D – Dewar; LS – light source; S – aperture; An – analyzer; F – light filter; Ch – chopper.

dicator records a maximum when the principal plane of the polarizer is parallel to the z axis, and a minimum when the principal plane is parallel to the x axis.

Proportionality of the light flux and the photocurrent i_F is observed for the photomultiplier over a wide range; therefore, we can write

$$\Delta = \frac{(i_F)_{\min}}{(i_F)_{\max}} = \frac{I_x}{I_z}.$$

It is well known [175] that the sensitivity of a photocathode is different for light of different directions of polarization. Therefore, it is quite necessary to depolarize the light, for example by means of a translucent scatterer or the already described method of converting the linearly polarized light into circularly polarized light before the light is incident on the photocathode.

A photoelectric polarimeter that is convenient for research was proposed by Tumerman [176] and used by him for the measurement of the degree of polarization in a study of fluorescence. The photometer based on this principle has also been used with success for the measurement of the depolarization factor of scattered light.

The working principle of this photometer is as follows: a beam of light from the scattering volume or other source of radiation passes through the polarizer, which is rotated with circular frequency O. The scattered light consists of a mixture of natural and linearly polarized light. Therefore, the light which passes through the rotating analyzer will be determined by the following expression:

$$I = \frac{1}{2}(I_x + I_z) + \frac{1}{2}(I_z - I_x)\cos 2Ot. \qquad (10.2)$$

If light of intensity I is directed on the photocathode of a photomultiplier, then the constant current at its exit will be proportional to $(I_X + I_Z)$, while the amplitude of the variable current is proportional to $(I_Z - I_X)$. Consequently, the ratio of the amplitude of the variable photocurrent to the value of the constant photocurrent is equal to

$$p = \frac{I_z - I_x}{I_x + I_z} = \frac{1-\Delta}{1+\Delta} \quad \text{or} \quad \Delta = \frac{1-p}{1+p}, \qquad (10.3)$$

where p is the polarization coefficient. The measurement of the constant and variable components of the photocurrent does not present much difficulty. Furthermore, with the help of an electronic potentiometer, for example an EPP-09, one can immediately record the ratio that is equal to p in (10.3).

Figure 19a shows a scheme for a photoelectric polarimeter, working on the principle described above. Figure 19b shows a scheme for a photoelectric polarimeter used with a laser.

Some Possible Errors in the Measurement of the Depolarization Factor of Scattered Light. The measurement of the polarization coefficient of scattered light is simple only from the viewpoint of principle. Actually, such measurements belong in the number of difficult optical measurements, not only because the intensity of the scattered light is very small, but principally because these measurements can be affected by a large number of different errors [165, 170, 174]. The possible errors of the measurement of the depolarization factor are given below.

1. Error Produced by the Finite Apertures of the Beams of the Incident and Scattered Light. We assume that the scattering volume is illuminated not by a collimated beam but, as is usually the case in reality, by a diverging beam of light incident on the scattering volume.

In this case, the measured depolarization Δ is larger than the real Δ_0 (Δ_0 is measured in parallel radiation).

The most careful investigation of the problem of the effect of a finite aperture on the measured value of the depolarization factor has been given by Gans [177]. He arrived at the following connection between the real and measured values of the depolarization factor and the aperture angle ϑ:

$$\Delta_0 = \frac{\Delta - \left(1 + \frac{\Delta}{2}\right)\Psi}{1 - \left(1 + \frac{\Delta}{2}\right)\Psi}. \tag{10.4}$$

Here $\Psi = (1 - \cos\vartheta)[(2 + \cos\vartheta)/3] = 2\sin^2(\vartheta/2)[1 - {}^2/_3 \sin^2(\vartheta/2)]$, ϑ being the aperture angle (angle between the generating cone of the light rays and the axis of this cone). If Ψ and Δ are small in comparison with unity, the following approximate formula follows from (10.4):

$$\Delta_0 = \Delta - \frac{1}{2}\sin^2\vartheta. \tag{10.5}$$

For a very small angle ϑ,

$$\Delta_0 = \Delta - \frac{1}{2}\vartheta^2. \tag{10.6}$$

Here we are concerned with the correction for the aperture of the incident light. It is quite clear that the same correction must be made for the finite aperture of the scattered light. The correction for rectangularly bounded apertures has also been carried out [227].

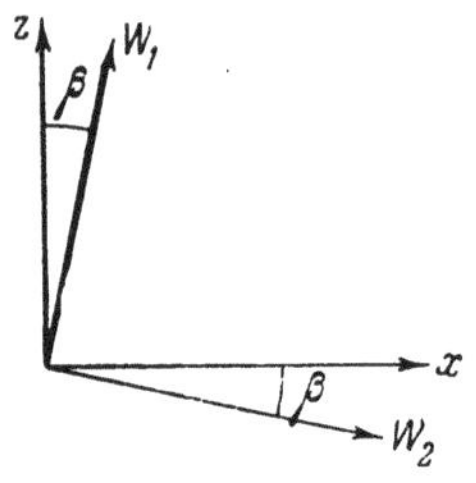

Fig. 20. Determination of the error in measurement of the depolarization factor of scattered light.

2. Error Resulting from Inaccurate Setting of the Polarizer. If one of the principal planes of the Wollaston prism W_1, W_2, or any other doubly refracting prism, does not pass through the vertical to the scattering plane z, but makes an angle β with it (Fig. 20), then, even if the scattering volume is illuminated by a parallel beam of light and a parallel beam of scattered light is radiated, the measured depolarization factor Δ will differ from the real Δ_0. It is not difficult to see (from Fig. 20)* that the connection between the angle β and the quantities Δ and Δ_0 is the following:

$$\Delta_0 = \frac{\Delta - \mathrm{tg}^2\beta}{1 - \Delta\,\mathrm{tg}^2\beta}. \tag{10.7}$$

If the inaccuracy in the location of the polarizer is small while Δ is large, then the correction will be very small. But, if $\beta \sim 1°$, and Δ is small (~0.02), then the error will be as large as 8%. This correction will be significant for the measurements of the depolarization factor in vapors and gases of atoms and molecules having a high degree of symmetry in their optical polarizability.

For polarimeters in which the analyzer turns rapidly and Δ is determined by (10.3), this error does not play an important role. If the Wollaston prism is replaced by a Glan prism or a polaroid, and the measurement is made at different settings, the correction (10.7) should be taken into account.

3. Errors Brought about by Parasitic Light. The imperfection of the "black background" in any setup for the study of light scattering leads to the entrance of parasitic light into the total flux of scattered light. As a rule, the state of polarization of the parasitic light is practically the same as the state of polarization of the incident light. It is therefore clear that when the incident

*It must be recalled that the scattered light is a mixture of linearly polarized (E || z) and natural light.

light is natural, the measured value of the depolarization is larger than this quantity in the absence of parasitic light.

For excitation by polarized light (E_z of the incident light perpendicular to the plane of scattering) the depolarization factor is smaller than in the absence of parasitic light. This is therefore an effective method for testing an apparatus for the presence or absence of parasitic light in it. Actually, from the definition of the depolarization factor in excitation by natural light Δ_u (4.10) and by polarized light Δ_v (4.9), it follows that these quantities are connected with one another by the relation

$$\Delta_u = \frac{2\Delta_v}{1+\Delta_v}. \tag{10.8}$$

The measurement of Δ in excitation by natural and polarized light allows us to test the equation (10.8). If it is not satisfied, then this indicates the presence of parasitic light in the scattered light studied.

It should be emphasized that if a Wollaston prism or similar device is used as a polarizer in the polarimeter, and the parasitic light is incident on the Wollaston prism not only from the central part of the "black background," but also from the peripheral parts and from the walls of the vessel, i.e., in other words, the parasitic light is incident on the Wollaston prism in a wide beam, then conditions arise in which the I_x^f component of the parasitic light is added to the I_z component of the scattered light and the I_z^f components of the parasitic light to the I_x component of the scattered light [165]. In this case,

$$\mathrm{tg}^2\alpha = \Delta = \frac{I_x - I_x^f}{I_z - I_z^f}. \tag{10.9}$$

We shall assume that the parasitic light is natural, i.e., $I_x^f = I_z^f$, and that $I_z^f \ll I_z$; the following approximate relation then follows from (10.9):

$$\Delta = \Delta_0 - \frac{I_x^f}{I_z}. \tag{10.10}$$

If it turns out that $I_x^f \sim \frac{1}{2} I_x$, then it follows from (10.10) that the measured depolarization will be half the true value.

4. Errors Resulting from an Inaccurate Knowledge of the State of Polarization of the Incident Light. The value of the depolarization measured in the excitation of scattered light by natural light differs from the value of the depolarization measured in the excitation by polarized light. In this case, when the incident light is partially polarized (a mixture of natural and linearly polarized light), a systematic error appears in the value of the depolarization.

If a Cornu photometer is placed in the path of the incident light passing through the entrance and exit windows for the vessel, and if the equality of fields is achieved in the photometer when the principal plane of the analyzer makes an angle β with the vertical, then the connection between Δ, Δ_0, and β will be the following:

$$\frac{\Delta_0}{1-\Delta_0}=\frac{\Delta}{1-\Delta}\,2\sin^2\beta. \tag{10.11}$$

Formula (10.11) is valid only when the primary light consists of mixtures of natural and linearly polarized light.

5. Error Produced by Raman Light Scattering. If special measures are not taken in the measurement of Δ to avoid Raman scattering, then this light has an effect on the value of the depolarization factor. The effect of this is especially marked if the depolarization factor of Rayleigh scattering is small and the intense lines of the spectrum of Raman scattering are strongly depolarized. For such a situation, a mixture of Raman scattering already small in intensity also leads to significant errors in the measured value.

Actually, if the total intensity of Raman scattering is represented by i, the intensity of the Rayleigh scattering by I, i/I by a, and i_x/i_z by ρ_{Ram}, and the rest of the notation is the same as before, then we have [170]

$$\Delta_0=\frac{\Delta-\dfrac{a}{\rho_{\text{Ram}}+1}(\rho_{\text{Ram}}-\Delta)}{1+\dfrac{a}{\rho_{\text{Ram}}+1}(\rho_{\text{Ram}}-\Delta)}. \tag{10.12}$$

It is seen from (10.12) that two cases are possible: (1) $\rho_{\text{Ram}} > \Delta$, whence $\Delta > \Delta_0$; and, (2) $\rho_{\text{Ram}} < \Delta$, whence $\Delta < \Delta_0$.

As an example, we shall show that by using the actual data of researches [178-181], we can find that $a \sim 3\%$ for pentane and hexane, and $a \sim 2\%$ for heptane and octane. Since most intense lines of Raman scattering in these liquids are depolarized, then, without great error, we can set $\rho_{\text{Ram}} \sim 0.8$; then the difference between Δ and Δ_0 amounts to 15%.

6. Errors Produced by Fluorescence. To eliminate fluorescent scattering of a medium is extraordinarily difficult and sometimes even impossible. Fluorescence distorts the value of the depolarization factor just as Raman light scattering does. To reveal the presence of fluorescent light in the medium under study, one usually employs the method of crossed light filters.

A light filter is placed in the beam of incident light which cuts a single spectral light or a narrow spectral interval intended for the excitation of light scattering. A light filter is then placed in the beam of scattered light which delays the light of the short wavelength part of the spectrum, including the portion of the spectrum passed by the first light filter. For the first entrance light filter, this is an especially fine interference light filter. If fluorescent light reaching the spectral region employed is excited in the scattering volume, then the photometer detects the presence of light in the crossed light filters. If there is no fluorescent light, or if it is negligibly small, then this is also established by the photometer.

When there is no possibility of reducing the effect of the fluorescence to negligibly small amounts, the relative intensity (with respect to the scattered light) and depolarization must be measured, and corrections determined by the formula given above (10.12) must be applied.

7. Errors Produced by Doubly Refracting Entrance and Exit Windows of the Test Cell. The entrance and exit windows of the test cell are made of optical glass or fused quartz of high quality. As a result of soldering and bad successive annealings, or as a consequence of tension in other methods of compression or, finally, in the investigation of the scattering medium under different pressures, double refraction can arise in the material, reducing the state of polarization of the exciting and scattered light.

The distortions of the value of the depolarization here are difficult to compute. Therefore, it is best to arrange the material and the conditions of compression of the cell window so that there is no necessity for corrections for double refraction of the window. Control of the final beams is best obtained by sensitive compensators [182].

8. Some Other Sources of Error. Among other sources of error in the measurement of the depolarization factor of scattered light, a major role is played by the different contaminations of the scattering medium, especially dust and other inhomogeneous impurities which appear as the result of poor cleaning.

Depending on the character and dimensions of the contaminations, these can either increase or decrease the depolarization factor.

In research with photographic or photoelectric photometers, when only one analyzer is used, the instability (drift) of the source of exciting light can lead to error if the measurements of the intensity of the I_x and I_z components of the scattered light are made at different times.

The instability of the temperature of the scattering volume is also a source of error.

In addition to the systematic errors outlined above, there are of course the random errors. They are different for different methods of photometry and range from a fraction of one percent to ten percent and higher. Analysis of the random errors in different photometers have received a considerable amount of attention [183-186].

11. Measurement of the Relative and Absolute Intensity of Scattered Light

Relative Measurements of the Intensity of Scattered Light in a Solid. Measurements of the relative intensity of scattered light are affected by fewer errors and are much simpler in experimental arrangement than the absolute measurements. For this reason, many relative measurements and comparatively few absolute measurements of the intensity of scattered light have been carried out.

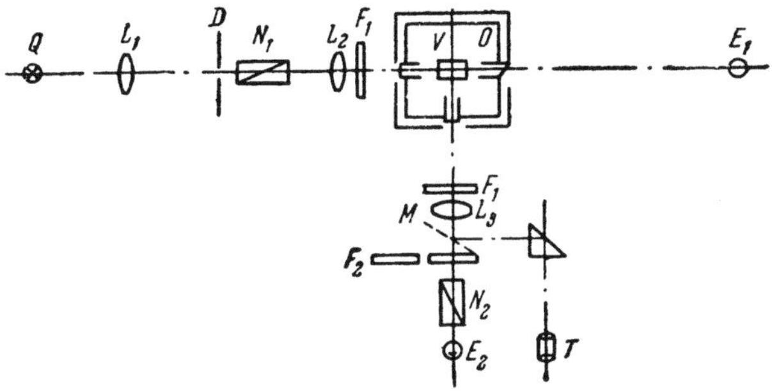

Fig. 21. Diagram of apparatus for the study of scattered light in a phase transition in a solid (Yakovlev, Velichkina, and Mikheeva [22]). V – quartz crystal; Q – mercury lamp SVD; L_1,L_2,L_3 – objectives; F_1 – heatproof light filter; F_2 – colored light filter; N_1, N_2 – polarizing prisms; E_1 – photocell; E_2 – photoelectric multiplier; M – mirror shutter; T – telescope; O – furnace.

If the absolute value of the scattering coefficient of the medium serving as the standard is well known, then, by carrying out the necessary comparison (relative measurements) of the medium studied with the standard, we can find the absolute values of the coefficient of scattering in any medium.

The first complete apparatus for the study of light scattered in crystals, and its temperature dependence, was constructed by Landsberg [17]. Apparatus of this type was used in the works of Vul'fson and Lombert [21], Motulevich [20], and a number of other investigations.

Yakovlev, Velichkina, and Mikheeva [22] studied the temperature dependence of the intensity of scattered light in the $\alpha \rightleftharpoons \beta$ transformation in crystalline quartz; their apparatus is pictured in Fig. 21. The light from a superhigh-pressure mercury lamp Q was directed by the objectives L_1 and L_2 inside the crystal V, placed in a special furnace O. A heat filter F_1 was placed in the path of the incident light, and a polarizer N_1 could be placed there also. The scattered light was observed at right angles to the incident light. The heat filter F_1 and the light filter F_2 were placed in its path. The scattered light was directed onto the photocathode of the photomultiplier by means of the objective L_3.

An analyzer N_2 could be placed in the path of the scattered light to the photomultiplier. Moreover, a mirror M could be introduced in this path, which made it possible to make visual observations with the help of the reversing prism and the telescope T. Replacement of the photomultiplier by a camera made it possible to obtain photographs of the scattered light.

A photocell was placed in the path of the passing light, allowing a check on the constancy of the light source. A great increase in the intensity in the region of the $\alpha \rightleftharpoons \beta$ transformation could be expected in a narrow temperature range in the vicinity of the transformation temperature, 573°C. Therefore, special attention was paid to the construction and preparation of the furnace in which the investigated sample of quartz was placed. It was possible to keep the vertical gradient of the temperature in the region near the crystal down to 0.01°C/mm in the oven used for these experiments (at a temperature of 573°C). The horizontal gradient in the direction of the incident light amounted to 0.03°C per mm.

In other variants of the apparatus, the intensity of the scattered light in the crystals was measured by its ratio to the intensity of a small fraction of the incident light, split off from the main beam. Such a detector is also used for measurement of scattering in other media (see the next section).

Apparatus for the Relative Measurement of Intensity of Light Scattered by a Volume of Liquid and by the Interface between Two Liquids. A convenient method for the measurement of the relative intensity of scattered light was proposed by Sweitzer [187], who equalized the intensity of the scattered light with the intensity of a small part of the incident light, specially split off from the incident beam. The Sweitzer method of relative measurements is widely used in various studies of light scattering. A variant of this scheme, used in the work of Velichkina [188, 189] in the study of the dependence of the intensity of the scattered light in the transition of a liquid to the glassy state, is shown in Fig. 22. Light from an ultrahigh-pressure mercury lamp Q is collected by the condenser L_1 on the diaphragm D_1 which is focused by the long-focal length objective L_2 inside the scattering volume V. F is a light filter which separates out the necessary mercury line or a narrow portion of the

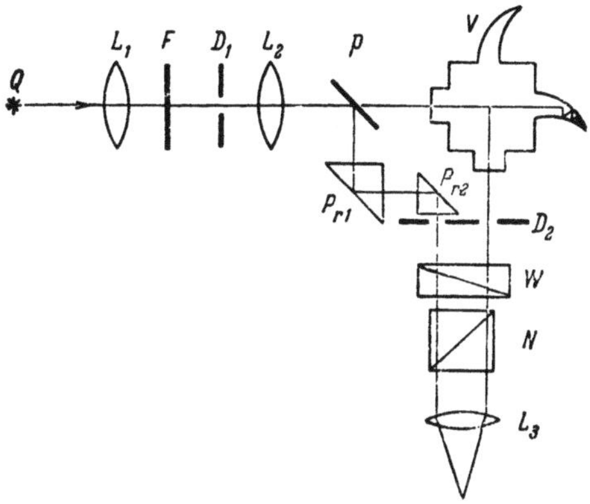

Fig. 22. Optical arrangement for the measurement of the relative intensity of light scattered in liquids at different temperatures (Velichkina [188, 189]). V – sample cell; L_1, L_2 – condensing lenses; L_3 – ocular; D_1, D_2 – diaphragms; F – light filter; P – separating plate; P_{r1}, P_{r2} – reversing prisms; W – Wollaston prism; N – Glan prism.

spectrum. Part of the light beam is split off by a plane-parallel glass plate P, and is then attenuated by a photometric wedge and with the help of reversing prisms is directed onto one of the apertures D_2, which is shielded by a translucent scatterer. The scattered light falls directly on the second aperture D_2.

A Cornu polarimeter was used in [189] and in Fig. 22 as a photometer, in which the single entrance aperture is replaced by two. One of these is illuminated by the comparison beam, the other by the scattered light.

In such a photometer, it is possible to measure only one of the polarized components of the scattered light.

If the photometer shown in Fig. 22 is replaced by an objective which reflects the illuminated diaphragms placed in the plane of D_2 onto the photographic plate, then the comparison light and the scattered light will be photographed on this plate simultaneous-

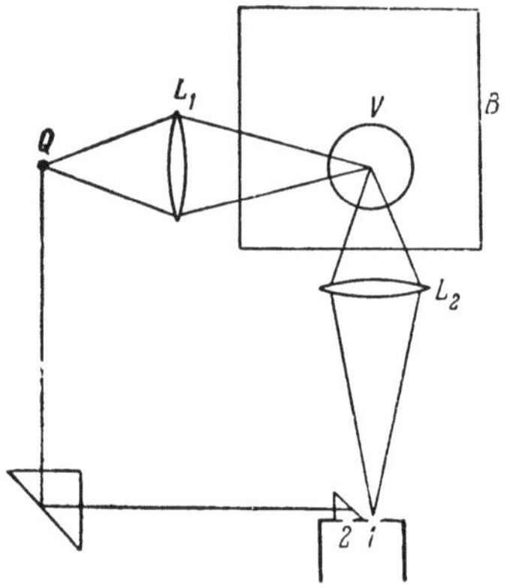

Fig. 23. Optical arrangement for the measurement of the intensity of light scattered by the interface between two liquids (Baryshanskaya [193]). V – spherical vessel for the medium under measurement; B – water bath; L_1, L_2 – objectives; 1,2 – slits of spectrophotometer.

ly. The relative intensity can be obtained by the usual rules of photographic photometry [190-192].

The photographic plate can, of course, be replaced by a photoelectric unit. Its alternating illumination by the comparison beam and the scattered light, with subsequent amplification of the photocurrent at the frequency of alternation, gives the relative intensity.

The inadequacy of this latter scheme for a photometer lies in the fact that such measurements fail to be simultaneous. However, even this difficulty can be overcome [176].

An apparatus using part of the light source for comparison with the scattered light was used by Baryshanskaya [193] for the study of light scattered by the interface between two immiscible liquids. The circuit arrangement, constructed by Baryshanskaya, is shown in Fig. 23.

The liquids under study were placed in the cell V of spherical shape with diameter 3 cm. The lens L_1 focused the image of the point light source Q on the interface between the two liquids. A second lens L_2 reflected the boundary surface of only one of the slits of a König–Martens spectral photometer. The second slit of the spectral photometer was illuminated by a beam of light split off from the source by a system of totally reflecting prisms. The spherical vessel with the liquids under study was placed in a large glass reservoir B, containing 12 liters of water. The walls of the reservoir were thermally insulated. The reservoir temperature could be changed, and kept with a constancy to within ±0.1°C.

By comparing the spectral composition of the scattered light with the spectral composition of the light of the source (knowledge of the energy distribution in the spectrum of the source used is not required), one can find the function $f(\lambda)$ which determines the spectral composition of the scattered light.

It is assumed that

$$f(\lambda)=\frac{a}{\lambda^n}. \quad (11.1)$$

Here n is the index of refraction, which must be measured.

Apparatus for Measuring the Angular Distribution of the Intensity of Scattered Light. The angular distribution of the scattered light intensity is a source of valuable information on the physical properties and characteristics of the scattering medium. Therefore, a great number of researches of this sort have been completed in the last twenty years, and many sets of apparatus and photometers have been constructed, primarily for the study of the angular distribution of the scattered light intensity in solutions of polymers, proteins, electrolytes, and other scattering media.

Figure 24 shows the apparatus of Quantie [194], which he used for the investigation of the asymmetry of the scattering indicatrices in solutions in the region of the critical temperature of mixing.

Light from a high-pressure mercury quartz lamp Q is directed by the quartz condenser L_1 onto the entrance slit S_1 of a monochromator, capable of operating in the spectral region from

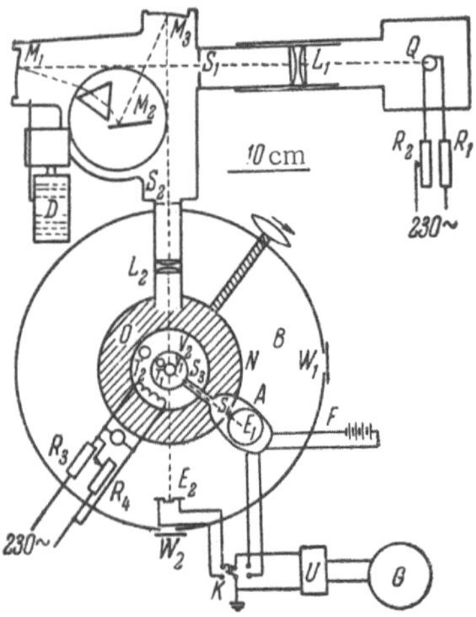

Fig. 24. Diagram of the apparatus for the photoelectric measurement of the angular distribution of scattered light (Quantie [194]). Q – quartz mercury lamp at ultra-high pressure; L_1 – quartz condenser; S_1 – entrance slit of monochromator (wavelength can be changed from 1850 to 10,000 Å); L_2 – condenser; M_1, M_2, M_3 – mirrors of monochromator; V_1 – scattering volume; S_3, S_4 – slits; E_1 – photomultiplier; E_2 – photoelement which establishes the constancy of the passing flux; A – housing of photomultiplier; T_2 – thermometer; T_1 – Beckman thermometer; W_1, W_2 – windows in protective housing; B – block; F – battery; K – switch; U – shunt; R_1, R_2, R_3, R_4 – resistors; D – drum monochromator; O – thermostat; N – base plate.

10,000 Å to 1850 Å. The condenser L_2 projects an image of the exit slit S_2 of the monochromator inside the quartz vessel V_1 in which the solutions under study are placed. The light scattered at a definite angle is directed into the photomultiplier E_1 (which is sensitive to ultraviolet light), enclosed in the housing A. Before the light is incident on the photocathode of the photomultiplier, it is first passed through the two slits S_3 and S_4, each of width 0.5 mm. The photocurrent is measured by the galvanometer G. The photomultiplier, together with the housing A, is fastened to a

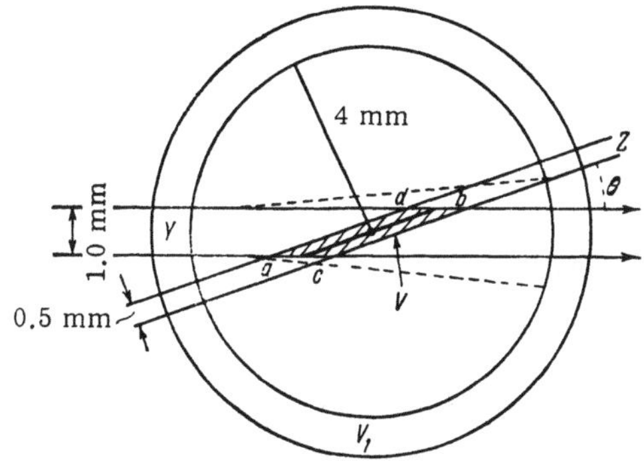

Fig. 25. Scheme for passing a narrow beam of light through the test cell V_1 (see Fig. 24) and the dimensions corresponding to this apparatus [194]. Y − beam of incident light; Z − beam of scattered light.

horizontal metallic bracket which can be turned by hand about a vertical axis passing through the center of the vessel V_1. The scattering angle is noted on a circular scale attached to the horizontal block B. The intensity of the directly transmitted light was measured by the photocell E_2 and by the same galvanometer G with shunt U before and after the measurement of the scattered light. The spectral sensitivity of both photodetectors was studied specially for the wavelength of the light used in the research. To obtain a constant temperature in the sample cell V_1, it was placed in a quartz cylinder V_2 which is filled with water and which is part of a large thermostat (not shown in the drawing). The temperature is checked by the Beckman thermometer T_2 and the thermometer T_1. The constancy of the temperature is verified with an accuracy to within a hundredth of a degree.

The entire apparatus is shielded from the surrounding light and heat by a special housing with windows W_1 and W_2 for observation of the direct light, and of the cell V_1. Most of these cells had the shape of cylinders of diameter from 2 to 18 mm. It is seen from Fig. 25 that the scattering volume inside the cell V_1 has the shape of the prism adbc and obviously depends on the angle θ.

Therefore, for purely Rayleigh scattering, the angular distribution of the intensity is determined not by the factor

$f(\theta) = 1 + \cos^2\theta$ (excitation by natural light), as shown in Sec. 1 et seq., but by the function

$$F(\theta) = \frac{1 + \cos^2\theta}{\sin\theta}, \tag{11.2}$$

which is valid for values of θ not too close to the values 0 and π.

In order to eliminate the purely Rayleigh dependence of the scattered light intensity on the angle of observation, and to reveal the other possible mechanisms which have an effect on the scattering indicatrix, one must divide the experimentally measured value of the scattering coefficient $R'(\theta)$ by $F(\theta)$:

$$R = \frac{R'(\theta)}{F(\theta)}, \tag{11.3}$$

where R for purely Rayleigh scattering is a constant independent of the angle θ. If R in (11.3) depends on θ, then this indicates the presence of more than the Rayleigh mechanism of light scattering. In the case of a very strong scattering (critical opalescence), secondary scattering must also be taken into account.

It is not difficult to see from Fig. 25 that the complete path length in V_1 of the scattered rays inside Z from entrance to exit is practically equal to the diameter of the vessel V_1, and does not depend on the angle θ for $\theta < \pi/2$. For $\theta > \pi/2$ we must replace (11.2) by the function [194]

$$F(\theta) = \frac{1 + \cos^2\theta}{hd}\,\mathrm{sh}\left(\frac{hd}{\sin\theta}\right), \tag{11.4}$$

where h is the extinction coefficient, and d is the width of the primary beam. For small extinction coefficients, (11.4) transforms into (11.2). In the apparatus described, the scattering volume and the illuminating ray are fixed, while the detector of the scattered light is rotated. Variants of the apparatus have been developed in which the light detector is fixed and the scattering volume, together with the ray of incident light or the direction of observation, are rotated [168, 198, 201-205]. The methodology of this problem is discussed in detail in the book by Stacey [43].

The great practical interest which attaches to measurements

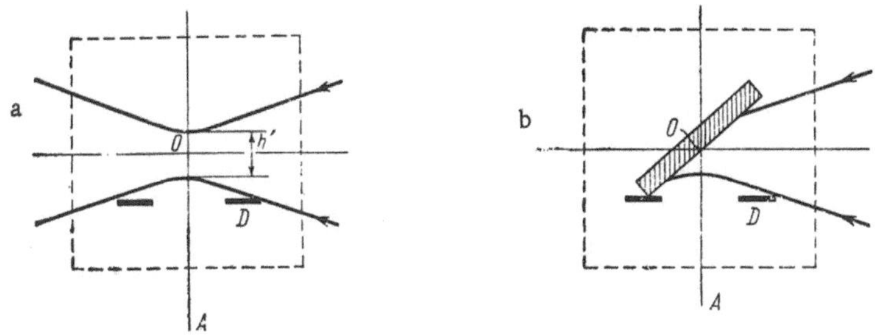

Fig. 26. Principle of the method of measurement of the absolute intensity of scattered light.

of the scattering coefficient and the angular distribution of the intensity in solutions of polymers, proteins, and in inhomogeneous media (suspensions, gels) have led to the appearance not only of a large number of different laboratory models, but also to commercial automatic or semiautomatic instruments that are quite simple to operate. Of practical importance is an apparatus intended for the determination of the dimensions, shapes, and distribution in size of suspended particles [195, 196]. One of these devices, described by Aughey and Baum [197], permits the determination of the size of particles from 0.1 to 100 microns. The angular distribution of the intensity can be measured in the range of scattering angles from 0.05° to 140° with an angular resolution of 0.02°.

The entire operation of the device is carried out automatically, and the scattering indicatrix is recorded on paper by a special recording pen for the angular range mentioned.

Measurement of the Absolute Intensity of Scattered Light. 1. Principle of the Method. The scattering coefficient R, defined by Eq. (1.68), can be rewritten in a different but entirely equivalent form:

$$R = \frac{I}{EV}, \tag{11.5}$$

where I is the light intensity scattered by 1 cm^3 of the medium in the direction OA (Fig. 26a), and E is the illumination per cm^2 produced by the incident light on an area oriented perpendicularly to the beam of incident light and located at the point O.

Thus, in order to determine R, it suffices to measure I, E, and the scattering volume V. For measurement of I and E, two methods are customarily used – photographic and photoelectric.

Let us first consider the photographic method of measurement of R when the scattering medium is a gas. Let the area of the diaphragm D be S and the depth of illumination of the scattering medium h'; then the scattering volume is V = Sh', and in place of (11.5), we must write

$$R=\frac{b}{h'E}, \tag{11.6}$$

where b is the brightness of the diaphragm D in the direction OA. If now a magnesium plate oriented at an angle of 45° to the direction of the incident light (Fig. 26b) is placed at the point O, then the brightness of the plate in the direction OA will be $b'' = (\sqrt{2}/2\pi) \cdot R'E$, where R' is the scattering coefficient of the magnesium plate. As a consequence of the fact that $R \ll R'$, the beam of light coming from the magnesium plate must be reduced by a factor of approximately 10^3 in order to compare the brightness of the two sources. By carrying out a reduction by a factor f', we have, for the brightness of the diaphragm D illuminated by the magnesium plate,

$$b'=\frac{\sqrt{2}}{2\pi}f'R'E. \tag{11.7}$$

Eliminating E from (11.6) and (11.7), we find

$$R=\frac{\sqrt{2}f'R'b}{2\pi h'b'}. \tag{11.8}$$

By projecting the diaphragm D by an objective with fixed aperture on the photoplate, once when D is irradiated by scattered light and once when it is irradiated by the light scattered by the magnesium plate, we get, in general, two different blackenings.

The photometric measurements allow us to find the illuminations of the photoplate to which these blackenings correspond. On the other hand, the illumination of the photoplate e is proportional to the brightness of the source (for fixed location of the objective and aperture); therefore, we can write e = mb and e' = mb', whence

$$\frac{e}{e'}=\frac{b}{b'}. \tag{11.9}$$

If the values of f', h', and R' are known, then it is clear from (11.8) and (11.9) that the determination of R reduces to the measurement of the two illuminations of the photographic plate.

It is clear from the foregoing that for the photographic method of determining R, it is appropriate to increase the brightness of the source of the excitation.

In the photoelectric method of measurement, a photoelement or photomultiplier serves as the radiation detector. The photocurrent here is, within broad limits [175], proportional to the light flux Φ incident on the photocathode. If we use the same experimental setup as in the photographic method, but replace the photoplate by a photomultiplier, then we can write the following in place of (11.8):

$$R = \frac{\sqrt{2} f' R' \Phi d\Omega'}{2\pi h' \Phi' d\Omega}, \tag{11.10}$$

where Φ and Φ' are the light fluxes in the direction OA from the diaphragm D, illuminated by the scattered light and by the light scattered from the magnesium plate, respectively; Ω and Ω' are the solid angles at which the photocathode is viewed from the center of the diaphragm D under the same conditions.

If the scattering material is a gas, then its index of refraction is virtually the same as the index of refraction of the surrounding air, and $\Omega'/\Omega = 1$. As has been shown, $i_F = m'\Phi$ and $i_F' = m'\Phi'$, whence

$$\frac{i_F}{i_F'} = \frac{\Phi}{\Phi'}. \tag{11.11}$$

Substituting (11.11) in (11.10), we find that the determination of R is reduced to the measurement of the ratio of photocurrents i_F/i_F'. This is the case only when one studies the absolute intensity of scattering in gases. When one passes to liquids, the experiment is complicated by the fact that the index of refraction of the liquid is materially different from unity. In this case, generally speaking, a number of corrections to the formulas given above are necessary. We shall discuss the character of these corrections below.

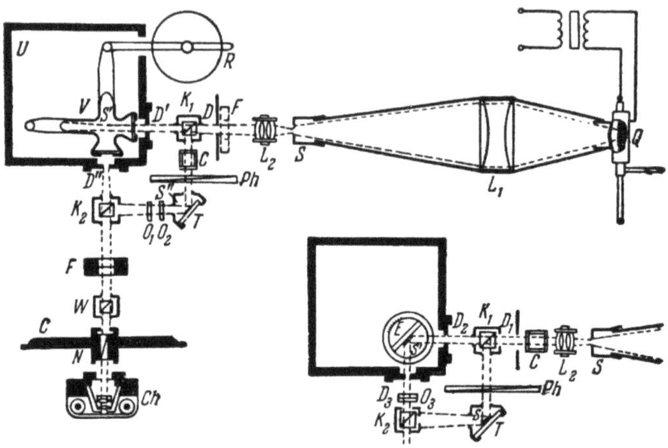

Fig. 27. Arrangement for the measurement of the absolute intensity of scattered light by the photographic method (Vaucouleurs [56]). Q – light source; L_1 – condenser; L_2 – objective which focuses the slit S at the center of the vessel V which is placed in the thermostat U; D, D', D", D_1, D_2, D_3 – diaphragms; K_1,K_2 – small cubes which separate the light beams; Ph – photometric wedge; T – diffusely scattering surface; O_1, O_2, O_3 – calibrated attenuators; E – standard diffuser; S' – image of slit; W – Wollaston prism; N – analyzer with limb C; Ch – photocamera; F – filter; R – reservoir of vessel V.

It is not easy to know when corrections are and are not needed. To date, it has not been possible to consider these questions as definitively settled. We shall return for discussion of this problem after a description of the photographic and photoelectric arrangements for the measurement of the absolute intensity of scattered light.

2. Apparatus for the Measurement of the Absolute Intensity of the Scattered Light. The principles of the arrangements for the measurement of the absolute intensity of scattered light differ but little from those for the measurement of the relative intensity. Figure 27 shows a setup for the measurement of R by a photographic method [56]. When the test cell is placed in the apparatus, blackening from the light scattered by the liquid and the light re-

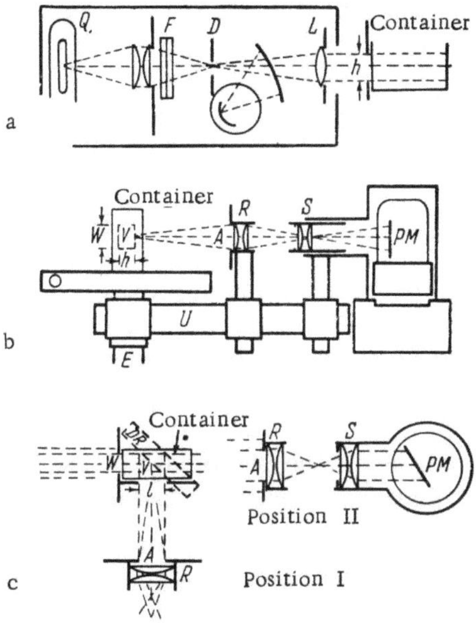

Fig. 28. Arrangement for the measurement of the absolute intensity of scattered light by the photoelectric method (Carr and Zimm [198]). Q – light source; V – test cell; L, R, S – lenses; A – aperture for illumination of photomultiplier PM; D – diaphragm; F – light filter; E, U – axes of rotation of the state O; h, W – cross section of illuminating beam; DR – magnesium plate (comparison body).

flected from the incident beam and attenuated by a known amount is recorded simultaneously by the photographic plate. In addition, a plate E is put in place of the sample cell at an angle of 45° with respect to the direction of the incident light. This plate is covered with magnesium oxide. The arrangement is that shown in Fig. 27 (lower right). Alternatively, a vessel is used with the same liquid in which a totally reflecting prism is immersed. This prism turns the direction of the incident beam by 90° [179]. This ray is strongly attenuated a known number of times before it is incident on the photoplate. A photograph is now made on this same photoplate of the direct attenuated light and the part of the light split off from the incident beam as in the photograph of the scattered light. The

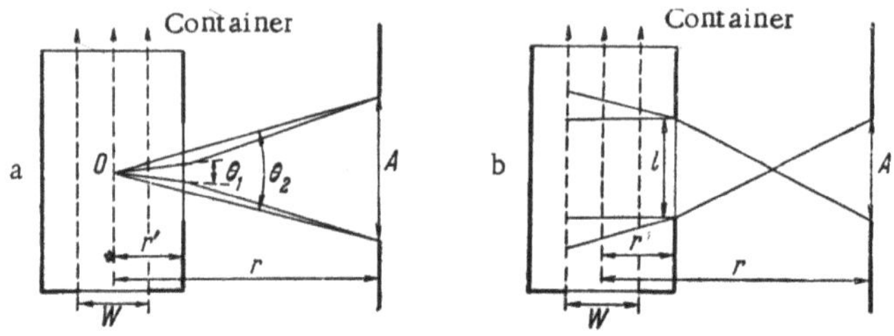

Fig. 29. For the corrections applied in the measurement of the scattering coefficient. (a) For corrections to the index of refraction, C_n; (b) for corrections to the effective volume, C_V.

ratio of the illuminations of the photoplate by the scattered and incident light can be obtained from these two photographs [242].

Figure 28 shows the arrangement of a typical photoelectric setup [198] for the determination of the scattering coefficient R. Figure 28a shows the illuminating part of the apparatus, and Fig. 28b the test cell, the diaphragm which limits the aperture, and the system of illumination of the photomultiplier PM by the condensers R and S. Figure 28c shows the plan of the same apparatus; it shows that the photometer can take the position I, for which one measures the scattered light or the light reflected from the magnesium plate DR, or the position II, for which the photometer records the passing light directly.

In apparatus for the measurement of the absolute intensity of the scattered light, the attenuation of the direct primary light produces great difficulties. For a reliable comparison of brightness or light fluxes that differ by factors of hundreds or thousands, it is necessary to know the damping coefficient f with sufficient accuracy. Various attenuation systems are used in the different arrangements.

In the arrangement of Fig. 27, the damping is due to a calibrated photometric wedge and the calibrated attenuators O_1, O_2, and O_3. In other cases, three polarizers are used for the attenuation of the light; they are located successively in the comparison beam as was done by Daure [199]. Attenuation by beams of multiple reflections [200] and a whole host of other techniques have also been used [43, 198, 201, 205] with success.

3. Corrections to the Measured Quantity. In explaining the principle of the method, it was assumed for simplicity that the index of reflection of the scattering material is equal to the index of refraction of the medium in which the photometer is located.

If the scattering substance is a liquid with an index of refraction n, while the photometer is located in air ($n \sim 1$) as before, then it is clear that the illumination at the point A (Fig. 26) will be less for otherwise equal conditions.

The reason is the change in the solid angle and the effective volume, as is clear from Figs. 29a and 29b. Therefore, if the light flux scattered by the liquid is compared with the light scattered by the magnesium plate located in air, then it is necessary to make corrections to the measured quantity. Corrections at various solid angles or, to put it another way, corrections to the index of refraction, C_n, and corrections for the effective volume, C_V, will be expressed in the following way [198]:

$$C_n = n^2\left(1 - \frac{r'(n-1)}{rn}\right), \tag{11.12}$$

$$C_V = 1 - \frac{r'(a+b)/(r-r')}{2[nl + r'(a+b)/(r-r')]}. \tag{11.13}$$

The meaning of the quantities entering into Eqs. (11.12) and (11.13) is clear from Figs. 29a and 29b.

To obtain the correct values of R measured by the photoelectric method, it is necessary to multiply the measured value by the product C_nC_V. The correction (11.12), under ordinary experimental conditions when $r' \ll r$, tends to the value $C_n \to n^2$, while under the same conditions $C_V \to 1$. Thus, in Eq. (11.6), under ordinary experimental conditions, $d\Omega'/d\Omega \to n^2$. The corrections (11.12) and (11.13) are usually made in photoelectric measurements, while such corrections are not made in the photographic measurements.

After the first photoelectric absolute measurements of R had been made, significant differences between the results of the measurement of R could clearly be seen, both photographically and photoelectrically. The divergence goes far beyond the limits of possible errors in both methods and amounts to ~50%.

Carr and Zimm [198] have shown that in the photographic method of measurement, account is not taken of the corrections (11.12) and (11.13), which have already been noted and, therefore, in their opinion, the photographic data are invalid.

Rousset and Lochet [206] have considered the photoelectric and photographic methods of measurement of R in detail, and reached the conclusion that the existing divergence cannot be explained by the fact that the correction C_n is not taken into account in the photographic measurement. In their opinion, it is not necessary to apply the correction C_n in the photographic method.

The discussions of these authors run approximately as follows. We shall assume that the scattering volume is bounded by the diaphragm D (see Fig. 26), placed in a liquid with index of reflection n. We denote the brightness of the diaphragm D in the direction OA by b. The brightness of the image of this diaphragm will no longer be b, but b/n^2, and the illumination of the photographic image will be reduced by the factor n^2. If now the magnesium plate is immersed in the scattering liquid, then the brightness of the diaphragm in the direction OA will be b'/n^2, and the ratio of the illuminations of the two images on the photoplate will be equal to b/b' as before. We then obtain the correct value of R from the previous expression (11.8). However, in some photographic arrangements, the comparison body (the magnesium plate) was located not in this same liquid, but in air. If, in place of the test cell, we place the magnesium plate in air, then the source will not be focused on it, and it is impossible to make definitive conclusions as to its illumination. For a focused source at the point O, it is necessary to change the location of the condenser or light source.

If, as the result of the adjustment, the solid angle is now shown to be the same as in the case in which the scattering liquid is introduced in place of the plate, then the brightness of the magnesium plate is reduced by the factor n^2 and the brightness of the diaphragm placed in air is increased by a factor of n^2 in comparison with the brightness of the image of the diaphragm, illuminated by the magnesium plate immersed in the liquid (see Fig. 26). Thus the result remains as before, even in this case. Analysis of the apparatus described in [56, 179, 207] led the authors [206, 208, 356] to the conclusion that the measurements obtained on these setups do not need revision, and one can refer to them in complete confidence. The discussion of this methodological question has, however, not yet been concluded.

In subsequent discussions, arguments have been raised in support of the "lower" value of R, just as in other researches, the correctness of the "upper" value of R have been maintained.

Probably, there is no need here of discussing the methodological value of the individual researches. We shall return to a discussion of the values of R and a comparison of the experimental values with the theoretical in Chapter V, Sec. 18, where there will also be a partial discussion of other researches in which the discussion of the "low" and "high" values of R is continued.

12. Methods of the Spectral Investigation of Molecularly Scattered Light

Let us consider some arrangements and apparatus used in the study of the fine structure and the wing of the Rayleigh line.

The wing of the Rayleigh line is a complicated spectrum with a maximum at a fixed frequency ν_0 which extends outward in both directions from ν_0, as a rule some tens or even hundreds of reciprocal centimeters. The fine structure of the Rayleigh line, even for such a substance as diamond, is confined to a spectral interval ± 3 cm^{-1}, while for liquids this interval does not exceed ± 0.3 cm^{-1}. Therefore, the apparatus for the study of these two phenomena must be essentially different in resolving power and range of dispersion of the spectral apparatus.

The difficulties of the experimental study of the spectral composition of scattered light are still greater than for the study of the spectrally unresolved scattered light.

In addition to specific difficulties arising in the solution of the different spectral problems, there is a general experimental difficulty, associated with the additional attenuation of the flux of the scattered light in the spectral equipment.

The basic sources of additional attenuation are the following:

1. Reduction of the aperture of the beam of incident light, necessary for the study of the fine structure and for polarization measurements.
2. Reduction of the aperture of the beam of the scattered light, necessary for efforts to bring to a minimum the effect of parsitic light.
3. The spatial spectral distribution of the scattered light, brought about by dispersion of the spectral apparatus.
4. Reflection and absorption of the light in the optical system as a whole.
5. Application of polarization optics.

An orienting estimate shows that the intensity decreases by an additional factor of 10^4-10^5 in the study of fine structure with the use of the polarization devices [53]. With such an enormous attenuation, the extremely weak scattered light must be taken into account in developing apparatus for the spectral investigation of scattered light and, in particular, for the study of the fine structure.

In the study of the fine structure on an apparatus without polarization optics, Venkateswaren [257] had to expose his photographs for various liquids from two to four days. Use of a polarizing prism required exposures of 10-12 days and made the experiment one of low productivity and of great difficulty in practice, inasmuch as, for example, under the conditions of the Moscow climate, it was necessary to take special measures, not only in carefully thermostatting the apparatus, but also in maintaining a constant barometric pressure in the volume where the spectral apparatus is placed; in particular, this care must be taken for the interference spectroscope. In order to avoid any part of these difficulties, it is necessary to increase the light intensity of the apparatus as much as possible. Simultaneously, of course, it is necessary to obtain a high quality of the image in order to minimize the possible distortions of the line shapes in the spectrum under study.

Apparatus and Methods of Study of the Wings of the Rayleigh Line. For the investigation of the wings of the Rayleigh line in low-viscosity liquids, in which the wings

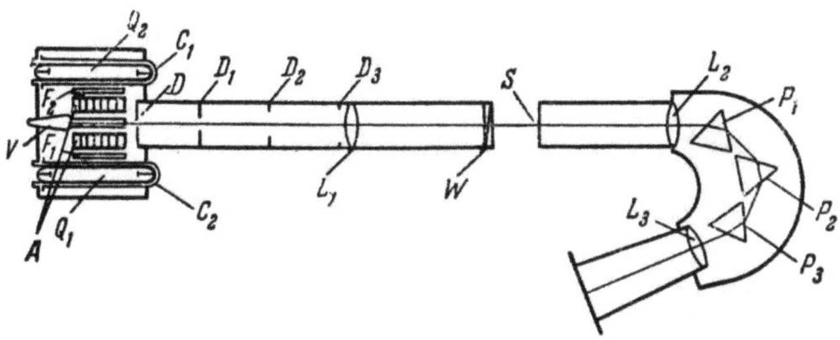

Fig. 30. Arrangement for the study of the wings of the Rayleigh line by the photographic method (Fabelinskii [73]). Q_1, Q_2 – mercury lamps; F_1, F_2 – light filters; V – test cell; C_1, C_2 – Dewars; D, D_1, D_2, D_3 – diaphragms; W – Wollaston prism; L_1, L_2, L_3 – objectives; P_1, P_2, P_3 – prisms; A – aperture limiter.

extend tens of reciprocal centimeters, prism spectrographs have been used successfully with high and medium resolving power and large linear dispersion (7-2 Å/mm in the region of 4358 and 4047 Å). In the different investigations, various types of spectral apparatus were used which satisfied these conditions.

Evidently, the use of diffraction spectrographs of the DSF-12 type for such investigations was especially useful; these allow one to obtain a photoelectric recording of the intensity distribution in the spectrum and possess sufficient dispersion and resolving power.

The principle of the apparatus with a prism spectrograph for the study of the wings of the Rayleigh line is demonstrated in Fig. 30. The test cell V is illuminated by two mercury lamps Q_1 and Q_2, which are shut off from V by the Dewar vessels C_1 and C_2, which keep the scattering medium from heating. The scattered light passes through a series of diaphragms D, D_1, D_2, and D_3. The objective L_1 focuses the diaphragm D through the Wollaston prism onto the entrance slit of the spectrograph S where two images of the diaphragm D are obtained in two polarizations. Two spectra of the scattered light are obtained simultaneously on the spectrogram in two mutually perpendicular states of polarization. An example of such a spectrum is shown in Fig. 31.

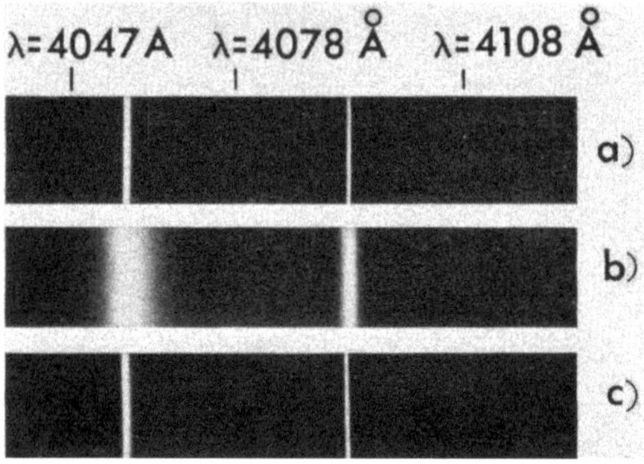

Fig. 31. Wings of the Rayleigh line obtained on the apparatus shown in Fig. 30 (Fabelinskii [73]). (a) Spectrum of the light source – mercury arc; (b) spectrum of the light scattering in carbon bisulfide; (c) spectrum of light scattered in water.

The diaphragms between the light filters F_1 and F_2 serve to limit the aperture for the incident light; this is necessary for measurements of the depolarization of the scattered light (see Sec. 10). If the polarization measurements are not carried out, the limits of the aperture can be removed.

We recall that a quarter waveplate should be placed after the Wollaston prism to transform the linearized polarized light into circularly polarized light, in order to create identical conditions of propagation in the optical apparatus for the x and z components of the scattered light. In the opposite case, the results of measurement will be strongly distorted. If a sufficiently wide range of the spectrum is investigated, then a slightly wedge-shaped plate of Iceland spar with its optic axis directed along the diagonal of the plate should replace the quarter wave plate. Such a plate is attached directly to the Wollaston prism by Canada balsam in such a way that its optic axis was directed at an angle of 45° to the directions of the principal planes of the Wollaston prism.* For po-

* Of course, the wedge-shaped plate is attached to that part of the Wollaston prism which is imaged in the slit of the spectrograph (see Fig. 30).

larization measurements, attention should also be paid to the removal of possible asymmetries in the vignetting of the beams of scattered light attenuated by the Wollaston prism on the collimator objective. In addition to the important methodological problems noted, there is also a series of other very important questions, which have already been discussed in the description of the methods of polarization measurements in the spectra of Raman scattering by Reitz [211], Bobovich and Vol'kenshtein [212], Sushchinskii [213, 214], Bobovich and Aleksandrov [215], Rank et al. [216], and also in the book by Kohlrausch [217].

On the photographic plate which detects the scattered light, the transmission of a graduated attenuator of known transmission is photographed with an exposure of the same order and with the same wavelength or interval of wavelengths. The distribution of the intensity over the spectrum is found according to the well-known rules of photometry [190, 210]. We emphasize that the photographs of the spectrum of the wings, obtained on spectrographs with ordinary linear dispersion, allow us to study the intensity distribution no closer than at a distance of 5-10 cm^{-1} from the maximum of the curve, because of the strong distortion of the intensity as the result of the light scattering in the emulsion of the photographic plate [186].

Figure 32 shows the arrangement for the photoelectric apparatus used in the study of the wings of the Rayleigh line. A DFS-12 spectrometer was used for the spectral apparatus.

The illuminating part of the apparatus is the same as in the apparatus shown in Fig. 33. The scattered light passes through a modulator which chops the light flux with a frequency of 400 cps, and enters the instrument through the entrance slit of the spectrometer. Its path inside the instrument is clear from Fig. 32. From the exit slit, the light falls on the photocathode of an electron multiplier. The photocurrent is amplified, rectified by the synchronous detector, and is fed to the electronic potentiometer which makes a pen recording of the signal.

The circuits, which are joined on a common stage, can be turned at different rates, by means of a special synchronous motor, which makes it possible to record the energy distribution in the spectrum of the scattered light.

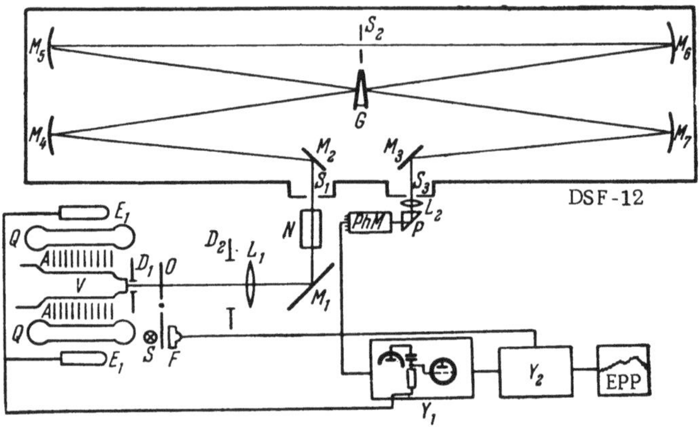

Fig. 32. Arrangement of photoelectric apparatus for the study of the wings of the Rayleigh line (Starunov [505]). V – test cell; A – aperture limiter; Q – low-pressure mercury lamp; E_1 – photoelements controlling the amplifier Y_1; D_1, D_2 – diaphragms; O – mechanical modulator of scattered light of frequency 415 cps; S – additional light source; F – photoresistor for obtaining the voltage controlling the synchronous detector in Y_2; L_1, L_2 – condensers; M_1, M_2, M_3, M_4, M_5, M_6, M_7 – mirrors; S_1, S_2, S_3 – spectrograph slits; G – two diffraction gratings of 600 lines/mm and of dimensions 150 × 140 mm; PhM – photomultiplier; P – reversing prism; N – polarizer; EPP – recording apparatus.

The controlling voltage on the synchronous detector is obtained from an additional light source which illuminates the photoresistance F through the modulator. Moreover, automatic regulation of the amplification factor of the electronic amplifier is turned on, compensating possible fluctuations of the intensity of the sources Q during the time of recording the distribution.

The width of the apparatus function of the DSF-12 for the 4358 Å line is about 1 cm^{-1}; therefore, it is possible to study, by means of the apparatus of Fig. 32, the parts of the wings that closely adjoin the fixed frequency ν_0, if the air gulls are weak and do not prevent such a study. In the apparatus with photoelectric recording, the measurement of the depolarization factor cannot, for

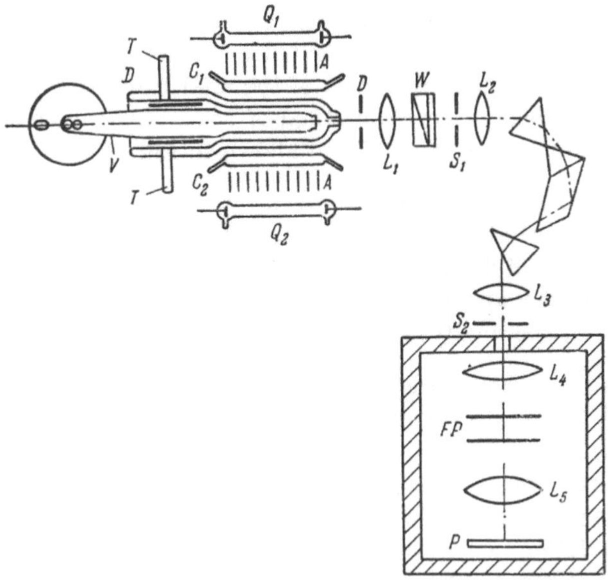

Fig. 33. Arrangement of the interference apparatus for the study of the fine structure of the Rayleigh line (Fabelinskii [53, 144]). V – test cell; W – Wollaston prism; D – Dewar; T – tubes in which the cool or hot air is introduced; C_1, C_2 – cylindrical lenses, separated from the low-pressure mercury lamps Q_1 and Q_2 by the aperture limiter A; D – diaphragm; L_1, L_2, L_3, L_4, L_5 – objectives; S_1, S_2 – entrance and exit of the ISP-51 spectroscograph; FP – Fabry–Perot interferometer; P – photoplate. The interferometer is placed in a separate, isothermal room or in a special, thermally insulated container.

well-known reasons, be made by means of the Wollaston prism in the same fashion as was done in the photographic method. In this case, it is necessary to insert a Glan prism or other polarizer of this type in the path of the scattered light, and record the distribution in the spectrum for different or for two mutually perpendicular positions of the polarizer. However, this method has the disadvantage that the intensity distribution is not obtained simultaneously for the different positions of the analyzer. It would be much better for the photoelectric method of recording to use the polarimeter described in Sec. 11 and shown in Fig. 23.

The resonance filter employed by Rosetti, and also by Landsberg and Mandel'shtam [219] was used by Vuks [220] for the study of the intensity distribution in the wings of the Rayleigh line, and by Chandrasekharan and Krishnan [222-226] for the study of the fine structures of the scattering line in quartz.

The principle of the resonance filter method consists of the following: the scattered light is excited by the resonance lines of the spectrum (in the Vuks apparatus, the resonance lines of the sodium spectrum; in the apparatus of Krishnan and Chandrasekharan, the resonance lines of the mercury spectrum λ = 2536.5 Å). The resonance filter, which is placed in the path of the scattered light at the slit of the spectrograph, is a cell filled with the vapor of the same element whose resonance line serves for the excitation of the scattered light. The cell can be heated by a special heater and thus the density of the vapors of the filter is altered; thereby, the width of the absorption lines can be changed.

The resonance filter absorbs part of the spectrum corresponding to the excitation line, while at a suitable temperature of the filter, the line of shifted frequency passes through the filter without being absorbed.

An appreciable increase in the temperature of the filter, as applied in [220], leads to a broadening of the part of the spectrum of scattered light absorbed by the filter. The intensity distribution in the wings of the Rayleigh line can be calculated from the quantity of light passed at the different temperatures and the law for change in transmission with filter temperature. One must be careful in the use of this indirect method of the investigation of the wings of the Rayleigh line, inasmuch as the theory of the method has been insufficiently developed to date and the reasons which lead to the distortion of the intensity distribution in the wings, studied by this method, are unclear [229-231]. The validity of the application of the method of the resonance filter to the study of the fine structure is also open to dispute [232, 233]. These methodological questions remain unresolved at the present time.

The direct interferometric method of investigation of the spectrum of the wings of the Rayleigh line has been proposed and developed by Shustin [234] and the author of this book [53, 144]. The method is based on the well-known phenomenon of interference in white light. The interference apparatus for the study of the wings by this method differs from the interference apparatus designed for the study of the fine structure only by the fact that there is no limitation on the spectrum that emerges from the monochromator, and the Wollaston prism is replaced by a polaroid or Glan prism.

The interference spectroscope in our case was a Fabry–Perot etalon, joined with a spectrograph (monochromator) in such a way that its linear dispersion acts in two mutually perpendicular directions.

The portion of the sum under study is unrolled spatially by virtue of the resulting dispersion. It is not difficult to see that the dispersion of such a system is equal to

$$D = D_{\text{int}}\left(1 + \frac{1}{2}\frac{D^2_{\text{sp}}}{D^2_{\text{int}}}\right), \tag{12.1}$$

where D_{sp} and D_{int} are the linear dispersions of the spectrograph and interferometer, respectively.

If the width of the entrance slit of the monochromator is so chosen that the width of its projection in the focal plane of the camera objective L, expressed in frequency units, is smaller than the dispersion range of the Fabry–Perot interferometer, then the spectra of neighboring orders will not overlap. Thus, the whole range of the investigated spectrum will be represented in each order of interference.

In order to avoid the superposition of neighboring orders of interference in the apparatus used in [53, 144, 234], the width of the entrance slit of the monochromator (the ISP-51 spectrograph)

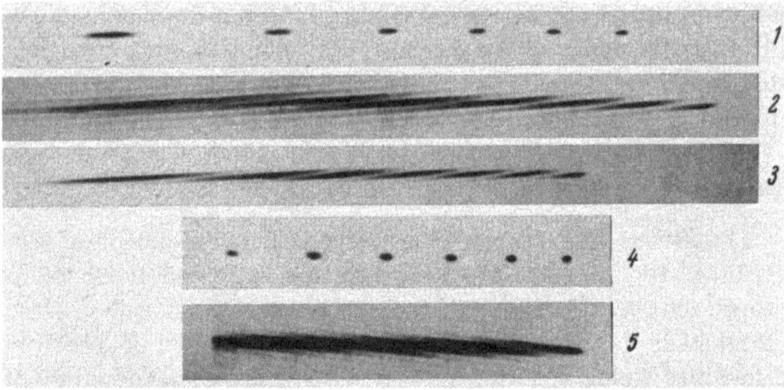

Fig. 34. Spectra obtained on an apparatus with a Fabry–Perot interferometer, joined with an ISP-51 spectrograph with a broad spectral interval, appreciably exceeding the range of dispersion of the interferometer (Fabelinskii [53, 144], Shustin [234]). 1,4 – exciting line of the mercury spectrum 4358 Å; 2 – spectrum of light scattered in carbon bisulfide; 3 – spectrum of light scattered in toluene; 5 – spectrum of light scattered in benzene.

was set at about 0.1 mm for a dispersion range of the etalon of 16.6 cm^{-1} (the width of the separating ring of the Fabry–Perot etalon was t = 0.3 mm). The range of the spectrum in each order for focal lengths of L = 270 mm, 120 mm, and 240 mm amounted to 0.2-0.3 mm (Fig. 34).

The dispersion of the Fabry–Perot etalon does not depend on the thickness t (see Appendix II) and appreciably exceeds the dispersion of the ordinary spectrograph; therefore, we can neglect $\frac{1}{2}(D_{sp}^2/D_{int}^2)$ in (12.1) in first approximation, and assume that the apparatus that has been described is characterized by a large constant dispersion, which is virtually equal to the dispersion of the interferometer. For the apparatus parameters that have been given, the linear dispersion in the region of 4358 Å varies from 0.2 to 1.3 Å/mm as a function of the order of the interference.

The theory of this method has been developed in detail in the work of Shustin [234]. Use of the apparatus has shown that it is suitable for quantitative measurements and that it has no unforeseen distortions.

This method permits the investigation of the wings of the Rayleigh line not only close to the unshifted line, but also over its entire extension. The method can naturally be used for the solution of other problems, which lie outside the framework of the questions associated with light scattering. However, it should be pointed out that the resolving power of the apparatus for an etalon thickness of 0.3-0.5 mm and for those reflection coefficients which we ordinarily use, amounts to 2500-30,000.

Further increase of the resolving power can be achieved by use of a monochromator with a larger angular dispersion than the ISP-51 (this allows us to increase the thickness of the etalon) and also by means of an increase in the reflection coefficient (use of a multilayered dielectric mirror).

In the study of the part of the wing close to the shifted line, significant difficulties arise which are connected with the high intensity of the Rayleigh triplet. By using the fact that both the central line and the MB components are completely polarized in low-viscosity liquids, the study of the wings is carried out on the x component of the scattered light. Even in this case, as a consequence of the finite aperture of the incident and scattered light, the superposition of a certain part of the z component on the x component is quite possible. Especially dangerous is the incidence of the z component in the region of the maximum of the wing.

Special experiments were carried out to make clear how important this effect is [30, 53]. The intensity was measured at the maximum relative to the intensity of the wing at a distance of 5 cm^{-1} from the maximum for different apertures of the exciting light (from 40 to 16° in the liquid). As the aperture angle is decreased, the relative intensity of the maximum falls off. In the range of angles ~16-15° the intensity ceases to fall off; consequently, the limitation of the aperture to 16° is sufficient for one not to have to take into account the insignificant fraction of the z component which falls on the x component.

The correction for the possible "incidence" of light of the z component on the x component can also be calculated from the dimensions of the limitation of the aperture and the law of change of the intensity as a function of the scattering angle. Far from the unshifted line, the effect of the components of the Rayleigh triplet is not seen. In this region, it is possible to carry out measurements with polarization apparatus, which greatly increases the optical power of the apparatus. Inasmuch as the intensity is negligibly small in the parts of the wing far from the undisplaced line, such an increase in the optical power is extremely desirable.

The duration of the exposure for the different parts of the wings and for different liquids is variable and lies within the limits 1-40 hours [53] for work with the photographic method. In the photoelectric variant of the method, the recording of the spectra on paper lasts from 30 min to 1.5 hours.

Apparatus and Methods for the Investigation of the Fine Structure of the Lines of Scattered Light. The change in the frequency for Rayleigh light scattering lies within the limits 0.05-3 cm^{-1} and, therefore, the study of this phenomenon requires the use of spectral apparatus of high resolving power.

Examples of such apparatus are interference spectroscopes with preliminary monochromators or diffraction lattices of high resolving power. In principle, any interference spectroscope is suitable for the study of the fine structure if only it is characterized by sufficient resolving power, and if the apparatus as a whole allows us to obtain photographs over an appropriate time interval. By an appropriate time interval is meant such an exposure during which the possible change in pressure and temperature does not distort the interference pictures (see Appendix II).

In the experimental researches that have been carried out, all the interference spectroscopes have been used which are generally employed in laboratory practice. Actually, in the first works of Landsberg and Mandel'shtam [131] a Lummer–Gehrcke plate was used; in the first work of Gross [27, 28], and later in the researches of Ramm [235] and Birus [236], the Michelson interferometer. In the investigations of Raman and his coworkers, both a Lummer–Gehrcke plate and a Fabry–Perot etalon were used, while in the later researches [53, 257, 479] a Fabry–Perot etalon was employed.

The variety of spectroscopes used in the research on fine structure is explained by the fact that there was no choice in the initial stages of the research, and each one used whatever instrument he had, and not that which he considered most suitable.

A comparison of the different spectroscopes, and the analysis of their operation [237, 238] and also the practical details of their use, show that, at the present time, the Fabry–Perot etalon is the most suitable for the study of fine structure, primarily because of its great relative aperture, comparative simplicity, well-identified apparatus functions, and its high, easily changed resolving power. The basic characteristics of the Fabry–Perot etalon are given in Appendix II.

Figure 33 shows the arrangement for a Fabry–Perot interferometer and an ISP-51 spectrograph as a preliminary monochromator. The illumination and interference parts of the instrument are located in different rooms. The illuminator of the scattering volume will be described in detail below. The objective L_1 with iris diaphragm focuses the exit diaphragm of the vessel D through the Wollaston prism W and the depolarizing wedge onto the entrance slit of the monochromator S_1. Two images of the diaphragm are obtained on the slit S_1, one above the other, one just touching the other, or else shifted slightly as in the apparatus for the study of the wings (Fig. 30). Thus, the upper and lower part of the monochromator slit are illuminated by scattered light with different polarizations. The exit slit S_2 of the monochromator lies at the focus of the objective L_4 which directs the parallel beam of light onto the interferometer FP. The objective L_5 also focuses the interference bands and the exit slit of the monochromator S_2 in the plane of the photoplate P. In such a system as Fig. 33, the image of the slit is speckled by the interference bands, half of which are due to the z component of the scattered light and the other half to the x component. Consequently, photographs of the fine structure are obtained simultaneously in such a setup in the two polarizations. A wedge of Iceland spar is attached to the Wollaston prism as in the apparatus of Fig. 30, guaranteeing identical conditions of light propagation in the optical apparatus. It is also necessary to track carefully in order that the collimator objective of the monochromator symmetrically divide the beams polarized along x and z; otherwise, the result will be highly distorted [211].

As an example, a reproduction of photographs of the fine structure of the Rayleigh line obtained on the apparatus shown for different liquids in Fig. 33 is reproduced in Fig. 35. The treatment of such photographs allows us to find the frequencies of the

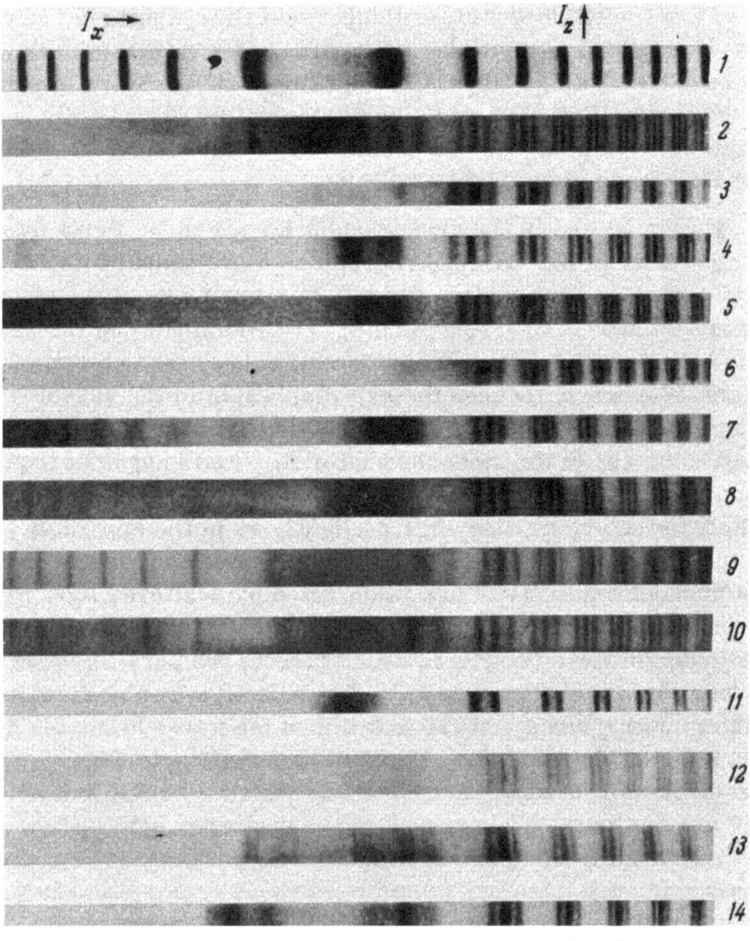

Fig. 35. Fine structure of the lines of Rayleigh scattering in liquids in the I_x and I_z components of polarized light. 1 – Exciting light λ = 4358 Å; 2 – carbon bisulfide (20°C); 3 – benzene (20°C); 4 – toluene (20°C); 5 – acetone (20°C); 6 – carbon tetrachloride (20°C); 7 – acetic acid (20°C); 8 – triacetin (72°C); 9 – glycerin (50°C); 10 – glycerin (150°C); 11 – water (20°C); 12 – methylene chloride (20°C); 13 – methylene bromide (24°C); 14 – chloroform (20°C).

components, the intensity distribution, and the depolarization factor.

Actually, as follows from the theory of the phenomenon (see Secs. 6 and 7) and experimental data (see below), the MB components and the central component in low-viscosity liquids are completely polarized; therefore, they are unweakened in the z component. To the component of the fine structure is added the intense and depolarized wing, which decreases the contrast of the picture and in some cases leads to the mixing of the position of the MB components and to a distortion of the intensity distribution in them. Use of a polarizer reduces the intensity of the wing in the z component almost to one half, and thus the conditions of measurement are significantly improved. In measurements of the frequencies of the MB components, appropriately oriented polaroids, Glan prisms, or Nicol prisms can be used as the polarizer.

The exposure varied from 10 min to 2 days for benzene, depending on the foci of the objectives L_4 and L_5. In the interference apparatus, a Fabry–Perot etalon was placed after the monochromator in the external parallel beam, if we use the terminology of Tolansky [239].

Of course, this arrangement of including an interferometer in the optical apparatus is not unique. The interferometer can also be located in the external parallel beam, but in front of the monochromator slit. The objective, placed between the interferometer and the monochromator, focuses the image of the interference band in the plane of the entrance slit of the monochromator in this case. Two positions of the interferometer in the internal parallel beam of rays are also possible [239]: (1) the interferometer is placed between the collimator objective and the dispersion system; (2) the interferometer is placed between the dispersion system and the camera objective.

An advantage of interferometer apparatus in the external parallel beam is that the apparatus is compact, not requiring additional objectives except those which are needed for operation of the monochromator. However, in the study of the fine structure, high requirements are also always necessary for the quality of the image of the components of the fine structure. From this viewpoint, the location of the interferometer in front of the monochromator or after the collimator objective of the monochromator is un-

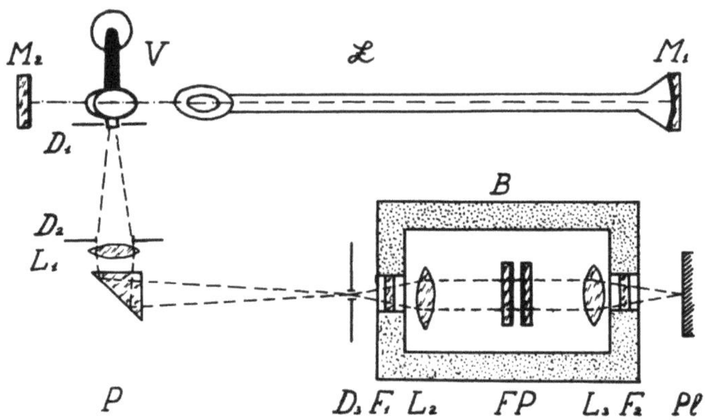

Fig. 36. Arrangement of interference apparatus for the study of the fine structure of the Rayleigh line with the use of a gas laser as the light source (Mash, Starunov, Tiganov, and Fabelinskii [246, 264]). $\mathscr{L}$ – gas laser; M_1 – spherical mirror; M_2 – plane mirror; V – test cell; D_1, D_2 – diaphragms; L_1, L_2, L_3 – objectives; F_1, F_2 – light filters; FP – Fabry–Perot interferometer; P – photoplate; B – isothermal and isobaric chamber in which the temperature did not vary more than 0.1°C over the time of the experiment, nor the pressure more than 1 mm Hg; R – reversing prism.

desirable, inasmuch as the image of the interference bands can be subject to notable distortion as a consequence of the imperfection of the dispersion system. Equivalent schemes of Fig. 33 and schemes in which the interferometer is located between the dispersion system and the camera objective of the monochromator can be shown. However, there will be a great deal of parasitic light scattered in the monochromator in the latter case. A detailed study of the different arrangements of the location of the interferometer, carried out by Sterin [241], has shown that the best image can be obtained with the location of the interferometer shown in Fig. 33 for the required choice of objectives L_4 and L_5. A discussion of some other advantages and disadvantages of the different systems of including the interferometer is given in [239].

A diffraction grating was used by Rank et al. for the study of the fine structure of the Rayleigh line in a liquid [243]. A spectrograph with a diffraction grating (focal length 12 m) was used by Benedek et al. [598]. The resolving power of their spectrograph was 900,000 and the linear dispersion 0.20209 ± 0.00014 $cm^{-1} \cdot mm^{-1}$. Recording of the spectrum was made by the photoelectric method. A Ne–He gas laser of 10-mW power and wavelength 6328 Å served as the light source.

For the study of the fine structure of the scattering line in fused quartz, Flubacher et al. [244] used an apparatus with a diffraction grating. The concave diffraction grating had a focal length of about 10.67 m (35 feet) and was based on the scheme of Igle [238, 244]. The linear dispersion amounted to 0.22 Å/mm (3.5 cm^{-1}–mm^{-1} in the region of 2500 Å) with a resolving power of about 300,000. Simpler and more effective is an apparatus with a Fabry–Perot interferometer, in which the light source was a laser.

Such a setup is shown in Fig. 36, in which a monochromator is not necessary and has been omitted. The light filters, placed at the entrance and exit apertures of an isothermal chamber, are more necessary for thermal insulation than for eliminating the light of spontaneous radiation. In the laser this latter is so much weaker than the light of stimulated emission that it can be neglected.

In the work of Mash, Starunov, and Fabelinskii [246], the stimulated emission of the $\lambda \approx$ 6328 Å line in a Ne–He gas laser was used. The radiated power of this line varied from 3 to 20 mW.

A special vessel [634] was placed inside the resonator of Fig. 36. In this case, the light power of the apparatus was so great that fine structure lines are seen by the eye (the focal lengths of the objectives L_2 and L_3 were respectively equal to 20 and 60 cm).

In the apparatus with the Fabry–Perot etalon, in addition to the photographic method of recording the fine-structure spectrum, a photoelectric method has also been used with success [237, 635, 602, 636, 637].

In the latter case, the interferometer is placed in a hermetically sealed chamber, such as B in Fig. 36, but is equipped with apparatus for evacuating and pressurizing the chamber with gas at the necessary rate.

As follows from (A.1), the change of the wavelength in the center of the interference picture is connected with a change in the index of refraction between the mirrors of the interferometer by the relation $\Delta\lambda = (2t/m)\Delta m$.

In a gas, the change in the index of refraction is directly proportional to the pressure change; therefore, the change in the wavelength in the center is proportional to the pressure change.

If, in place of Pl (Fig. 36), a small diaphragm is placed at the center of the interference picture, and behind it is placed a photomultiplier with amplifying and recording system, then a change in the gas pressure in the chamber allows us to record a change in the photocurrent in correspondence with the distribution of the intensity in the spectrum investigated.

One can select a gas with a suitable index of refraction and necessary rate of input leakage in order to ensure the scanning of the investigated spectral interval over a definite period of time. One can easily see that the finite dimension of the diaphragm contributes an additional apparatus halfwidth [see Eq. (A.7)], $\delta\nu'_a \approx (\nu/2)(r_0^2/f^2)$, where r_0 is the radius of the diaphragm and f is the focal distance of the camera objective L_3 (Fig. 36).

Hanes et al. [638] considerably improved this method by using a phase-sensitive detector to eliminate the fluctuations of intensity of the laser over the time of recording of the spectrum.

For the study of the width of the central line of the fine structure under different conditions, Cummins et al. [593], and Alpert et al. [639] used the method of light heterodyning. Ford and Benedek [640] developed a new method, based on light beats which arise on the quadratic photodetector between the different spectral components lying inside the line.

The resolving powers of both methods lie far beyond the limits of possibility of the usual optical methods and achieve values of $5 \cdot 10^{13}$ to 10^{14}.

Apparatus for Temperature Experiments and the Illuminator. Figure 37 shows a diagram of apparatus for the measurement of the effect of temperature on the intensity

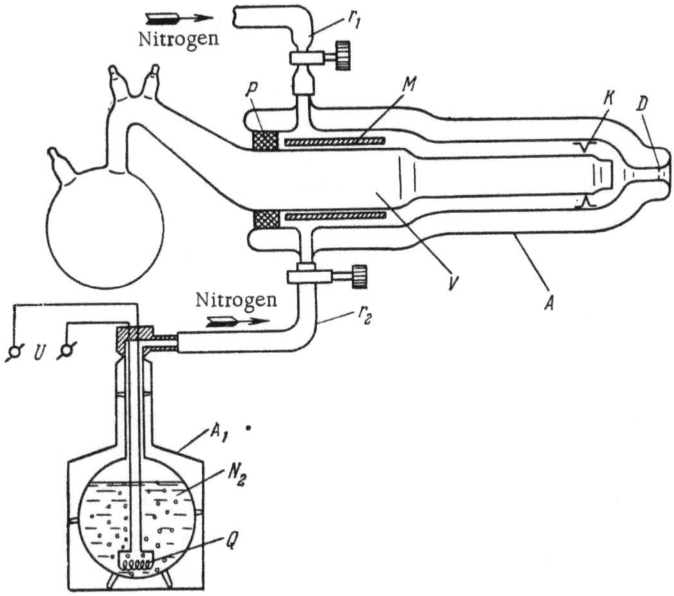

Fig. 37. Arrangement for the cooling of the scattering liquid (Fabelinskii [53]). A – glass Dewar; N_2 – liquid nitrogen; Q – heater; A_1 – metallic Dewar; r_1, r_2 –tubes with stopcocks; V – test cell; P – Dewar condenser; M – metallic ring; K – support for front part of vessel; U – pressure control voltage; D – aperture.

and spectrum of the scattered light. The vessel with the scattering medium V is placed inside a specially made Dewar A. Air is evacuated between the walls of the Dewar. Cold or hot air can be circulated through the tubes r_1 and r_2 of the Dewar. The temperature of the vessel V is controlled by two thermocouples (not shown in the diagram). A metallic ring M protects the vessel V from a direct flow of hot or cold air on it. For heating experiments, hot air which was first passed through a variable power heater is blown through the Dewar (not shown in Fig. 37).

For a definite current and power of the heater, the variation of the temperature of the liquid in V did not exceed 0.5-1.0°C during the time of exposure. In such a way, the temperature of the liquid could easily be changed from room temperature up to 300°C.

For experiments in the Dewar A at a lower temperature, cold air was circulated. The method of introducing the cold air is shown in the lower part of Fig. 37 (see also Fig. 38). A heating coil, the heating current of which was controlled by an LATR transformer, was placed at the bottom of a large metallic Dewar A_1 filled with liquid nitrogen or liquid air. A strong flow of nitrogen enters the Dewar A. By a change in the heating current of the coil, the strength of the flow of cold air could be regulated and, hence, the temperature of the vessel V also. Such an arrangement made it possible to change the temperature in a vessel of capacity 350-400 cm^3 from room to −150°C and the system could be operated with the temperature kept constant to within 0.5-1°C. In maintenance of the temperature of a 400-cm^3 vessel at −70°C for 15-20 hr, the expenditure of 15-20 liters of liquid nitrogen was required.

In optical investigations at low temperatures, a difficulty always arises in which water droplets from the atmosphere condense at the cold exit window (in our case, the exit window of the vessel V) and the scattered light in practice is not incident in required direction. There are different ways of eliminating this difficulty [245]. In the case of the apparatus of Fig. 37, we proceed in the following fashion: the Dewar A is filled from the horn-shaped part of the vessel V. Consequently, the nitrogen admitted into it can go only forward, where a nozzle is located with an opening D of diameter 5-10 mm. At the opening, there is thus an appreciable excess pressure, which prevents the appearance of moisture from the atmosphere in the vessel V. Consequently, work can be carried on in such a system until there is a flow of cold and dry air or nitrogen. Therefore, there are two tubes (r_1 and r_2) in the Dewar, through which one can force air or nitrogen either immediately from the two Dewars A_1 or, after the liquid air in one Dewar reaches the outlet, by joining the other to the tube r_2, without interrupting the work.

Various lighting arrangements have been used in the investigations on scattered light [43, 53, 66, 75]. Figure 38 shows a photograph of the lighting arrangement with which a large part of the investigations of the author on the fine structure of the Rayleigh line was carried out (see also Fig. 33). The light from the gas discharge source was guided through the stopping aperture and a cylinder lens to the inside of the test cell. Glass tubes were used as

Fig. 38. Photograph of the illuminating part of the apparatus with two illuminating lamps without a reflector (Fabelinskii [53]). 1 – Illuminator; 2 – condensor; 3 – Wollaston prism; 4 – ICP-51 spectrograph; 5 – barograph; 6 – metallic Dewar.

the cylindrical lenses. Flowing water was circulated inside the lenses. Thus, the cylindrical lenses simultaneously protected the liquid in the vessel from heating.

The relative locations of the light sources Q_1 and Q_2, the cylindrical lenses C_1 and C_2, and the vessel could be varied. Figure 39a shows such arrangements when the line from the sources is focused in the scattering vessel. The angular opening can easily be computed beforehand.

Figure 39b shows such an arrangement of the lenses, sources, and cell when practically parallel beams of light are introduced into the cell. Figures 39a and 39b represent the arrangement in vertical cross section. In the plane, the cylindrical lenses form a flat plate and, therefore, the stopping of the aperture takes place parallel to the fixed plates. In the measurement of the depolarization factor, it is necessary to see to it that the angular aperture, given by the stops, is approximately equal to the angle given by the cylindrical lens C and the vessel V. The correction for a finite aperture can be made on the basis of Eq. (10.4), which

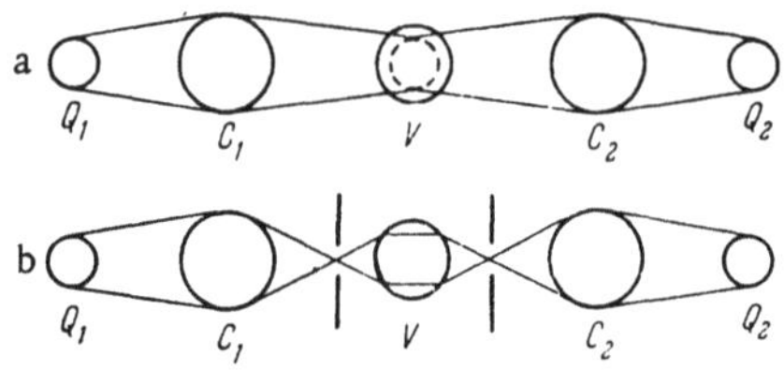

Fig. 39. Arrangements of the positions of the cylindrical lenses and the test cell shown in Fig. 33. Q_1, Q_2 – gas discharge lamps; C_1, C_2 – cylindrical lenses; V – sample cell.

applies to illumination by a symmetric cone of rays and the formulas obtained in the researches [27, 36], which take into account the correction for plane parallel stops.

Figure 40a shows a scheme of illumination developed by Mikhailov [247] for the study of the spectra of scattering in gases at pressures up to 450 atm. The lighting apparatus which he used makes it possible to take the work up to pressures of 1500 atm. The light-source housing, together with the sample cell, was made of chrome steel alloy and consisted of three steel hollow cones, pressed against one another to create the preliminary pressure of the metal of the housing.

In the arrangement of Fig. 40b, three light sources were used simultaneously, located at angles of 120° with respect to one another. Such lighting arrangements were used successfully for the study of the Raman scattering and can also be used for the study of the wing of the Rayleigh line (not the polarization studies). For the study of the fine structure or polarization measurements, it is necessary to limit oneself to only one lamp or to change the scheme of lamp locations or, finally, to put a special aperture stop in the scattering volume, similar to that used for illumination with an elliptical reflector [53].

A four-mirror instrument was placed inside the scattering volume of this apparatus, making it possible to use the scattered light more effectively [248].

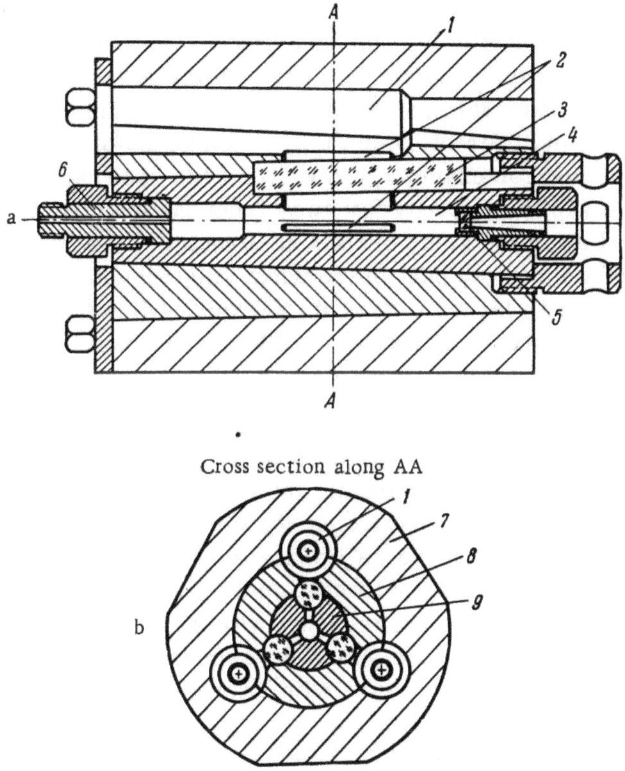

Fig. 40. Illumination scheme for the study of the spectrum of scattered light in gases at pressures up to 450 atm (Mikhailov [247]). 1 – Housing for the light; 2 – windows for illuminator; 3 – cylindrical lenses; 4 – scattering volume; 5 – exit window; 6 – closing structure for gas feed; 7, 8, 9 – external, central, and inner framework.

Light Sources. For the light source used in the excitation of the scattering spectrum, two rather opposing requirements are necessary, as has already been noted: the highest possible intensity and the sharpest possible line of the exciting light.

It is well known that the intensity growth occurring either because of the increase in the density of the vapors in the gas discharge lamp, or because of the increase in the strength of the discharge current, inevitably leads to a growth in the line width [238, 239]. Consequently, the problem of the choice of the source is a question of a reasonable compromise between the line width and its intensity. In different cases, this problem is solved in different ways. Thus, for example, the width of the exciting line for the study of the wing, especially far away from the maximum, is not as essential as in the study of the fine structure. Therefore, in the study of the remote part of the wing, one can use more intense light sources with larger widths of the exciting line than in the study of fine structure or the parts of the wing close to its maximum.

In the first experiments of Gross on the study of the fine structure, a mercury lamp was used at low pressure. The discharge took place in a cylindrical tube half immersed in flowing water. To decrease the absorption of radiation, the gas discharge was compressed against the wall of the tube by a magnetic field. A mercury lamp with internal water cooling was used by Komarov. Several variants of the mercury lamp with slight changes were also put forward by the author [53]. The successful construction of a mercury lamp of low pressure with internal water cooling was achieved by Sosinskii [251].

The reproduction of interference photographs of some lines of the visible part of the mercury spectrum (discharge at low pressure) is shown in Fig. 41. All the lines which are suitable in intensity for excitation of scattered light are accompanied by a significant amount of hyperfine structure components which naturally makes their application complicated. However, for very intense visible lines of the mercury spectrum $\lambda = 4358$ Å, the hyperfine structure components are much weaker than the fundamental line [252]. Approximately the same thing can be said for the line 4078 Å. But the green line $\lambda = 4358$ Å of the mercury spectrum is more intense than the violet 4078 Å and, therefore, we have preferred to study the fine structure with the green mercury line [53], although in other researches almost all the other lines of the mercury spectrum have been used. A much weaker hyperfine structure with low line intensity is observed in the visible triplet of the zinc spectrum (Fig. 41), although any of the lines of this triplet is less intense than the line $\lambda = 4358$ Å of the mercury spectrum.

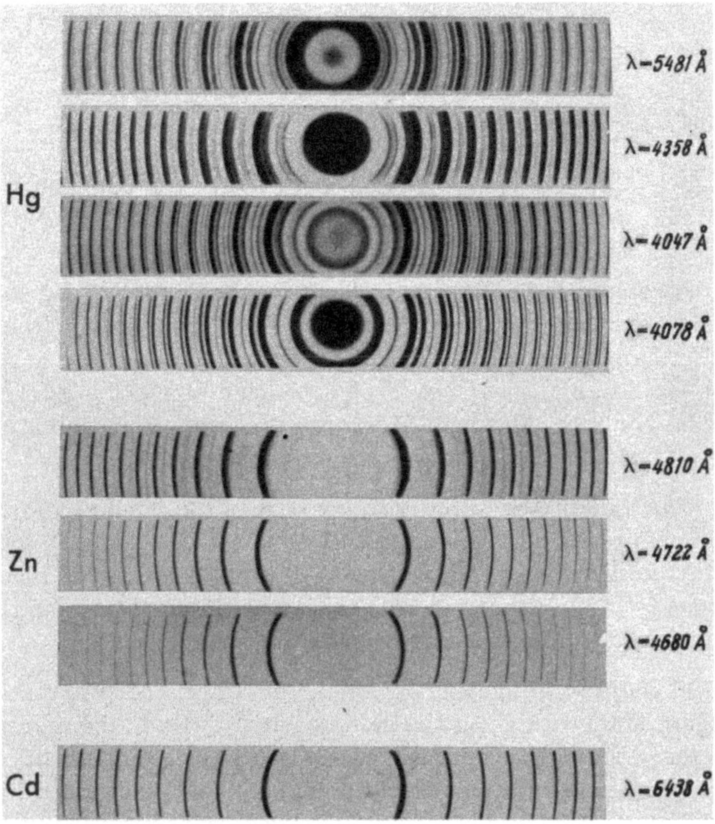

Fig. 41. Interferograms of some lines of the spectra of mercury, zinc, and cadmium, obtained on the apparatus shown in Fig. 36 (Fabelinskii).

The lines of the visible triplet of zinc have already been used in some studies for the excitation of the fine structure of the scattering line. E. H. Meyer and Ramm [253] constructed a special gas discharge zinc quartz lamp. The corrosiveness of zinc relative to glass and quartz did not permit them to use this zinc source more than once. The zinc settling on the walls of the tube after the lamp was turned on destroyed the walls.

More long-lived and, therefore, more practical, were tubes operating with an amalgam. Such light sources have long been known [254], but for the study of scattered light, lamps with mercury–zinc amalgam were first used by Venkateswaren [172, 257] and Sunanda Bai [258].

The amalgam used in these lamps contained 12-15% Zn by weight. The discharge current amounted to 2 A. The lamps had a curved shape. Ablekov, Zaitsev, and Pesin [255] developed a relatively simple technology for the preparation of amalgam light sources generally.

Amalgam light sources of low pressure with Zn–Hg and Cd–Hg were constructed by them. The content of zinc in the Zn–Hg lamp amounted to 10-20% by weight. The operating part of the lamp was a linear cut of pipe. Liquid electrodes of the lamp could be cooled by flowing water or by air blast. The cooling part of the electrodes of the lamp was the same as in [251].

For a diameter of the working part of the lamp of 20 mm, the discharge current varied from 6 to 25 A. By changing the temperature of the water cooling the condenser, it is easy to change the density of the vapors in the discharge; the intensity and the spectral line widths are also changed in this case.

In addition to the Zn–Hg amalgam lamps, there is a Cd–Hg amalgam lamp with about the same content of Cd in the mercury as in the zinc case; it operates under about the same conditions [255]. For the excitation of scattering, the well-known red line of the cadmium spectrum $\lambda = 6438$ Å is used (this line was chosen by Michelson as a standard of length [256]).* This line is the sharpest of the lines shown in Fig. 41. The intensities of the lines of its hyperfine structure are so low that they are practically invisible in the exposition time for the scattered light; therefore, we have used this light source for the study of the fine structure [36]. The lines of the zinc spectrum that have been mentioned and, in particular, the red line of cadmium, are much sharper than the lines of the mercury spectrum used in the study of fine structure;

*Another line is now used as the international standard of length, namely the line in the spectrum of krypton ^{86}Kr corresponding to the transition between the $2p_{10}$ and $5d_5$ levels [260]. It can be expected that the best standard of length in the future will be the wavelength of radiation in a gas laser [261].

they are, therefore, preferable, but their intensity in the amalgam is much less than the intensity of the mercury line (4358 Å) and the intensities of the zinc and cadmium lines in the amalgam source fall off with time. This great weakness of amalgam sources of light led to a search for another way. The solution was found when Zaitsev, Starunov, and Fabelinskii [263] constructed a powerful cadmium discharge tube with liquid electrodes.*

The difficulty in the construction of such a tube lies, first of all, in the fact that the melting point of cadmium is 320.9°C and, therefore, a temperature no lower than this is needed in order that discharge take place in the tube with liquid electrodes. Therefore, the working part of the tube consisted of a double tube (Fig. 42). The space between the tubes was evacuated to a high vacuum and sealed off. Cadmium was loaded into the branches for the electrodes according to the method described in [255]. Neon or argon under a pressure of 5-6 mm Hg was introduced into the tube. Such a light source can be used in any arrangement in the same way as was the mercury low-pressure tube. Upon connection of the tube in a constant current circuit of 110 V, discharge begins in the inert gas, which heats the entire tube to the melting temperature of cadmium, after which discharge begins in the cadmium vapor. In order to avoid contraction of the discharge (too much line width but also large intensity), the electrodes were cooled by running water.

The vacuum jacket on the working part of the tube guaranteed a sufficiently poor thermal conductivity, so that the tube was ready for operation within 5-7 min after connection.

A comparison of the intensity of emission of the cadmium line $\lambda = 6438$ Å in the tube shown in Fig. 42 with the intensity of emission of the line $\lambda = 4358$ Å of the mercury spectrum in low-pressure tubes [251] with the help of a neutral bolometer showed that when a discharge current of 5 A passed through both tubes, the spectral line $\lambda = 6438$ Å of cadmium was about twice as in-

*The discharge in pure cadmium usually takes place in low power tubes with an oxide cathode. Such tubes are small in their dimensions and power output and, therefore, they are not suitable for the study of light scattering.

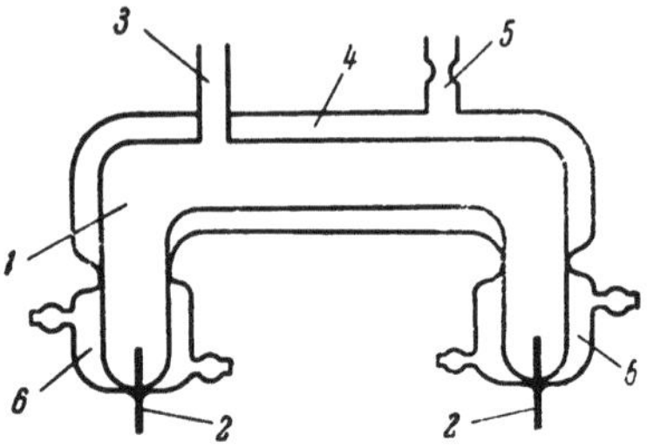

Fig. 42. Diagram of a cadmium tube with liquid electrodes (Zaitsev, Starunov, and Fabelinskii [263]). 1 – Discharge space (inner tube); 2 – electrodes; 3 – port for filling and emptying tube; 4 – vacuum space between inner discharge tube and outer tube; 5 – port for evacuation of vacuum jacket; 6 – refrigerator.

tense as the λ = 4358 Å line of mercury. For a discharge current of 10 A, both these lines had the same intensity. The intensity of the 6438 Å line in the lamp of Fig. 42 was about an order of magnitude higher than for the amalgam tube.

In the work of Flubacher et al. [244], a tube was employed which works on the pure ^{198}Hg isotope. The construction of the tube was the same as for the tube used in the study of Raman light scattering; about 2 mg of ^{198}Hg was used in the tube in place of the ordinary mercury. The halfwidth of the 2537 Å line was so small that the visual shape of it was determined principally by the apparatus halfwidth of the spectrograph. For an even better source of excitation in the study of the fine structure, it is necessary to use the laser, which is characterized by a very small beam opening and a limitingly small halfwidth of the emission line. With such a source, a series of qualitatively new experiments is possible. In our first experiments, we measured the damping of the thermal wave by means of the halfwidth of the MB components. The accuracy of the measurement of the hypersonic velocity is an order of magnitude greater. Measurements were carried out on the intensity of the fine structure components [246, 264]. Gas lasers were used in [593, 594, 598, 602] and other researches on light scattering.

Treatment of the Results of Spectral Measurements. In the spectrum of the fine structure line of scattering, the frequencies (locations of the maxima) are measured of the discrete components as well as the intensity distribution over the frequencies. In the spectrum of the wing of the Rayleigh line, the distribution of the intensity is measured over the frequencies. Of course, the depolarization factor can be measured in both spectra. The method of obtaining the spectra in two different states of polarization has been described earlier, while the obtaining of the value of the depolarization at any point of the spectrum reduces to a measurement of the intensities at the corresponding points in the two polarizations of the divided spectra.

For the measurement of the frequencies of the fine-structure components, it suffices to determine the location of the maxima of the components by a microphotograph or to measure the location of the components by a comparator and, using one of the methods of interpreting the interference spectra described in Appendix II, to translate the linear distances between the components into frequency or wavelength. After the corresponding values have been obtained for $\Delta\nu$ – the difference in wave number between the unshifted frequency (the central line) and the wave number of any component – it is easy to find the hypersonic velocity. Actually, we get from Eqs. (5.9) and (1.26) for the hypersonic velocity

$$v = c\Lambda\Delta\nu \frac{\mathrm{m}}{\mathrm{sec}}, \tag{12.2}$$

where c is the velocity of light in m/sec, Λ is the wavelength of the hypersound, determined from (1.27) in cm, and $\Delta\nu$ is in cm^{-1}.

The relative error of measurement of v is obtained from the relative errors in measurement of the quantities entering into (12.2). Taking (1.27) into account, we find that the relative error in the determination of the hypersonic velocity is

$$\frac{\Delta v}{v} = \frac{\Delta c}{c} + \frac{\Delta(\Delta\nu)}{\Delta\nu} + \frac{\Delta\nu}{\nu} + \frac{\Delta n}{n} + \frac{1}{2}\frac{\Delta\theta}{\mathrm{tg}\,\frac{\theta}{2}}. \tag{12.3}$$

The velocity of light c, the wave number of the exciting light ν, and the index of refraction n are known with high accuracy; therefore, one can discard the terms corresponding to them in (12.3). Then there are only two terms left in (12.3) – the relative errors in the determination of $\Delta\nu$ and the scattering angle. The finding of the intensity in the spectrum of the fine structure is generally a complicated problem. The complication here is generally produced by the fact that the real spectral apparatus always distorts the true intensity distribution.

In the case of scattered light, the location is additionally complicated by the fact that the exciting line is not an infinitely sharp spectral line, but is characterized by a finite halfwidth and by a certain shape.

We first assume that the scattered light is excited by monochromatic radiation. Then, what we see on the photographic plate or by photoelectrical measurement is, as it is said, the "visible" contour of the intensity distribution in the spectrum. Let the visible contour be described by the function $w'(\nu)$, the contour of the apparatus function $a(\nu)$, and the real intensity distribution in any Mandel'shtam–Brillouin component or in any other part of the scattering spectrum by $I(\nu)$.

The spectral apparatus converts any infinitely sharp band chosen within the contour $I(\nu)$ into the contour $a(\nu)$. Therefore, the visible contour is connected with the apparatus function and the real distribution by the well-known integral equation (see for example [267])

$$w'(\nu) = \int_{-\infty}^{\infty} I(x)\, a(\nu - x)\, dx. \qquad (12.4)$$

The integral equation (12.4) is valid, as has been pointed out, under the condition that the excitation takes place by monochromatic light. But actually this is not so. The contour of the exciting line can also be characterized by a certain function $B(\nu)$. We guide the flux of the exciting light into the Fabry–Perot interferometer. The visible contour which will now be obtained is expressed, similar to (12.4), by

$$A(\nu) = \int_{-\infty}^{\infty} B(x)\, a(\nu - x)\, dx. \qquad (12.5)$$

If B(x) is a δ function, i.e., the exciting light is monochromatic, then the visible contour is identical with the apparatus function. On the other hand, if $a(\nu)$ has the form of a δ function (ideal apparatus) then the visible contour $A(\nu)$ is identical with $B(\nu)$. But in the real case, neither $B(\nu)$ nor $a(\nu)$ is a δ function. Therefore, $A(\nu)$ will be some function, generally speaking, different from $B(\nu)$ and $a(\nu)$ which, acting on any infinitely sharp portion of $I(\nu)$ converts it into a curve of the form $A(\nu)$, and we must write

$$w(\nu)=\int_{-\infty}^{\infty} I(x)\,A(\nu-x)\,dx. \tag{12.6}$$

Now the problem reduces to the solution of Eq. (12.6) relative to $I(\nu)$, with account of (12.5). Exactly the same problem arises in the study of the shape of the lines of Raman light scattering [214]. The solution of integral equations of the form (12.4)-(12.6) relative to the kernel for arbitrary functions appearing under the integral is a rather complicated problem which, however, has been solved both approximately and exactly [267, 269]. By solving this problem exactly, Sushchinskii [270] showed that for finding $I(\nu)$, it suffices to know the visible contour of the studied line $w(\nu)$ and the visible contour of the exciting line $A(\nu)$, which plays the role of an effective apparatus function. This conclusion simplifies the determination of the true distribution for any spectral apparatus.

When both integrands are either Gaussian functions or functions corresponding to the dispersion (Lorentzian) distribution of the intensity, the integral equations of the type (12.4)-(12.6) are solved relatively simply [267].

Here we are interested in the simplest case, for which the integrands are Lorentzian. This case relates directly to the intensity distribution in the central line and in the MB components of the fine structure (5.19), (5.36), and (5.37). The apparatus function for the Fabry–Perot interferometer has a Lorentzian character for sufficient resolving power [240] [see Eq. (II.14) of Appendix II]. The visible contour of the exciting line $A(\nu)$ (Fig.43) is also close to Lorentzian in character. Inasmuch as it is known that, if the visible distribution has a Lorentzian character, one of the integrand functions is dispersive, then this means that the

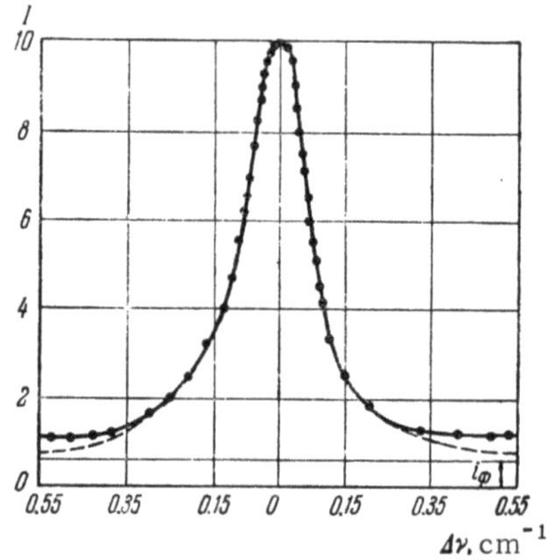

Fig. 43. Contour of the line of the exciting light λ = 4358 Å of the discharge of mercury at low pressure (Fabelinskii [53]). i_Φ — Background intensity; the "apparatus function" is taken into consideration in the dashed curve and the superposition of neighboring orders.

other integrand function, in our case I(ν), is also a Lorentzian function.

If the function I(ν) has a halfwidth $\delta\nu_I$, and A(ν) has the halfwidth $\delta\nu_A$, then, in the visible distribution,

$$\delta\nu_w = \delta\nu_A + \delta\nu_I.$$

Therefore, the halfwidth of the Lorentzian function I(ν) is equal to

$$\delta\nu_I = \delta\nu_w - \delta\nu_A. \qquad (12.7)$$

By taking into account the Lorentzian character of A(ν) and $a(\nu)$, we can draw the conclusion, as in the previous case, that B(ν) is a Lorentzian function, and

$$\delta\nu_A = \delta\nu_B + \delta\nu_a. \qquad (12.8)$$

Consequently, we get the halfwidth of the true distribution from (12.7) and (12.8)

$$\delta\nu_I = \delta\nu_w - \delta\nu_A = \delta\nu_w - (\delta\nu_B + \delta\nu_a). \qquad (12.9)$$

By $\delta\nu_I$ we understand both the real width of the MB components $\delta\nu_{MB}$ and the halfwidth of the central component $\delta\nu_c$ as well as the halfwidth of the central line produced by fluctuations of the concentration $\delta\nu_{conc}$.

Determination of the Ratio of the Integrated and Maximum Intensities in the Fine Structure Components. The discrete components of the fine structure of the scattered line in liquids are resolved against a more or less intense background (Fig. 44).

The intensity of the background is due to several factors: the basic part of the intensity of the continuous background is due to the wing of the Rayleigh line. In some liquids, the background is very intense; in others it is quite weak corresponding to the intensity of the wing of the Rayleigh line in these liquids.

In addition to the wing of the Rayleigh line, the scattering of the continuous spectrum of the light source and the nonvanishing value of I_{min} in the apparatus function of the etalon have an effect on the increase in the "background." (See Appendix II.) In the determination of the ratio of the integrated or maximum values of the intensity of the fine-structure components, a difficulty arises in obtaining the "background line," from which one can calculate the intensity. This question is important, inasmuch as the intensities that we found will depend significantly on the location of the "background" line. In order to eliminate any arbitrariness in drawing the "background line," we proposed the following method [29, 53], which we demonstrate by the example of the treatment of the results for benzene. Photographs are made simultaneously of the fine structure in the x and z components of the polarization according to the method described above (Fig. 33). We shall denote the total intensity of the scattering in the polarized components by $\mathcal{I}_x$ and $\mathcal{I}_z$ and the corresponding components of the anisotropic scattering by i_x and i_z.

The treatment of the photograph of the fine structure in the x and z components yields the necessary material for avoiding arbitrariness in drawing the "background line." The intensity distribution in the exciting line $\lambda = 4358$ Å of the mercury spectrum is shown in Fig. 43, and one order of interference is shown in the drawing (region of dispersion $\langle\Delta\nu\rangle = 1$ cm^{-1}). The intensity at a distance of 0.5 cm^{-1} from the center of the exciting line

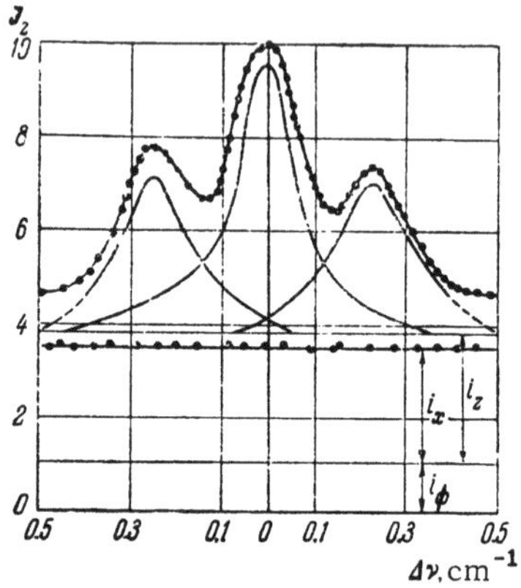

Fig. 44. Observed and interpreted contour of the fine structure of the Rayleigh line in benzene (Fabelinskii [53]). i_Φ – Background intensity; i_x, i_z – intensities of the components of anisotropic scattering.

is determined by the following two factors: (1) by the scattering of the continuous background of the light source, and, (2) by the minimum of the apparatus function.

The separate determination of the background intensity and the "apparatus intensity" is completed in the following fashion: in the same apparatus, pictures are taken of the exciting line and its halfwidth $\delta\nu_A$ is determined. By using the fact that the contour of this line in its best approximation is described by the Lorentzian distribution

$$\mathcal{I}(\Delta\nu) = \frac{\mathcal{I}(0)}{1+\left(\frac{\Delta\nu}{\delta\nu_A}\right)^2},$$

we find what the intensity will be at the point $\Delta\nu = 0.5$ cm^{-1} for the given halfwidth $\delta\nu_A$, while $\mathcal{I}(0)$ corresponds to the intensity at the center of the exciting line. By determining the part attributable to the apparatus function, and assuming that the background of the light source is uniform at a distance of 1 cm^{-1} (within the limits of error of the experiment this is unquestionably so), we determine (by means of a planimeter) what part of the area (in percent) is attributable to the fraction of scattering of the source background.

It is clear that the same percent of the intensity of these components lies in the fraction of background in the $\mathcal{I}_x$ and $\mathcal{I}_z$ components as in the excitation line, inasmuch as the line and the background of the exciting light are unpolarized. Planimeter measurement of the intensity distribution curves for the $\mathcal{I}_x$ and $\mathcal{I}_z$ components makes it possible to note this area in the $\mathcal{I}_x$ and $\mathcal{I}_z$ components in Fig. 44, which belongs to the background fraction. The corresponding ordinate in Fig. 44 is denoted by i_Φ. Furthermore, the intensity i_x of the anisotropic scattering is measured in the $\mathcal{I}_x$ component. The determination of i_z by the minimum of the intensity in the $\mathcal{I}_z$ component is inconvenient in practice, since this minimum depends not only on i_Φ and i_z, but also on the superposition of the "tails" of the displaced components of the neighboring orders and of the "apparatus intensity" of the same orders; therefore, the determination of i_z is more accurately and simply obtained in the following way: from an independent experiment, carried out by us for benzene and toluene, it is known that the depolarization factor of the anisotropic scattering close to the triplet is equal to $^6/_7$. Therefore, multiplication by $^7/_6$ gives the corresponding ordinate for i_z. Inasmuch as $\mathcal{I}_x$ and $\mathcal{I}_z$ are plotted in the same arbitrary intensity units, one can simply add $^7/_6 i_x$ to i_Φ in $\mathcal{I}_z$ in order to obtain the ordinate determining the anisotropic scattering and the scattering brought about by the continuous background of the light source. The "excess" intensity in the region between the orders in the z component is the result of the superposition of the apparatus functions of the neighboring orders. Dividing the ordinate at the point corresponding to the midpoint between the orders in the z component, we find the part of the "apparatus intensity" belonging to the order under study. From what has been said above, it is seen that now all the necessary data are at hand, found experimentally, in order to determine unambiguously the origin from which to measure the intensity in the fine-structure components.

The next stage of the analysis consists in the determination of the integrated intensity of each of the three components of the fine structure. Inasmuch as these components are close to each other ($\Delta\nu$ = 0.24 cm^{-1} in benzene) while the exciting line of the mercury spectrum λ = 4358 A possesses a finite width ($\delta\nu$ = 0.17 cm^{-1}), one is always observing some superposition. The problem of the interpretation of the complicated contour of spectral lines when the components making up the contour are clearly resolved can be solved uniquely [210].

After the procedure of plotting the "background line" given above, the left side of the contour of the left satellite and the right part of the contour of the right satellite are shown to be (in the first or zeroth approximations) undistorted superpositions of the neighboring lines. Therefore, one could reflect the left and right parts of the contours symmetrically relative to $\mathcal{I}$ (0); however, this produces some error, inasmuch as the exciting line is somewhat asymmetric. As is clearly seen in Fig. 43, the asymmetry is not large, but in our photographs it is clearly evident and must therefore be taken into account. In particular, the mentioned asymmetries of the exciting lines are determined by the somewhat different values of the maxima of the displaced components.

Account of the asymmetry of the exciting line in the interpretation of a complicated contour is achieved in our method inthe following way: to the right of the line of the maximum for the left component, the right part of the right satellite is constructed in the corresponding scale, and conversely. After such a construction, it is seen what part of the "tails" of the lateral components is superimposed on the maximum of the central line. The systematic operation is as follows: the contours of the lateral components are "subtracted" from the general contour. The contour of the central component is obtained in this way. As a result of this operation, all three contours are obtained in first approximation. In such an operation, we have not considered the parts of the intensity of the central component which are superimposed on the laterals and which exaggerate their intensity.

Therefore, the central component is now "subtracted" from the total contour in first approximation, as the result of which the lateral components are obtained in second approximation; these are again subtracted from the total contour, after which the contour of the central component is obtained in second approximation, etc. Figure 44 shows the final contours. After two or three such operations, the contours no longer change and we consider the result as final. Figure 44 shows the final contours drawn with the fine lines. The determination of the intensities of the components is completed by measurement of their areas with a planimeter.

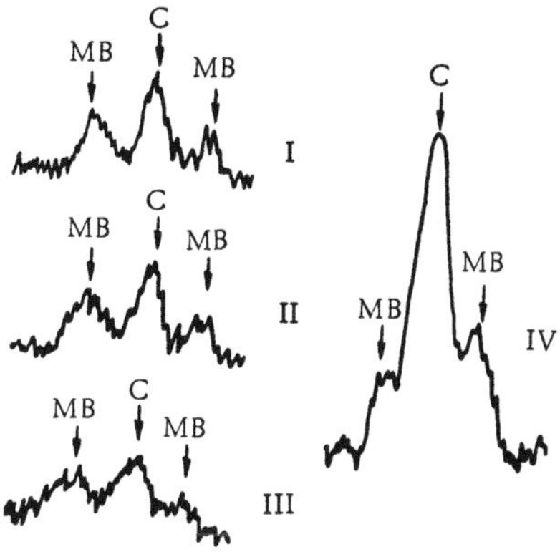

Fig. 45. Microphotographs of the fine structure of pure benzene (I), mixtures of benzene and toluene (II), toluene (III), and a badly purified benzene (IV) (Sterin [241, 444]). MB – Mandel'shtam–Brillouin components; C – central component.

Figure 44 shows the result of this treatment for the fine structure of benzene. The molecule of benzene contains a significant optical anisotropy and, therefore, a wing of appreciable intensity (²⁄₃ of the total intensity of the scattered light) is contained in the scattered spectrum. The Rayleigh triplet in benzene is therefore raised above the intense spectrum of the "anisotropic" scattered light. If polarization resolution is not applied in the photographs of the fine structure, then the background intensity increases by a factor of almost two and the possibility of uniquely determining the zero point of the intensity scale in the components of the triplet is lost. It is not difficult to see in Fig. 44 that various intensity ratios of the components of the triplet can be obtained by locating the zero point in different places.

In some liquids with less anisotropic molecules than in benzene, there is a less intense "background," against which the Rayleigh triplet stands out; the determination of the intensities of the fine-structure components is therefore less arbitrary. In this sense the data of Rank [479] for liquids with isotropic molecules should be closer to the true values than the data of other authors for liquids with anisotropic molecules.

The measurement of the integrated intensities can be used directly in the determination of the relation of the intensities in the triplet. Measurement of the maximal intensities is generally insufficient to compare theory with experiment. One must also know the widths of the shifted and unshifted lines (see Sec. 5). However, if we consider a real case, for example benzene, then one can reach the conclusion that the calculation of the ratio of the intensities by the integrated and maximal values differ slightly from one another. Actually, if the width of the exciting line $\delta\nu_A = 0.17$ cm^{-1} and the width of the MB component and the central component are $8.5 \cdot 10^{-3}$ cm^{-1} and $6.5 \cdot 10^{-4}$ cm^{-1} (see below), respectively, then the difference of the integrated and maximum intensities is determined by the factor

$$\frac{\delta\nu_A + \delta\nu_{MB}}{\delta\nu_A + \delta\nu_c} = 1.046,$$

i.e., it differs at most by 4.6%, which is much less than the photometric errors of the experiment. The validity of such a conclusion has been shown by the author in a direct experiment [53] (see also Sec. 25).

This appears even more strongly in the ratio of intensities for a poorly purified liquid. Sterin [241, 444] carried out a special study, comparing the fine structure components in well-purified benzene, in a mixture of benzene and toluene, and in poorly purified benzene (Fig. 45).

His results demonstrate very clearly how distortions of the actual distribution can be expected for a poorly purified preparation.

Cummins and Gammon [637] have confirmed that for the wing, $\rho_u \neq 6/7$; therefore, they found $I_c/2I_{MB}$ [Eq. (5.45)] from measurements of the maximum intensity and the halfwidth of the components of the visible contour. To eliminate the distorting effect of the superposition of the components, corrections were made by means of a numerical calculation under the assumption that the form of all the components is Lorentzian. We note, however, that the method of unique determination of $I_c/2I_{MB}$ carried out by us above is suitable not only when $\rho_u = 6/7$, but also for any known ρ_u.

13. Measurement of Some Auxiliary Parameters Necessary for the Computation of the Intensity of Molecular Light Scattering

Calculation of the scattering coefficient R or the absolute intensity of molecular scattering and the ratio of the intensities of the components, fine structure, as is evident from Eqs. (1.93), (1.94), and (5.40), requires the knowledge of the adiabatic and isothermal values of the derivative $(\rho\,\partial\varepsilon/\partial\rho)$, the adiabatic and isothermal values of the compressibility β, the temperature coefficient of the dielectric constant $(\partial\varepsilon/\partial T)_\rho$, and a number of other quantities. Up to now, it has not been possible to compute these values for condensed media, and measurements are necessary. The static measurements of the isothermal value of $(\rho\partial\varepsilon/\partial\rho)$ were carried out in the work of Roentgen and Zehnder [273], Eisele [274], Himstedt and Wertheimer [275], and Raman and Venkateswaran [59]. However, the determination of the intensity of scattered light by adiabatic fluctuations of the density (the integrated intensity of the MB components) requires a knowledge of the adiabatic, and not the isothermal values of $(\rho\partial\varepsilon/\partial\rho)_S$. On the other hand, the scattering of light by adiabatic density fluctuations is a process which is characterized by frequencies of about 10^{10} Hz.

Consequently, a determination of the corresponding quantities by a dynamic method is desirable. Two dynamic methods for determining the adiabatic value of $(\rho\partial\varepsilon/\partial\rho)_S$ were developed by Motulevich and Fabelinskii [277-279]. For the study of high-viscosity liquids, difficulties are possible in the determination of the compressibility β_S throughout the entire range of variation of the viscosity from a fraction of a poise up to the glassy state. A special method was developed by Velichkina and Fabelinskii [280] for the determination of the compressibility (sound velocity) under these conditions [188, 189]. Methods of measuring $(\rho\partial\varepsilon/\partial\rho)_S$ and β_S developed for the study and discussion of measurements in scattered light are essential also for a whole series of other problems, not directly connected with the problem of light scattering (see, for example, [281]).

A Dynamical Method of Finding $(\rho\partial\varepsilon/\partial\rho)_S$ from the Diffraction of Light by Ultrasound. Great interest attaches to the measurement of $(\rho\partial\varepsilon/\partial\rho)_S$ at frequencies corresponding to hypersound. Although such high frequencies have now been generated artificially, it has not yet been possible to study the propagation of waves at such a frequency in the volume of a liquid because of the large damping in the latter. Therefore, it has been necessary to make the measurements in the ultrasonic range of frequencies and use the well-known phenomenon of the diffraction of light by ultrasound for this purpose [282-285]. The phenomenon of optical diffraction by ultrasound differs in no essential way from optical diffraction by an elastic thermal wave. The only (and insignificant) difference lies in the fact that in the first case the sound wave is generated artificially, while in the second it arises from the thermal motion of the molecules of the body. The phenomenon of optical diffraction by ultrasound thus gives an adequate method for the determination of $(\rho\partial\varepsilon/\partial\rho)_S$ by the phenomenon of the diffraction of light at ultrasonic frequencies.

The theory of the diffraction of light by ultrasound was first considered in detail by Rytov [282]. The case which is realized in practice corresponds most closely to the condition of light scattering corresponding to a weak sound field (Brillouin approximation). The requirements of low intensity of the acoustic field means physically that the intensity of the second and higher orders of the diffraction maxima are small not only in comparison with the intensity of the zero maximum, but also in comparison with the intensity of the first diffraction maximum. This condition brings us close to the corresponding conditions of light scattering and is easily achieved and controlled in experiment.

In the work of Rytov [282] it was shown that for a weak sound wave $[|(2\Delta n/n)(\Lambda^2/\lambda^2)| < 1]$, the intensity of the first diffraction maximum is determined by the condition

$$I_{\pm 1} = I_0 \frac{l^2\pi^2}{4\varepsilon\lambda^2}\left(\rho\frac{\partial\varepsilon}{\partial\rho}\right)_S^2\left|\frac{\Delta\rho}{\rho}\right|^2\left\{\frac{\sin\left[\frac{\pi l}{2\Lambda\cos\varphi}\left(\frac{\lambda}{\Lambda}\pm 2\sin\varphi\right)\right]}{\frac{\pi l}{2\Lambda}\left(\frac{\lambda}{\Lambda}\pm 2\sin\varphi\right)}\right\}^2, \quad (13.1)$$

where $I_{\pm 1}$ are the intensities of the first diffraction maxima; I_0 is the intensity of light which passes through the test cell in the absence of ultrasound; $(\rho\partial\varepsilon/\partial\rho)_S$ is the adiabatic value of the derivative; l is the thickness of the ultrasonic column traversed by the light; $\Delta\rho_0$ is determined from the relation $\Delta\rho = \rho_0 + \Delta\rho_0$.

$\cos(\Omega t - kx)$ as the amplitude of the density of the medium, which varies as the result of the propagation of the sound wave; Λ is the acoustic wavelength; and φ is the angle of incidence of the light on the sound wave.

At the angle of incidence of the light on the sound wave corresponding to the Bragg condition (1.27), Eq. (13.1) is simplified and takes the form

$$I_{\pm 1} = I_0 \frac{\pi^2 l^2}{4\varepsilon\lambda^2} \left(\rho \frac{\partial \varepsilon}{\partial \rho}\right)_S^2 \left(\frac{\Delta\rho_0}{\rho}\right)^2. \qquad (13.2)$$

Equations (13.1) and (13.2) show that $(\rho\, \partial\varepsilon / \partial\rho)_S^2$ can be determined from experiment by the diffraction of light by ultrasound.

For the determination of $(\rho\partial\varepsilon/\partial\rho)_S^2$ at an ultrasonic frequency on the basis of the relation (13.2) measurements are necessary of the ratio of the intensity of the lines of the plus one and minus one diffraction orders to the intensity of the incident light, as well as the amplitude of the oscillation of the density $\Delta\rho_0/\rho$ and the thickness of the ultrasonic column l in the direction of light propagation. In the case of an amplitude-inhomogeneous acoustic field, the problem becomes one of the effective values of l_{eff} and $(\Delta\rho_0/\rho)_{\text{eff}}$.

Measurement of $|\Delta\rho_0|^2$ was carried out by a radiometric method [277, 286, 287].

In accord with the hydrodynamic theory of sound [157], $(\Delta\rho_0)^2$ is connected with the pressure p of a plane ultrasonic wave on a fixed obstacle by the relation

$$p = \overline{E}_1 \cos^2\theta_1 + \overline{E}_1' \cos^2\theta_1 - \overline{E}_2 \cos^2\theta_2, \qquad (13.3)$$

where $\overline{E}_1$, $\overline{E}_1'$, and $\overline{E}_2$ are the energy densities in the incident, reflected, and transmitted waves, respectively. Choosing the detecting radiometer in the form of a prism with the dimensions a, h', and vertex angle α such that all the sound waves incident on the face of the prism undergo total internal reflection, we set $\overline{E}_2 = 0$ and $\overline{E}_1 = \overline{E}_1'$ in Eq. (13.3), and thus $p = 2\overline{E}_1\cos^2\theta_1$. Taking into account the dimensions of the prismatic detector and the expression for the energy density $\overline{E}_1$, we obtain for the force acting on the prism from the side of the sound wave,

$$F = 2ah' \cos^2\left(\frac{\pi}{2} - \frac{\alpha}{2}\right) \frac{1}{2} \rho v^2 \left(\frac{\Delta\rho_0}{\rho}\right)^2. \qquad (13.4)$$

Combining (13.2) and (13.4) we get, finally,

$$\left(\rho \frac{\partial\varepsilon}{\partial\rho}\right)_S^2 = \frac{4ah'\varepsilon\lambda^2\rho v^2 \frac{I_1}{I_0} \cos^2\left(\frac{\pi}{2} - \frac{\alpha}{2}\right)}{\pi^2 l^2 F}. \qquad (13.5)$$

In the apparatus used in [53, 277], an aluminum prism was employed with a vertex angle of 120° and dimension a exceeding the width of the sound column l; therefore, the effective dimensions of the prism are $a = l_{\text{eff}}$. With this in mind, we obtain for the chosen dimensions of the radiometer detector,

$$\left(\rho \frac{\partial\varepsilon}{\partial\rho}\right)_S^2 = \frac{3.02h'\lambda^2\varepsilon\rho v^2 \frac{I_1}{I_0}}{\pi^2 l_{\text{eff}} F}. \qquad (13.6)$$

A wide prism is suitable as a radiometer detector in two respects: first, in the final relation (13.6), l appears to the first power and the errors in the measurement of l have less effect on the result; second, in the case of an inhomogeneous field, there is a better averaging over the cross section of the sound beam. The principle of this apparatus is evident from Fig. 46.

As a source of light, we used the PRK-2 mercury lamp. Light of $\lambda = 4358$ Å is emitted from the monochromator UM-2. A photographic plate and step attenuator are placed in the focus of the lens L_2, the attenuator covering only the zero diffraction maximum. In another case, the photoplate was replaced by a photomultiplier; then, instead of the step attenuator, the zero maximum was attenuated by a continuous narrow band of metal deposited on glass. The generator of the high-frequency oscillations was made in the usual way, but special attention was paid to the stability of its operation.

The acoustical transducers were quartz crystals chosen with special care from the viewpoint of optimal homogeneity of their acoustical radiation. In the work, x-cut quartz crystals were used in the shape of circular cylinders of 50 mm diameter and 5 and 10

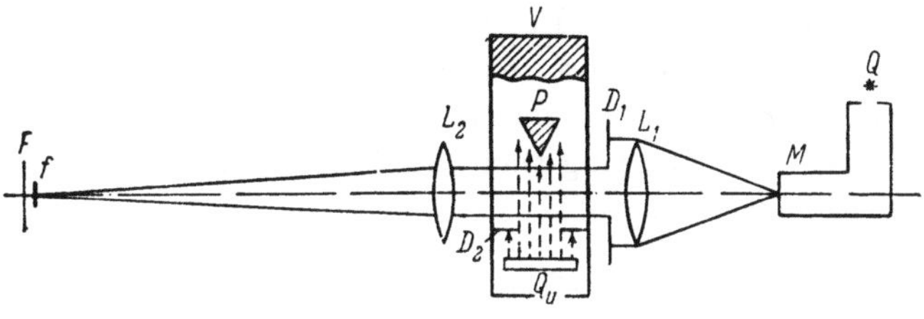

Fig. 46. Arrangement of ultrasonic apparatus for the determination of $(\rho\partial\,\varepsilon/\partial\rho)^2_S$ (Motulevich and Fabelinskii [277]). Q – light source; M – monochromator; L_1, L_2 – objectives; D_1, D_2 – diaphragms; Q_u – sound radiator; P – pressure receiver; V – test cell; F, f – photographic plates with step attenuator or photomultiplier.

mm thickness. The research was performed on the 21st and 91st harmonics.

The test cell in which the piezoelectric radiator and the radiometer detector were placed had a horn-shaped form.

The errors of measurement of $(\rho\partial\,\varepsilon/\partial\rho)^2_S$, as is clear from (13.6), consist of the errors of measurement of the sound velocity v, the ratio I_1/I_0, the pressure on the radiometer F, and l_{eff}.

Estimates show that the errors from measurement of v, F, and I_1/I_0 do not exceed 1.5-2%. A very serious error is connected with the inhomogeneous radiation of the sound [272]. The inhomogeneous radiation makes the determination of l_{eff} more difficult. For inhomogeneous radiation, l_{eff} is not equal to the geometric cross section of the beam. Moreover, the inhomogeneously radiating quartz transmits oblique beams onto the detector and they exert a pressure on it. While the oblique beams of sound either do not give their contribution to the intensity of the diffraction maximum everywhere, or they give a very small contribution, inasmuch as they do not satisfy the Bragg condition. Therefore, complete elimination of the oblique rays and a reliable method of determination of l_{eff} are necessary.

Elimination of the Oblique Sound Rays and Control of the Homogeneity of the Acoustic Field. Waves of oblique directions can be removed or weakened to a significant degree by "filtering them out" by means of acoustic interference "direction" filters.

Actually, we put a plate of thickness t = m(Λ/2) (where m is an integer) in the path of the sound waves emitted by the piezoquartz. This plate will be almost "transparent" for sound waves falling perpendicularly on its surface. Waves which fall on the plate at nonzero angle will be appreciably attenuated. Thus, a wave of practically one direction emerges from the acoustic interference filter. The interference filter can be prepared of metal, glass, or other material with a low sound absorption coefficient. Filters whose thickness corresponds to many half wavelengths filter out the oblique rays better, but they require more accurate preparation (the planes of the filter and the wave front must be parallel). However, it must be remembered that incorrect alignment of the filter does not improve, but can significantly worsen the character of the acoustic field. Therefore, this use of interference filters destined for equalizing the acoustic field makes sense only when one can exert control over their action. One such method control, proposed by Motulevich and Fabelinskii [288], is described here.

This method is based on the investigation of the intensity distribution of the first diffraction maximum of light as a function of the direction of propagation of the ultrasonic wave. A brief theory of the proposed method of control is as follows. If a collimated beam of light is incident in the x direction on a plane sound wave propagating in the xy plane at an angle φ to the y axis (Fig. 47), then the intensity of the light diffracted at an angle θ to the x axis, for small sound amplitudes, is determined by the relation

$$I = C\frac{\sin^2\left[\frac{\pi l}{2\lambda}\theta(\theta-2\varphi)\right]\sin^2\left[\frac{\pi a}{\lambda}\left(\theta\pm\frac{\lambda}{\Lambda}\right)\right]}{\left[\frac{\pi l}{2\lambda}\theta(\theta-2\varphi)\right]^2\left[\frac{\pi a}{\lambda}\left(\theta\pm\frac{\lambda}{\Lambda}\right)\right]^2}. \tag{13.7}$$

Here, l is the width of the ultrasonic column transversed by the light, a is the dimension of the sound column along the z axis, and C is a constant for a given liquid, a definite wave length, and a definite intensity of

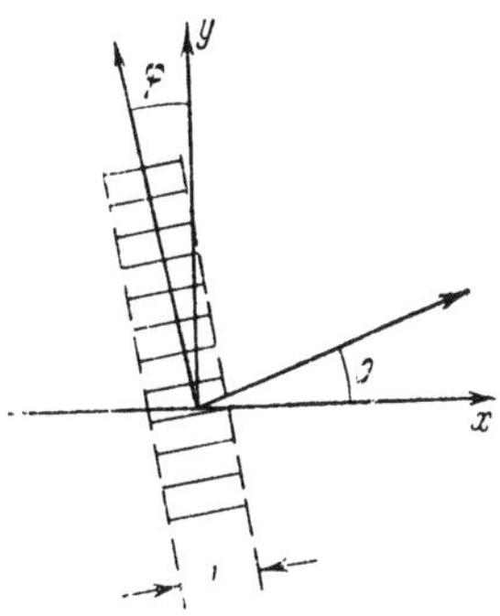

Fig. 47. Theory of method of control of acoustic field. A light wave is directed along the x axis. The first diffraction maximum is observed at the angle θ to the x axis. The ultrasonic wave is propagated at an angle φ to the y axis; l is the width of the ultrasonic column.

the incident light. For $\theta = \pm\lambda/\Lambda$, which corresponds to the direction of the plus (minus) one diffraction maximum, the intensity distribution will be

$$I(\psi) = C\frac{\sin^2\frac{\pi l}{\Lambda}\psi}{\left(\frac{\pi l}{\Lambda}\psi\right)^2}, \quad (13.8)$$

where $\psi = \varphi \pm {}^1\!/_2\lambda/\Lambda$. Figure 48 shows the graph of Eq. (13.8). By turning the radiator in the xy plane (see Fig. 47), we can obtain the experimental relation between the intensity of the first diffraction maximum and the angle of rotation of the normal to the radiator. For the zero angle of this rotation we choose the direction of the normal of the radiator corresponding to the maximum intensity of the diffraction maximum.

By normalizing the experimental and theoretical curves so that their maxima are equal to unity, we obtain a coincidence of both curves if the plane wave of one direction is used, and a divergence if the acoustical field has a different character. Measurements carried out on the intensity distribution as a function of angle have shown that this distribution frequently differs substantially from the theoretical (such a situation is shown by the crosses in Fig. 48). For a quantitative estimate of the energy of a sound beam, whose direction of propagation differs from the chosen by an angle α, we consider two cases.

1. Let $\alpha > \psi_0 = \Lambda/l$. Then an additional maximum appears in the experimental curve $I = I(\psi)$ corresponding to $\psi = -\alpha$. The area bounded by the curve corresponding to this maximum will be proportional to this beam. In [288] one could easily discover the oblique rays whose energy amounted to 0.5% and more of the ener-

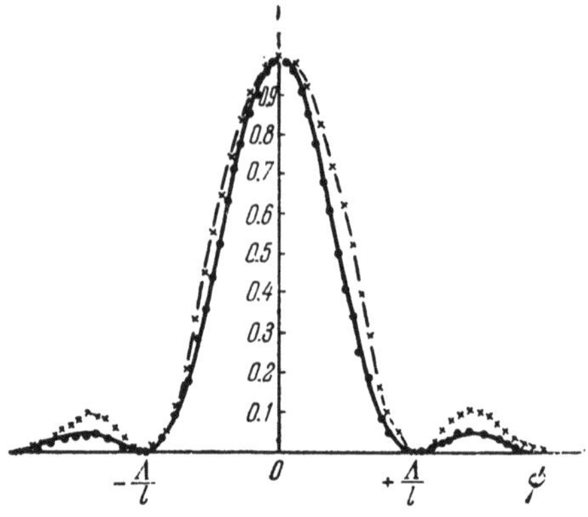

Fig. 48. Dependence of the intensity of the first diffraction maximum $I(\psi)$ on the angle of rotation $\psi = \varphi + \frac{1}{2}\lambda/\Lambda$ (Motulevich and Fabelinskii [288]). Continuous line – theoretical curve; crosses – experimental points without filter; circles – the same, with filter.

gy of the main beam. Since the research was carried out at a frequency of 10^7 Hz with a width of the ultrasonic column $l = 1.5$ cm, then $\psi_0 = 0.001$.

2. Let $\alpha < \psi_0$. Then the ratio of the areas bounded by the experimental and theoretical curves is equal to the ratio of the total energy of the sound field to the energy of the "effective wave," i.e., the wave which is propagated along the direction corresponding to the maximum experimental curve with some "effective amplitude." The "effective amplitude" in the given case is determined by the condition of equal optical action of the actual and "effective" sound beams.

In fact, let us consider two plane sound waves, which are characterized by the amplitudes A_1 and A_2, and by the directions of propagation α_1 and α_2 (α_i is the angle between the direction of propagation of the beam and the direction corresponding to the maximum of the experimental curve). The intensity of the first-order diffraction maximum is proportional to the expression $A_1^2 f(\psi - \alpha_1) + A_2^2 f(\psi - \alpha_2)$, where f is a function of the form (13.8).

The experimental curve is represented by the formula

$$I_{\exp}(\psi)=\frac{1}{A_1^2 f(-\alpha_1)+A_2^2 f(-\alpha_2)}\left[A_1^2 f(\psi-\alpha_1)+A_2^2 f(\psi-\alpha_2)\right]. \qquad (13.9)$$

The theoretical curve is represented by the equation

$$I_{\text{theor}}(\psi)=f(\psi).$$

The corresponding areas will be

$$S_{\exp}=\int I_{\exp}(\psi)\,d\psi, \quad S_{\text{theor}}=\int I_{\text{theor}}(\psi)\,d\psi.$$

It is not difficult to see that

$$\frac{S_{\exp}}{S_{\text{theor}}}=\frac{A_1^2+A_2^2}{A_1^2 f(-\alpha_1)+A_2^2 f(-\alpha_2)}=\frac{A_1^2+A_2^2}{A_{\text{eff}}^2}. \qquad (13.10)$$

The corresponding conclusion can be extended to the case of any number of waves.

If it is necessary to know more accurately the character of the ultrasonic beam in case (2), then we must expand the experimental curve to a curve of the form (13.8). This gives the amplitude and the direction of propagation of all the rays entering into the composition of the field under study.

The method outlined here of control of the acoustic field makes it entirely possible to control the action of the interference sound filter. Figure 48 clearly shows to what degree an interference filter of thickness Λ improves the character of the emission field of the quartz. In most cases, correct location of the interference filter virtually eliminates the oblique rays; however, it is sometimes impossible to obtain such results. In the latter case, it is entirely valid to make the correction to the oblique rays and the result of the measurement becomes free from their distorting action. Furthermore, from a measurement of the intensity of the first diffraction maximum as a function of the angle, it is possible to determine quantitatively the value of l_{eff}, the effective width of the sound beam in the direction of propagation of the light. l_{eff} is determined from the data of Fig. 48, and the corrections to the measured quantities, distorted as the result of the noncollimated character of the sound beam, from the ratio $S_{\exp}/S_{\text{theor}}$ [Eq. (13.10)].

As is clear from Eq. (13.8), $I(\psi)$ first falls to zero [in the case of incomplete elimination of the oblique rays one should rather speak of the minimum value of $I(\psi)$] under the condition that

$$\frac{\pi l_{\text{eff}}\,\Delta\psi}{\Lambda}=\pi,$$

where $\Delta\psi$ is the angular distance from the maximum of the curve $I(0)$ to the first zero of the function $I(\psi)$, whence it follows directly that

$$l_{\text{eff}}=\frac{\Lambda}{\Delta\psi}\,. \tag{13.11}$$

Here, $\Delta\psi = \Delta\psi'/n$, and $\Delta\psi'$ is the angle of rotation of the cell, measured from the position corresponding to the Bragg angle, n is the index of refraction of the liquid. The correction for the oblique rays, which are not completely eliminated by the interference filter, is determined by means of a planimeter measurement of the area of the theoretical curve and the experimental curve. The ratio of the areas gives the ratio of the energy of the sound waves passing in the given direction and obliquely to it. In this case, the ratio has been shown to be equal to

$$\delta=\frac{S_{\text{exp}}}{S_{\text{theor}}}=1.026.$$

An important question arises in the introduction of a correction for the sound absorption into the result of the determination of $(\rho\partial\varepsilon/\partial\rho)_S$ from (13.6). The fact is that the distance from the emitter to the axis of the sound beam is different than to the radiometer receiver. Therefore, it would appear that it is necessary to take into account the attenuation of the sound over the path d. Such an account has been undertaken [146, 278].

However, for the radiometer used, such a calculation led to a decrease in the values of $(\rho\partial\varepsilon/\partial\rho)_S$ for the following reason: for a propagating sound wave, one must satisfy the laws of conservation of energy and momentum. In an absorbing medium, part of the energy of the sound wave goes over into heat (into the energy of Debye waves), raising the temperature of the liquid. There is also some decrease in the momentum of the wave, which creates the "acoustic wind" in the liquid, i.e., the current of liquid, directed from the source to the receiver. As a result of the fact that the liquid is not ideal, the "acoustic wind" transfers momentum to the receiving radiometer and consequently creates an additional pressure. Ultrasonic investigations which appeared after our researches [289, 290] make it possible to estimate roughly this excess pressure, and to introduce a correction to the already completed measurements of $(\rho \partial \varepsilon / \partial \rho)_S$. The complication brought about by the "acoustic wind" can be avoided if one uses such a radiometer (piezoelectric detector with a tuned amplifier) which will record the pressures at the frequency of the sound wave, and which does not respond to the acoustic wind.

However, at the time when the method just described was being developed, there was no possibility of carrying out a sufficiently accurate calibration of the radiometer with a piezoelectric sensing element.

Dynamic Method of Determination of $(\rho \partial \varepsilon / \partial \rho)_S$ at Low Sound Frequencies. The method of measurement of $(\rho \partial \varepsilon / \partial \rho)_S$ at low frequencies was developed by Motulevich and Fabelinskii [279]. A special vessel was placed between the plates of a Jamin interferometer (Fig. 49), made of a single piece of stainless steel, in which two parallel channels were made. Two beams from the interferometer passed along them. In channel 2 of the vessel, an electromagnetic sound emitter 6 produced a pressure that varied with a frequency of 50 Hz. The length of the channel of the vessel is $1/100$ of the sound wavelength ($\Lambda = 30$ m). Consequently, it can be assumed with excellent approximation that a uniform pressure is generated along the entire length of the channel.

The distribution of the light intensity in the interference picture will be determined by the expression [291]

$$I = I_1 + I_2 + 2\sqrt{I_1 I_2}\cos\delta, \tag{13.12}$$

where δ is the path difference, equal to $(2\pi l \Delta n/\lambda)\cos(\Omega t + \varphi_0)$; l is the length of the vessel, and φ_0 is the phase difference in the absence of excitation. The time average $\overline{\cos[\varphi_0 + (2\pi l/\lambda)\Delta n \cos \Omega t]} = \cos\varphi_0 J_0[(2\pi l/\lambda)\Delta n]$ (here J_0 is the Bessel function of zero order). Consequently, for visual observation, the interference picture is determined by the expression

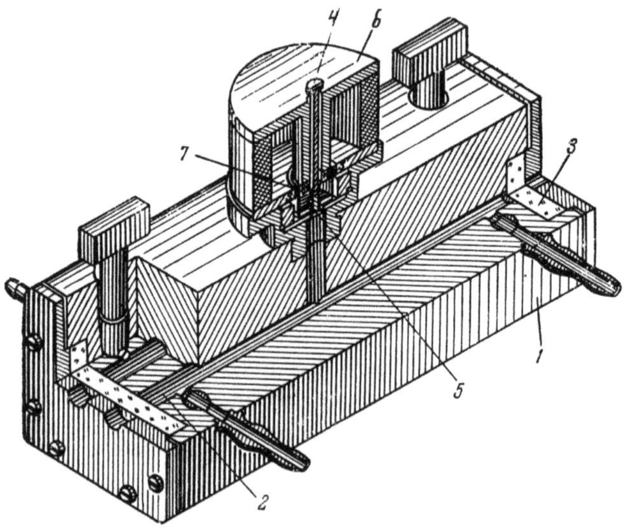

Fig. 49. Vessel for the measurement of $(\rho\partial\varepsilon/\partial\rho)_S$ at low sound frequency (Motulevich and Fabelinskii [279]). 1 – Steel block; 2 – channel for liquid; 3 – glass window; 4 – mirror fastened to membrane; 5 – piston of radiator; 6 – electromagnetic sound source; 7 – framework (antenna loop).

$$I = I_1 + I_2 + 2\sqrt{I_1 I_2}\cos\varphi_0 J_0\left[\frac{2\pi l}{\lambda}\left(\rho\frac{\partial n}{\partial\rho}\right)_S\frac{\Delta\rho}{\rho}\right]. \quad (13.13)$$

The change of amplitude of the vibration of the piston 5 (Fig. 49) of the electromagnetic radiator produces a change in the density of the liquid and, consequently, a change in its index of refraction by Δn.

If the amplitude of the vibration of the piston is changed continuously, then the visibility of the bands will be changed in the visual observation of the interference picture. When the Bessel function $J_0(x)$ vanishes, the interference picture changes to uniform illumination. Upon further change of amplitude of the radiator, the interference picture again becomes visible, and then again disappears when $J_0(x)$ passes through its second zero, etc. Thus, for an amplitude of vibration of the piston of the radiator for which

$J_0(x)$ vanishes, $x = (2\pi l/\lambda)(\rho\partial n/\partial\rho)_S(\Delta\rho/\rho)$ is a root of the function $J_0(x)$. Consequently, in order to determine $(\rho\partial n/\partial\rho)_S$, one must only determine $\Delta\rho/\rho$. It is only necessary to know the value of the pressure $\Delta p = (1/\beta_S)(\Delta\rho/\rho)$ accurately at the moment when $J_0(x)$ vanishes. In other words, an accurate calibration of the transmitter is necessary. The method of calibration for obtaining p is as follows. A rod (Fig. 49) is attached to the radiator, on the upper end of which the mirror is placed. This mirror is one of the mirrors of a Michelson interferometer. The change in the position of the piston is measured by the shift of the interference pattern. A pressure is transmitted to the liquid in channel 2 of the vessel through a valve. This pressure can be measured exactly. The displacement of the piston of the radiator under the action of the given pressure is compensated by the flux through the loop 7 of the electromagnetic sound radiator. The moment of compensation is established with great accuracy by the position of the interference pattern. A small additional effect, produced by the elasticity of the membrane, was also measured by the interference method and is taken into account in the calibration. Thus, the pressure was determined from the flux through the coil of the radiator.

Consequently, if Δp, determined by calibration [291, 292] is known, then

$$\left(\rho\,\frac{\partial n}{\partial \rho}\right)_S = \frac{x_i\lambda}{2\pi l \beta_S \Delta p}\,. \tag{13.14}$$

Here, x_i is the i-th root of the function $J_0(x)$. By this method, $(\rho\partial n/\partial\rho)_S$ was determined for water and benzene and converted to $(\rho\partial\varepsilon/\partial\rho)_S$; the latter were shown to be equal to 0.90 ± 0.02 and 1.61 ± 0.06, respectively. These values agree with the static values within the limits of error.

Measurement of the Adiabatic Compressibility. For the calculation of the intensity I or the scattering coefficient R from the Einstein formula (1.94), it is necessary to know the isothermal compressibility with great accuracy. For the calculation of R by Eq. (1.93), it is necessary to have at one's disposal the adiabatic value of the compressibility.

The adiabatic compressibility β_S is connected by the well-known relation $\beta_S = 1/\rho v^2$ with the sound velocity, and this latter can be determined with high accuracy [280, 283]. Therefore, even

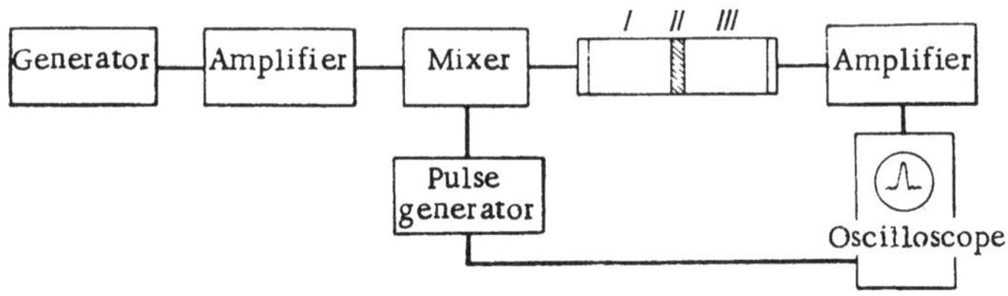

Fig. 50a. Electrical block diagram of apparatus for the measurement of velocity and absorption of ultrasound (Velichkina [189]).

for the determination of the isothermal compressibility, it is obviously better to use the measurements of the sound velocity to find β_S and to use the well-known thermodynamic relation between β_S and β_T (1.47).

The coefficient of volume expansion and c_p, which enter into Eq. (1.93), are determined with high accuracy. Even if these quantities are known with not too large an accuracy in some cases, then they nevertheless do not lead to large errors in the determination of β_T, inasmuch as the term $T\sigma^2/c_p\rho$ in (1.47) amounted to at most 25-35% of β_S.

Many different methods have been proposed for the measurement of the sound velocity, which are based on the phenomenon of optical diffraction by ultrasound, the interference of sound (Pierce interferometer) and on the pulse method of the measurement of ultrasonic velocity [283]. All these methods give high accuracy in the determination of the sound velocity under ordinary conditions, but if it is necessary to know the sound velocity at an arbitrarily low temperature, where the viscosity of the liquid, and hence its absorption, both increase, the application of the known methods of measurement of the sound velocity becomes very difficult or even, in some cases, impossible [280]. However, in the study of the scattering of light in liquids, in the transition from ordinary viscosities to the glassy state, it is necessary to know the adiabatic compressibility at any temperature. Therefore, Velichkina and Fabelinskii [280] developed a special method of measurement which, as it seems to us, is better suited than the other methods for these conditions. Of course, it can also be used in other cases. This method of measurement of the ultrasonic velocity is based on

Fig. 50б. General view of the apparatus for the measurement of the velocity and absorption of ultrasound.

the phenomenon of the interference of sound waves in a plane-parallel layer of finite thickness. The apparatus which operates on this principle is a Fabry–Perot interferometer for sound waves. The schematic diagram of the apparatus and the block diagram of the electronic part are shown in Fig. 50a, while a general view of the apparatus is given in Fig. 50b. The radiating quartz sends out an ultrasonic pulse which is propagated in medium I (glass, quartz, or metallic rod); on the boundary of media I and II it is partially reflected back into medium I and partially passes into medium II, the plane-parallel layer of the liquid to be studied. Here, the sound

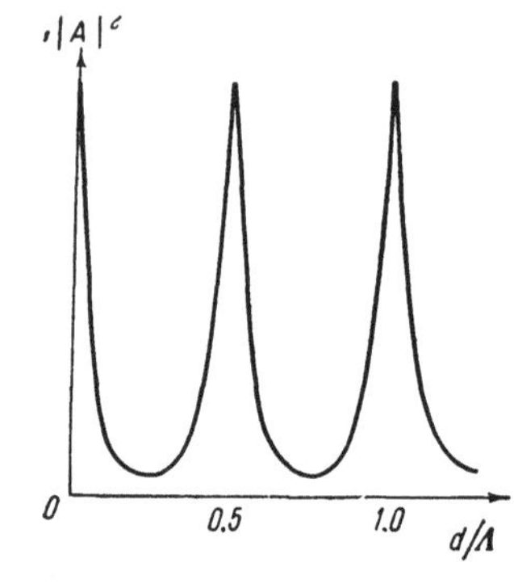

Fig. 51. Dependence of the square of the amplitude of the transmitted wave on the ratio d/Λ.

wave undergoes multiple reflections from the surface of separation bounding the layer. The reflected rays interfere with one another and, depending on the relative phases, give a maximum either in medium I or in medium III (a second rod). We assume that the media I and III are the same, and are characterized by the density ρ_1 and the sound velocity v_1. The density and the sound velocity in the layer of liquid sample are ρ and v, respectively.

The square of the amplitude of sound which passes through the layer of liquid (medium II) of thickness d is expressed by the formula [137]:

$$|A_0|^2 = \frac{1}{\cos^2 q_0 d + \frac{1}{4}\left(\frac{\rho_1 v_1}{\rho v} + \frac{\rho v}{\rho_1 v_1}\right)^2 \sin^2 q_0 d}. \qquad (13.15)$$

Here it is assumed that the amplitude of the incident wave is equal to unity and that there is no absorption in all three media; q_0 is the wave number and d is the thickness of the layer.

The dependence of $|A_0^2|$ on d/Λ is shown graphically in Fig. 51. It is seen from the graph and Eq. (13.15) that the sound intensity which passes through the liquid layer is a maximum when the thickness of the layer is equal to an even number of half wavelengths, and a minimum when the thickness of the layer is equal to an odd number of half wavelengths. Consequently, if the media I and III are the same, and the gap between them can be continuously changed, then, as the gap distance d changes, a maximum in the sound intensity will be noted whenever d = nΛ/2, and a minimum in intensity when d = (2n + 1)Λ/4. Here, it follows from (13.15) that

$$|A_0|^2_{max} = 1 \quad \text{and} \quad |A_0|^2_{min} = \frac{1}{4\left(\frac{\rho_1 v_1}{\rho v} + \frac{\rho v}{\rho_1 v_1}\right)^2}.$$

The measurement procedure thus reduces to a continuous change in the distance d, for example from zero to some value of d_1, and to a counting of the number of maxima (or minima) detected by the receiver in the change of the gap by an amount d_1. If we count the maxima, then

$$d_1 = n\frac{\Lambda}{2} = n\frac{v}{2f}, \tag{13.16}$$

where f is the sound frequency.

From (13.16), we get for the sound velocity

$$v = \frac{2d_1}{n}f. \tag{13.17}$$

It is clear from (13.16) and (13.17) that, for the determination of v, we can either change the thickness of the layer, keeping the frequency constant, or keep the thickness of the layer constant and change the frequency f. The second method is more convenient in a number of cases and is possibly more accurate than the first, but it is used only when it is known beforehand that there is no dispersion in the medium.

Two different methods of sound detection were employed in the work of [280]. In one of these, a receiving quartz was attached on the end face of the rod III (Fig. 50a), the signal of which is fed through an amplifier to an oscilloscope. Maxima and minima in the signal are observed on the oscilloscope. In the second method, a variant of the Toepler method was used that was proposed by Raiskii. In this variant, the glass rod III had a polarized cut, through which the light passed. At the moment when the sound (maximum) passed through the rod III in the focal plane of the Raiskii apparatus, a bright flash of light was observed. In this method, it is necessary to count the number of flashes.

Velichkina [189] developed a theory of the proposed method for the case in which the studied medium in the gap d absorbs sound, while the media I and III (Fig. 50a) are the same and, as before, do not absorb the sound.

In this theory, it was assumed that the sound propagation in the gap d obeys Eq. (5.10) while in the media I and III it obeys Eq. (5.5). As a result of the solution of the boundary problem, it was found that the amplitude of the sound in medium III is

$$|A|^2 = \frac{|C|^2}{\left[\frac{1}{r^2} e^{2\alpha d} - 2\cos 2(k_0 d + \psi + \varphi) + r^2 e^{-2\alpha d}\right]}, \tag{13.18}$$

where C is a constant, independent of d. Here, we have introduced the notation

$$\varphi = \operatorname{arctg} \frac{\rho_3 \Gamma \omega}{v_1 \rho_1 v_2 \left(1 + \frac{v_2 \rho_2}{v_1 \rho_1}\right)},$$

$$\psi = \operatorname{arctg} \frac{\rho_3 \Gamma \omega}{v_1 \rho_1 v_2 \left(1 - \frac{v_2 \rho_2}{v_1 \rho_1}\right)},$$

and Γ is determined from (5.11).

In the range of applicability of (5.10), i.e., for $\alpha\Lambda \ll \pi$, the analysis of (13.18) shows that the distance between the extrema remains the same as in the absence of absorption. It follows from (13.18) that

$$|A|_{\max} = \frac{C}{2 \operatorname{sh} (\alpha d + \beta)},$$

$$|A|_{\min} = \frac{C}{2 \operatorname{ch} (\alpha d + \beta)}.$$

Consequently, the measurement of the ratio of the minimum to the maximum of the amplitude of the transmitted sound for different d allows us to determine the absorption coefficient α from the relation

$$\operatorname{arcth} \frac{|A|_{\min}}{|A|_{\max}} = \alpha d + \beta, \tag{13.19}$$

where β is a constant.

The apparatus of the type described can be used for the measurement of absorption by the usual pulse method [283-296].

This method of the determination of the ultrasonic velocity does not require large amounts of liquid and makes it possible to obtain a quick result with sufficiently high accuracy (about 0.1%). In a great number of measurements, the accuracy can be signifi-

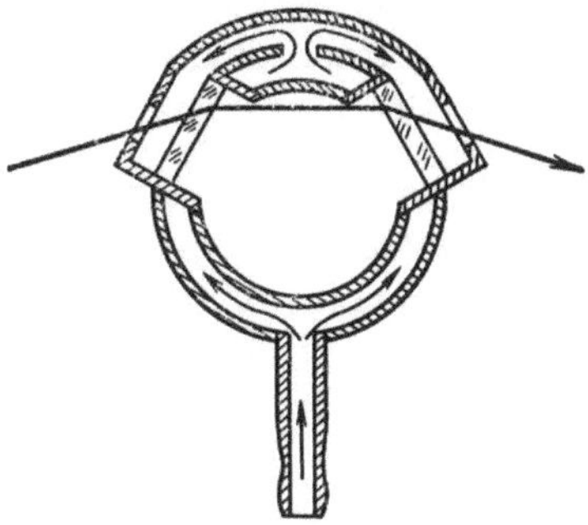

Fig. 52. The method of measurement of the temperature dependence of the index of refraction (Velichkina [189]). The entrance and exit windows of the vessel are parallel to the faces of a 60° prism. Cold air is blown on the windows and prevents condensation on them of moisture from the atmosphere.

cantly improved with excellent thermostatting. The method is especially convenient when measurements are required over a wide range of temperatures.

Measurement of the Temperature Dependence of the Index of Refraction and the Dependence of the Index of Refraction on the Concentration. The measurement of the quantity $(\partial \varepsilon/\partial T)_p \equiv 2n(\partial n/\partial T)_p$ is important not only because it is necessary to know $[(1/\sigma)(\partial \varepsilon/\partial T)]_p$ for the calculation of R [(1.92)] or the ratio of the intensity of the fine structure components [145, 53], but even more because, in a number of cases, when there are not directly measured values of $(\rho\partial \varepsilon/\partial \rho)_T$ it can be replaced with excellent approximation by the quantity $[(1/\sigma)(\partial \varepsilon/\partial T)]_p$ [see (18.8)].

Actually, when the light scattering is measured in liquids in the transition from the liquid to the glassy state, the application of the dynamic methods described above for the measurement of $(\rho\partial \varepsilon/\partial \rho)_S$ becomes difficult. Under these same conditions, one can carry out the measurement of $(\partial \varepsilon/\partial T)_p$ quite simply. There are many methods for the measurement of the temperature dependence or the index of refraction. The most accurate of these are based on the application of interference methods [298-302].

Here, we shall describe briefly only that method which was used in the course of our investigations of liquids cooled to low temperatures, where the viscosity increases strongly.

The principle of this method has long been known and it has been used with success in different cases [297]. Essentially, it amounts to the following: a prism is "prepared" of the sample substance and the index of refraction is measured from the angle of minimum deviation of the light beam in this prism, from the relation

$$n = \frac{\sin \frac{1}{2} (\varphi + \delta)}{\sin \frac{1}{2} \varphi}, \qquad (13.20)$$

where φ is the refracting angle of the prism and δ is the angle of minimum deviation.

For the operation of this method, a cylindrical metallic vessel was made with two windows (Fig. 52). Two entrance windows of finely polished plane-parallel glass plates are made in the vessel.

The glass plates are so set that their planes face each other with an angle of approximately 60°. The windows and upper covers of the vessel are hermetically sealed. Cold or hot air flows in the gap between the walls, as was done in the heating or cooling of the liquid in the scattering vessel (see Fig. 37). For the protection of the entrance and exit windows from moisture condensation at low temperatures, the outgoing cold and dry air was directed onto the diaphragms put on the windows. The back pressure of the dry air prevents the appearance of moisture of the atmosphere on the glass window. In the research of Velichkina [189], the apparatus described was placed on the stage of a Schmidt and Hensch goniometer, capable of measuring the angle with an accuracy to within 6". Measurements of the angle in the temperature range from +20 to −30°C were carried out with an accuracy to within 2', and for temperatures below −30°C, with accuracy to within 4'. The total error in the measurement of n because of inaccuracy in the determination of φ, δ, and the temperature amounted to about 10^{-3}.

Chapter IV

Molecular Scattering of Light in Gases

14. Results of Measurement of the Absolute and Relative Intensity of Light Scattered in Gases

By using the photographic method of measurement described in Chapter III, Sec. 11, Cabannes [303] first made measurements of the absolute intensity of scattered light in argon and measurement of the relative intensity in some other gases and vapors of organic substances. Later, Daure [199], in Cabannes' laboratory, made absolute measurements in vapors of ethyl chloride, and Vaucouleurs [56] made measurements in argon, air, and ethyl chloride.

The small number of studied substances is explained by the experimental difficulties, but the results are sufficient to give a

Table 6. Absolute Intensities of Light Scattered in Gases

Gas	t, °C	Press., mm Hg	λ, Å	$\Delta_u \cdot 10^2$	$R_{90} \cdot 10^8$, cm^{-1}	$N_A \cdot 10^{-23}$	Author
Argon				$\leqslant 5 \cdot 10^{-2}$			Massoulier [581]
	27	760	4358	0	1.34	6.90	Cabannes [303]
	0	760	4350	0.00	1.668 ± 0.015	5.99 ± 0.07	Vaucouleurs [56]
Ethyl chloride	19	750	5200—5700		10.6	6.50	Daure [199]
	0	760	4350	0.0145 ± 0.0005	26.25	6.09 ± 0.10	Vaucouleurs [56]
Air	0	760	4350	0.0310 ± 0.0005	1.818 ± 0.016	5.98 ± 0.7	Vaucouleurs [56]

convincing proof of the complete agreement of theory with experiment.

The results of the measurement of the absolute scattering coefficient and the computed values of Avogadro's number N_A are shown in Table 6.

Equation (4.4) has been used for the calculation of $N_A = N_L V$; it was obtained from the Einstein formula (1.94) and the equation of state of the ideal gas. In the calculation of N_A, carried out by Vaucouleurs [56], a correction was introduced for the departure of a real gas from the equation of state of an ideal gas.

Obviously, the values of N_A calculated by Cabannes [303] and Daure [199] from their measurements of the absolute intensity of the scattered light involve some errors. The data of Vaucouleurs give excellent values for N_A. The mean value of Avogadro's number computed by him from his data for argon, air, and ethyl chloride, $N_A = (6.02 \pm 0.07) \cdot 10^{23}$, is in excellent agreement with the most accurate value of this quantity, $N_A = (6.0254 \pm 0.0002) \cdot 10^{23}$. The determination of Avogadro's number by the measurement of the extinction coefficient h in the atmosphere of the earth, worked up by Tien Kiu [305] and Vassy [306] from the data of many authors, also gives quite satisfactory results, namely, $N_A = (6.10 \pm 0.08) \cdot 10^{23}$.

From the coincidence of N_A determined by measurements of the absolute intensity of the scattered light with the most accurate value of this quantity, it follows that the propounded theory of molecular light scattering in gases is a valid one.

Table 7. Relative Intensity of Light Scattered in Gases

	Relative intensity	Gas	Relative intensity
Air	1	Nitrous oxide N_2O	3.8
Nitrogen N_2	1	Xenon Xe	5.5
Hydrogen H_2	0.23	Ethane C_2H_6	6.3
Argon Ar.	0.8	Propane C_3H_8	12
Oxygen O_2.	0.9	Ethyl chloride C_2H_5Cl.	16
Krypton Kr	2.0	Butane C_4H_{10}.	22
Methane CH_4	2.4	Butyl alcohol $CH_3(CH_2)_3OH$.	26
Carbon dioxide CO_2.	2.6		

The measurement of the intensity of the light scattered in gases, permitted Gill and Heddle [218] [with the help of Eq. (4.4)] to determine the index of refraction for ultraviolet (λ = 1216 Å) for argon, krypton, xenon, hydrogen, and nitrogen.

Table 7 gives the results of the measurements of the relative intensity of scattered light in various gases and vapors, obtained by Cabannes [75, 315].

If the light intensity scattered by air is taken to be unity, then, by using the data of Vaucouleurs (Table 6), we obtain for argon 0.92 and for ethyl chloride 14.4, while from the data of Cabannes (Table 7) one obtains 0.8 and 16, respectively, for these same gases. Evidently, this means that the data of Table 7 cannot be very accurate; they are suitable only as estimates.

Equation (4.4) does not contain an explicit dependence of the scattering coefficient on the temperature [307]. For a change in temperature, R_{90} can change only in the range of the change of $(n^2 - 1)^2$.

For atmospheric optics, the discussion of the problem of the temperature dependence of R_{90} for the case of a real gas is essential for the case in which the temperature is changed [308] at constant pressure, and when the change in temperature at constant pressure is brought about as the result of the departure of the behavior of the real gas (air) from the equation of state of an ideal gas [309].

For a consideration of the first part of the problem, following Barrell [310], we write down the dependence of the index of refraction on the temperature and pressure of the air in the following fashion:

$$\frac{n-1}{n_0-1} = F(t,\ p)\frac{p(1+\sigma t_0)}{p_0(1+\sigma t)}, \tag{14.1}$$

where n_0 is the index of refraction measured at a temperature t = 15°C and p_0 = 760 mm Hg, $\sigma = 1/273$ (°K)$^{-1}$, and

$$F(t,\ p) = \frac{1+f_t \cdot p}{1+f_{15} \cdot p_0}. \tag{14.2}$$

Here $f_t = (1.049 - 0.0157t) \cdot 10^{-6}$ cm^2/dyne. The measurement of the scattered light in the lower atmosphere was carried out by Crosby and Koerber [271].

By using the equation of state of an ideal gas, one can put (14.1) in the form

$$n - 1 = c_\lambda \cdot F(t,\ p)\, \rho, \tag{14.3}$$

where $c_\lambda = (n_0 - 1)/\rho_0$. On the other hand, one can set $n^2 - 1 = 2(n - 1)$ in Eq. (4.4) with sufficient accuracy, and, by using (14.3), rewrite

$$R_{90} = \frac{2\pi^2}{\lambda^4 N_L} c_\lambda^2 F^2 \rho^2 f(\Delta). \tag{14.4}$$

For the extinction coefficient h [see (1.72)], we get

$$h = \frac{32\pi^3}{3\lambda^4 N_L} c_\lambda^2 F^2 \rho^2 f(\Delta). \tag{14.5}$$

Thus, by measuring the extinction coefficient at different altitudes, one can determine the density of the air ρ at the corresponding altitude.

The dependence of R_{90} or h on the temperature is included in F(t, p) and can easily be estimated [309].

Actually, for a change of temperature from −40°C to +40°C, the factor F^2 changes from 0.9994 to 1.0012, i.e., it changes at most by 0.18%.

For an account of the departure of a real gas from the equation of state of an ideal gas, we write the Einstein formula (1.94) for gases in the following form:

$$R_{90} = \frac{2\pi^2}{\lambda^4} (n - 1)^2 \frac{R_g}{N_A} T \beta_T f(\Delta), \tag{14.6}$$

where R_g is the gas constant per gram molecule. For an ideal gas, $\beta_T = 1/p$, and (14.6) transforms into (4.4).

If we start out from the equation of a real gas,

$$p(V - B') = R_g T, \tag{14.7}$$

where B' is the second virial coefficient, equal to

$$B' = b - \frac{a}{R_g T} \tag{14.8}$$

(a and b are the Van der Waals constants), then we get for the compressibility [from (14.7) and (14.8)]:

$$\beta_T = \frac{1}{p}\left(1 - \frac{p}{R_g T} B'\right). \tag{14.9}$$

In the same temperature interval ±40°C, we find the correction for β_T. B' (40°C) = 0.9 · 10^{-4} liters, while B' (−40°C) = 349 · 10^{-4} liters; therefore, the correction to the compressibility changes from 0.035 · 10^{-4} atm^{-1} at t = 40°C to 18.3 · 10^{-4} atm^{-1} for t = −40°C. Thus, if p = const ≈ 1 atm, the total change in the compressibility ~0.18%, i.e., the same amount as the change in F^2 in the interval ±40°C.

In the consideration of the temperature dependence of R_{90}, a large range of temperature change is chosen; this can be realized in practice under atmospheric conditions.

Under laboratory conditions, a small change in the temperature of about ±1°C has a trivial effect on the measured value of N_A.

15. The Results of Measurement of the Depolarization Factor and the Kerr Constant in Gases and Vapors

It has already been shown, in Sec. 10, that great difficulties accompany the measurement of the depolarization factor of scattered light in gases.

Although all the necessary precautions were taken for the measurement of Δ, and the necessary corrections carried out, the results of the measurement are filled with errors, in particular when the depolarization factor is small. This applies especially to measurements in inert gases and in vapors with optically symmetric molecules.

Table I (see Appendix III) shows the results of the measurement of the depolarization factor and the Kerr constant in gases and vapors. The measurements of the depolarization factor are rather completely represented, but only some measurements of the Kerr constant are listed. The measurements of Δ_u for the inert gases were made principally by Rayleigh [316], Cabannes [303, 314, 315], Parthasarathy [317], and Vaucouleurs [56]. The data for helium and neon have only an orientational character.

So far as the data for argon, krypton, and xenon are concerned, in the opinion of Cabannes, there exists a finite depolarization of the scattered light in these gases; it is equal to Δ_u = 0.0055 for all three gases. Parthasarathy found $\Delta_u = 0.0056$ for argon, while later Vaucouleurs could show only that Δ_u was equal to zero, with accuracy to within 10^{-3}. Still later, Massoulier [581] found that for argon $\Delta_u \leq 5 \cdot 10^{-4}$. It was shown that as the techniques of measurement of Δ_u for argon improved, one always came closer to the expected value of Δ_u, which was equal to zero.

Very recently, Bridge and Buckingham [592] used a gas laser (λ = 6328 Å) as a light source and found for argon at a pressure of 1 atm and 20°C, $\Delta_u = 0.041 \pm 0.005$. The authors [592] show that their high values for argon and xenon (Table I) are explained by defects of the apparatus. Evidently, one must regard the measurement of the depolarization factor in inert gases as an unsolved experimental problem, and use must be made of the modern photoelectric method, which guarantees a high accuracy of measurement.

The depolarization factor found in methane, carbon tetrachloride, and certain other symmetric molecules is also too large.

The results for methane are probably connected with inaccuracies of the experiment, inasmuch as the attempts to find rotational structure in the scattering line in methane gives a negative result [319-322]. Evidently, this negative result shows that the molecule of ammonia is optically isotropic and that its measured depolarization factor cannot characterize the molecule, but is rather connected with the uncertainties of the method of observation.

Experiments on the vapors of carbon tetrachloride have not succeeded in revealing the rotational structure of the scattering line; therefore, such conclusions which appear to be convincing relative to the methane molecule cannot be drawn here.

Stuart [115, 116] assumes that some anisotropy should be expected, assuming that there are different isotopes of chlorine in one and the same molecule. However, even this effect cannot explain the large depolarization factor that is found in the vapors of CCl_4. It is also impossible to attribute it to experimental errors.

In all the cases named, further careful experimental investigations are necessary.

Measurements of the depolarization factor and the Kerr constant in vapor and gases allow us to determine the principal polarizabilities of the molecules on the basis of the formulas of Sec. 4 and, by using the qualitative considerations of Silberstein [117], to draw important conclusions on the structure of the molecules. Many investigations have been devoted to this problem, which have been summarized adequately by Stuart [115, 116].

Table II gives the data for the mean and principal polarizabilities of a sufficiently varied group of molecules. The last column gives the directions of the maximum polarizability, the dipole moment, and data on the character of the molecule.

In Sec. 4, a calculation of the depolarization of scattered light in gases was carried out under the assumption that the anisotropic molecules are randomly distributed, but are not moving and an averaging is carried out over all possible positions of the molecule in space relative to the laboratory system of coordinates. Here, of course, one has not taken into account the vibrational and rotational combination (Raman) light scattering.

The intensity of the vibrational Raman scattering in gases is extremely weak, $a \sim 10^{-3}$ [see Eq. (10.12)] and, therefore, if the lines of vibrational Raman scattering are strongly depolarized, then the correction to the depolarization will be negligible in accord with (10.12). A much more serious consequence would be expected for neglect of the rotational Raman scattering.

Raman and Krishnan [323] discovered that the rotational wing of the Rayleigh line grows in intensity with increase in the anisotropy of the molecule, and found a larger depolarization than for the unshifted scattering line. In order to make it clear whether one could continue to use the measurements obtained on the depolarization factor and the theory developed in Sec. 4 which gives the connection between Δ and the polarizabilities used for the interpretation of the molecular structure, Cabannes and Rocard [324], and later Bhagavantam [325], carried out a calculation that takes the rotation of the molecule into account. They treated the case of an elongated diatomic or polyatomic molecule, to which we also limit ourselves. The same derivation can be made for a molecule without a symmetry axis, as was shown by Born [124].

In order to take the rotation of the molecule into account in the calculation made in Sec. 4, by following [75, 115], it is necessary to assume that the system of coordinates attached to the molecule rotates relative to the laboratory system of coordinates with a frequency ω_r and, consequently, the Eulerian angles become functions of time.

In this case, in place of (4.5), we get

$$\left.\begin{aligned} P_z &= \left[b+(b_1-b_2)\left(\frac{\cos^2\theta}{2}-\frac{1}{3}\right)\right]E_{0z}\cos\omega_0 t+ \\ &\quad+\frac{(b_1-b_2)}{4}\cos^2\theta E_{0z}\{\cos[(\omega_0+2\omega_r)t+\delta]+ \\ &\quad+\cos[(\omega_0-2\omega_r)t+\delta]\}; \\ P_x &= \frac{b_1-b_2}{2}\sin\theta\cos\theta\cos\varphi E_{0z}\cos\omega_0 t+ \\ &\quad+\frac{b_1-b_2}{4}\sin\theta\cos\theta\cos\varphi E_{0z}\{\cos[(\omega_0+2\omega_r)t+\delta]+ \\ &\quad+\cos[(\omega_0-2\omega_r)t-\delta]\}-\frac{b_1-b_2}{4}\cos\theta\sin\varphi E_{0z}\{\sin[(\omega_0+ \\ &\quad+2\omega_r)t+\delta]-\sin[(\omega_0-2\omega_r)t-\delta]\}. \end{aligned}\right\} \tag{15.1}$$

It is natural that not only ω_0 is present in Eq. (15.1), but also $\omega_0 \pm 2\omega_r$, corresponding to the modulation of the light frequency ω_0 by the rotational frequency of the molecules ω_r.

Averaging over all orientations of the molecules in space, and assuming that dN molecules have frequencies of rotation lying in the range from ω_r to $\omega_r + d\omega_r$, we find for the intensity of the Rayleigh line

$$\left.\begin{aligned} I_z &= \text{const}\left[b^2+\frac{(b_1-b_2)^2}{45}\right], \\ I_x &= \text{const}\,\frac{(b_1-b_2)^2}{60} \end{aligned}\right\} \tag{15.2}$$

and for the intensity of the rotational wing for the Stokes and anti-Stokes parts,

$$\left.\begin{aligned} I_z^W(\omega_r) &= \text{const}\,\frac{(b_1-b_2)^2}{30}dN\left(1\pm\frac{2\omega_r}{\omega_0}\right)^4, \\ I_x^W(\omega_r) &= \text{const}\,\frac{(b_1-b_2)^2}{40}dN\left(1\pm\frac{2\omega_r}{\omega_0}\right)^4. \end{aligned}\right\} \tag{15.3}$$

The total intensity of the rotational wing (Stokes and anti-Stokes parts), distributed according to the Maxwell–Boltzmann law, will be

$$\left.\begin{aligned} I_z^W &= \text{const}\,\frac{(b_1-b_2)^2}{15},\\ I_x^W &= \text{const}\,\frac{(b_1-b_2)^2}{20}. \end{aligned}\right\} \tag{15.4}$$

Equation (15.4) shows that the wing exists in only those cases where there is an anisotropy $b_1 \neq b_2$ and, in accord with experiment [115], the intensity of the wing is the greater the larger the anisotropy.

It also follows from (15.4) that the depolarization factor of the rotational wing $\rho_V = I_X^W / I_Z^W = 3/4$ for illumination by linearly polarized light, and $\rho_u = 6/7$ for illumination by natural light.

The depolarization factor with account of the rotation of molecules by (15.2) and (15.4) is equal to

$$\Delta_v = \frac{I_x + I_x^W}{I_z + I_z^W} = \frac{\frac{1}{15}(b_1-b_2)^2}{b^2 + \frac{4}{45}(b_1-b_2)^2} = \frac{(b_1-b_2)^2}{3b_1^2 + 4b_1b_2 + 8b_2^2}. \tag{15.5}$$

Consequently, the account of rotation in no way changes the expression just obtained for the depolarization factor (15.5), which is entirely identical with (4.7). The same identity is obtained for Δ_u, and for molecules which do not have an axis of symmetry [115].

For the calculation of the intensity of the individual rotational lines and the distribution of these lines, one must use the quantum mechanical method [249]. However, in the calculation carried out, it was required to find only the integrated intensity of the entire rotational wing relative to the integrated intensity of the Rayleigh scattering.

In the latter case, a classical consideration is entirely possible.

The reason for the identity of the calculations of Δ which take the rotation into account and those which do not is evidently found in the following [115, 116]: the completed calculation, with account of rotation of the molecules, is suitable for any strongly increasing moment of inertia. The wing always gives the same depolarization factor $^3/_4$ or $^6/_7$ for irradiation by linearly polarized and natural light, respectively. Upon increase in the moment of inertia, the rotational lines contract toward the unshifted line and merge with it for a moment of inertia equal to infinity; then the depolarization remains the same.

But the case of the moment of inertia equal to infinity is also the case in which the molecules do not rotate, but are randomly oriented. It is necessary to assume that the identity of (15.5) with (4.7) occurs for just this reason.

16. Selective Light Scattering in Mercury Vapor

The fundamental study of selective scattering of light made by Landsberg and Mandel'shtam [76] belongs in the number of delicate and extremely complicated experimental investigations. As a result of this research, the phenomenon of selective scattering was discovered in mercury vapor and a radical departure from the Rayleigh law ($I \sim 1/\lambda^4$) was observed.

It was shown in Sec. 4 that one must expect a significant departure in the formula of Rayleigh–Cabannes (4.4) from the Rayleigh law when the wavelength of the incident light lies in the region of the characteristic absorption of the scattering material. Landsberg and Mandel'shtam [76] turned their attention to this fact.

However, for the region of characteristic absorption, not only do we have an increase in the factor $(n-1)^2$, which leads to the selective light scattering, but also a very strong increase in the absorption of the line whose wavelength lies in this region. Moreover, in mercury vapor, in which all the investigations were carried out, molecular formations appear which are capable of luminescence and which threaten to mask the sought-after phenomenon completely.

For the calculation of the scattering coefficient R_{90}, we shall use Eq. (4.4). Figure 53 shows the variation of $(n-1)^2$ as a function of the wavelength in the vicinity of the resonance line 2537 Å of the mercury spectrum, computed from the data of [328]. Landsberg and Mandel'shtam chose the lines 2558 Å and 2502 Å of the zinc spectrum as lines to excite the light scattering.

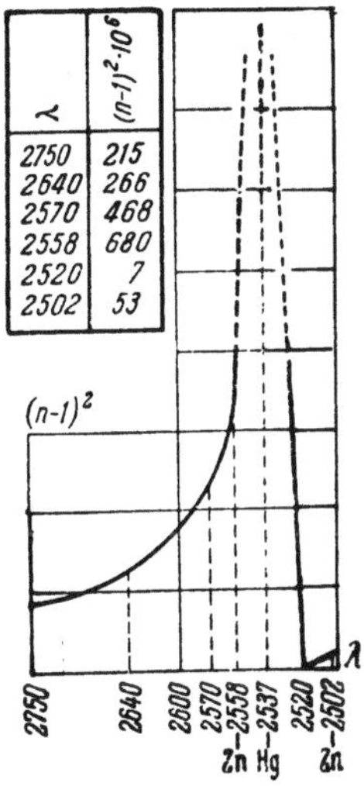

Fig. 53. Dependence of $(n-1)^2$ on the wavelength λ (Landsberg and Mandel'shtam [76]).

It is clear from the shape of the curve in Fig. 53 that for the lines of the zinc spectrum that were mentioned, $(n-1)^2$ changes by a factor of 12.4, while the change in $1/\lambda^4$ for these two lines amounts to no more than 9% (in the opposite direction). Therefore, $R_{\lambda=2558}$ is 12.4 times larger than $R_{\lambda=2502}$ and, consequently, the light scattering should have a purely selective character. According to the estimates of Landsberg and Mandel'shtam [76], the intensity of the scattering in the mercury vapor is 2.5 times greater than in ether and 70 times greater than in air, for the same number of scattering centers.

The exciting line 2558 Å can appear in the region of the molecular absorption band which, with increase in pressure of the mercury vapor, occupies the region of the spectrum from 2540 to 3000 Å.

It was made clear by a special experiment that for t = 280°C, the absorption of the line 2558 Å in the test cell does not exceed 10% and, for t = 270°C, and a pressure of saturated mercury vapor of 123 mm Hg, the absorption is so small that it cannot be observed. Therefore, the experiments were carried out under these favorable conditions.

But under these same conditions, mercury molecules amounting to 1% of the total number of atoms appear in the scattering volume. This is very unfavorable for the experiment, since some of the molecules emit the 2558 Å line, which lies inside the band of molecular fluorescence.

Before embarking on an account of the methods of eliminating these difficulties, and an account of the experimental results, we shall show that the experimental arrangement is the same as for the observation of light scattering in ordinary cases. In this experiment, a condensed spark between disk-shaped rotating zinc electrodes served as the light source [327].

The mercury vapor was placed in a horn-shaped vessel with plane polished walls for the entrance and exit of the primary light and for the exit of the scattered light; the vessel had a diameter of 40 mm and a length of 60 mm. The light of the spark was focused inside the volume and the scattered light was collected on the slit of a quartz spectrograph. The branch of the vessel in which the drop of mercury was placed was put in a bath of molten tin. The temperature of the branch was controlled by a thermocouple. All the rest of the test cell was placed inside the oven. The resonance line 2537 Å was almost completely absorbed by a resonant filter. The exposure time for photographing the scattered light varied from 2 to 6 hr. First of all, it was computed by means of (4.4) that the ratio of the scattering coefficient of the mercury vapor* at $t = 270°C$ to the scattering coefficient in saturated ether vapor at 35°C amounts to 0.185. The experimental determination of this ratio gave the figure 0.72 for the line 2558 Å.

From this the natural conclusion was drawn that about $^3/_4$ of the experimentally observed light arises as the result of molecular fluorescence. Under such conditions, it is impossible to count on the uniqueness of the experimental results and an effective method is needed which permits us to make the results unique and beyond doubt. Such a method, or rather two methods, were found which serve one and the same purpose. First, the property of molecular hydrogen to quench the fluorescence of the molecules of mercury was used; second, the different characteristics of the dependences of the light intensity of atomic scattering and resonance luminescence on the pressure of the mercury vapor were employed. The radiation of the continuous spectrum of fluorescence, which served as an obstacle in this experiment, corresponds to the transition from the molecular state (1^1S-2^3P_1) to the state (1^1S-1^1S). Moreover, the absorption transforms the molecule from the normal state to the state (1^1S-2^3P_1) and, finally, as a result of collisions, transitions of the molecule are possible to the state (1^2S-2^3P_0).

*The depolarization factor Δ is unknown for mercury vapor; therefore, it was assumed to be the same as for ether, which obviously gives an insignificant error.

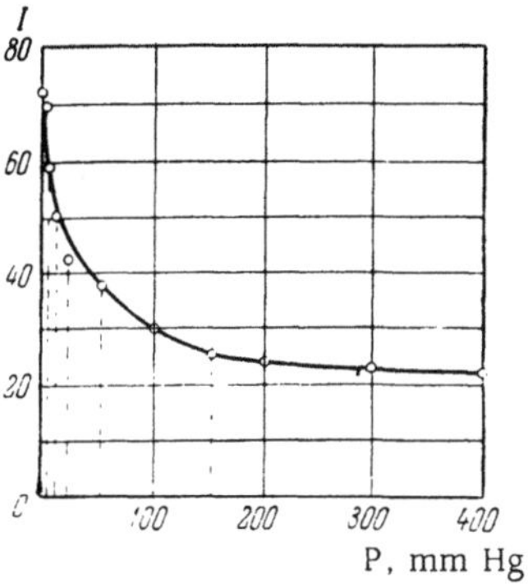

Fig. 54. Dependence on the hydrogen pressure of light intensity scattered in mercury vapor at a temperature of 270°C (Landsberg and Mandel'shtam [76]).

Radiative transitions from the three named states to the state (1^1S-1^1S) determine the radiation of the molecular bands in the region of wavelengths close to 2537 Å. By the assumption made in [76], one of the circumstances leading to the illumination of mercury vapors is the absorption of the 2558 Å line lying in the limits of the continuous spectrum of absorption of the mercury molecules. The state (1^1S-2^3P_1) that arises in this case determines the radiation of the mercury molecules and the line 2558 Å is radiated here with particular intensity. Landsberg and Mandel'shtam turned their attention to the fact that stimulated emission, proportional to the density of the incident radiation, is added. Thus the light scattered in mercury vapor contains: (1) the light of atomic selective scattering, and (2) the light of resonance fluorescence from the comparatively small number of mercury molecules.

For the suppression of the fluorescent light in the experiments of Landsberg and Mandel'shtam, molecular hydrogen was added to the mercury vapor. The energy of dissociation of the molecule of hydrogen (4.44 eV) is close to the energy of the mercury terms 2^3P_1 and 2^3P_0; therefore, one can suppose that the effect of the hydrogen appears in the disruption of the excited states of the mercury molecule, while the energy of excitation will be consumed in the dissociation of the hydrogen molecule. Thus, it is expected that the intensity of emission of luminescence will be strongly reduced, while the impurity gas has practically no effect on the selective atomic scattering. Such a significantly different effect on the resonance radiation of fluorescence and on the selective atomic scattering can be made clear by the following classical consideration.

The intensity of the scattered light $I \sim (n^2 - 1)^2 \sim A/[(\omega - \omega_0)^2 + \delta^2]$, where ω is the frequency of the exciting light, ω_0 is the characteristic frequency of the scattering center, and δ is the damping. The quantity δ is small and is determined by the damping, and amounts to about a thousandth or a hundredth of an Angstrom. The addition of an impurity gas increases δ, but in the case considered, the addition of a quenching gas increases δ much more, to several tenths of an Angstrom. Therefore, if $(\omega - \omega_0)^2$ is large in comparison with δ^2, then the change of δ^2 noted does not vary the intensity. In this case, the difference of the wavelengths corresponding to $(\omega - \omega_0)$ is equal to 21 Å and $\delta^2/(\omega - \omega_0)^2$ is a quantity of the order of several hundredths. In the case of resonance scattering, the intensity depends exclusively on δ and, consequently, the change of δ by tens can change the intensity by hundreds. The results of the experiment with the addition of molecular hydrogen are shown in Fig. 54. The light intensity scattered by mercury vapor at a temperature of 270°C is plotted along the ordinate, and the hydrogen pressure in mm Hg is plotted along the abscissa. The intensity of light scattered by saturated ether vapor at 35°C was chosen as 100 on the scale.

The circles on the curve of Fig. 54 denote the average value of the results of measurement. Taking into account the assumptions made above, that the total molecular (I_m) and atomic (I_a) scattering is observed in the experiment, and that the intensity of the atomic scattering depends on the hydrogen pressure, while the intensity of the molecular scattering falls off with increase in pressure, it can be assumed that I is a function of the pressure p.

For unlimited increase in $pI_m = f(p) \to 0$; for $p = 0$, $f(p) =$ const; therefore, I_m can be approximated by the Stern–Follmer function

$$I_m = \frac{I_0}{1+bp},$$

where I_0 is the intensity of molecular radiation without the quenching gas, and b is a constant depending on the nature of the quenching gas and the lifetime of the excited state. The total intensity of scattered radiation can then be written in the form

$$I = I_a + \frac{I_0}{1+bp}.$$

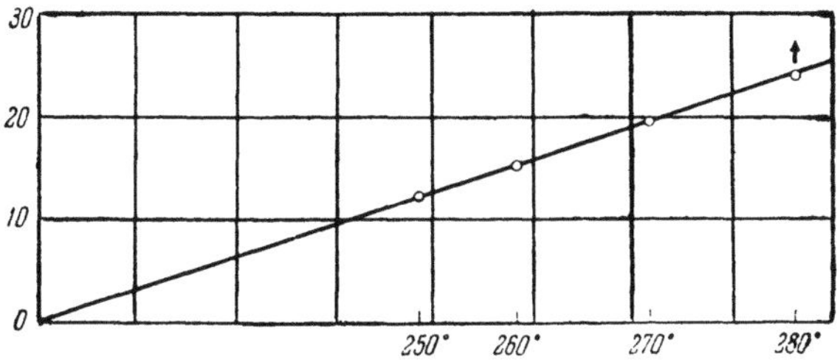

Fig. 55. Dependence of the intensity of the atomic scattering I_a in mercury vapor on the density of this vapor (Landsberg and Mandel'shtam [52, 76]).

The curve in Fig. 54 corresponds to

$$I_a = 18, \quad I_0 = 52, \quad b = 0.04.$$

Of course, the value of I_a is obtained from the curve $I = f(p)$, independent of the form of the function $f(p)$, as the asymptotic value for unlimited increase in p. After the method of separation of the atomic scattering from the molecular scattering had been found, one could conduct another control experiment. Inasmuch as the intensity of the molecular radiation is proportional to the square of the density of the mercury vapor, while the intensity of the atomic scattering is proportional to the first power of this density, measurements of I_a were made as a function of the density of the mercury vapor.

The results of the experiments are shown in Fig. 55. Taking what has been shown earlier into account, we see that the intensity at 280°C is somewhat reduced (by 5-10%) in comparison with what should be the case for this temperature as a consequence of the initial absorption of the 2558 Å line. The straight line of Fig. 55 actually shows that linear dependence exists between I_0 (in arbitrary units) and the density of the mercury vapor.

Moreover, this straight line passes through the origin, as it should. Consequently, in the experiments that have been described, the possibility of the measurement of the intensity of the selective scattering in mercury vapor is demonstrated, and the intensity of atomic scattering I_a is thus measured.

In the presence of molecular hydrogen, the ratio of the intensity of the lines $\lambda = 2558$ Å and 2502 Å in atomic scattering was measured. According to the theoretical calculations already mentioned, it follows from (4.4) that the 2558 Å line should scatter 11.5 times more strongly than the 2502 Å line. If it is taken into account that the exciting line 2558 Å in the source is 1.8 times as intense as the 2502 Å line, then the expected ratio of intensities is about 20. Measurements have shown that just such a ratio is actually observed for these intensities.

Thus, the presence of selective scattering was discovered in mercury vapor and its complete agreement with the conclusions of the theory has been shown. In this research of Landsberg and Mandel'shtam, polarization measurements were carried out, although these, it is true, have only qualitative character. It was shown by these measurements that while the molecular radiation has significant depolarization, the atomic scattering is practically completely polarized.

17. Fine Structure and Line Width of Rayleigh Scattering of Light in Gases

The only experimental investigation of the fine structure of the Rayleigh line, carried out for H_2 at a pressure of 100 atm, for N_2 and O_2 at a pressure of 80 atm, and for CO_2 at 50 atm, has been made by Venkateswaran [330]. The scattered light was observed at an angle of $\theta = 180°$. The gap between the mirrors of the Fabry–Perot etalon $t = 7.5$ mm. No discrete fine structure of the Rayleigh line was ever observed in one of the investigated gases. Only some broadening of the line-scattered light was observed in H_2; in the other cases, the spectrum of the scattered light reproduced the spectrum of incident radiation.

This result is surprising to us, and its interpretation in [330] is incorrect [335]. Taking into account that the problem of the fine structure of the Rayleigh line in gases is an important problem for understanding the scattering mechanism in general, we shall consider it and attempt to explain under what conditions one should expect a discrete fine structure of the scattering line in gases, and under what conditions it is impossible to expect it and, finally, what we must expect in specific cases of the experiment [330].

In the opinion of Venkateswaran [330], there should not be a discrete fine structure in the spectrum of scattered light in all the gases studied by him, but a broadening of the lines of scattering should be observed as a result of the Doppler effect on the molecules moving with a thermal velocity u. If the velocity u is directed along the bisectrix of the angle θ, then the halfwidth determined by the Doppler effect should be equal to $\delta\nu_D = \nu(u/c)2\sin(\theta/2)$, where c is the speed of light.

The thermal velocity corresponding to the temperature T is equal to $u = \sqrt{2\ln(2kT/m)}$ and, consequently, the halfwidth is

$$\delta\nu_D = 2\nu\sin\frac{\theta}{2}\sqrt{2\ln 2\frac{kT}{mc^2}}, \tag{17.1}$$

where m is the mass of the molecule.

According to (17.1), for H_2, with an exciting line 4047 Å and at room temperature, $\delta\nu_D = 0.22\ \text{cm}^{-1}$, and for three other gases, $\delta\nu_D \sim 0.06\ \text{cm}^{-1}$. The broadening of the scattering line in hydrogen is shown to be equal to $0.18\ \text{cm}^{-1}$, which the author assumes to be satisfactory agreement with the calculation of $\delta\nu_D$ from Eq. (17.1).

It seems to us that the assumption that the line broadening in a gas at a pressure of 50-100 atm is brought about by the Doppler effect on individual molecules moving with thermal velocities amounts to a misunderstanding. This misunderstanding has already been discussed in part in the Introduction.

Actually, if there is a medium in which one can separate volumes that are small in comparison with the wavelength, which contain within them molecules whose number is proportional to the size of such a volume, then, in this case, one must consider the collective action of the centers of scattering and the thermal motion of the molecules of the medium can be represented by Debye elastic waves. Change of frequency of the scattered light in such a consideration is determined by the reflection from the elastic thermal wave as from a mirror moving with velocity $\pm v$. In the direction satisfied by the Bragg condition, the maximum intensity of the scattered light will be observed, and the change of frequency will be due to the Doppler effect, but not by molecules moving with thermal velocities, distributed according to Maxwell, but by reason of reflection from the elastic Debye wave traveling with the velocity $\pm v$. Such a Doppler effect, as is well known, leads to the discrete fine structure of the scattering line (see Sec. 5) [24, 25, 134].

Landsberg [134] published a detailed critique of the research devoted to finding the effect of the Doppler shift of the light frequency in reflection from a metallic mirror. In these researches, it was assumed that the frequency shift came about because of the thermal motion of the atoms of the material of the mirror. As analysis, carried out by Landsberg [134], Lennuier [331], and Cabannes [332] has shown, incorrect conclusions are obtained even when the collective action of the atom or molecules is replaced by the action of the set of isolated atoms or molecules.

It follows from what has been said (see also Sec. 5) that the theory of the fine structure of the scattering line in a gas is no different from the theory of this phenomenon in liquids, so long as the mean free path length $\bar{l}$ of the molecules of the gas is much less than the length of the thermal elastic wave Λ which is important for scattering of light at the given angle. When the damping of the hypersound of wavelength Λ, which is propagated in the gas, is very small, a clearly delineated fine structure will be present in the spectrum of the light scattered by the gas. When the wavelength of the hypersound becomes equal to or less than the mean free path length of the molecules of the gas, the fine structure ceases to exist; in addition, the hydrodynamic consideration of the problem of sound absorption becomes impossible.

Consequently, for the case $\bar{l} \ll \Lambda$, as was shown in Sec. 5, the sharpness of the fine-structure lines will be determined by the criterion (5.28). We now calculate what will be the value of $\alpha\Lambda$ in gases.

For the sound absorption coefficient (5.21), we have

$$\alpha = \frac{\Omega^2}{2\rho v^3}\left[\frac{4}{3}\eta + \eta' + \varkappa\left(\frac{1}{c_V} - \frac{1}{c_p}\right)\right]. \tag{17.2}$$

The process of the scattering of sound in gases at pressures used in [330] is adiabatic. Actually, the frequency for which the sound propagation goes from adiabatic to isothermal is

$$\Omega_{\text{iso}} \sim \frac{v^2}{\chi}, \tag{17.3}$$

where χ is the coefficient of temperature conductivity ($\chi = \varkappa / c_p \rho$).

The frequency which is important for light scattering (5.8) is

$$\Omega_{\text{l.s.}} = \frac{2\pi v}{\Lambda} = \frac{4\pi n \cdot v \cdot \sin\frac{\theta}{2}}{\lambda}. \tag{17.4}$$

It is not difficult to see that even in the case of gases at the pressures shown in [330],

$$\frac{\Omega_{\text{iso}}}{\Omega_{\text{l.s.}}} = \frac{v \cdot \Lambda}{2\pi \chi} = \frac{\lambda \cdot v}{4\pi \cdot n\chi \cdot \sin\frac{\theta}{2}} \gg 1, \tag{17.5}$$

and, consequently, even in this case, the process of the hypersonic scattering is actually an adiabatic process. Therefore, one must use the expression for the Laplace sound velocity in the gas in (17.2), i.e.,

$$v = \sqrt{\gamma \frac{p}{\rho}}. \tag{17.6}$$

Taking (17.2) and (17.6) into account, we get the following expression for the sound absorption per wavelength:

$$\alpha\Lambda = \frac{2}{\gamma} \pi^2 \frac{f}{p} \left[\frac{4}{3}\eta + \eta' + \varkappa\left(\frac{1}{c_V} - \frac{1}{c_p}\right)\right]. \tag{17.7}$$

It is difficult to compute η'; therefore, we set $\eta' = 0$ for the time being. Later, we shall return to this problem and show that such a neglect is completely justified.

For $\eta' = 0$, we get from (17.7)

$$\alpha\Lambda = \frac{2\pi^2}{\gamma} \frac{f}{p} \left[\frac{4}{3}\eta + \varkappa\left(\frac{1}{c_V} - \frac{1}{c_p}\right)\right]. \tag{17.8}$$

Taking into account the well-known gas-kinetic relations

$$\varkappa = b c_V \eta; \quad \eta = 0.5 \rho \bar{u} \bar{l}, \tag{17.9}$$

where the average thermal speed of the molecule $\bar{u}$ is related to the speed of sound by the equation

$$v = \bar{u}\sqrt{\gamma \frac{\pi}{8}}\,. \qquad (17.10)$$

Also taking into account that b is a constant equal to 1.9 for diatomic and 1.75 for triatomic gases, we find

$$\alpha\Lambda = A\frac{\bar{l}}{\Lambda}\,. \qquad (17.11)$$

Here,

$$A = \left[\frac{2\pi}{\gamma}\right]^{3/2}\left[\frac{4}{3}\gamma + b(\gamma - 1)\right]. \qquad (17.12)$$

From (17.12), we get A = 25 for a diatomic and A = 24 for a triatomic gas.

On the basis of Eq. (17.11) and the computed values of A, one can draw the conclusion that at atmospheric pressure ($\bar{l} \sim 10^{-5}$ cm), $\alpha\Lambda \gg 1$ and, consequently, it is impossible to observe any fine structure. But at a pressure of about 20 atm, $\alpha\Lambda < 1$, and under the conditions of the experiment [330], $\alpha\Lambda = 0.1$ (Table 8), and thus one can expect that a sharp fine structure will be observed in the spectrum of scattered light.

Table 8 also gives the values of $\Delta\nu$, the shift of the MB components computed from Eq. (5.9). The halfwidths of the MB components $\delta\nu_{MB}$ are computed from (5.20) and the halfwidth of the central component from (5.37):

$$\delta\nu_c = \left(\frac{2\pi}{\Lambda}\right)^2 \frac{\chi}{\pi c} = \frac{4\pi\varkappa v^2}{c\gamma c_p p \Lambda^2}\ \mathrm{cm}^{-1}. \qquad (17.13)$$

The confirmation that a sharp fine structure Eq. (17.11) will be observed for pressures higher than 20 atm is valid only when the bulk viscosity coefficient η', which we neglected in (17.7), actually plays no significant role in the phenomenon being studied. The fact that η' relaxes for much lower values of f/p than the quantity characterizing the fine structure is justification for such a neglect. In confirmation of what has been pointed out, we plot the experimental data for $\alpha\Lambda$ as a function of f/p for carbon dioxide and hydrogen [333, 334]. Actually, it is seen from Fig. 56 that even for $f/p \sim 1$ MHz/atm, the bulk viscosity η' ceases to give any appreciable contribution to the absorption, and all the absorption is de-

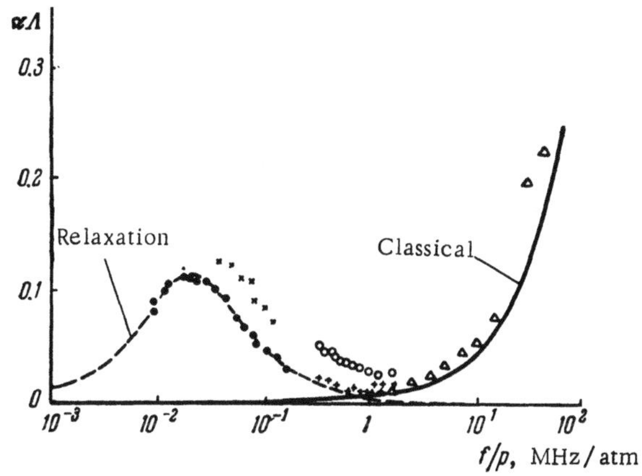

Fig. 56. Dependence of the sound absorption coefficient per wavelength on the ratio of the sound frequency to the pressure for gaseous CO_2 (Zartman [333]). The different symbols correspond to measurements of different authors.

termined by η. For hydrogen and for diatomic gases generally, the maximum of the relaxation of η' lies at much higher values of f/p than for triatomic gases, but even in the case of diatomic gases η' does not play a role for light scattering (Fig. 57).

Table 8 shows the values of $\alpha\Lambda$ taken from experiment for the values of f/p corresponding to the experimental conditions [330]. Comparison of the computed and measured values of $\alpha\Lambda$ shows that they are in excellent agreement with one another, and this means that for the phenomenon of light scattered in these gases η' actually equals zero with good accuracy. Thus, it follows from the data of Table 8 that in all cases considered in [330], a sharp fine structure should be observed.

Table 8 also gives some parameters which characterize liquids, in the scattered light spectrum of which a clear fine structure is observed [29, 30, 53]. The form of the fine structure in carbon bisulfide and benzene is especially sharp. The fine structure of carbon tetrachloride is poorer but still clearly evident. Comparison of the data given in Table 8 for hydrogen and benzene shows that both these substances are characterized by similar

Table 8. Some Parameters Which Characterize Gases in Which the Spectrum of the Rayleigh Light Scattered Has Been Studied (Scattering Angle $\theta = 180°$, Wavelength $\lambda = 4047$ Å, Temperature 20°C), and Liquids in Which the Discrete Fine Structure Has Been Observed (Scattering Angle $\theta = 90°$, Wavelength $\lambda = 4358$ Å, Temperature 20°C)

	Scattering medium	Pressure, atm	f/p, MHz/atm	Velocity of sound* v, m/sec	Index of refraction,† n	Wavelength of hypersound $\Lambda \cdot 10^5$, cm	Freq. of hypersound $f \cdot 10^{-9}$, sec^{-1}	Abs. coeff. $\alpha \cdot 10^{-3}$, cm^{-1}	$\alpha\Lambda$ computed	$\alpha\Lambda$ measured	$\Delta\nu$, cm^{-1}	$\delta\nu_{MB} \cdot 10^3$, cm^{-1}	$\delta\nu_c \cdot 10^3$, cm^{-1}
Gas	Nitrogen N_2	80	21	339	1.0232	1.97	1.69	6.35	0.12	0.1	0.054	2.28	3.4
	Hydrogen H_2	100	73	1460	1.0132	2.0	7.3	8.75	0.17	0.1	0.243	12.8	29
	Oxygen O_2	80	20.6	327	1.0205	1.98	1.65	6.30	0.12	—	0.055	2.94	3.8
	Carbon dioxide CO_2	50	33.2	332	1.0124	2.0	1.67	5.8	0.11	0.1	0.055	2.04	2.9
Liquid	Benzene C_6H_6	1	—	1470	1.522	2.02	7.26	8.05	0.163	0.1	0.25	8.5	6.5
	Carbon tetrachloride CCl_4	1	—	1040	1.472	2.1	5	2.6	0.545	0.32	0.16	28.7	6.3
	Carbon bisulfide CS_2	1	—	1265	1.674	1.85	6.8	2.64	0.049	0.05	0.23	3.53	3.1

*Velocity of sound in a real gas depends on the pressure in approximately the following way: $v = \sqrt{\gamma \frac{(R_g T + 2B'p)}{M}}$, where B' is the second virial coefficient, equal to two; R_g is the gas constant; M, the molecular weight; and P, the gas pressure. For 50-100 atm, v increases relative to the velocity at atmosphere pressure by ~2%, which was not taken into account.

†The index of refraction was found from the formula $r_0 = (n-1)\frac{760(1+\alpha t)}{p(1+\beta t)}10^{-6}$, where $r_0 = n-1$ for atmospheric pressure, $\alpha = 1/273\ deg^{-1}$, we have set β equal to zero.

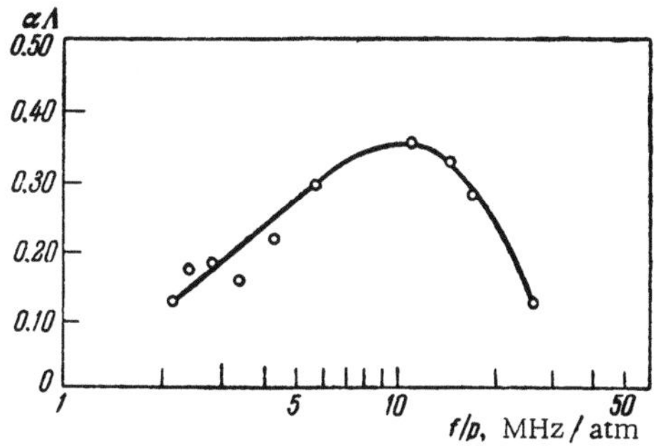

Fig. 57. Dependence of the sound absorption coefficient per wavelength on the ratio of the sound frequency to the pressure for gaseous H_2 (Markham, Beyer, and Lindsay [422]).

values of v, α, $\alpha\Lambda$, $\Delta\nu$, and $\delta\nu_{MB}$; only $\delta\nu_c$ is appreciably different, as a consequence of the large thermal conductivity of hydrogen. However, if it is taken into consideration that the width of the exciting line in the experiments of [330] is much greater than the true widths $\delta\nu_c$ and $\delta\nu_{MB}$, then this difference is shown to be unimportant.*

Thus, practically the same fine-structure picture should be observed in liquid benzene and in gaseous hydrogen at 100 atm and room temperature.

In nitrogen, oxygen, and carbon dioxide, the observation of the fine structure is accompanied by great difficulties, inasmuch as the shift of the MB components in the spectrum of the light scattered by these substances (taking into account the width of the exciting line), does not exceed the limits of the halfwidth of the central component $\delta\nu_c$. Here, therefore, only the width of the central line could be observed.

Meanwhile, in the experiment of Venkateswaran [330], neither broadening of the central line nor discrete components of the fine structure were observed in the last three gases.

*In [330], the visible width of the incident mercury line 4047 Å was $0.08\,cm^{-1}$.

The broadening of the central line was observed by him only in the spectrum of light scattered by hydrogen where, according to the analysis given above, there should be a clear discrete line structure. It is difficult to say how such a conflict between theory and experiment can be explained.*

The classical phenomenological theory of the fine structure in a gas, developed in this section, can scarcely be shown to be invalid. On the other hand, the only experiment [330] is in qualitative contradiction with the theory.

Therefore, one must regard the experimental problem of discovering the fine structure of the Rayleigh line for light scattered by gases as unresolved.

The solution of such a problem is important to us in principle, although it is undoubtedly very difficult. One can hope that the new powerful lasers with extremely sharp exciting lines will considerably facilitate the solution of this experimental problem.†

At low pressures (isolated molecules) of the gas, the scattering spectrum consists of the Rayleigh line and the rotational wing. This phenomenon is described with extraordinary completeness by the theory of Placzek and Teller [249].

---

*We are inclined to assume that great experimental errors entered into such results. In particular, the extremely small quantity of light used requires very long exposures. In the work of Venkateswaran [330] the exposures lasted 15 days and while measures were taken in his experiment to thermostat the region where the interference part of the apparatus was located, it is not said whether pressure control precautions were also taken. Barostatting is no less essential than thermostatting (see Appendix II). If the pressure could change by 2 mm Hg during the 15 days of exposure, then the absence of the discrete components in the spectrum of the light scattered by the hydrogen with an etalon thickness $t = 7.5$ mm and reflection coefficient ~85% could be explained.

†At the end of 1965, investigations were published in which stimulated MB scattering was discovered in compressed gases (see Sec. 34). Thus, the presence of fine structure under these conditions was demonstrated.

For an increase in the pressure to 60 atm, a change in the rotational wing in the gas has been studied experimentally by Weiler [164]. At high pressures the "rotational" wing is described by an intensity distribution which is characteristic for liquids rather than for gases.*

We also turn our attention to still one more possible mechanism of the line broadening of Rayleigh scattering in gases, as the result of an increase in pressure, which was pointed out by Ginzburg [141].

This reason is that, in the collision of the scattering molecules, a change in their direction of motion takes place. Consequently, the component of the velocity in the direction of observation will be changed. This effect will depend on the pressure and, as a consequence of the Doppler effect, will lead to a broadening of the Rayleigh line.

It has been possible [141] to show that for the case of pressures below atmospheric ($\bar{l} \gg \Lambda$), the intensity of the "wing," produced by such an effect will depend strongly on the frequency, to wit:

---

*Attempts to describe the wing of the Rayleigh line by the same mechanism which leads to the broadening of the emission line in nonquenching collisions [72, 324], as Ginzburg has pointed out [141], cannot lead to the desired explanation. The scattered radiation stimulated by the incident line of frequency ω does not change its phase and frequency because of the collision. Only the phase and amplitude of the characteristic vibrations of the oscillator change, whose frequency ω_0 is far from ω. Therefore, only at the instant of collision can the frequency of the scattered waves be changed somewhat, but the duration of the collisions is short ($\Delta\tau/\tau_0 \ll 1$) and therefore such a frequency change can be neglected.

The collision halfwidth is $\delta\omega = 1/\tau_0$ (where τ_0 is the time of free flight) only in those cases in which the collision time $\Delta\tau$ is much smaller than τ_0, which is the case neither in compressed gases nor (even less) in liquids. Broadening of the emission line as a result of nonquenching collisions takes place because of the phase change of the characteristic emission of the oscillator in the collision.

$$I \simeq \frac{Czb^2}{(\Delta\omega)^6}, \tag{17.14}$$

where $\Delta\omega$ is calculated from the maximum of the Rayleigh line, z is the number of collisions in the scattering volume per second, b is a quantity dependent on kT, λ, and θ (the scattering angle), and C is a constant.

For the case $\bar{l} \simeq \Lambda$, it has not yet been possible to solve the problem, while for the case $\bar{l} \ll \Lambda$, the theory was set forth at the beginning of this section.

An estimate made according to (17.14) showed that this effect is already negligible at a distance of about 1 Å from the center of the unshifted Rayleigh line.

As was shown above, the classical consideration of the problem [141] leads to the conclusion that the collision of molecules at low gas density should not affect the width of the scattering line.

It is true that Weisskopf [250], on the basis of quantum mechanical considerations, reached a different conclusion.

A molecule in a state with a finite lifetime τ is energetically fixed with an accuracy to within h/τ; therefore, the energy of the scattered quantum can differ by $\Delta E \sim h/\tau$.

Problems of the width of the Rayleigh scattering line in gases have been considered in detail from the quantum mechanical viewpoint by Sobel'man [304], who showed that quantum theory also leads to the results which are identical with the classical for the Rayleigh scattering.

Chapter V

Molecular Scattering of Light in Liquids

18. Results of Measurement of the Absolute and Relative Intensity of Scattered Light in Liquids

Absolute Measurements. The measurement of the absolute intensity of scattered light in liquids, begun more than forty years ago by Martin and Lehrman [336], continues to the present day. Especially great attention is paid to such measurements today, when they have not only a scientific but also a practical value.

The practical value of the absolute measurements became evident after the researches of Debye [42] and his coworkers showed that the molecular weight of macromolecules and other of their important characteristics could be determined from the absolute intensity of the scattered light in solutions of polymers and proteins.

The method of light scattering in such problems becomes especially convenient if the complicated absolute measurements can be replaced by the comparatively simple relative measurements. But in the latter case, one must have at one's disposal a standard, the absolute intensity of the light scattering in which is extremely well known. As such standards, use has been made of a number of substances [43, 337], but liquid benzene has been used most frequently. It has therefore been investigated most frequently.

Table III (Appendix III) gives all the absolute measurements of R_{90} known to the author at the present time. It can be seen immediately that the divergence between the different measurements in benzene (and also in certain other liquids) is very great. Some of the measurements in benzene group about the value $R_{90} \cdot 10^6 \sim 32\ cm^{-1}$, while others group about the value $R_{90} \cdot 10^6 \sim 46\ cm^{-1}$.

Consequently, the difference between the two groups of measurements is greater than 40% and obviously cannot be explained by random errors in the experiment. It must be noted that the divergences apply only to the absolute values. The relative values give complete consistency with one another.

At first [29] it appeared that the photoelectric method always gave a "high" value, while the photographic method led to a "low" value of R_{90} (if we exclude the data of Raman and Rao [338], which are subject to great errors). However, special photoelectric measurements made by Harrand [342] according to a suggestion of Cabannes [356] gave a "low" value of R_{90}.

The discussion of the photographic and photoelectric methods is still being carried on, chiefly from the viewpoint of the necessity of introducing various corrections into the measured values, in particular, corrections for the index of refraction c_n [351, 356, 358].

In the discussion of the methods of measurement of the scattering coefficient in Sec. 11, some of the arguments on the side of the "high" and "low" values of R_{90} were given, and we shall not go further into the details of the problem, the more so in that the discussion has not yet been completed. We shall show only that a rather complicated problem has been produced here.

The usual and natural method of checking a theory is the comparison of its conclusions with the results of experiment. Here it is impossible to check the theory, because there is no agreement between the two different groups of experiments, although there is agreement within each of the groups. Possibly that is why attempts have been made to support the correctness of the various measurements by their agreement with the theory [198, 341, 351]. However, even here, the situation is complicated by the fact that different values are given in the different sources for the parameters which enter into the formulas of the theory. Therefore, one and the same formula gives satisfactory agreement with the "low" and "high" values of R_{90}, depending on what values of the parameters are taken.

The problem of the choice of the parameters for the calculation according to the formulas of the theory of light scattering cannot be regarded as simple, but evidently it can be solved in an entirely satisfactory way (see Sec. 19).

Discussion of the General Formulas and Comparison of Calculated and Measured Values. It was shown in Sec. 7 that if there is dispersion, then, under known conditions, one can compute the scattering coefficient according to Eqs. (1.93) and (1.94), using the corresponding dynamic values of the parameters. For convenience of discussion we shall again write Eqs. (1.93) and (1.94):

$$R_{90}=\frac{\pi^2}{2\lambda^4}\left[\left(\rho\frac{\partial\varepsilon}{\partial\rho}\right)_S^2\beta_S kT+\left(\frac{1}{\sigma}\frac{\partial\varepsilon}{\partial T}\right)_p^2\frac{\sigma^2 kT^2}{c_p}\right]\frac{6+6\Delta_u}{6-7\Delta_u}\,. \qquad (1.93)$$

As has been shown, under the condition (1.66), Eq. (1.93) transforms into the Einstein–Cabannes formula:

$$R_{90}=\frac{\pi^2}{2\lambda^4}\left(\rho\frac{\partial\varepsilon}{\partial\rho}\right)_T^2\beta_T kT\left(\frac{6+6\Delta_u}{6-7\Delta_u}\right). \qquad (1.94)$$

Equation (1.94) has frequently been compared with the experimental results.

Here we shall speak about the comparison of (1.94) only with the "low" values of R_{90}.

In the calculation of R_{90} from (1.94) or (1.93), there arises the problem as to what values of the parameters β_T, β_S, $(\rho\,\partial\varepsilon/\partial\rho)_T$, and $(\rho\,\partial\varepsilon/\partial\rho)_S$ must be used. It is clear from what was said in Sec. 7 that, strictly speaking, all these parameters should be measured at a frequency of 10^{10} Hz, which defines the phenomenon of light scattering.

Since the dispersion of the sound velocity had not been discovered (in spite of the attempts made to discover it, see Sec. 23), one could have used the static values of β and the thermodynamic formula (1.94); in any case, it appeared that this is so from the very meaning of the theory. But the use of the static values of the parameters for the calculation of R_{90} from (1.94) led to a divergence of the measured and computed values [29, 53]. This divergence of theory and experiment forces us to assume that the values measured under static conditions of the quantities β_T, β_S, $(\rho\partial\varepsilon/\partial\rho)_S$ and $(\rho\partial\varepsilon/\partial\rho)_T$ cannot be used for the calculation of the intensity of the scattered light [29], which corresponds to processes taking place at frequencies of about 10^{10} Hz. The optical dielectric constant ($\varepsilon \equiv n^2$) can change appreciably at frequencies of about 10^{15} Hz. Therefore, the static value of $(\rho\partial\varepsilon/\partial\rho)$ can scarcely be different from this quantity measured at a frequency of 10^{10} Hz if there are no processes now known to us which relax at a frequency much less than 10^{10} Hz.

This problem has been little discussed [53], although one can refer in particular to the note of Cabannes [75], that the static measurement of $(\rho\partial\varepsilon/\partial\rho)$ carried out in a large volume, cannot be suitable for the characteristics of $(\rho\partial\varepsilon/\partial\rho)$ at a point (volume of fluctuations).

The meaning of this remark can be seen in the fact that in principle the values of $(\rho\partial\varepsilon/\partial\rho)$ measured statically and dynamically will be different if we assume that this quantity is a function of a certain region [279]. This region should have the order of magnitude determined by the intermolecular interaction. Taking this circumstance into account, we can write the result that, in the measurement of $(\rho\partial\varepsilon/\partial\rho)$ at the sound frequency corresponding to the wavelength Λ,

$$\rho\frac{\partial\varepsilon}{\partial\rho}=\left(\rho\frac{\partial\varepsilon}{\partial\rho}\right)_{0,S}\left\{1+k\left(\frac{a}{\Lambda}\right)^2\right\}, \tag{18.1}$$

where $(\rho\partial\varepsilon/\partial\rho)_{0,S}$ is the static adiabatic value of this quantity, a is a constant having the dimensions of length and equal in order of magnitude to 10^{-7} cm, k is a dimensional constant ~1, the sign of which is not determined beforehand. For the frequencies of 50, 10^7, and 10^{10} Hz, $k(a/\Lambda)^2$ is of the order of 10^{-20}, 10^{-10}, and 10^{-4}, respectively. Thus, account of this circumstance does not explain the divergence, and the problem needs further consideration.

Cabannes [75] and a number of other authors [54, 56, 179], after a comparison of R, computed on the basis of the static data, with the experimental value, gave up the use of the static values of $(\rho\partial\varepsilon/\partial\rho)$, and turned to the calculation of this quantity from a formula of the Lorentz–Lorenz type [89] or that of Gladstone and Dill [72].

The central point in the solution of the problem of the connection between ε and ρ, or, what amounts to the same thing, of $(\rho\partial\varepsilon/\partial\rho)$, lies in finding the values of the internal field acting on the molecule. It is well known how this problem was solved by Lorentz [89]. The connection that he found between ρ and ε is expressed by the formula which now bears the name Lorentz–Lorenz:

$$\frac{\varepsilon-1}{\varepsilon+2}=C'\rho, \tag{18.2}$$

where C' is a constant.

Einstein [14] used (18.2) for the calculation of $(\rho\,\partial\varepsilon/\partial\rho)_T$, and obtained

$$\left(\rho\frac{\partial\varepsilon}{\partial\rho}\right)_T=(\varepsilon-1)\left(\frac{\varepsilon+2}{3}\right). \tag{18.3}$$

Substituting (18.3) in (1.94), we get

$$R_{90}=\frac{\pi^2}{2\lambda^4}(\varepsilon-1)^2\left(\frac{\varepsilon+2}{3}\right)^2 kT\beta_T\left(\frac{6+6\Delta_u}{6-7\Delta_u}\right). \tag{18.4}$$

The scattering coefficients computed from (18.4) are shown to be appreciably larger than those measured by experiment (the "low" values), and in many cases larger than those computed from (1.94) with the help of the static values of $(\rho\partial\varepsilon/\partial\rho)_T$.

Rocard [357] subjected the Einstein method of the calculation of $(\rho\partial\varepsilon/\partial\rho)_T$ from Eq. (18.2) by simple differentiation to a criticism. He showed that the expression $\varepsilon+2$ in Eq. (18.3) appears because of the consideration of the action of molecules located outside the physically infinitely small sphere (the Lorentz sphere) surrounding the molecule under study. The fluctuations of ε in the material outside the chosen sphere is shown to have a very small effect on the change of the field inside the sphere; therefore, ε can be regarded as virtually constant in the term $(\varepsilon+2)$. Consequently, only the factor $(\varepsilon-1)$, which is due to molecules within the small sphere, where the density fluctuations are considerable, need be differentiated with respect to ρ.

In such a case, setting $(\varepsilon+2)=\text{const}$, we get, in place of (18.2),

$$\varepsilon-1=C''\rho. \tag{18.5}$$

Then

$$\left(\rho\frac{\partial\varepsilon}{\partial\rho}\right)_T=\varepsilon-1 \tag{18.6}$$

and Eq. (1.94) takes the following form:

$$R_{90}=\frac{\pi^2}{2\lambda^4}(\varepsilon-1)^2 kT\beta_T\frac{6+6\Delta_u}{6-7\Delta_u}\ . \tag{18.7}$$

The computations according to (18.7) for certain substances are found in satisfactory agreement with experiment; however, for a whole series of other substances, the divergence is as great as before. The divergence of the experimental data from those computed from (18.7) has been interpreted until very recently as an inadequacy of the Einstein theory of light scattering in a liquid. This problem has already been discussed in Sec. 1; it was shown that there is no basis for such conclusions.

This is a case of light scattering by the fluctuations $\Delta\varepsilon'$ which, in accord with Ginzburg [45] (see Sec. 1), could take place in principle and not be reduced to fluctuations of the anisotropy and the density.

However, one could hardly draw any conclusions from the divergence of Eqs. (18.7) or (18.4) from experiment on the imperfection of the thermodynamic theory of light scattering in liquids, most of all because the application of Eqs. (18.2) and (18.5) to the calculation of $(\rho\partial\varepsilon/\partial\rho)$ in a liquid is completely without foundation, in spite of several improvements introduced in Eq. (18.2) [357]. The inapplicability of Eqs. (18.2) and (18.5) in liquids follows from the inadmissibility of the assumption that the molecules of the liquid are located entirely randomly. Even more incorrect is the neglect, in the case of liquids, of the dimensions of the molecules, as is done for the derivation of (18.7), because the distance between the molecules in the liquid is of the order of their dimensions. It is interesting to note that in his "Theory of Electrons," Lorentz [89], comparing his formula and the empirical formula $n-1=C'''\rho$ with experiment, saw that there was no satisfactory agreement, and pointed out the incorrect account of the internal field as the reason for the observed disagreement.* So far as Eq. (18.5) is concerned, although today one calls it the formula of Gladstone and Dill, it was obtained theoretically by Laplace, and Lorentz called it the law of Laplace. Lorentz wrote about this law: "So far

*We have in mind Eq. (18.2) and the formula $n-1=C'''\rho$. Equation (18.2) was established by Lorentz for a cubic crystal.

as Laplace's law is concerned, we can say no more, since in all cases it leads to much less satisfactory results than either of the other formulas."

Therefore, the divergence or agreement of Eqs. (18.4) and (18.7) with experiment cannot serve either as proof or disproof of the static theory of light scattering in the liquid that we have just considered. There also exists a whole series of relations of an empirical character, of the type of (18.2) and (18.5); however, they are at least as unsatisfactory as those already considered and, therefore, we shall say nothing in detail about them.

The modern position of the theory of the liquid state is such that it is not possible to solve the problem of the connection of the optical dielectric constant with the density of the medium. On the other hand, Eqs. (18.2) and (18.5), which are needed to compute the dielectric constant empirically for a liquid, are unsatisfactory in application to the problems of light scattering.

It is therefore seen that at the present time it is best to use the experimental values of $(\rho \partial \varepsilon / \partial \rho)$ choosing, however, a method of determination of this quantity which is based on phenomena whose nature is close to the phenomenon of light scattering.

Starting from this position, a method has been developed for determining $(\rho \partial \varepsilon / \partial \rho)_S^2$ from the phenomenon of the diffraction of light by ultrasound, described in Sec. 13. The results found by this method are given in Table IV. Unfortunately, an inaccurate account of the "acoustic wind" makes the dynamic value of $(\rho \partial \varepsilon / \partial \rho)$ unreliable. From the results of measurements given in Table IV, it follows that for benzene and toluene, the difference between the static and dynamic values of $(\rho \partial \varepsilon / \partial \rho)$ is practically nonexistent. In the other cases studied, there is some difference between these values. It is possible that it is produced by the inexact account of the "acoustic wind," inasmuch as there is no physical basis to expect dispersion of $(\rho \partial \varepsilon / \partial \rho)$ at frequencies of about 10^7 Hz, as has been repeatedly pointed out [29, 53, 288]. In those cases in which there are no values of $(\rho \partial \varepsilon / \partial \rho)_S$ measured in the dynamic regime, one can, with sufficiently good approximation, replace $(\rho \partial \varepsilon / \partial \rho)_T$ or $(\rho \partial \varepsilon / \partial \rho)_S$ by another measured quantity, namely $2[(n/\sigma)(\partial n/\partial T)]_p$. Actually, if we set $[(\rho \partial \varepsilon / \partial \rho)_T - (\rho \partial \varepsilon / \partial \rho)_S] = 0$, then it follows from (1.61) that we neglect a quantity equal to $(\partial \varepsilon / \partial T)_\rho (T\sigma / c_V \beta_T \rho)$, which for benzene, for example, $\sim 5 \cdot 10^{-2}$. If such a quantity can actually be neglected, then we get from (1.65):

$$\left(\rho \frac{\partial \varepsilon}{\partial \rho}\right)_T = \left(\frac{1}{\sigma} \frac{\partial \varepsilon}{\partial T}\right)_p \equiv 2\left(\frac{n}{\sigma} \frac{\partial n}{\partial T}\right)_p, \tag{18.8}$$

and the Einstein–Cabannes formula (1.94) can be written approximately as

$$R_{90} = \frac{2\pi^2}{\lambda^4}\left(\frac{n}{\sigma} \frac{\partial n}{\partial T}\right)_p^2 \beta_T kT \left(\frac{6+6\Delta_u}{6-7\Delta_u}\right). \tag{18.9}$$

Equation (18.9) has been used in a number of cases for the calculation of the scattering coefficient, although in other cases (18.7) has been used up to the present time. At this time, when there is no acceptable approach to the calculation of the internal field in a liquid, one must use (18.9), (1.94), and (1.93) for the calculation of R_{90}, substituting the experimentally determined parameters in these formulas [45, 129, 146]. Unfortunately, the tabulated values of the parameters in the different comparisons and papers do not have the same values of the same quantities. Therefore, some difference is always possible in the computed values of R_{90} found by the different authors using somewhat differing parameters.

Table IV (Appendix III) gives the values of the constants which we used for the calculation of the scattering coefficients, taken from different sources. Table III (Appendix III) gives the results of the measurement of the scattering coefficient in liquid benzene by different authors.

Table 9 lists the scattering coefficients averaged separately for the "high" and "low" values, after reduction to a temperature of 20°C by use of the temperature coefficient [179] $[(1/R_{90})(\Delta R_{90}/\Delta T) = 0.007]$* and to a wavelength of 4358 Å by use of the Rayleigh law $(1/\lambda^4)$. Table 9 gives the results of the calculation of R_{90} according to the different formulas and with different parameters. In the calculation of the scattering coefficient, it is necessary to keep in mind that in a number of cases it is not possible to use the static value of the compressibility, as was done earlier in all cases. And

* From the data of Vaucouleurs [56], $(1/R_{90})(\Delta R_{90}/\Delta T) = 0.005$.

Table 9. Comparison of Measured and Computed Values of R_{90} in Benzene

$R_{90} \cdot 10^6$, cm^{-1}, averaged over "low" and "high" values shown in Table III	$R_{90} \cdot 10^6$, cm^{-1}, computed from the formulas					
	(1.94)*		(1.93)†		(18.9)‡	
	1	2	3	4	5	6
32.0 46.1	37.6	36.0	31.8	47.7	38.6	34

*1. The static values of $(\rho\,\partial\varepsilon/\partial\rho)_T$ and β_T were used. 2. The dynamic values of $(\rho\partial\varepsilon/\partial\rho)_T$ and the static value of β_T were used.

†The dynamic values of $(\rho\,\partial\varepsilon/\partial\rho)_T$ and the dynamic value of β_S, computed from the hypersonic velocity in benzene, were used.

‡For an explanation, see the text.

this is especially important to keep in mind. The fact is that, in a number of our researches [29-36], dispersion of the sound velocity was discovered in such liquids as benzene, carbon bisulfide, carbon tetrachloride, chloroform, etc., while the velocity of hypersound at a frequency of 10^{10} Hz for different liquids is larger than the ultrasonic velocity by 10-20%, which decreases the adiabatic compressibility by 20-40%. Such a difference in the compressibility obviously ought to be taken into account in the calculation of R_{90}.

In the fourth column of Table 9, the results are given of the calculation of R_{90} carried out by Carr and Zimm [198], and reduced by us to 20°C by the same temperature coefficient according to which the experimental values were reduced. It is immediately necessary to turn our attention to the fact that the authors of [198] did not have available data on the dispersion of sound velocity and, therefore, used a value of β_T which is unsuitable for the calculation of R_{90} in benzene.

The calculation which is carried out in [198] can be corrected if we find β_T, which is necessary for the calculation of R_{90} in benzene.

Table 10. Comparison of the Scattering Coefficient $R_{90} \cdot 10^6$, cm^{-1}, Measured in Certain Liquids, with the Calculation for 20°C and λ = 4358 Å

Liquid	Exptl. value at various temperatures (see Table III Appendix III)	Eq. (1.94) static	Eq. (1.94) dynamic	Eq. (1.93)	Eq. (18.9)
Water	2; 3.08; 2.89; 4.3	2.3	2.0	2.0	
Ether.	10.1	11.6	9.75	10.3	
Carbon tetrachloride. . .	15.5; 15.4; 14.9			9.3	
Carbon bisulfide. .	151	166	—	147.1	
Toluene	32.2; 60; 55.6; 60	41.9	44.1	46.0	
Acetone	14.8				10

Table 11. Coefficients of Light Scattering by Adiabatic and Isobaric Density Fluctuations and Fluctuations of the Anisotropy for Several Liquids

Substance	$(R_{90})_{ad} \cdot 10^6$, cm^{-1}	$(R_{90})_{is} \cdot 10^6$, cm^{-1}	$(R_{90})_{an} \cdot 10^6$, cm^{-1}	$R_{90} \cdot 10^6$, cm^{-1}
Water	1.70	0.02	0.365	2.08
Methyl alcohol	3.68	0.96	0.65	5.19
Ethyl alcohol	3.90	1.105	0.79	5.80
Ether	6.04	2.59	1.67	10.30
Benzene.	7.05	4.39	20.4	31.8
Toluene	9.66	4.15	32.6	46.48

The isothermal compressibility at a frequency of 10^{10} Hz can be found from (1.47). This formula can be rewritten by considering the connection between β_S and the sound velocity in the following way:

$$\beta_T = \frac{1}{\rho}\left(\frac{1}{v^2} + \frac{\sigma^2 T}{c_p}\right), \tag{18.10}$$

where v is the speed of hypersound, found from the components of the fine structure (see Sec. 23), while the remaining parameters can be taken from static experiments (Table IV). The basis of such a calculation can be found in Sec. 7.

By determining β_T from (1.94), one can find the ratio g = β_T/β_T' (where β_T' is the compressibility used in the calculation of [198]), by which it is necessary to multiply the R_{90} computed in [198] in order to obtain the correct values of this quantity. Such a correct value is reduced to 20°C by (18.9) on the basis of the constants given in Table IV and the β_T computed from (18.10) by the method already shown. The result of such a calculation is given in the last column of Table 9.

A comparison of the measured and computed values of R_{90} for benzene lead to the conclusion that the best agreement is observed with the "low" values of R_{90} in the calculation by (1.94), while calculation by (18.9) gives either the best agreement with the "high" values of R_{90} or the best agreement with the "low" values of R_{90}, depending on the tabular values of the constants used. Therefore, if, in general, one uses the results of the calculation for choosing the correct measurements, then this choice falls on the "low" values. However, it seems to us that the theory cannot serve for the discrimination of the experimental values and, therefore, new absolute measurements of the scattering coefficient are necessary, as well as careful analysis of the conditions of measurement and a critical review of the parameters for the calculation of R_{90}. Comparison of the measured and computed values of R_{90} for six other liquids is given in Table 10.

By using Eq. (1.93) and the data of Table IV, one can compute separately the absolute scattering coefficient of light, which is produced by the adiabatic $(R_{90})_{ad}$ or isobaric $(R_{90})_{is}$ density fluctuations and the fluctuations of the anisotropy $(R_{90})_{an}$. In correspondence with (1.53), (1.54), and (1.93), these quantities can be expressed in the following fashion:

$$(R_{90})_{ad}=\frac{\pi^2}{2\lambda^4}\left(\rho\,\frac{\partial\varepsilon}{\partial\rho}\right)_S^2\beta_S kT, \tag{18.11}$$

$$(R_{90})_{is}=\frac{\pi^2}{2\lambda^4}\left(\frac{1}{\sigma}\,\frac{\partial\varepsilon}{\partial T}\right)_p^2\frac{\sigma^2 T}{\rho c_p}\,kT, \tag{18.12}$$

$$(R_{90})_{an}=[(R_{90})_{ad}+(R_{90})_{is}]\,\frac{13\Delta_u}{6-7\Delta_u}\,. \tag{18.13}$$

The calculations of the absolute intensity according to (18.11)-(18.13) for several liquids are given in Table 11.

The last column of Table 11 lists the value of R_{90} corresponding to the sum of (18.11)-(18.13) and is equal to R_{90} in (1.93).

Relative Intensity Measurements. The relative measurements of the scattering coefficient, made by various authors, are internally consistent and those complications with which the absolute intensity measurements were connected do not arise here.

The fundamental results of measurement of the relative intensity of scattered light were obtained by Martin [171, 366], Martin and Lehrman [367], Krishnan [368], Raman and Rao [338], Ramanathan [91], Ananthakrishnan [369], Parthasarathy [370], Peyrot [179], and Sweitzer [187]. The largest number of measurements, carried out very carefully, belong to Peyrot [179]. Table V lists the results of most of his measurements, supplemented by data of other authors. The data were taken from [66] and converted to the coefficient R_{90} for benzene, the scattering coefficient of which is taken as unity.

The dependence of the intensity of the scattered light on the temperature was investigated in a wide temperature range for many liquids by Rao [371].

In these liquids, the compressibility and the depolarization factor were also measured over a wide temperature range and calculations of the relative intensity $I_t/I_{30°C}$ were carried out by means of a formula similar to (18.7). All these results are given in Table VI. Excellent agreement of the computed and measured values indicate that practically the entire temperature variation of the intensity is derived from the temperature variation of β_T and Δ_u. Some departure of the computed values from those measured at high temperature is due to the approach to the critical temperature, where (18.7) ceases to be valid (see Sec. 2).

The qualitative observations and the relative measurements of the intensity of light scattered in liquid helium is of undoubted interest, especially considering the history of the problem.

Even before the well-known theory of Landau appeared [372] (which correctly describes the properties of liquid helium), London [373] considered the λ transition in helium (the transition of the He I into He II) as a direct analog of the condensation of an ideal Bose–Einstein gas.

If the situation were the case that London proposed, then, as was shown by Galanin [374] and Schiff [375], a strong increase in the intensity of scattered light would be observed at the λ point.

According to the estimate of Galanin and Schiff, the intensity of scattering at the λ point should increase by a factor of 10^4–10^5.

We now estimate R_{90} according to the classical formula (1.94). Considering that $(\rho \partial \varepsilon/\partial \rho)_T$ has not been determined in liquid helium, one can start out for the estimate of the order of magnitude of R_{90} from Eq. (18.7), setting $\Delta_u \sim 0$ without great error. The compressibility β_T was determined experimentally, and it can be found at the λ point from the value of the sound speed $v = 225$ m/sec, the density $\rho = 0.147$, and $c_p/c_V = 1.0008$ [147, 376]. From these data, $\beta_T = 1.35 \cdot 10^{-8}$ cm^2/dyne.

The index of refraction [376] close to the λ point is $n = 1.027$. Substituting these values in (18.7) and assuming that $T = 2.19$°K and $\lambda = 4358$ Å, we obtain $R_{90} = 1.6 \cdot 10^{-8}$ cm^{-1}. This value of R_{90} for liquid helium is about 100 times smaller than R_{90} for water at room temperature (see Table 10) and of the same order as for air (see Table 6).

It follows from the conclusions of [374, 375] that the intensity of the scattered light at the λ point should be larger than in liquid water at room temperature by a factor of 50 or even 500. In other words, in the transition from He I to He II, a very bright flash should be observed.

In the first experimental investigations of light scattering in liquid helium by McLennan, Smith, and Wilhelm [377], definite results were not obtained, and no anomalously large light scattering was detected.

In the work of Yakovlev [378], the intensity of the light scattered in air and in liquid He II was estimated visually; it was established that the intensity of scattering in liquid He II has the same order of magnitude as in air at room temperature, and thus it was demonstrated that the alleged great increase in the intensity actually does not exist.

Starting from the theory of Landau [372] which describes the properties of liquid helium, Ginzburg [147] showed that the intensity of the scattered light in liquids (down to temperatures ~0.1°K)

Table 12. Relative Intensity of Light Scattered in Liquid Helium at Various Temperatures

T_1, °K	p, atm	Phase	$R_l(T)/R_g(T_{b.p.})$, measured	$R_l(T)/R_g(T_{b.p.})$, computed
4.2	1	Gas	1.0	–
4.2	1	Liquid	3.8	–
4.2	1	Liquid	1.0	1.0
4.2	1	Liquid	0.75	0.7

is described by the Einstein–Cabannes formula (1.94) (see also Secs. 1 and 5) and this, as we see immediately, is in agreement with experiment.

Lawson and Meyer [379] again turned to the problem of the measurement of the intensity of the scattered light in liquid He II and completed quantitative measurements of the scattering coefficient in liquid helium at different temperatures, relative to the scattering coefficient of gaseous helium at the boiling temperature (Table 12). In their experiment, the scattered light was recorded by a photomultiplier, cooled to a temperature of 50°K.

The experimental results thus obtained were compared with the computed. The calculation was carried out by means of Eq. (18.7) under the assumption that $\beta_T = 1/p$, and that the Δ_u are identical in a gas and in its liquid. Beginning with this, one can write

$$\frac{R_l(T)}{R_g(T_{b.p.})} = \frac{(\varepsilon_l - 1)^2}{(\varepsilon_g - 1)^2} \frac{T}{4.2} \beta_T P,$$

where R_l is the scattering coefficient in the liquid, R_g is the scattering coefficient in the gas at the boiling point, and ε_l and ε_g are the optical dielectric constants in the liquid and the gas, respectively. The results of the measurements and the calculations from the last formula are given in Table 12.

Within the limits of error of the experiment, which amount to 10%, the calculated and measured values are in excellent agreement with one another, which confirms quantitatively the previous observation [378] and the conclusions of the theory [147].

19. Depolarization of Light Scattered in Liquids

It has been established that the measurement of the depolarization of scattered light in vapors and gases can be used along with the measurement of the Kerr constant and the molecular refraction for a quantitative determination of the optical anisotropy of molecules and an explanation of the problems of geometric structure of various compounds (Sec. 4).

The use of the depolarization factor of scattered light measured in liquids for this purpose encounters considerable difficulties, inasmuch as it has not been possible to date to obtain a rigorous quantitative relation between the measured value of the depolarization factor and the optical anisotropy of the molecules of the liquid.

The chief difficulty in the development of the theory of the depolarization of scattered light in liquids is connected with the necessity, in the first place, of considering the internal field acting on the molecule and, in the second place, of finding the distribution function for the calculation of the mean values of the square of the time derivative of the polarization components, which determine the intensity of the x and z components of the scattered light.

The internal field acting and the distribution function of the molecules are not yet accurately known; therefore, there is no possibility of carrying out a proper calculation of Δ, as was done in the case of gases, where the molecular distribution is assumed to be random and the field acting on the molecule is virtually identical with the external field of the incident light wave (see Sec. 4).

In the situation that has arisen, it is rational to connect the depolarization factor, measured in the liquid, with parameters characterizing other phenomena. Such a connection can be found by computing the different effects in a liquid on the basis of phenomenological theory. Using the Maxwell representation of the viscosity of a relaxing liquid (with a single relaxation time), Leontovich [39] computed the degree of depolarization of scattered light [see Sec. 6, Eq. (6.28)].

Taking into account that the intensity distribution in the wing of the Rayleigh line is characterized by two and even more relaxation times τ, (6.28) can scarcely be used for an accurate calculation of Δ_u.

Rather, (6.28) will be useful for the estimate of the value of some τ_{eff} determined from the measured depolarization factor and the other parameters entering into (6.28).

In the correlation theory of light scattering, Rytov [40] found a formula for the depolarization factor in low-viscosity liquids (7.35), which takes into account two relaxation times: τ_1 and τ_2.

We derive an estimate of Δ_u for benzene from Eq. (7.35) by taking $\beta_T = 82 \cdot 10^{-12}$ bar^{-1}, $\eta = 6.5 \cdot 10^{-3}$ poise, $\varepsilon = 2.32$, $(\rho\, \partial\varepsilon/\partial\rho)_T = 1.62$, $M = 9 \cdot 10^{-11}$, $\tau_1 = 2.16 \cdot 10^{-13}$ sec, $\tau_2 = 3 \cdot 10^{-12}$ sec, and $\beta = 0.385$; this gives $\Delta_u = 0.42$, which is in excellent agreement with the measured $\Delta_u = 0.43$. Although it is necessary to assume that such good agreement comes about from accidental circumstances, inasmuch as τ_1 and τ_2 are determined experimentally with considerable inaccuracy, nevertheless, Eq. (7.35) gives encouraging results and should be tested on a large number of liquids, where τ_1 and τ_2, and the other necessary parameters, have been reliably established. In previous theoretical investigations, a number of attempts have been made to compute the depolarization factor of scattered light in liquids on the basis of different molecular theories.

In particular, Raman and Krishnan [65] considered the anisotropy of the effective field, replaced the Lorentz sphere by an ellipsoid, considered the depolarization in liquid pentane, hexane, heptane, and octane, and obtained results which agree well with experimental data [170]. But Stuart and Volkmann [380] showed that the results of Raman and Krishnan have an accidental character. Exactly the same model, as well as other models of molecular theory, developed by Stuart and his coworkers [380-382] and Müller [383], have not led to desirable results and confirm the serious criticism on the part of Ansel'm [69] and Vol'kenshtein [384]. All these theories now have an historical interest. Their exposition and criticism can be found in the book by Vol'kenshtein [384]. Of interest is the attempt of Ansel'm [69] to carry out a calculation of the depolarization factor and the Kerr constant in liquids by applying the correlation function of Prins and Zernike [100], used by them in the study of the x-ray scattering in liquids (see Sec. 18).

The basis of the theory of Ansel'm is the very well established assumption that the polarizability tensor of the molecule does not change at all in the transition from the gas to the liquid, or so very little that this change can be neglected. This means that all the changes in the depolarization factor in the transition from a gas to a liquid are explained by the change in the internal field acting, and by the statistical effect. A second, much less substantiated assumption is that the effective field is taken to be the Onsager [385] expression for this quantity. The intensities I_x and I_z of the scattered light are computed from the scheme of calculation adopted in [125] and the orientation interaction is accounted for by means of the correlation functions. Furthermore, an expression is found for the depolarization factor in terms of the polarizability of the molecule in the gas and in terms of some integral parameter $\Gamma_{\Delta u}$ which must be calculated. But it is impossible to do this, since the distribution function is unknown. On the other hand, the Kerr constant is computed for the liquid on the basis of these assumptions, and with the same correlation functions. The Kerr constant is expressed in terms of the same integral parameter Γ_K.

The parameter Γ is not computed in [69], but is determined from the measured values of the depolarization factor or the Kerr constant. Excellent agreement of the computed values Γ_K and $\Gamma_{\Delta u}$ for two different phenomena would have signified excellent agreement of theory with experiment. The calculation carried out by Ansel'm [69] for the case of carbon bisulfide gave the values $\Gamma_{\Delta u} = -0.14$ and $\Gamma_K = -0.31$, and for the case of benzene, $\Gamma_{\Delta u} = -0.32$ and $\Gamma_K = -0.45$.

The quantitative agreement cannot be regarded as good, but the author of the theory only assumed agreement of Γ_K and $\Gamma_{\Delta u}$ in order of magnitude, inasmuch as the Onsager relation was used for the effective field on the molecule. The conclusion in [69] that Γ varies negatively from 0.2 to 0.4 is in contradiction with the model representations of Stuart and his coworkers [380-382], who assume that the axes of symmetry of the molecules should orient themselves parallel to one another. For such a model, even for pair association, $z = 2$, $\Gamma \approx {}^2\!/_3 z$.

On the basis of the values of Γ found in [69], they confirmed that the correlation of the orientations of the molecular axes in benzene and carbon bisulfide lead to the formation of large angles between the molecular axes. However, such a unique comparison is still insufficient for the final conclusions. On the basis of the Ansel'm theory, it is impossible accurately to compute the depolarization factor of light in a liquid by determining Γ from the Kerr effect, or conversely, but one can agree with his conclusion that a change in the depolarization or the Kerr constant in the transition from a gas or vapor to a liquid is brought about principally by a change in the forces of interaction of the molecules in the liquid, as has already been shown [384].

In the work of Buckingham and Stephen [388] a new attempt was made to construct a statistical theory of the depolarization of scattered light in a condensed medium. It seems to us that the work of Buckingham and Stephen [388] is in idea rather close to the work of Ansel'm [69], but of course it differs essentially in the method of obtaining the final results, and in the result itself. Following [388], let us choose a system of cartesian coordinates x, y, z with unit vectors along the axes of **i**, **j**, **k**, respectively. Let the incident light be directed along the x axis, the observation of the scattered light be carried out along the y axis, and the electric vector of the incident, linearly polarized light be parallel to the z axis. The linear dimensions of the molecules are much less than λ. The field acting on the i-th molecule is equal to $\mathbf{E}^{(i)} + \mathbf{F}^{(i)}$, where $\mathbf{E}^{(i)}$ is the field of the incident light wave

$$\boldsymbol{E}^{(i)} = E^{(i)}\boldsymbol{k} = E_0 e^{-i\frac{2\pi}{\lambda}ct}\boldsymbol{k}. \tag{19.1}$$

$\mathbf{F}^{(i)}$ is the field at the point where the i-th molecule is located, brought about by the distribution of charges of the neighboring molecules; consequently, it is dependent on E and on t. Furthermore, we can write

$$\boldsymbol{E}^{(j)} = E^{(j)}\boldsymbol{k} = \boldsymbol{E}^{(i)} e^{\frac{2\pi}{\lambda} i\,(r^{(j)} - r^{(i)})\,i}, \tag{19.2}$$

where $\mathbf{r}^{(i)}$ is the radius vector drawn from some arbitrary but fixed point O inside the scattering volume to the place where the i-th molecule is located; $\mathbf{r}^{(j)}$ is the radius vector drawn from the point O to the i-th molecule.

The observation of the scattered light is carried out from points located on the y axis. The distance from the point of observation to the point O is equal to L, where $L \gg r^{(i)}$ for all i. The instantaneous value of the electric field of the scattered light wave at the point of observation will be [390, 391] (if we neglect the small electric quadrupole and the magnetic dipole moments of the molecule)

$$E' = \frac{1}{c^2}\sum \frac{1}{L^{(i)}}\left(\frac{\partial^2 P^{(i)}}{\partial t^2}\right)_{t-\frac{L^{(i)}}{c}} = -\frac{4\pi}{\lambda^2} E \sum \left(\frac{\partial P^{(i)}}{\partial E}\right)_{t-\frac{L^{(i)}}{c}}. \quad (19.3)$$

Here $\mathbf{L}^{(i)} = \mathbf{L} - \mathbf{r}^{(i)} + O(1/L)$. Such a description is possible because the dipole moment $\mathbf{P}$ is a function of $\mathbf{E}$ and t. The dipole moment of the molecule is described in [388] in the following way:

$$P_\alpha = \mu_\alpha + b_{\alpha\beta}E_\beta + \frac{1}{2}\beta_{\alpha\beta\gamma}E_\beta E_\gamma + \frac{1}{6}\gamma_{\alpha\beta\gamma\delta}E_\beta E_\gamma E_\delta + \ldots. \quad (19.4)$$

Here, μ_α is a component of the constant dipole moment of the molecule, $\beta_{\alpha\beta\gamma}$ and $\gamma_{\alpha\beta\gamma\delta}$ are the hyperpolarizabilities characterizing the nonlinearity in the dependence of P on E. In Eq. (19.3), the derivative is taken at the instant of time $t' = t - L^{(i)}/c$. The components of the intensity I_x and I_z in the scattered light are determined as the mean squares $\overline{[\mathbf{E}(t) \cdot \mathbf{i}]^2}$ and $\overline{[\mathbf{E}(t) \cdot \mathbf{k}]^2}$. Considering this and Eq. (19.3), we get

$$\Delta_v = \frac{\overline{\sum_{ij} \frac{1}{L^{(i)}L^{(j)}}\left(\frac{\partial P_\alpha^{(i)}}{\partial E}\right)^*_{t-\frac{L^{(i)}}{c}}\left(\frac{\partial P_\beta^{(j)}}{\partial E}\right)_{t-\frac{L^{(j)}}{c}} i_\alpha i_\beta}}{\overline{\sum_{ij} \frac{1}{L^{(i)}L^{(j)}}\left(\frac{\partial P^{(i)}}{\partial E}\right)^*_{t-\frac{L^{(i)}}{c}}\left(\frac{\partial P_\beta^{(j)}}{\partial E}\right)_{t-\frac{L^{(j)}}{c}} k_\alpha k_\beta}}, \quad (19.5)$$

where the starred quantities represent the complex conjugate values and the bar indicates averaging over all configurations inside the scattering volume. As before, the repeated Greek indices in (19.4) and (19.5) denote summation.

With account of the fact that the field acting on the i-th molecule is determined by the expression $[\mathbf{E}_\alpha^{(i)} + \mathbf{F}_\alpha^{(i)}]$, the moment $\mathbf{P}_\alpha^{(i)}$ should be written

$$P_\alpha^{(i)} = \mu_\alpha^{(i)} + b_{\alpha\beta}^{(i)}(E_\beta^{(i)} + F_\beta^{(i)}) + \frac{1}{2}\beta_{\alpha\beta\gamma}^{(i)}(E_\beta^{(i)} + F_\beta^{(i)})(E_\gamma^{(i)} + F_\gamma^{(i)}) +$$
$$+ \frac{1}{6}\gamma_{\alpha\beta\gamma\delta}^{(i)}(E_\beta^{(i)} + F_\beta^{(i)})(E_\gamma^{(i)} + F_\gamma^{(i)})(E_\delta^{(i)} + F_\delta^{(i)}) + \ldots. \quad (19.6)$$

Information on the tensor of the hyperpolarizability $\gamma_{\alpha\beta\gamma\delta}$, as has been shown in the work of Buckingham and Stephen [392, 393], can be obtained from a study of the dependence on the value of the intensity of the field of that part of the Kerr constant which is temperature independent. Measurement of the Kerr constant in spherically symmetric molecules can be used for finding the quantity $\gamma_{\alpha\beta\gamma\delta}$. So far as the quantity $\beta_{\alpha\beta\gamma}$ is concerned, it has not yet been found anywhere, although this quantity must enter into the Kerr constant [392] and the molecular refraction [393] of a substance with polar molecules.

For the calculation of Δ_V, we write

$$\left(\frac{\partial P_\alpha^{(i)}}{\partial E_\varepsilon^{(i)}}\right)_{E=0} = \pi_{\alpha\varepsilon}^{(i)} = \pi_{\varepsilon\alpha}^{(i)} = \left\{ b_{\alpha\beta}^{(i)} + b_{\alpha\beta\gamma}^{(i)} F_\gamma^{(i)} + \right. \\ \left. + \frac{1}{2}\gamma_{\alpha\beta\gamma\delta} F_\gamma^{(i)} F_\delta^{(i)} + \ldots \right\} \left\{ \delta_{\alpha\beta} + \left(\frac{\partial F_\beta^{(i)}}{\partial E_\varepsilon^{(i)}}\right) \right\}_{E=0}, \tag{19.7}$$

where

$$\delta_{\alpha\beta} = \begin{cases} 1 & \text{for} \quad \alpha = \beta; \\ 0 & \text{for} \quad \alpha \neq \beta. \end{cases}$$

For $\mathbf{L} \gg \mathbf{r}^{(i)}$ for all i, (19.5), with consideration of (19.7), transforms into

$$\Delta_v = \frac{\sum\limits_{ij} \overline{\left(\pi_{\alpha\gamma}^{(i)}\right)^* \pi_{\beta\delta}^{(j)} \exp(i\psi_{ij})\, i_\alpha i_\beta k_\gamma k_\delta}}{\sum\limits_{ij} \overline{(\pi_{\alpha\gamma})^* \pi_{\beta\gamma} \exp(i\varphi_{ij})\, k_\alpha k_\beta k_\gamma k_\delta}}, \tag{19.8}$$

where

$$\psi_{ij} = \left(\frac{2\pi}{\lambda}\right)(\boldsymbol{r}^{(j)} - \boldsymbol{r}^{(i)})(\boldsymbol{i} - \boldsymbol{j}) = \psi_{ijx} - \psi_{ijy}. \tag{19.9}$$

The latter expression characterizes the phase difference of the light scattered by the i-th and j-th molecules. After averaging over all possible orientations and using the results of [392], we get

$$\overline{i_\alpha j_\beta k_\gamma k_\delta} = \frac{1}{30}\{\delta_{\alpha\beta}\delta_{\gamma\delta} - \delta_{\alpha\gamma}\delta_{\beta\delta} - \delta_{\alpha\delta}\delta_{\beta\gamma}\}; \tag{19.10}$$

$$\overline{k_\alpha k_\beta k_\gamma k_\delta} = \frac{1}{15}\{\delta_{\alpha\beta}\delta_{\gamma\delta} + \delta_{\alpha\gamma}\delta_{\beta\delta} + \delta_{\alpha\delta}\delta_{\beta\gamma}\}. \qquad (19.10)$$

In the case in which all the molecules are identical, (19.8) goes over into the formula

$$\Delta_v = \frac{3A}{4A+B}, \qquad (19.11)$$

where

$$A = \frac{1}{N}\sum_{ij}\overline{\left\{\pi_{\alpha\beta}^{(i)*}\pi_{\alpha\beta}^{(j)} - \frac{1}{3}\pi_{\alpha\alpha}^{(i)*}\pi_{\beta\beta}^{(j)}\right\}\exp(i\psi_{ij})} \qquad (19.12)$$

and

$$B = \frac{10}{3N}\sum_{ij}\overline{\pi_{\alpha\alpha}^{(i)*}\pi_{\beta\beta}^{(j)}\exp(i\psi_{ij})}. \qquad (19.13)$$

If the scattering volume is irradiated not by linearly polarized light but by natural light, then one must write

$$\Delta_u = \frac{6A}{7A+B} \qquad (19.14)$$

in place of (19.11) (see Sec. 4). On the basis of this general theory of Buckingham, Stephen considered the case of an ideal gas and obtained the results already developed in Sec. 4. He also considered the depolarization of scattered light in a real gas, where it is shown that the depolarization in this case can be expressed in terms of the second and third "virial coefficients of depolarization" which can be determined experimentally in principle. Chief interest is centered on the attempts to calculate the depolarization factor in liquids. For a spherically symmetric isolated molecule (inert gas), $\mu_\alpha = \beta_{\alpha\beta\gamma} = 0$, as also for all molecules with a center of inversion and

$$\left.\begin{aligned} b_{\alpha\beta} &= b\delta_{\alpha\beta}, \\ \gamma_{\alpha\beta\gamma\delta} &= \frac{1}{3}\gamma\,\{\delta_{\alpha\beta}\delta_{\gamma\delta} + \delta_{\alpha\gamma}\delta_{\beta\delta} + \delta_{\alpha\delta}\delta_{\beta\gamma}\}.\end{aligned}\right\} \qquad (19.15)$$

Expressing $\pi_{\alpha\beta}$ in terms of b and γ, and assuming that the moment of the molecule is produced only by the electric field of the light wave, we get

$$\frac{\partial F_\alpha^{(i)}}{\partial E_\beta^{(i)}} = -\sum_{k \neq i} \epsilon_{\alpha\gamma}^{(ik)} \pi_{\beta\gamma}^{(k)} \exp\left\{i \frac{2\pi}{\lambda} (\boldsymbol{r}^{(k)} - \boldsymbol{r}^{(i)})\, \boldsymbol{i}\right\} = -\sum \epsilon_{\alpha\gamma}^{(ik)} \pi_{\beta\gamma}^{(k)} \exp(i\psi_{ikx}). \tag{19.16}$$

Using the results obtained by Pople [395], we write

$$\epsilon_{\alpha\gamma}^{(ik)} = r_{ik}^{-5} \left\{r_{ik}^2 \delta_{\alpha\gamma} - 3r_{ik\alpha} r_{ik\gamma}\right\}, \tag{19.17}$$

where r_{ik} is the distance between the i-th and k-th molecules. We obtain from Eqs. (19.7) and (19.6),

$$\pi_{\alpha\varepsilon}^{(i)} = \left\{\alpha_{\alpha\beta}^{(i)} + \frac{1}{2} \gamma_{\alpha\beta\gamma\delta}^{(i)} F_\gamma^{(i)} F_\delta^{(j)} + \ldots\right\} \left\{\delta_{\beta\varepsilon} - \sum_{k \neq 1} \epsilon_{\beta\zeta}^{(ik)} \pi_{\zeta\varepsilon}^{(k)} \exp i\psi_{ikx}\right\} =$$

$$= \left(b + \frac{1}{6} \gamma F_\gamma^{(i)} F_\gamma^{(i)} + \ldots\right) \delta_{\alpha\varepsilon} + \frac{1}{3} \gamma F_\alpha^{(i)} F_\varepsilon^{(i)} - b^2 \sum_{k \neq 1} \epsilon_{\alpha\varepsilon}^{(ik)} \exp(i\psi_{ikx}). \tag{19.18}$$

For molecules with constant polarizability $\gamma = 0$, and on the basis of (19.11), (19.13), and (19.18), we obtain

$$A = b^4 \sum_j \sum_{k \neq 1} \overline{\epsilon_{\alpha\beta}^{(ik)} \epsilon_{\alpha\beta}^{(jl)} \cos(\psi_{klx} - \psi_{ijy}) + \ldots}, \tag{19.19}$$

$$B = 30b^2 \sum_j \overline{\cos \psi_{ij} + \ldots}. \tag{19.20}$$

Small quantities have been neglected in the latter formulas. It is possible to compute B by following the method used by Benoit and Stockmayer [394]:

$$\sum_j \cos \psi_{ij} = 1 + \int \cos \psi \left\{g(r) - \frac{N}{V}\right\} dV, \tag{19.21}$$

where g(r) is the radial density distribution function and g(r)dV is the probability that the molecule is located in the volume dV located at a distance r from a certain i-th molecule. Integration is carried out over the entire scattering volume. It is assumed that g(r) differs essentially from N/V only in the region close to the i-th molecule, and that λ is much greater than this region; then we get

$$\sum_{j} \cos\psi_{1j} = 1 + \int_0^{\infty} \left\{ g(r) - \frac{N}{V} \right\} r^2\, dr = \beta_T \frac{R_g T}{V}, \qquad (19.22)$$

where V is the molar volume. On the other hand, the molecular refraction MR of a substance can be represented in the following form:

$$MR = \frac{4\pi}{3} N \overline{\left(\frac{\partial p_{\alpha}^{(i)}}{\partial E^{(i)}} k_{\alpha} \right)} = \frac{4\pi N}{9} \overline{\pi}_{\alpha\alpha}. \qquad (19.23)$$

The real part of (19.23) is

$$MR = \frac{4\pi N}{3} \left\{ b + \frac{b^3}{3} \sum_{j} \sum_{k \neq 1,\, j} \epsilon_{\alpha\beta}^{(ik)} \epsilon_{\alpha\beta}^{(jk)} \cos\psi_{1,\, jx} + \dots \right\}. \qquad (19.24)$$

In the latter formula, it is assumed that $\gamma = 0$. If the chief contribution to (19.19) is made by the term corresponding to $l = k$, then one can write

$$A = 3b^2 \left(\frac{MR - MR_0}{MR_0} \right), \qquad (19.25)$$

where $MR_0 = (4\pi/3)Nb$ is the molar refraction for vapors or gases.

From Eqs. (19.11), (19.20), (19.22), and (19.25), we obtain

$$\Delta_v = \frac{3\left(\frac{MR - MR_0}{MR_0} \right)}{4\left(\frac{MR - MR_0}{MR_0} \right) + \frac{10\beta_T R_g T}{V}}. \qquad (19.26)$$

If the principal contribution to A is made by the term corresponding to $j = 1$, $k = l$, then

$$A = 6b^4 \sum_{k \neq 1} \overline{r_{ik}^{-6}}, \qquad (19.27)$$

therefore,

$$\Delta_v = \frac{3 \sum_{k \neq 1} \overline{r_{1k}^{-6}}}{4 \sum_{k \neq 1} \overline{r_{1k}^{-6}} + \frac{5\beta_T R_g T}{b^2 V}} \qquad (19.28)$$

In [388] the radial distribution functions g(r) obtained by experiments on the scattering of x rays are used for the calculation of $\Sigma \overline{r_{ik}^{-6}}$. In this case,

$$\sum_{k \neq 1} \overline{r_{ik}^{-6}} = \int r^{-6} g(r)\, dV = 4\pi \int_0^\infty r^{-4} g(r)\, dr. \tag{19.29}$$

By using the experimental curve for g(r) as a function of r, found for liquid argon at −188.8°C and 0.8 atm, Eisenstein and Gingrich [396] found that

$$\sum_{k \neq 1} \overline{r_{ik}^{-6}} = 4.2 \cdot 10^{-45}\ \mathrm{cm}^{-6}.$$

For $\beta_T = 1.8 \cdot 10^{-10}$ dyne/cm^2 and $b = 1.63 \cdot 10^{-24}$ cm^3, $\Delta_V = 0.12$ is obtained for liquid argon; the experimental data for the depolarization in liquid argon are not yet available.

For tetragonal molecules $\beta_{\alpha\gamma\delta}$ is not equal to zero and the depolarization factor Δ_V should be written in the following fashion:

$$\Delta_v = \frac{1}{5b^2} \left\{ 3b^4 \sum_{k \neq 1} \overline{r_{ik}^{-6}} + \beta^2 \overline{F^2} \right\} \frac{V}{\beta_T R_g T}. \tag{19.30}$$

By means of a quantum mechanical calculation, Buckingham and Stephen found that for methane $\beta = -0.21 \cdot 10^{-30}$ cgs units, while $\overline{F^2}$ was determined by them by setting the energy of polarization $\frac{1}{2} N b \overline{F^2}$ equal to the change in the internal energy in evaporization, which is equal in turn to the latent heat of vaporization. On the basis of these assumptions, the calculation gives the result that for liquid nitrogen, $\overline{F^2} = 1 \cdot 10^{11}$ cgs units and $\beta^2 \overline{F^2} = 5 \cdot 10^{-51}$ cm^6 – a quantity that is small in comparison with the first component of Eq. (19.30), computed in the same way as in the previous case.

From the discussions thus set forth, it follows that the term containing β^2 in (19.30) is small and can be dropped from consideration in the calculation of the depolarization of tetragonal molecules. If this is so, then one can also use the formulas (19.26) and (19.28) in this case. For polar molecules, in the opinion of the authors of the theory, $\beta_{\alpha\beta\gamma}$ plays an essential role and cannot be omitted from the calculation.

In the case of chloroform considered in [383], one can determine the difference $\beta_1 - 0.8\beta_2 = 40 \cdot 10^{-30}$ cgs units, if we assume that the depolarization factor is known. β_1 and β_2 corresponding to the hyperpolarizability are parallel and perpendicular to the major axis of polarizability of the molecule.

Actually, if we assume that β_1 and β_2 vanish, then the depolarization factor thus calculated turns out to be equal to $\Delta_u = 0.42$, while, in reality, $\Delta_u = 0.24$. In the calculation, Buckingham and Stephen [388] took it into account that the internal field is determined according to Onsager, while the average polarizability is determined from the Lorentz–Lorenz formula. Therefore, it is not known to what extent the calculated and measured values of Δ_u diverge from one another because of the neglect of $\beta_{\alpha\beta\gamma}$ or because of the inexact account of the internal field.

For axially symmetric molecules with a center of inversion, the result is obtained (after averaging) in [388] that

$$\left.\begin{aligned} A &= \frac{2}{3}(b_1 - b_2)\left\{b_1 - b_2 + \frac{7}{18}(\gamma_a - \gamma_b)\overline{F^2}\right\}, \\ B &= 10b\left\{3b + \frac{5}{9}(\gamma_a - 2\gamma_b)\overline{F^2}\frac{\beta_T R_g T}{V}\right\}. \end{aligned}\right\} \tag{19.31}$$

It is shown in that work that the terms containing γ make a small contribution to the calculation of the depolarization and therefore they were neglected, while the equations (19.31) are simplified. By using tabular data for the quantities entering into (19.31), under the condition $\gamma = 0$ for carbon bisulfide, one obtains the result that $\Delta_v = 0.49$, while the experimental value is $\Delta_v = 0.45$. From the calculations and comparisons made by the authors [388], it follows that only for a completely clear case of polar molecules do the hyperpolarizabilities β and γ play a negligibly small role. The calculation of the mean values with the use of the density radial distribution function obtained from x-ray investigations appears to be expedient to us, but the final conclusion can only be made after calculation of a large number of cases. A comparison of the results of measurement of the depolarization factor in gases and vapors (see Table I, Appendix III) with experiments in liquids (see Table VII) shows that, roughly speaking, the depolarization in liquids is an order of magnitude larger than in gases.

If we assume that the ellipsoid of polarizability of the molecule undergoes virtually no change in the transition from one state of aggregation to another, then one must draw the conclusion that the value of the depolarization in the liquid is determined by the internal field and statistics to a greater degree than by the anisotropies of the molecule. Therefore, it appears that there is no possibility of drawing any conclusion on the anisotropy of the molecules from the value of Δ_u measured in a liquid. So far as quantitative calculations of the anisotropy is concerned, they actually cannot be made while there is no reliable theory. But in several cases, qualitative conclusions on the anisotropy of the molecules still appear to be possible from a measurement of Δ_u in a liquid [170, 174].

Above all, we note that with a few exceptions, the depolarization of the scattered light in a liquid is the greater, the larger the anisotropy of this liquid. One can assume that the noted regularity could be especially sharply marked in normal paraffins and their isomers. With this aim, the author [174] undertook a special measurement of Δ_u in 25 normal paraffins and their isomers. In the transition from normal paraffins to the isomers, the geometric structure of the molecules changes very substantially, and, consequently, there ought to be a decrease in the depolarization of the scattered light. At the same time, there is no special reason to assume that the character of the intermolecular forces in this transition is changed so much that the change of depolarization brought about by the structural change is masked. The results of the measurements are given in Table VII.

The measured values show that Δ_u in the normal paraffins becomes somewhat larger as the length of the carbon chain increases, from 7.3% for pentane to 8.8% for octane.

Much more sharply etched is the significant decrease in the depolarization factor in the transition from the normal paraffin to its branched isomer. This general rule is clearly expressed in all the cases studied, while it can also be noted that the depolarization factor for paraffins with three branches is less than for the corresponding paraffins with two branches.

From the results obtained for the paraffins, it follows directly that the branched molecules possess more optical symmetry and,

Table 13. Depolarization Factor of $SnCl_4$, $SnBr_4$, and Their Mixtures

Substance and mixture	$SnCl_4$	$SnBr_4$	$\frac{3SnCl_4}{1SnBr_4}$	$\frac{1SnCl_4}{1SnBr_4}$	$\frac{1SnCl_4}{3SnBr_4}$
$\Delta_u \cdot 10^2$	6.5±0.1	9.6±0.1	12.8±0.1	12.7±0.1	12.7±0.1

consequently, in the light of the theory of induced dipoles, differ by a higher geometric symmetry than the normal hydrocarbons. Such regularities were observed earlier by Wolf, Briegleb, and Stuart [386] only for certain normal and isomeric molecules in the vapor state.

It has been shown to be possible to decide on the change of the optical anisotropy of molecules in the liquid state in another special case [174]. It was shown in the investigations of Trumpy [387] (by the method of Raman scattering) in mixtures of $SnCl_4$ and $SnBr_4$, that intermediate molecules of the type $SnCl_2Br_2$, $SnCl_3Br$, and $SnClBr_3$ are formed in such mixtures.

The intensities of the lines of Raman scattering of the original products in all the mixtures studied, having molecular ratios of the components of 1:3, 1:1, and 3:1 were small; on this basis, the conclusion was drawn [387] that the concentration of the initial products in these mixtures is small and hence almost the entire substance consists of molecules of the intermediate forms, mentioned above.

The initial products possess a high degree of symmetry and are characterized by a small depolarization of the scattered light, while the assumed intermediate products are much less symmetric and, if they actually exist, then the same assumption is valid for them as for the paraffins. Consequently, one can expect an appreciable increase in the depolarization in the transition from the individual components to the mixtures. The results of measurement given in Table 13 actually show that the depolarization factor for the mixtures is much greater than for the initial substances.

Although the departure of the depolarization factor from additivity is sometimes observed in mixtures, this departure has a

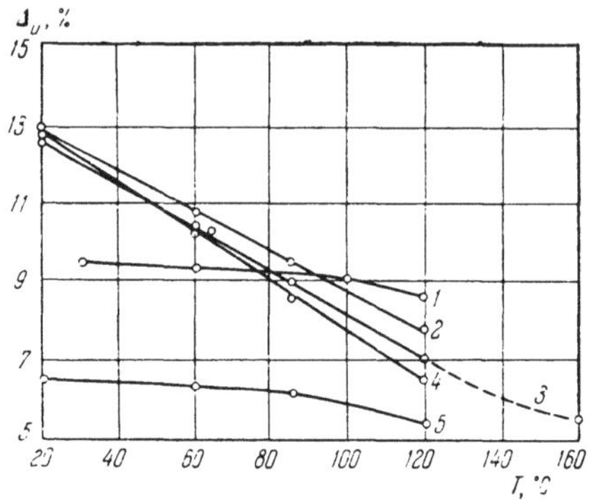

Fig. 58. Dependence of the depolarization factor of scattered light on temperature (Fabelinskii [174]). 1 – Temperature dependence of depolarization in $SnBr_4$; 2 – temperature dependence for the mixture $1SnCl_4 + 3SnBr_4$; 3 – temperature dependence for the mixture $1SnCl_4 + 1SnBr_4$; 4 – temperature dependence for the mixture $3SnCl_4 + 1SnBr_4$; 5 – temperature dependence for $SnCl_4$.

different character. Nowhere did the depolarization observed in mixtures exceed the depolarization of the most anisotropic component. In the case studied by us, the depolarization observed in the solution is almost one and a half times the depolarization in $SnBr_4$.

The second circumstance forcing us to reject the explanation of the observed phenomenon as a departure from additivity lies in the fact that the result obtained does not depend on the composition of the mixture. It is difficult to explain this if the reason is the departure from additivity. Conversely, the explanation of the variation of the depolarization by the formation of asymmetric molecules allows us to expect one and the same result for different mixtures, because the character of the anisotropy of the assumed products is sufficiently similar. The strong decrease in depolarization with increase in temperature (Fig. 58) should evidently not be regarded as proof of the decomposition of the intermediate products upon heating; because the no less strong decrease in the depolarization is observed in the individual substances, as is seen for example from Table VI.

The degree of depolarization of light scattered in liquids in the case of $SnCl_4$, $SnBr_4$, and the intermediate molecules falls off with increase in temperature. The same law is observed in the case of n-pentane, isopentane, and ethyl ether (Table VI). Such a variation of the temperature dependence of the depolarization is natural, inasmuch as the depolarization factor should decrease, roughly speaking, by an order of magnitude in the transition from the liquid to the gas.

The character of the decrease in the depolarization with increase in temperature apparently can be different if one takes into account the individual features of the intermolecular interactions in the liquids. The study of the depolarization factor for a decrease in temperature is of considerable interest in liquids which allow appreciable supercooling and transform continuously into the solid amorphous state. Such investigations were completed by Velichkina [188, 189] for triacetin, salol, and β chloroethyl ether benzenesulfo acid (β-ether). The data obtained by her are listed in Table VIII and in Fig. 86.

The data for salol and β-ether show that Δ_u slowly increases with a decrease in temperature; the growth practically ceases in the temperature range 20-30°C, and a rather broad plateau is formed.

At the glass temperature or even at much lower temperatures, an increase in the depolarization is observed. The reason for the growth of the depolarization in this temperature interval remains unexplained. Possibly the growth of the depolarization is produced by the splitting of the specimen or by much finer defects which arise at low temperature. In triacetin, the plateau is more poorly delineated but the increase in the depolarization at low temperatures is clearly seen.

In concluding this section, we recall the interesting phenomenon of anomalous depolarization, observed by Krishnan [262] in the region of the critical temperature of mixing of solutions. This phenomenon has essentially not yet found satisfactory theoretical explanation. According to definition, the depolarization factor

[Sec. 4, Eqs. (4.9) and (4.10)], upon irradiation of the scattering volume by light polarized in a plane perpendicular to the plane of scattering, is equal to

$$\Delta_v = \frac{I_{zx}}{I_{zz}}. \tag{19.32}$$

For irradiation by light polarized in the plane of scattering,

$$\Delta_h = \frac{I_{yx}}{I_{yz}}. \tag{19.33}$$

For irradiation by natural light,

$$\Delta_u = \frac{I_{zx} + I_{yx}}{I_{zz} + I_{yz}}. \tag{19.34}$$

On the basis of the principle of reversibility [262], we get

$$I_{zx} = I_{yz}. \tag{19.35}$$

From the relations (19.32)-(19.34), we obtain

$$\Delta_u = \frac{\Delta_v (1 + \Delta_h)}{1 + \Delta_v}. \tag{19.36}$$

Krishnan [262] discovered that at the critical temperature of mixing Δ_h slightly exceeds unity in all eight critical mixtures studied by him, while far from this temperature $\Delta_h = 1$. As an example, we give the results of measurement for two mixtures of phenol–water for different concentrations of the phenol and different temperatures, and for the critical mixture of carbon bisulfide–methyl alcohol in the ratio 3 : 1 for different temperatures (Table IX).

This phenomenon has also been observed in such individual liquids as hydrochloric and acetic acids, where Δ_h is more than unity even at room temperature, and falls to unity when the temperature is reduced to 90°C. In propionic and n-butyric acids, such a phenomenon was not observed.

Krishnan [262] showed that the phenomenon which he observed was brought about by the formation of "clusters" or "swarms" of molecules at the critical temperature. The dimensions of such molecular formations can be comparable or even larger than the wavelength of the incident light; therefore, in the calculation of the intensity and polarization, it is necessary to take into account the phase difference between the waves emitted from different points of the "cluster," as is done in the theory of Mie.

One of the attempts at a theoretical explanation of the Krishnan effect was made by Gans, who assumed that the volumes of the clusters or swarms were not too small in comparison with λ^3. The formulas of the Gans theory give $\Delta_h > 1$ in some cases but, in others, in contrast with experiment, $\Delta_h < 1$. Thus, although the Gans theory shows in principle the possibility of explaining $\Delta_h > 1$, his quantitative conclusions are not supported by experiment. As possible reasons for the anomalous depolarization, Vladimirskii pointed out the second scattering by the concentration fluctuations at the critical point, or by the correlation fluctuations of the anisotropy at distances $\sim\lambda$ for the explanation of the phenomenon in acetic and butyric acids.

All these explanations of the anomalous depolarization have a qualitative character at the present time. A satisfactory quantitative theory of the phenomenon has yet to be constructed.

20. Experimental Investigation of the Light Scattering at the Interface between Two Media

The basic theories of light scattering at an interface, developed by Mandel'shtam [15] and generalized by Andronov and Leontovich [111], and by Gans [113], were set forth in Sec. 3. We shall give here the fundamental results of the experimental investigations and comparisons of the theory with experiment.

Intensity and Polarization of Light Scattered at an Interface. As was shown earlier, the first experimental results belong to Mandel'shtam [15], who studied qualitatively the scattering of light by a molecularly inhomogeneous surface between carbon bisulfide and methyl alcohol.

The methyl alcohol and carbon bisulfide mix in any proportions at temperatures above 40.5°C. Depending on the ratio of the two components of the mixture, the temperature at which only a single phase exists is far removed from the critical temperature. In the experiments of Mandel'shtam, such a ratio of components was used that the mixture became of single phase at a temperature of 39°C. Special measures were taken to purify the interface by distillation of the components of the mixture. The vessel with the sealed liquid could be kept a long time at a constant temperature with accuracy to 0.1°C. Light was incident on the interface at an angle of 80°. At room temperature, mirror reflection and a very feeble molecular scattering of light by the interface were the case. Upon increase of the temperature to 37°C, the intensity of the molecular scattering increased appreciably, although it was still weaker than the specular reflection. In the experiment, it was noted that the intensity of the scattered light falls off if the direction of observation departs from the direction of specular reflection, keeping in the plane of incidence. The intensity of the scattered light falls off much more rapidly if the direction of observation comes out of the plane of incidence. Upon irradiation by natural light, strong polarization of the scattered light was not observed.

Upon further increase in temperature, the intensity of the scattered light increased until the scattered and specularly reflected light could not be distinguished.

At a temperature of about 39°C, the liquid became murky and then again became transparent. The experimental results given by Mandel'shtam [15] are in qualitative agreement with the theory. Actually, upon approach to the critical temperature of mixing, the surface tension γ^* decreases; therefore, as is evident from Eqs. (3.4) and (3.5) [the intensity of the scattered light is inversely proportional to γ^* (see Sec. 3)], the intensity of the scattered light increases sharply. A dependence of the intensity on the angle of observation of the scattered light was also observed, in complete agreement with the theory. Subsequently, quantitative measurements of the characteristics of light scattered by a liquid surface were completed by Raman and Ramdas [397, 398], Ramdas [399], and Baryshanskaya [193].

Moreover, Ramdas [400, 401] investigated the scattering of light by a fresh surface of slip mica and glass, by a surface of polished gold, and by gold sputtered on mica.

The theory was compared with the very complete results of measurement of scattering of light by an interface between liquid mercury and methyl alcohol [111, 399]. The sunlight was projected onto the clean surface of mercury normal to the surface. The scattered light was observed at an angle of 45° and compared with the light scattered under the same conditions by a surface of gypsum.

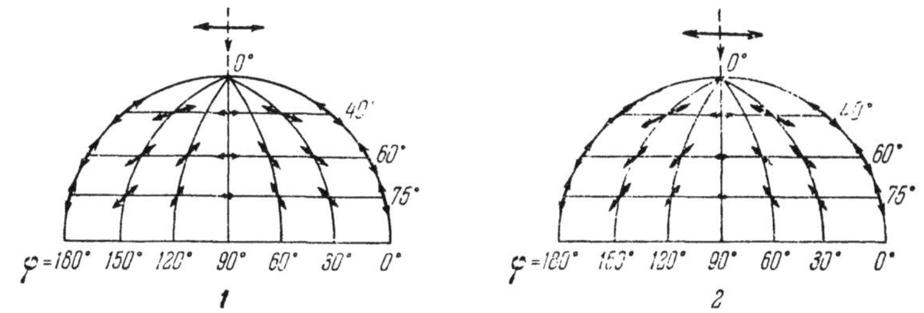

Fig. 59a. Diagram of the light scattered by a mercury surface (Andronov and Leontovich [111]). Scattering angle $\theta = 0°$. (1) Experiment; (2) theory.

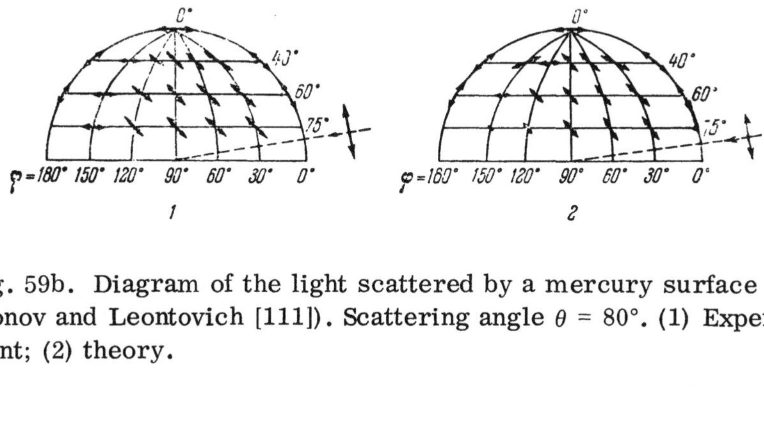

Fig. 59b. Diagram of the light scattered by a mercury surface (Andronov and Leontovich [111]). Scattering angle $\theta = 80°$. (1) Experiment; (2) theory.

For the yellow part of the sun's spectrum [397] the results

$$\frac{I}{I_g} = 5.7 \cdot 10^{-7},$$

were obtained, where I_g is the intensity of the light scattered by the gypsum. By assuming that the gypsum scatters according to Lambert's law, we get [111]

$$I_g \, d\Omega = \frac{I_0 S \cos\theta \, d\Omega}{\pi}.$$

When the surface possesses total reflection, one must use the index of refraction $n = \infty$, $\theta' = \theta_0' = 0$ in Eqs. (3.5) and (3.10)–(3.12), we then get from (3.5) by taking $\lambda = 5460$ Å and $\gamma^* = 540$ dynes/cm,

$$\frac{I}{I_g} = 6.71 \cdot 10^{-7}.$$

If it is taken into account that the reflection coefficient of the mercury surface is not equal to unity, while the gypsum plate is not an equal scatterer, as was assumed in the calculation, then the agreement of the theory with experiment must be regarded as excellent. Ramdas came to the same conclusion [399].

In the researches of Raman and Ramdas [397], where they irradiated the mercury surface by polarized light, no absolute numbers came into the picture, but their experimental data are represented in the form of diagrams, similar to those of Fig. 59a, where the length of the arrows give the intensity and the direction of the arrows, the direction of the oscillations of the electric vector in the scattered light [66]. The dashed arrow shows the direction of the light incident on the surface, while the solid arrow perpendicular to it gives the direction of the oscillations in the incident polarized radiation.

For a comparison of the experimental data with the theories of Andronov and Leontovich [111], a calculation of the corresponding diagrams was made.* Figure 59a shows the diagram calculated for normal incidence of polarized light on the mercury surface. Figures 59b, (1) and (2), represent the experimental and theoretical data for the case of light incident on the mercury surface at an angle $\theta_0 = 80°$,also by linearly polarized light. From a comparison of the experimental and calculated diagrams, one can see that there is excellent agreement between them.

The experimental data of Raman and Ramdas [398] for light scattered by a surface of methyl alcohol refer to the case in which the surface was irradiated by natural light from the liquid side at the angle of total internal reflection ($\theta = 49°$) for different values of the azimuth φ. The theoretical calculation, according to the formula (3.5) ($n = 1/1.328$, $\lambda = 4200$ Å, and $\gamma^* = 23$ dynes/cm), completed by Andronov and Leontovich [111], is represented in the graph of Fig. 60 for three angles θ and for φ changing from 0 to

---

*Andronov and Leontovich [111] noted that the method of obtaining the diagrams was not shown in [397].

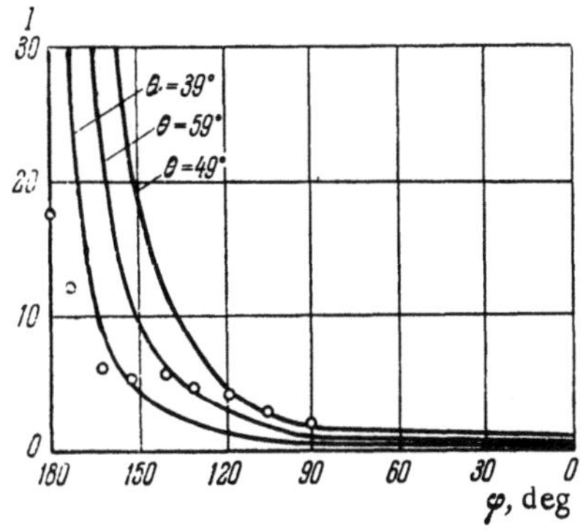

Fig. 60. Dependence of the intensity (in arbitrary units) of the light scattered by the methyl alcohol surface on the azimuth φ for three different angles θ. Solid curve – theoretical [111]. Experimental data are represented by the circles [398].

180°. The experimental data are represented by the circles. Up to azimuths of $\varphi \sim 130°$, the agreement of experiment with theory is quite complete, while for larger azimuths, a serious divergence is evident. It has been thought that the reason for the observed divergence is an imperfection of the theory. It is pointed out in [402] that the theory cannot pretend to be a rigorous, quantitative description of the phenomenon, inasmuch as it does not take into account the damping of the surface waves, the diffraction of light by these waves is calculated in this theory. However, it is considered in none of the calculations or estimates and therefore the reason for the divergence continues to be unknown.

The results of measurement and calculation of the relative intensity of some of the liquids studied are given in Table X (Appendix III) for the case of normal incidence of the exciting unpolarized beam from the liquid side and observations of the scattered light at the angle of total internal reflection, also from the liquid side. The measurements were made relative to water as a standard. In the last graph of Table X, the intensity of the scattering in the volume is given relative to the intensity in water.

The agreement between the computed and measured values of the relative light intensity at the surface cannot be regarded as satisfactory in all cases. Taking into account the difficulty of the experiment, however, it would be hard to require better agreement.

It is much more important that the general law of the dependence of the intensity of the scattered light on the value of the surface tension is clearly evident in the results given in Table X. Water possesses the largest surface tension (with the exception of mercury), while the intensity of the scattering by the water surface is less than for all other liquids, in complete correspondence with the prediction of the theory (see Sec. 3).

Ramdas [399] completed measurements of the intensity of the surface scattering in liquid CO_2 in the transition to the critical state. His data are given in Table XI. Comparison of the experimental data with the calculation showed that there is excellent agreement between them.

Dependence of the Intensity of the Light Scattered by the Interface between Two Liquids on the Wavelength of the Incident Light. The theory of molecular scattering of light by the surface of a liquid, given in Sec. 3, predicts that the intensity should be inversely proportional to λ^2 of the incident light, while in the volume scattering the intensity is inversely proportional to λ^4. Experimental test of this conclusion of the theory was carried out by Baryshanskaya (Landsberg) [193] for two critical mixtures: phenol–water and isobutyl acid–water. The first mixture has a critical temperature of 65°C for a concentration of phenol of 32%. For the second mixture, the critical temperature is close to 24° for a concentration of isobutyl acid of 36%. Measurements were carried out according to the method described in Sec. 11.

In the mixture of phenol–water, measurements were made at a temperature of 69.3°C and graphs were constructed for the dependence of $\log I$ on $\log \lambda$. The corresponding slope of the straight line gives the value of the exponent in the law $I \sim 1/\lambda^n$. The mean value of n obtained in this experiment is approximately equal to two. The dependence of the light intensity scattered by the surface on the temperature is plotted in Fig. 61. The results of the treatment of the intensity measurements at different wavelengths and

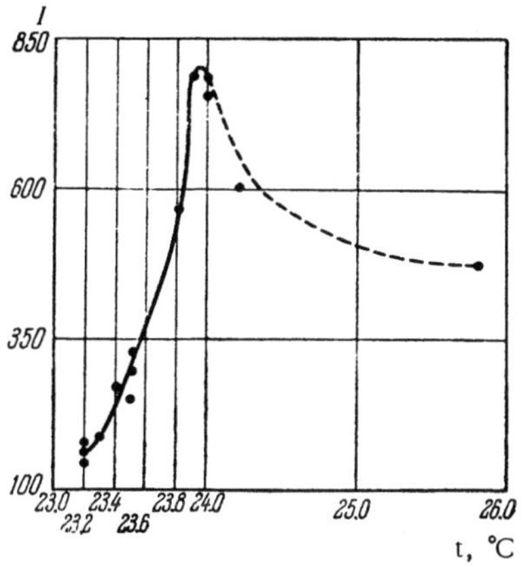

Fig. 61. Temperature dependence of the light intensity scattered by the interface between isobutyl acid and water (Baryshanskaya [193]).

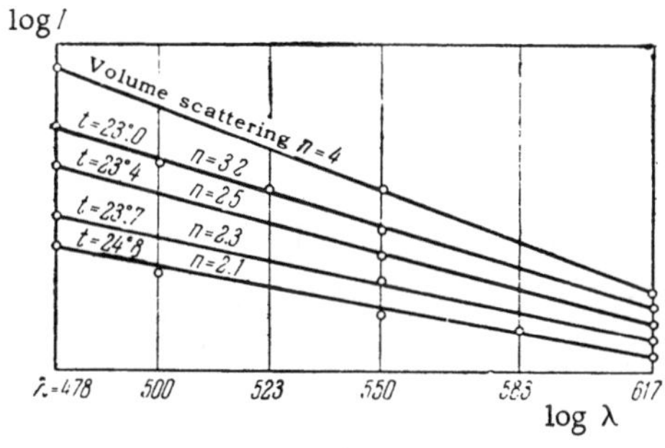

Fig. 62. Dependence of the logarithm of the intensity of light scattered by the interface between isobutyl acid and water on the logarithm of the wavelength of the incident light (Baryshanskaya [193]).

temperatures are shown in Fig. 62. In the case of the mixture of isobutyl acid–water, the exponent n is close to two for elevation of the temperature, and becomes practically equal to two when the temperature becomes close to critical. But the intensity of the volume scattering at the critical temperature is also inversely proportional to λ^2.

At low temperature, far from the critical point, n is much larger than expected (Fig. 62). For an explanation of the results, Baryshanskaya [193] showed that, in the measurements, along with the light scattered by the surface, some part of the light scattered by the adjacent volume is also unavoidably incident on the measuring apparatus. The relative fraction of the light scattered by the surface and by the volume depends strongly on the temperature, since the intensity of the scattering by the surface increases more rapidly, upon approach to the critical temperature of mixing, than the intensity of the light scattered by the volume. Therefore, the fraction of the light scattered by the volume at low temperatures is greater than at high temperatures.

The approximate calculation of the effect of the addition of volume scattering on the exponent n in the total flux of scattered light shows that if the addition of volume scattering is 5%, then n = 2.1; for an addition of 20%, n = 2.4, and for equal amounts of surface and volume scattering, n = 3.2.

Thus, it is possible to give a satisfactory explanation of the departure of the exponent n from two at a temperature of 23.0°C.

21. Application of the Method of Light Scattering to the Study of Solutions of Polymers, Proteins, and Electrolytes

General Remarks. The Einstein formulas, introduced in Sec. 1 for the scattering by density and concentration fluctuations, are valid for the case in which the dimensions of the optical inhomogeneities (fluctuations) are much less than the wavelength of the incident light, and their index of refraction differs but slightly from the index of refraction of the surrounding medium. Not only the density or concentration fluctuations, but the very molecules of polymers, proteins, or ions can be such optical inhomogeneities.

If the conditions given above are satisfied, then Eq. (1.82) or (1.83) can be used for the determination of the mean [43] molecular weight $\overline{M}$ of the dissolved substance. In this case, the experimental value of Kc/R_{90} or Hc/h, plotted on a graph as a function of c, should be a straight line which intercepts the ordinate at the value $1/\overline{M}$, while the slope of the line determines a quantity equal

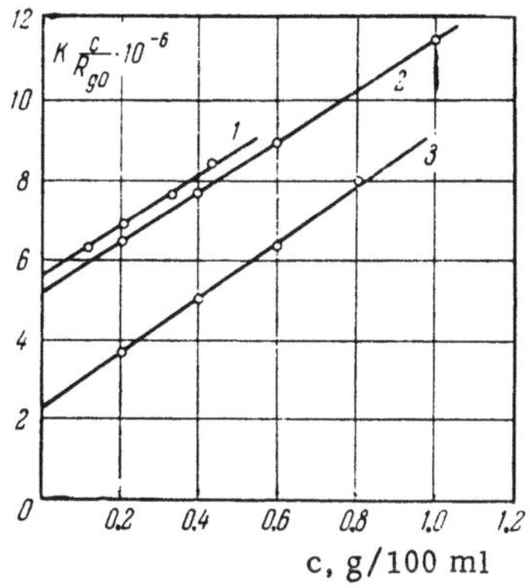

Fig. 63. Dependence of $K(c/R_{90})$, the fraction of polystyrene dissolved in methylethylketone, on the concentration (Doty, Zimm, and Mark [404]). (1) M = 178,000; (2) M = 190,000; (3) M = 445,000.

to 2B'. Such a linear dependence is actually observed in a number of cases, for example, for a solution of some fraction of polystyrene in methylethylketone [404] (Fig. 63) and for some other substances [43]. Evidently, in a large part of the practically important cases, the dimensions of the polymers or protein molecules are comparable with the wavelength of the incident line or even larger. In such a case, it is already impossible to apply the equations of Sec. 1.

In the case in which the dimensions of the scattering inhomogeneity (dimensions of the macromolecule) become comparable with the wavelength of the incident light, while the difference in the index of refraction of the dissolved molecules and the solvent remains small, the scattering problem becomes similar to the problem of the scattering of x rays at small angles [43, 48]. When the dimensions of the inhomogeneities become comparable with the wavelength, it is necessary to expect a significant departure from the Rayleigh indicatrix of scattering (see Fig. 2).

Figure 64 shows schematically the polymer molecule in the form of a disordered ball of thread: the distance between two selected points on the molecule, A and B, is equal to l; the angle

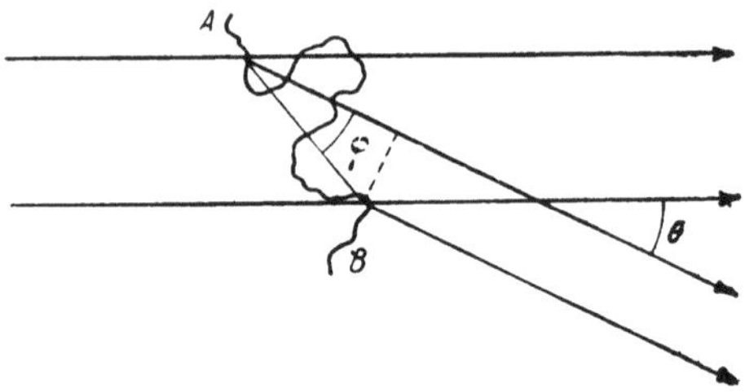

Fig. 64. The asymmetry of the scattering by a ball of molecules, the dimensions of which are comparable with the wavelength of the incident light.

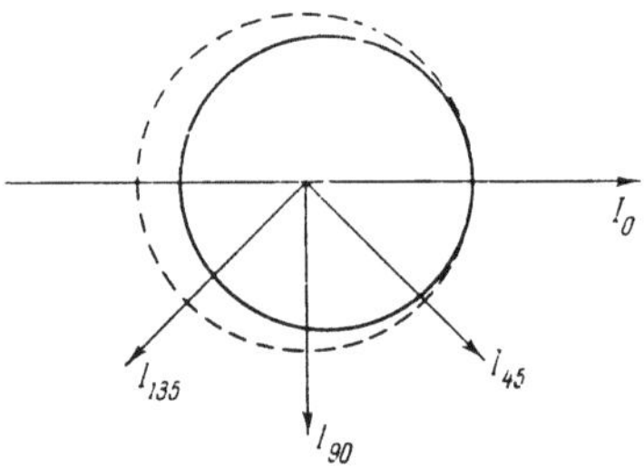

Fig. 65. Indicatrices of scattering for a particle with $l/\lambda = 0.2$ (solid curve) and for a particle with the same molecular weight, but with $l/\lambda = 0.067$ (dashed line) (Stacey [43]).

between the line connecting AB and the direction of propagation of the incident light is equal to φ; the scattering angle is equal to θ. The phase difference of the secondary waves emitted by the points A and B is equal to

$$\delta_\varphi = 2\pi n \frac{l}{\lambda}[\cos\varphi - \cos(\varphi - \theta)]. \tag{21.1}$$

It is clear from (21.1) that for any dimensions and configurations of the thread, the forward scattering ($\theta = 0$) leads to $\delta_\varphi = 0$ while, with increase in θ, the phase difference, which reaches a maximum value for $\theta = 180°$, is equal to

$$\delta_{\varphi_{max}} = 4\pi n \frac{l}{\lambda}.$$

If $1/\lambda > 1/20$, then the interference already plays an important role and less light will be scattered in the direction $\theta = 180°$ than in the direction $\theta = 0$, while the intensity of the scattering will be asymmetric relative to $\theta = 90°$. Figure 65 gives the indicatrices of scattering by spherical particles with $l/\lambda = 0.2$ and by spherical particles with the same molecular weight, but with $l/\lambda <$ 0.067. If in the case considered, $l/\lambda > 0.05$, the small difference in the index of refraction and the scattering coefficient of the particle is characterized by the relation

$$P(\theta) = \frac{R_\theta}{R_0}, \tag{21.2}$$

(R_θ and R_0 are the scattering coefficients for the angles θ and 0, respectively), then it is evident that $P(\theta)$ will depend on the shape and dimensions of the inhomogeneities. Debye [403] showed that for polymer molecules in the form of a ball, for which the mass distribution of the chain relative to the center of mass of the molecule is determined by a Gaussian function, the function $P(\theta)$ is expressed in the following fashion:

$$P_c(\theta) = \frac{2}{u^2}[e^{-u} - (1-u)], \tag{21.3}$$

where $u^2 = \frac{1}{6} q^2 \overline{l^2} = \frac{1}{6}[(4\pi n/\lambda)\sin(\theta/2)]^2 \overline{l^2}$. Here $\overline{l^2}$ is the mean square length between the ends of the polymer chain. For a sufficiently small u,

$$\left.\begin{aligned} P_c(\theta) &= 1 - \frac{u}{3}, \\ P_c^{-1}(\theta) &= 1 + \frac{u}{3}. \end{aligned}\right\} \tag{21.4}$$

For scattering particles which are thin rigid rods [403, 406],

$$P_r(\theta) = \frac{1}{v}\int_0^{2v} \left(\frac{\sin v}{v}\right) dv - \left(\frac{\sin v}{v}\right)^2, \tag{21.5}$$

where $v = (2\pi/\lambda) l \sin(\theta/2)$. Here l is the rod length. For small values of v,

Table 14. Order of Magnitude of the Scattering Coefficients of Water and Solutions of Various Proteins and Polymers [337]

Substance	R_{90}, cm^{-1}
Water.	$2 \cdot 10^{-6}$
Solution of polystyrene ($c = 10^{-3}$ g/ml, $M = 5 \cdot 10^5$)	
a. in toluene.	$8 \cdot 10^{-5}$
b. in methylethylketone	$2 \cdot 10^{-4}$
Protein in water solution ($c = 10^{-3}$ g/ml, $M = 5 \cdot 10^5$). . . .	$2 \cdot 10^{-4}$
Nucleic acid in water solution ($c = 10^{-3}$ g/ml, $M = 8 \cdot 10^6$). . . .	$6 \cdot 10^{-3}$

$$P_r^{-1}(\theta) = 1 + \frac{1}{9}v^2 + \frac{7}{2025}v^4 + \ldots. \tag{21.6}$$

Finally, for a spherical particle [403, 407],

$$P_{sph}(\theta) = \left[\frac{3}{w^3}(\sin w - w\cos w)\right]^2, \tag{21.7}$$

where $w = (2\pi/\lambda)d\sin(\theta/2)$. Here d is the diameter of the sphere. For small values of w,

$$P_{sph}^{-1}(\theta) = 1 + \frac{1}{5}w^2 + \frac{4}{175}w^4 + \frac{47}{23{,}625}w^6 + \ldots. \tag{21.8}$$

In all cases in which u, v, and w tend to zero, $P(\theta)$ approaches unity, either because of the fact that the ratio of the dimensions to the wavelength approaches zero, or because the scattering angle approaches zero. Figure 66 shows the dependence of $P^{-1}(\theta)$ on u, v, w for particles of spherical shape [339]. Taking the relation (21.2) into account, we can write another expression in place of (1.82), which is valid for large molecules, namely,

$$\frac{Kc}{R_\theta} = \frac{1}{\overline{M}_w P(\theta)} + \frac{2B'C}{R_g T}. \tag{21.9}$$

It is clear from (21.9) that the determination of the molecular weight is possible when $P(\theta)$ is known, or else the measurements must be obtained under the condition that $P(\theta) = 1$.

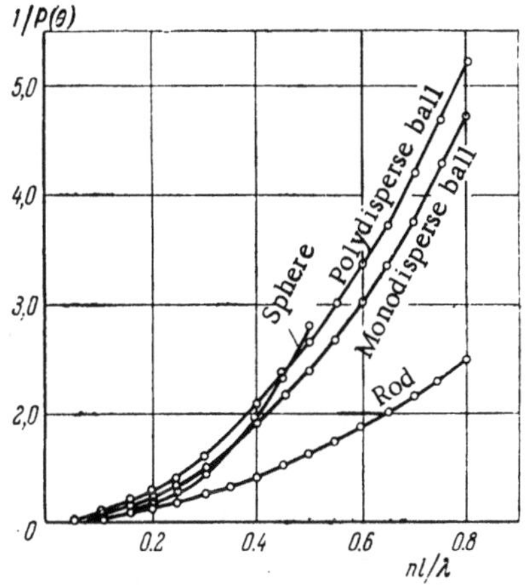

Fig. 66. Dependence of the transmission coefficient $P^{-1}(\theta)$ on the dimensions (nl/λ) and the shape of the scattering molecules (Doty and Steiner [339]).

The scattering by solutions of polymer and protein molecules is much more intense than the scattering by pure liquids with small molecules. Table 14 gives information on the order of magnitude of the scattering coefficients of proteins and polymers.

The measurement of the molecular weight requires the determination of the excess scattering by the solution over and above the scattering by the solvent; therefore, the determination of R should be carried out with every precaution which, however, is not too difficult to do with a reliable standard (see Sec. 18).

Methods of Treatment of the Experimental Data. Two methods of treatment of the experimental data are given. The method proposed by Debye is based on the determination of the asymmetry of the scattering $z = I_\theta / I_{\pi-\theta} = R_\theta / R_{\pi-\theta}$, which arises when the dimensions of the molecule become comparable with the wavelength of the incident light (Fig. 64). The asymmetry of z or other characteristic of the asymmetry of $P(\theta)$ or $P^{-1}(\theta)$ depends on the shape and dimensions of the scattering molecules. Figures 66 and 67 give a representation of the character of this dependence [339]. Consequently, if the shape of the molecule is known beforehand, then its dimensions can be deter-

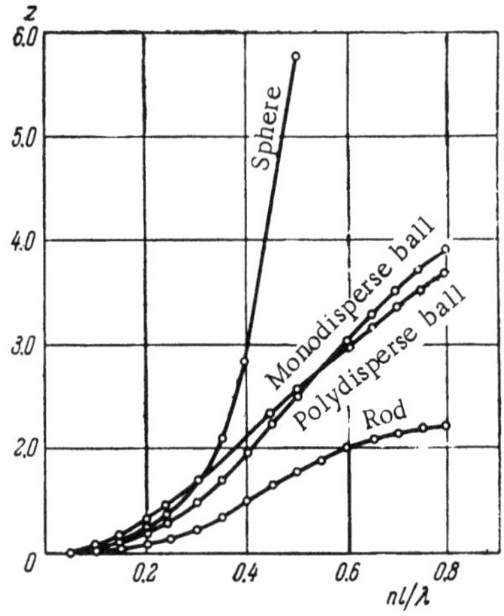

Fig. 67. Dependence of the asymmetry $z = I_{35}/I_{145}$ on the dimensions (nl/λ) and shape of the scattering molecules (Doty and Steiner [339]).

mined from the measured z. The coefficient of asymmetry depends on the concentration c; therefore, a value of z is used in the calculation of $\overline{M}$ which is extrapolated to zero concentration. For the purpose of greater convenience in extrapolation, a graph is constructed of the dependence of $1/(z-1)$ on the concentration c. Such a graph frequently has the shape of a straight line. The value of $z_{c=0}$ is used for the calculation of $P^{-1}(\theta)$, which is necessary for the transition from $(Kc/R_{90})_{c=0}$ to $(Kc/R_0)_{c=0}$, and $\overline{M}$ is found without difficulty from this latter measurement.

Definite conclusions can be made on the dimensions of the scattering molecules, found by the method of asymmetry, if one makes some assumption on the shape of these molecules.

The method of treatment of the experimental data used to determine the molecular weight and the dimensions of the molecules without having recourse to assumptions on the shape of the molecules was proposed by Zimm [201]. This method is based on the fact that at $\theta = 0$, $P^{-1}(\theta) = 1$.

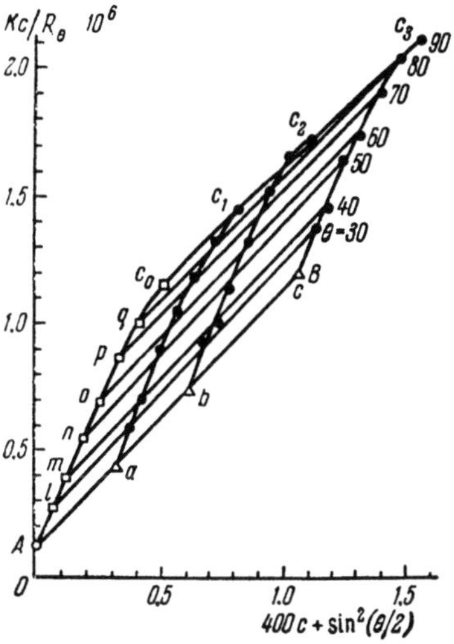

Fig. 68. Dependence of Kc/R_θ on $\sin^2(\theta/2)$ + 400c, constructed for the treatment of experimental results according to the Zimm method [201].

It is virtually impossible to measure the scattering at the angle $\theta = 0$, but one can find it by means of extrapolation. Zimm [201] showed that the best way to obtain such an extrapolation is to construct a graph of the dependence of Kc/R_θ on $\sin^2(\theta/2) + Ac$, where A is an arbitrary constant, chosen so as to satisfy the experimental data. With the use of this method, one must measure the scattering coefficient at several scattering angles for each concentration. The experimental data omitted on the graph in the coordinates given above form a grid, a typical example of which is shown in Fig. 68. The line AB on this drawing corresponds to the extrapolation to zero angle, while the line Ac_0 represents a line of extrapolation to zero concentration. Both the lines AB and Ac_0 intersect at the point which cuts off a section on the ordinate whose value is numerically equal to $(Kc/R_\theta)_{c=0,\,\theta=0}$, while the molecular weight can be found from this quantity without resorting to any assumption as to the shape of the scattering molecules. If it is assumed that Ac_0 on Fig. 68 can be represented by a linear law, then one can write

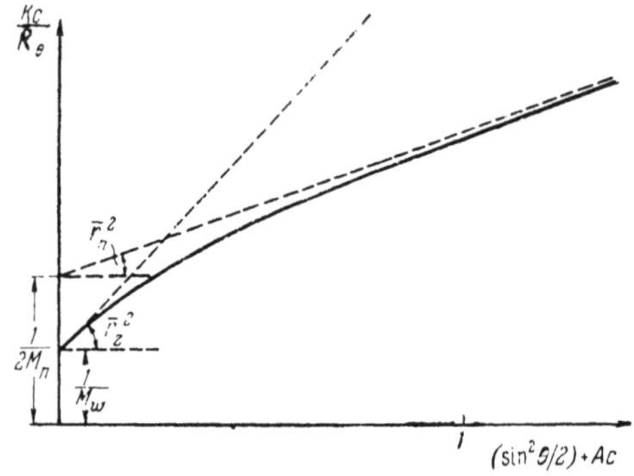

Fig. 69. Dependence of Kc/R_θ on $\sin^2(\theta/2)$ + Ac and the characteristics of a molecule which can be obtained in principle from this curve.

$$\left(\frac{Kc}{R_\theta}\right)_{c=0}=\left(\frac{Kc}{R_\theta}\right)_{c=0,\ \theta=0}+k'\sin^2\frac{\theta}{2}. \tag{21.10}$$

Dividing both parts of Eq. (21.10) by $(Kc/R_\theta)_{c=0,\ \theta=0}$ = OA (see Fig. 68), we obtain

$$P^{-1}(\theta)=1+\frac{k'}{OA}\sin^2\frac{\theta}{2}, \tag{21.11}$$

where $k' = \operatorname{tg}\beta$ is the slope of the line (21.10). By equating (21.11) to (21.4), we find

$$\frac{\overline{l^2}}{\lambda^2}=\frac{9}{8\pi^2n^2}\frac{\operatorname{tg}\beta}{OA}. \tag{21.12}$$

Zimm [201] and Benoit [408, 409] have shown that the mean square of the radius of the scattering center r_z can also be determined when the shape of the scattering molecules does not correspond to one of the shapes considered earlier (see Figs. 66 and 67).

It is shown in Fig. 69 how the characteristics of the molecule can be determined from the line Ac_0, i.e., by the dependence of $(Kc/R_\theta)_{c=0}$ on $\sin^2(\theta/2)$. The quantities with the index n are the mean numerical values, while the quantities with the index w are the mean weight values.

If the molecule weight is defined as

$$\overline{M}_n = \frac{\sum_i N_i M_i}{\sum_i N_i},$$

or also as the mass of the entire sample, divided by the number of molecules in the specimen, then we obtain the mean numerical molecular weight $\overline{M}_n$. In such a method of determination, the statistical weight of the molecules of each size is equal to its mole fraction. If the molecular weight is defined as

$$\overline{M}_w = \frac{\sum_i N_i M_i^2}{\sum_i N_i M_i},$$

then, in this case, the statistical weight is determined by the weight fraction of the molecules of given size. From the two latter expressions, one can see that they are equal to each other if the molecules are homogeneous. In the case of heterogeneous molecules, these quantities are more or less different, and $\overline{M}_w > \overline{M}_n$, and their ratio characterizes the width of the distribution curve of the molecular weights.

In all cases, the depolarization factor of the light must be taken into account in such a way as has already been shown in Sec. 1.

However, if the depolarization factor of the dissolved molecules is larger than that of the solvent, then it must be considered that z and $P(\theta)$ become functions of the anisotropy [43, 410]. It is also necessary to take it into account that for large molecules, depolarization arises which does not depend on the anisotropy [406].

Goldstein and Michelin [96] developed a theory of light scattering applicable to glasses, helium, and liquid crystals. According to this theory, the dimensions of the fluctuation regions can be determined from the curve of the dependence of the depolarization on the asymmetry.

The simple but very important cases of scattering in solutions of large molecules were considered above. In this book, we shall not discuss the general problems associated with light scattering in solutions of molecules which are balls of distributed links that do not obey Gaussian statistics, or the case in which the index of refraction of the inhomogeneity is large in comparison with the index of refraction of the medium surrounding it (Mie scattering). Also, we shall not consider questions of the anomalous polarization in solutions of large molecules.

All these interesting and important questions are treated with notable completeness in the books of Vol'kenshtein [384], Shifrin [411], van de Hulst [44], Stacey [43], and the review articles of Stuart [413] and Eskin [600].

Some Results of the Application of the Light Scattering Method to the Investigation of the Molecules of Proteins and Polymers. The comparison of the results of the determination of the molecular weight by the method of light scattering and by the method of osmotic pressure or viscosity shows that entirely satisfactory agreement of the results is obtained. Values of $\overline{M}_w$ determined from light scattering, and of $\overline{M}_n$ from experiments on the osmotic pressure, are given in Table XII (Appendix III). In this case, one can assume that these quantities are close to one another because very narrow fractions of the polymers have been used.

The excellent agreement of the data of light scattering with the data of other methods of determination of molecular weights is apparently not exhausted by the example given (Table XII). The method of light scattering has now become a far-reaching method of determination of the molecular weights of polymers, proteins, and polyelectrolytes, and also their shapes and dimensions. Table XIII gives the molecular weights of some proteins, determined by the method of light scattering, and by the method of measurement of the attenuation of a passing beam of light as a result of scattering. The data actually show the comparatively excellent agreement in the quantities obtained by the different methods. In connection with Table XIII, one must make the following observation. For the calculation of the molecular weight, a correction to the depolarization factor of about 5% was made, while later data show that the depolarization is less than the quantity assumed. New data for Δ_u are given in the next to the last column of Table XIII.

In the literature, attention is paid to the fact that the molecular weight of proteins depends on the time between the preparation

Table 15. Molecular Weights and Dimensions of Fibrinogens, Found by Different Methods

Method of measurement	Fibrinogen	
	Human	Animal
	$M \cdot 10^{-3}$	
Osmotic pressure.	580	440
Light scattering.	440	540
		407
		340
	Length of molecule, Å	
Double refraction in flow	700	670
Sedimentation and viscosity	700	–
Light scattering		840
Light scattering		520-650
Electron microscope		700
Electron microscope		500

of the material and the measurement of scattering in it. In the worst case, it was observed that the molecular weight of serum albumin varied over 18 months from 99,000 to 237,000 although, in the method of osmotic pressure, the growth was only up to 104,000 [417]. In other cases, it was shown that the prepared solution of protein practically did not change its properties during the passage of time [418]. Therefore, special attention must be paid to the method of proper preparation of the solutions.

Table 15 demonstrates the agreement of the molecular weights of fibrinogen, determined by different methods, and the excellent agreement of the lengths of these molecules, found by four different methods.

It is especially important that the method of light scattering give results very close to the results of the direct electron microscope method of the determination of the dimension of the macromolecules. It has been noted [337] that when molecules are not too large (~1000 Å), and the relative index of refraction ~1.5, the method of light scattering [Eq. (21.12)] and the calculations by the Mie theory, completed by Blumer [419] for particles with $nl/\lambda = 0.33$, give an error of about 10%. In this case, the method of light scattering is applicable also for the study of emulsions of polymers in water and for polymers precipitated in organic solvents [337].

Chapter VI

The Study of the Fine Structure of the Line of Scattered Light in Liquids with Large Bulk Viscosity and Small Shear Viscosity

22. The Fine Structure of the Scattering Line. Classical and Relaxational Theories of Sound Propagation in Viscous Liquids

After the fine structure of the Rayleigh line had been discovered in such liquids as benzene, carbon tetrachloride, and some others, and the ultrasonic absorption coefficient had been measured in them [79, 429], Leontovich [138] and Mandel'shtam [420, 421] turned their attention to a difficult point of principle which arises in the interpretation of the fine structure from the viewpoint of the hydrodynamic theory of sound propagation. The difficulty is the following: experimentally, the fine structure of the scattering line in the liquids studied is quite clearly observed. This means, as was shown in Sec. 5 [see (5.27)] that $\alpha\Lambda \ll \pi$. On the other hand, the experimental determination of the absorption coefficient α for the ultrasonic frequency range, and the extrapolation of this coefficient by the square law (5.21) to frequencies of about 10^{10} Hz, is essential for the scattering of light; (5.29) gives, for example for carbon tetrachloride, benzene and carbon bisulfide, values of $\alpha\Lambda$ respectively equal to 5, 11, and 70.

If this were actually so, then the fine structure could not be observed. Therefore, the extrapolation of the ultrasonic absorption coefficient by the square law up to hypersonic frequencies is not permissible and, consequently, the hydrodynamic theory is not applicable in this case. It must be recalled that Eq. (5.21) is valid only when $\alpha\Lambda$ is small. If $\alpha\Lambda$ is not small, extrapolation by the square law, which follows from (5.21), becomes inconsistent.

The way out of these difficulties was found by Mandel'shtam and Leontovich [421] in the relaxation theory of sound propagation in a liquid created by them.

Their theory was developed under the assumption that the liquid state is determined not only by the pressure p and the density ρ, but also by a certain parameter ξ, or by several such parameters, which characterize the internal state of the liquid.

In the equilibrium state, ξ is a function of p and ρ, and in the nonequilibrium case (propagation of sound through the liquid) ξ obeys a kinetic equation or reaction equation. In such a case, the equilibrium expansion leads to viscous forces. If the sound frequency is low (slow processes), then the viscous forces can be taken into account by the second viscosity coefficient; in other words, for such slow processes, the Stokes equation is valid with $\eta' \neq 0$. For rapid processes (hypersound), the effect of viscosity is not exhausted by the account of the second viscosity coefficient, which plays a small role or no role at all at high frequencies. From the formulas obtained in the relaxation theory of Mandel'-shtam and Leontovich [421], it follows that the absorption coefficient due to the second viscosity coefficient generally ceases to depend on frequency at the high frequencies. In the simplest case, the formula expressing the dependence of the absorption on the frequency becomes identical in form with the Kneser formula for the sound absorption in a monatomic gas.

After the publication of the relaxation theory of Mandel'-shtam and Leontovich, a number of variants of the relaxation theory, both phenomenological and molecular, appeared. The development of this part of modern acoustics is set forth in the review of Markham, Beyer, and Lindsay [422], in special detail in the book of Herzfeld and Litovitz [424], and in other books and reviews.

As a result of the development of the relaxation theory for the simplest case, in which a single parameter ξ and a single reaction time τ corresponding to it are introduced, Mandel'shtam and Leontovich obtained an expression for the square of the complex sound velocity

$$w^2 = v_0^2 \left\{ 1 + \frac{i\Omega\tau \left[\left(\frac{v_\infty}{v_0} \right)^2 - 1 \right]}{1 + i\Omega\tau} \right\}. \tag{22.1}$$

Here v_0 is the Laplace value of the velocity, toward which w tends as $\Omega \to 0$; v_∞ corresponds to the limiting value of w which is approached as $\Omega \to \infty$; and the relaxation time $\tau = \tau'(c_V/c_{V_\xi})$ (here c_{V_ξ} is the specific heat at constant volume V and constant ξ and is not equal to c_V measured in the static experiment).

The actual values of the sound velocity v and the amplitude absorption coefficient per unit length $\alpha_{\eta'}$, determined by the bulk viscosity η', can be found from the relations $1/v = \mathrm{Re}(1/w)$ and $\alpha_{\eta'} = \Omega \,\mathrm{Im}(1/w)$.

By application of these operations, and under the assumption that w differs slightly from v_0, the following formulas are obtained in [421]:

$$\frac{v_0}{v} = 1 - \frac{1}{2}\,\frac{\Omega^2\tau^2\left[\left(\frac{v_\infty}{v_0}\right)^2 - 1\right]}{1+\Omega^2\tau^2}, \text{*} \tag{22.2}$$

$$\alpha_{\eta'} = \frac{1}{2v_0}\,\frac{\Omega^2\tau\left[\left(\frac{v_\infty}{v_0}\right)^2 - 1\right]}{1+\Omega^2\tau^2}. \tag{22.3}$$

It follows from the theory of Mandel'shtam and Leontovich that

$$\left(\frac{v_\infty}{v_0}\right)^2 - 1 = \frac{1}{\rho^2 v_0^2}\left\{\frac{\left(\frac{\partial \mathcal{P}}{\partial \xi}\right)^2}{\frac{\partial^2 \psi}{\partial \xi^2}} + \frac{T\left[\frac{\partial p}{\partial T} - \frac{\left(\frac{\partial \mathcal{P}}{\partial \xi}\right)\left(\frac{\partial E}{\partial \xi}\right)}{T\,(\partial^2\Psi/\partial\xi^2)}\right]^2}{c_V - \frac{1}{T}\left(\frac{\partial E}{\partial \xi}\right)^2 \Big/ \frac{\partial^2 \Psi}{\partial \xi^2}} - \frac{T\left(\frac{\partial p}{\partial T}\right)^2}{c_V}\right\}. \tag{22.4}$$

The left side of Eq. (22.4) or twice the value of the dispersion $\Delta v/v$, which is approximately equal to it, is determined not only by the equilibrium values of the derivative of the pressure with respect to temperature $\partial p/\partial T$ and the specific heat c_V, but also by the quantities $\partial\mathcal{P}/\partial\xi$, $\partial E/\partial\xi$, and $\partial^2\Psi/\partial\xi^2$, which determine the change in the pressure, energy, and free energy for changes in ξ. It is shown in [421] that the first two quantities can be computed

* Translator's note: This formula, which is taken directly from [421], contains approximations which become invalid as $\omega \to \infty$.

only if one introduces special molecular representations of the liquid. The last quantity $\partial^2\Psi/\partial\xi^2$ could be found if one could somehow determine $\overline{(\Delta\xi)^2} = kT/(\partial^2\Psi/\partial\xi^2)$. In the research discussed [421], a general formula was written for w in the case in which it is not possible to use a single ξ and a single relaxation time τ, but one must introduce many internal parameters.

The formulas (22.2) and (22.3) written above were obtained without the restrictions which are characteristic of the hydrodynamic formula (5.21) for the absorption coefficient.

For low frequencies $\Omega\tau \ll 1$, it follows from (22.3) that $\alpha_{\eta'} \sim \Omega^2\tau$. If it is taken into account that the quantity τ is inversely proportional to the reaction rate, while the latter is proportional to the viscosity (at least for a number of phenomena), then $\alpha_{\eta'} \sim \Omega^2\eta'$ which is identical with the conclusions of hydrodynamic theory. In those cases in which $\Omega\tau \gg 1$, Eq. (22.3) transforms into

$$\alpha_{\eta'} = \frac{\left[\left(\frac{v_\infty}{v_0}\right)^2 - 1\right]}{2v_0\tau}, \tag{22.5}$$

whence it follows that beyond the region of relaxation, i.e., when $\Omega \gg 1/\tau$, $\alpha_{\eta'}$ no longer depends on the frequency and remains a constant quantity at a constant temperature. The damping per wavelength for $\Omega\tau \gg 1$ is equal to

$$\alpha_{\eta'}\Lambda = \frac{1}{2v_0}\frac{\left[\left(\frac{v_\infty}{v_0}\right)^2 - 1\right]}{\tau}\Lambda \tag{22.6}$$

and tends to zero upon increase in frequency.

Thus the relaxation theory of sound propagation in a liquid developed by Mandel'shtam and Leontovich [421] gives a natural qualitative explanation of the existence of the fine structure of the Rayleigh line in liquids in which $\eta' \gg \eta$.

Moreover, it follows from (22.5) that $\alpha_{\eta'}$ increases with increase in temperature and decrease in τ, while, according to the formulas of hydrodynamic theory, one should expect the opposite effect [see (5.21)].

Experiments with the fine structure confirm the validity of the conclusions of relaxation theory.

In the region of applicability of the classical expression (5.21) ($\alpha\Lambda \ll 1$) in the ultrasonic region of frequencies ($\Omega\tau \ll 1$), one can equate (22.3) to that portion of (5.21) which is determined by η'. As a result, we obtain

$$\eta' = \tau\rho\,(v_\infty^2 - v_0^2). \tag{22.6a}$$

The maximum difference is determined by the quantities η' and τ in the velocities $v_\infty - v_0$, as is seen from (22.6a). Actually, it is not difficult to obtain the result that

$$\frac{\Delta v}{v} = \frac{1}{2\rho v^2}\frac{\eta'}{\tau} = \frac{v}{\Omega^2}\frac{\alpha_{\eta'}}{\tau}\left[\frac{\alpha}{\alpha_\eta} - 1\right], \tag{22.7}$$

from (5.21) and (22.6), where $\Delta v = v_\infty - v_0$, $v = \frac{1}{2}(v_\infty - v_0)$, α is the experimentally determined value of the absorption, and α_η is the absorption coefficient determined by the shear viscosity and computed according to (5.21), where it is assumed that $\eta' = \varkappa = 0$.

The total jump in the velocity from sonic frequencies to frequencies $\sim 10^{10}$ Hz, which is of importance for the fine structure of the Rayleigh line, is determined with excellent accuracy by the formula (22.7) in liquids with a large bulk viscosity and small shear viscosity.

The relaxation theory was extended by Isakovich [153] to the case in which it is necessary to consider the relaxation of η' and the relaxation of η at the same time. Since that paper, this theory has been reproduced in different variants [424].

The quantitative conclusions of this theory will be used below, and we note for the present that the corresponding expressions for the velocity and damping coefficients consist of two components of the same form as (22.2) and (22.3). The qualitative conclusions, drawn from the formulas of relaxation theory already described, also follow from the formulas which contain the relaxation of both η and η'.

We turn our attention to the characteristic feature of the formulas of relaxation theory (22.1)-(22.4). In all these formulas, v_0 and v_∞ enter in significant fashion, while it is assumed that

$v_\infty > v_0$. In other words, a positive dispersion of the sound velocity is assumed. If it were to turn out that there were no velocity dispersion, then this would mean that the relaxation theory did not correspond to the real law of sound propagation.

23. Dispersion of Sound in Liquids

Preliminary Remarks on the Previous Investigations of the Dispersion of the Sound Velocity. The study of the dispersion of the sound velocity is a problem in itself, having physical meaning in principle. The experimental study of the dispersion of the sound velocity was begun long ago, in any event, before relaxation theory appeared, and produced especially great interest in this problem.

Here it is not intended to give anything like a full description of the investigations of the dispersion of the sound velocity that have been carried out to date. A sufficiently complete review of the investigation of the dispersion of the sound velocity in the ultrasonic frequency region can be found in the reviews already mentioned [283, 422, 424]. However, we shall recall the results of some of the investigations in order to make the situation clear.

The basic difficulty of the measurement of the dispersion of the sound velocity in the sonic and ultrasonic frequency regions lies in the smallness of the effect itself.

Zachoval [433], Bär [425], Krishnan [426], and others, investigated water, ethyl acetate, xylol, benzene, carbon tetrachloride, carbon bisulfide, and other organic liquids and in the limits of the rather high accuracy of measurement (1%) did not find any sound velocity dispersion at frequencies up to 50 MHz.

The investigations of Randall [427] revealed sound velocity dispersion in some alcohols of from 0.01 to 0.15%. But such small values of the dispersion were not easily established experimentally.

In a number of cases, it appeared that sound velocity dispersion had been found; however, in reality, it turns out that the reported velocity measurements are explained by local heating of the liquid or other reasons [283]. The first proof of the possible real existence of sound velocity dispersion was obtained in the investigation of the frequency dependence of the sound absorption coefficient by Bazhulin [79] and Biquard [429] in acetic acid, by Bär [430] in methyl acetate, and by Beyer and Smith [431] in ethyl acetate. For the substances mentioned, the absorption coefficient, divided by the square of the frequency, α/Ω^2, slowly decreased, as a function of Ω, with increase in Ω. Such a rule is inexplicable by the hydrodynamic theory, since it follows from this theory that $\alpha/\Omega^2 = \text{const}$, and does not depend on the frequency [see (5.21)], while the relaxation theory explains the dependence of α/Ω^2 on the frequency in a natural fashion [Eq. (22.3)]. In these researches, however, the dispersion of sound was not measured; one would have expected to find it if the observed phenomena actually had a relaxational character.

The first dispersion of the sound velocity in the ultrasonic range was discovered in acetic acid by Shpakovskii [432], whose results were confirmed by Lamb and Pinkerton [434]. Dispersion in propionic acid was found by Lamb and Huddart [435]. The magnitude of the dispersion of the sound velocity in these substances was $\Delta v/v \sim 1\%$.

Wong T'ieh-Ch'ao [283] discovered a decrease of 2.2% in the sound velocity in carbon bisulfide for a frequency change from $4 \cdot 10^5$ Hz to $1.36 \cdot 10^6$ Hz. Negative dispersion was also found by Schallamach [283] for geraniol and di-dihydrocitronellyl ether in the range of temperatures from +15 to −80°C. At frequencies from $6.89 \cdot 10^5$ to $2.943 \cdot 10^6$ Hz, negative dispersion of $\Delta v/v \sim 0.65\%$ at −50°C was also found. Upon further decrease in temperature, the dispersion changes sign.

Measurements of the sound velocity at ultrasonic frequencies, as has already been said, have been carried out in great number and with great accuracy. Only for acetic and propionic acids were small, positive dispersions of the sound velocity observed. In all other cases, the results of the experiments were negative or produced doubt and required careful confirmation. It is therefore expedient to repeat the investigations of the dispersion of the sound velocity in the region of very high frequencies, where the effect of dispersion ought to be strongly marked.

The generation of waves of high frequency or hypersound has recently met with significant success. It has been possible to obtain the oscillation of a quartz radiator at a frequency of 10^{10} Hz [576].

However, the absorption of sound at such a frequency in a liquid is so great that the investigation of hypersonic propagation in ordinary liquids has not been possible to date, although there already exist methods which promise progress in the future [596, 597].

Until recently, there was only one direct acoustic measurement of hypersonic absorption [437]. Recently, however, E. and J. Stewart [596, 597, 605] have completed direct acoustic measurements of the hypersonic velocity (acoustic impedance ρv at a frequency of $3 \cdot 10^9$ Hz in water, acetone, and carbon tetrachloride. The authors [605] hope to extend their measurements to other liquids and to frequencies up to 10^{10} Hz.

Now, however, fundamental results on the measurement of the velocity and absorption of hypersound in liquids have been obtained by an optical method, involving measurements on the fine structure of the Rayleigh line.

The first attempt to discover dispersion in the sound velocity by the shift in the Mandel'shtam–Brillouin (MB) component was made by Rao [438]. It followed from his measurements that a negative dispersion was observed in acetone ($\partial v/\partial \Omega < 0$), and $\Delta v/v \sim 20\%$, while in carbon tetrachloride, a positive dispersion was found ($\Delta v/v > 0$) of the same order. Rao and Samaya [439] attempted to explain the negative dispersion by the fact that at a frequency essential for light scattering, the adiabatic velocity went over to the isothermal. But the inadequacy of this explanation was pointed out by Ginzburg [440], and this has already been noted in Sec. 7.

The basic formulas of relaxation theory, given in the previous section, show that as a result of the relaxation of η', the velocity of sound should increase with increase in frequency. Exactly the same effect should be observed for the relaxation of η.

Estimate of the Possible Negative Dispersion of the Sound Velocity. Vladimirskii [46] and Ginzburg [440] have shown that, in principle, negative dispersion of the sound velocity in liquids can be observed.

In [46], the dispersion of the sound velocity, not connected with relaxations of η and η', is considered. This is assumed to be due to intermolecular interactions. Here the sound velocity is calculated by use of the generalized thermodynamic functions introduced in [46].

In the research of [440], the Navier–Stokes equation was studied in second approximation, where it is assumed that the absorption is very small; that is, essentially, the effect of viscosity on the dispersion is not considered and other reasons for dispersion appear.

Both calculations lead to a result which can be expressed by the following formula:

$$v = (v_0^2 + f^* q^2)^{1/2}, \tag{23.1}$$

where v_0 is the Laplace sound velocity at low frequency, f^* is a quantity which characterizes the intermolecular interaction, the same as in Eq. (2.3). Equation (23.1) can be written in the following fashion (for $\Delta v \ll v_0$):

$$\frac{\Delta v}{v_0} = \frac{1}{2v_0^2} f^* q^2 = \frac{2\pi^2 f^*}{v_0^2 \Lambda^2}, \tag{23.2}$$

where $\Delta v = v - v_0$. The sign of the dispersion of the sound velocity in (23.2) will be determined by the sign of f^*, the latter depending on the character of the intermolecular interactions. Thus, for example, in a one-dimensional model of a solid—a chain, which consists of pairwise interacting particles— $f^* < 0$, and the velocity of the elastic vibrations of the chain falls off with increase in frequency [46]. In the molecular aggregate, between the particles of which elastic forces act, $f^* < 0$ and the dispersion according to (23.2) will be negative [440]. These discussions lead to the conclusion that negative dispersion of the sound velocity is possible in principle in a liquid.

For an estimate of the amplitude of the dispersion, it is obviously necessary to assume that $|f^*| \sim v_0^2 r^2$, where r is the distance at which the molecular interaction becomes appreciable— the correlation radius or the dimensions of the associated aggregate. By taking into account the last expression for f^*, (23.2) can be written as:

$$\left|\frac{\Delta v}{v}\right| \sim 2\pi^2 \left(\frac{r}{\Lambda}\right)^2. \tag{23.3}$$

Far from the critical temperature, $r \sim 10^{-7}$ cm, and the considered mechanism for acetone, under the conditions of Rao's experiment [438], gives $\Delta v/v \sim 0.05\%$.

To explain a dispersion of 20%, it is necessary to assume that the dimensions of the aggregates in acetone are $r \sim 3 \cdot 10^{-6}$ cm under ordinary conditions, which is highly unlikely, inasmuch as such a correlation radius appears only at temperatures close to the critical temperature of mixing of some liquids.

Under such conditions, the scattering indicatrix should be completely changed, which in reality is not observed in the study of the scattering in liquid acetone at room temperature. However, the study of the dispersion of the sound velocity in strongly associated liquids and solutions is of interest from the viewpoint of the considerations set forth.

The mechanism considered here of negative dispersion, as has been shown by Ginzburg, reduces to the consideration of spatial dispersion in acoustics.

The natural rotation of the plane of polarization in acoustics, considered by Andronov [265], is also determined by spatial dispersion.

Continuation of the Search for Dispersion of Sound Velocity from the Fine Structure. The study of the fine structure with the aim of finding the dispersion of the sound velocity was later undertaken by Venkateswaran [441]. He studied sixteen pure liquids, including acetone, benzene, chloroform, and carbon tetrachloride. In all these liquids, Venkateswaran found no difference between the hypersonic and ultrasonic velocities. Errors in the determination of the velocity of hypersound from the data of [441] were ±25 m/sec and for some cases still less; i.e., they amounted to about 2%.

The research of [441] disproved the data of [438]. No negative dispersion was observed in acetone, or positive dispersion in carbon tetrachloride. The works of Rao and Venkateswaran just described were specially devoted to the determination of the hypersonic velocity from the components of the fine structure, but antedating these works, investigations of the fine structure were published which had the aim of testing Eq. (5.9). Here the measured $\Delta\nu$ was compared with the computed $\Delta\nu$. In the calculation of $\Delta\nu$, the velocity of sound measured in the ultrasonic region was used.

One could do otherwise – i.e., use the $\Delta\nu$ measured in these researches for the calculation of the hypersonic velocity according to (5.9). The results of such a calculation from the data of earlier researches are given in Table XIV (Appendix III). The velocity of the hypersound computed from the data of the researches of Gross [27], Gross and Khvostikov [442], and Meyer and Ramm [253] exceeds in many cases the value of the ultrasonic velocity, but in those cases in which the accuracy of the determination of $\Delta\nu$ is given, the excess of the velocity lies within the limits of roughness of the experiment. According to the data of other authors (Table XIV), the velocity of hypersound in benzene is identical with the velocity of ultrasound, while the velocity of hypersound in carbon tetrachloride in the researches of Ramm [235] and Venkateswaran [441] is also identical with the values of the ultrasonic velocity. The conclusion can then apparently be drawn that in a whole series of investigated substances, including benzene, carbon tetrachloride, and other liquids, up to frequencies $\sim 10^{10}$ Hz, no dispersion of the sound velocity is observed, with an accuracy to within 1.5-2%.

If this is actually so, then the alternative arises: either the relaxation theory and the measurements of the dependence of the absorption coefficient on the frequency are incorrect or, at frequencies $\sim 10^{10}$ Hz (fine structure), an appreciable positive dispersion of the sound velocity should be observed in a number of liquids; that such an alternative exists is not difficult to prove [29, 30, 146].

Estimate of the Possible Positive Dispersion of the Sound Velocity as a Result of the Relaxation of η'. One can estimate the dispersion of the sound velocity on the basis of the formula of relaxation theory (22.7). The velocity of sound v_0 for different liquids varies little; consequently, the dispersion will be determined chiefly by the ratio η'/τ. The larger η' and the smaller the relaxation time τ, the larger will be the dispersion. An estimate of the expected dispersion in benzene can be made from Eq. (22.7) by using the definition of η' and the fact that α increases with increase in frequency f proportional to the square of the frequency, up to $f = 1.65 \cdot 10^8$ Hz [96]. This means that the relaxation time found from the condition $\Omega\tau \ll 1$ will be

$$\tau \ll \frac{1}{2\pi f} = 9.6 \cdot 10^{-10} \text{ sec.}$$

On the other hand, as has already been shown, a definite fine structure ($\alpha\Lambda < 1$) is observed in benzene; consequently, the frequency 10^{10} Hz corresponding to the formation of the fine structure, lies beyond the region of relaxation of η', i.e., $\tau \gg 1/(2\pi \cdot 10^{10}) \sim 10^{-11}$ sec. Taking into account that for benzene, $\eta' = 0.866$ poise, we find $\Delta v/v > 3\%$ or $\Delta v > 40$ m/sec. Consequently, the measurements of the sound velocity from the fine structure of the scattering line should show a dispersion of the sound velocity in each case greater than 3%.

A still more reliable estimate of the dispersion of the sound velocity can be made for carbon bisulfide. According to the data of [424, 448], the dependence of α on f in the interval of frequencies $f = 10^7$-$2 \cdot 10^8$ Hz, departs from the square law and follows formula (22.3) with $\tau = 1/2\pi f_c = 2.2 \cdot 10^{-10}$ sec ($f_c = 7.2 \cdot 10^7$ Hz). Taking into account that for carbon bisulfide $\eta' = 5.5$ poise, we find from (22.7),

$$\frac{\Delta v}{v} \sim 9\% \quad \text{or} \quad \Delta v \sim 104 \text{ m/sec.}$$

A significant dispersion must also be expected in the case of carbon tetrachloride. In the same way, one can obtain an estimate of the dispersion of the sound velocity for a number of other liquids. However, it is necessary to remark that among the liquids studied few possess a large coefficient of bulk viscosity and a small relaxation time. Therefore, the investigation of the dispersion of the sound velocity from the components of the fine structure can be carried out in a limited number of cases, even though this method gives a comparatively low accuracy.

The estimate carried out above showed that in each case, in carbon bisulfide and benzene, the expected effect is so large that it should be observed from the components of the fine structure if the relaxation theory is correct, and if the measurement of α on which that estimate is made is also correct.

The negative result of [441] is evidently explained by purely experimental conditions which have been discussed in part in the literature [53, 241, 243, 444]. The circumstances given above led the author of this book and his coworkers to turn again to the study of the dispersion of the sound velocity from the fine structure of the Rayleigh line.

Measurement of the Hypersonic Velocity from the MB Components; Dispersion of the Sound Velocity. Preliminary estimates of the dispersion of the sound velocity have shown that it is first necessary to measure the velocity of hypersound in such liquids as benzene, carbon bisulfide, carbon tetrachloride, chloroform, methylene chloride, and methylene bromide. These have also been measured by the author and his coworkers [29-34, 36, 53]. The method described above of estimating the size of the dispersion showed that in such liquids as toluene, acetone, and acetic acid, the dispersion of the velocity is so small that it cannot be observed from the components of the fine structure [30, 36, 53]. Photographs of the fine structure for certain cases, in the excitation of λ = 4358 Å spectrum of mercury (for which ordinary light sources were used) are given in Fig. 35. Certain spectra of the fine structure in the excitation of scattering by irradiation from a Ne–He gas laser λ = 6328 Å are given in Fig. 70.

The photographing of the fine structure is carried out at somewhat different temperatures of the scattering liquid depending on the experimental conditions in different experiments. Therefore, it is essential to know the temperature coefficient of the hypersonic velocity. The temperature coefficients of the ultrasonic and hypersonic velocities were assumed to be equal, and the well-known coefficients for ultrasound were used.

The direct measurements of the dependence of the velocity of hypersound on the temperature in acetone, reported in [32, 445], serve as a strong confirmation of such an assumption.

The results of the measurements are shown in the graph of Fig. 71. The data of the two researches are plotted along the one straight line, and therefore allow us to determine the temperature coefficient of the hypersonic velocity for acetone with sufficient

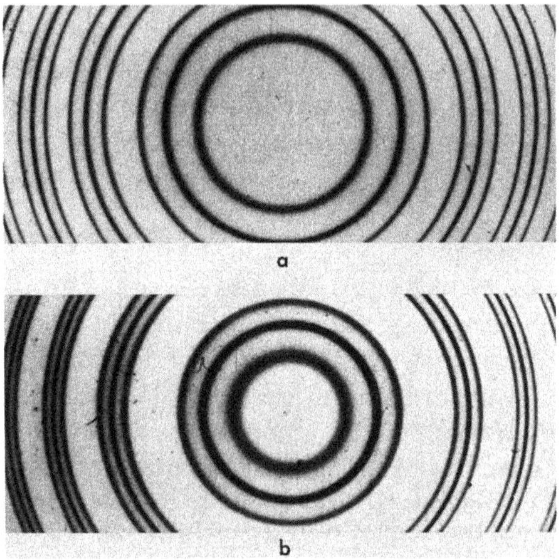

Fig. 70. Fine structure of the lines of Rayleigh scattering in some liquids with the excitation by the stimulated emission of $\lambda = 6328$ Å from a Ne–He gas laser (Mash, Starunov, Tiganov, and Fabelinskii [246, 264]). (a) – Benzene; (b) – carbon tetrachloride.

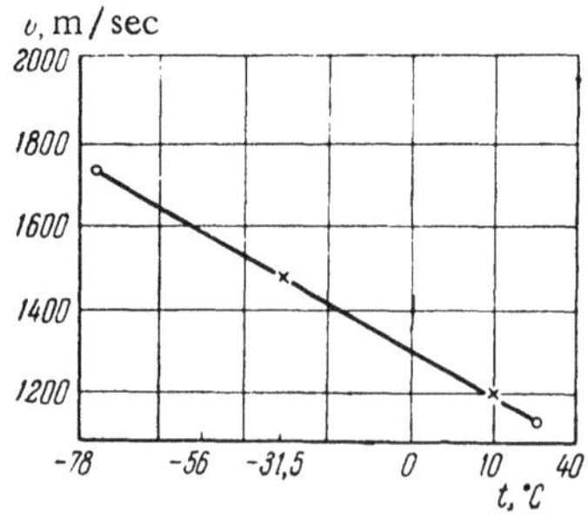

Fig. 71. Temperature dependence of the velocity of hypersound in acetone. Circles – measurements of Rank, Shull, and Axford [445]; crosses – measurements of Molchanov and Fabelinskii [32].

Table 16. Velocity of Hypersound and Ultrasound in Various Liquids

Liquid	Temp., °C	Frequency of hypersound, f, 10^9 Hz	Velocity of hypersound, v, m/sec	Velocity of ultrasound, v, m/sec	Dispersion of sound velocity $(\Delta v/v) \cdot 10^2$	Reference
Benzene	20	6.6	1470 ± 20	1324	10	[31, 53]
	20		1497 ± 30	1324	11	[446]
	20	4.9	1470 ± 8	1324	11	[264]
	21.5	4.9	1964 ± 8	1317	11	[589]
	20	4.9	1434 ± 15	1324		[630]
	20	5.01	1500	1324	11.5*	[637]
	25.0	6.75	1546 ± 13	1304	15.5*	[641]
Carbon bisulfide	20	6.8	1265 ± 22	1158	9	[32]
	20		1253.0	1158	8.2	[602]
		0.123	1223 ± 24			[611]
		4.5	1232 ± 15			[630]
	20	4.53	1250	1158	7.5*	[637]
	27.6	4.48	1217.0 ± 30			
	24.7	4.7	286 ± 42	923	6.5*	[641]
Carbon tetrachloride	20	5.0	1040 ± 27	920	12	[31]
	20	3.2	1015 ± 6	920	9.5	[264]
	2.4	3.25	991 ± 11	940	8.2	[589]
	22	2.9	1007 ± 7			[599]
	20	2.977	972 ± 5	938	4	[605]
	20	3.27	1000	920	8*	[637]
	27.3	3.21	965.4 ± 3.0			[647]
	25.0	5.93	1236 ± 23	1158	6.3*	[641]
Methylene chloride	20	3.53	1113 ± 6	1079	3.2	[264]
	22.3	4.55	1105	1071	3.1*	[642]
	19.1	4.67	1142	1080	5.4*	[642]
	25.7	3.53	1106 ± 5	1072	3.2	[638]
	24.1	4.33	1028 ± 17	1077	−4.7*	[641]
Methylene bromide	24	5.5	1099 ± 30	971	12(?)	[33]
Chloroform	20	5.7	1211 ± 43	1005	17(?)	[33]
	20	3.4	1055 ± 5	1005	5	[264]
	24.1	4.60	1112 ± 6	987	11*	[641]
Pyridine	20	7.2	1549 ± 31	1434	7.4	[446]

*The dispersion was calculated from the velocity of hypersound and ultrasound; three significant figures are given in [637] for the hypersonic velocity.

Table 16 (continued)

Liquid	Temp., °C	Frequency of hypersound, f, 10^9 Hz	Velocity of hypersound, v, m/sec	Velocity of ultrasound, v, m/sec	Dispersion of sound velocity $(\Delta v/v) \cdot 10^2$	Reference
Furan	20	6.3	1272 ± 37	1124	10.9	[446]
Quinoline	70		1572 ± 29	1360	13.5	[446]
Salol	20	6.9	1544 ± 15			[630]
Nitrobenzene	20	6.9	1546 ± 15	1473	6	[630]
	20	5.31	1535 ± 3	1473	4.3	[644]
	85		1467 ± 23	1263	14	[446]
Toluene	20	6.5	1314 ± 34	1324	0	[31, 53]
	21.5	4.39	1317 ± 7.6	1317	0*	[598]
	20		1380.1	1324	4	[602]
	20	4.50	1350	1324	1.9*	[637]
	22.2	6.4	1379 ± 1.1	1317	4.5*	[636]
	27.4	4.48	1316.2 ± 3.0			[647]
	24.7	5.85	1354 ± 19	1324	2.2*	[611]
Acetone	20	5.3	1190 ± 40	1192	0	[53]
	22.0	3.47	1144 ± 6.5	1190	−3.8*	[598]
	22		1174 ± 7			[599]
	20	2.951	1180 ± 20	1190		[605]
	20	3.6	1190	1190		[637]
	20	3.60	1190 ± 2.5	1190 ± 2.0	~0	[643]
	27.4	3.51	1136.8 ± 3.0			[647]
Acetic acid	20		1140 ± 35	1144	0	[53]
	24		1212.1	1144	5.6*	[602]
	20	3.57	1160	1144	1.3*	[637]
Water	20	6.4	1480 ± 20	1485	0	[36]
	22.9	4.33	1457 ± 10	1491	−2.8*	[598]
	20	2.951	1490 ± 20	1490	0	[605]
	22	1	1471 ± 8		0*	[599]
	20		1470.4	1492	−1.4*	[602]
	20	4.41	1480	1489-1492	−0*	[637]
	27.1	6.31	1504 ± 1.4	1505	0*	[636]
	26.6	4.47	1468 ± 10			[647]
	24.7	5.66	1466 ± 37	1492		[641]
n-Heptane	20		1186 ± 36	1180	0	[446]

Table 16 (continued)

Liquid	Temp., °C	Frequency of hypersound, f, 10^9 Hz	Velocity of hypersound, v, m/sec	Velocity of ultrasound, v, m/sec	Dispersion of sound velocity $(\Delta v/v)\cdot 10^2$	Reference
Methanol	22.1	3.21	1102 ± 6	1114	0	[598]
	22		1100 ± 11		0	[599]
	20	3.30	1110	1105		[537]
Aniline	22		1699 ± 8	1564	8.6	[599]
Ethyl acetate	24		1273.5	1176	7.6*	[602]
Ether	14		1051.8			[602]
	20	3.03	1000	985		[637]
Ethyl alcohol	20	3.54	1160	1150		[637]
	27.3	3.46	1112.3 ± 3.0			[647]
n-Hexane	20	3.42	1110	1113		[637]
Methylene iodide	24.1	5.07	986 ± 10	965		[691]

accuracy. It is shown to be equal to $\Delta v/\Delta T = -5.6$ m/sec-deg, and is identical with the value for ultrasound [283], which is equal to -5.5 m/sec-deg. Table XV gives some parameters of the liquids studied.

The determination of the hypersonic velocity is carried out by the formula (12.2), where $\Delta\nu$ is the frequency separation between the central line and each of the MB lines. In some cases, a notable difference is observed between the Stokes and the anti-Stokes values of $\Delta\nu$ (benzene, carbon tetrachloride). The reason for the observed difference of the Stokes and anti-Stokes shifts remains unexplained. In a recent investigation, such a difference was not observed for benzene.

Table 16 gives the measurements of the hypersonic velocity carried out by the author and his coworkers [31-34, 36, 53], data of other measurements [446, 447], in which dispersion of the sound velocity was discovered, as well as measurements of the hypersonic velocity carried out with the use of a He–Ne gas laser and a ruby laser [264, 598, 599].

The study of the fine structure of the light scattering showed the existence of the dispersion of the sound velocity up to frequencies (3-6) · 10^9 Hz in such low-viscosity liquids as benzene and carbon tetrachloride (see Table 16). From certain measurements of the hypersonic velocity (Table 16), it follows that negative dispersion is observed in water and acetone. Taking into consideration the importance of this result, Tiganov [643] carried out special experiments on the measurement of the hypersonic velocity in water and acetone. From his results, it follows that negative and positive dispersion of sound velocity is absent with accuracy up to several tenths of a percent in these liquids.

The dispersion of the sound velocity, established from the fine structure in a number of liquids, is shown in Table 16, and the dispersion of sound found in the ultrasonic region made the relaxational theory reliable, in each case, in a qualitative way. With regard to those liquids for which dispersion of the sound velocity has not been observed from the fine structure (Table 16), it is still impossible to say that dispersion of the sound velocity is completely absent from them. The result which was obtained for them means only that the method of fine structure is insufficiently sensitive for the small dispersion of the sound velocity to be found.

24. Discussion of the Results of Measurement of the Dispersion of the Sound Velocity in Liquids with a Large Bulk Viscosity

Summary of the Characteristics of the Propagation of Hypersound. A very rough variant of the relaxation theory with a single characteristic parameter ξ and a single relaxation time (see Sec. 22) plays an enormous role in the understanding of the nature of the fine structure of the Rayleigh line and the prediction of significant dispersion of the sound velocity in a whole series of liquids. The investigation of the liquids in which dispersion had not previously been found, while the relaxation theory clearly showed that appreciable dispersion should be present, indicated that the dispersion of the sound velocity is actually found up to frequencies $\sim 10^{10}$ Hz. In spite of such heuristic successes of the relaxation theory, it could scarcely be expected that the variant of the theory with a single relaxation time quantitatively describes the variation of the absorption with frequency in a

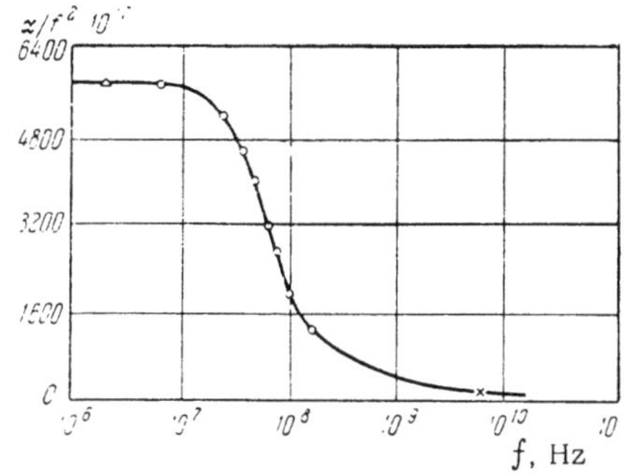

Fig. 72. Dependence of α/f^2 on f in carbon bisulfide. Experimental points of [448].

number of liquids. And, for all that, it was shown that in carbon bisulfide the dependence of the absorption coefficient on the frequency, measured in the ultrasonic frequency range, is very accurately described by the formula

$$\frac{\alpha}{f^2} = \frac{\alpha_\eta}{f^2} = \frac{\alpha_{\eta'}}{f^2} = B + \frac{A}{1+\left(\frac{f}{f_c}\right)^2}\,. \tag{24.1}$$

Here,

$$\left.\begin{aligned} \frac{\alpha}{f^2} = B\,\frac{8\pi^2}{3\rho v_0^3}\,\eta, \quad A = \frac{2\pi}{2v_0^3 f_c}\,[v_\infty^2 - v_0^2], \\ f_c = \frac{1}{2\pi\tau}\,. \end{aligned}\right\} \tag{24.2}$$

The second component in (24.1) is identical with (22.3). Lamb and Andreae [448, 449] in their measurements and in the measurements of Rapuano [450] constructed a curve of the dependence of α/f^2 on f (Fig. 72) and found that f_c = 72.1 MHz.

Later, Heasell and Lamb [451] carried out new measurements of the absorption in carbon bisulfide in the frequency range from 10 to 190 MHz and found that f_c = 78 MHz.

These results correspond completely to the formulas of relaxation theory with a single relaxation time. Neglecting the small difference between the values of f_c for CS_2 obtained in the different experiments, we obtain the relaxation time $\tau = 21 \cdot 10^{-10}$ sec,

Table 17. Some Characteristics of Liquids and the Propagation of Hypersound in Them

Liquid	t, °C	f_H, 10^9 Hz	$\eta \cdot 10^3$, poise	η', poise	$\tau \cdot 10^{10}$, sec	$f_c = \frac{1}{2\pi\tau}$, 10^8 Hz	$\alpha_{\eta'} \cdot 10^{-3}$, cm^{-1}	$\alpha_\eta \cdot 10^{-3}$, cm^{-1}	$\alpha \cdot 10^{-3}$, cm^{-1}	$\alpha\lambda$	References
Benzene	20 20	7.2 4.9	6.5	0.87	2.4 2.5	6.5 7.1	3.65 2.5	4.4 2.0	8.05 4.5	0.16	[29, 31, 53] [246, 264]
Carbon bisulfide	20	6.9	3.6	6.01 5.52	21.8	0.72	0.38	2.26	2.64	0.04	[29, 32, 53]
Carbon tetrachloride	20 20	5.0 3.3	9.5	0.[illegible]	0.8 0.93	20.5 21	19.5 14	6.4 2.0	26 16	0.54	[29, 31, 53] [246, 264]
Methylene chloride	20	5.8	4.03	0.98	3.3 17	4.8	4.3	1.8	6.1 0.91	0.13	[29, 33, 53] [264]
Methylene bromide	24	5.5	1.4	0.65	1.2	13.3	12.6	0.5	13.1	0.26	[29, 33, 53]
Chloroform	20	5.7	5.65	0.30	0.6 1.68	26.5	38.3 9	3.2 1.15	41.5 4.5	0.87	[29, 33, 53] [264]
Pyridine	20	7.4	9.58	0.658	1.95	8.2	2.95	4.73	7.68	0.16	[446, 447]
Thiophene	20	7.2	6.62	1.860	4.0	4.0	2.46	3.84	6.30	0.13	[446, 447]
Furan	20	5.8	3.72	0.176	0.58	27.0	16.45	2.49	18.94	0.41	[446, 447]
Quinoline	70	8.1	13.5	0.199	0.305	52.0	28.87	8.86	37.73	0.73	[446, 447]
Naphthalene	85	7.6	8.94	0.876	1.61	9.8	8.57	6.86	15.43	0.30	[446, 447]

in accord with (24.2). On the other hand, this quantity can be computed by (22.7) for the use of the measured.velocity $\Delta v/v$ and $\eta' =$ 6.01 poise. The relaxation times τ, found by these two methods, are identical within the limits of accuracy of the experiment. The application of the stimulated emission of the line $\lambda = 6328$ Å of the Ne–He gas laser for the excitation of the fine structure made it possible to increase the accuracy of the measurement of hypersound by an order of magnitude and, which is especially important, made it possible to measure the absorption coefficient of hypersound.

Therefore, for such liquids as benzene, carbon tetrachloride, and some others, one can now simultaneously and independently measure the absorption coefficient α and the hypersonic velocity v (and, consequently, the dispersion $\Delta v/v$) by the width and relative position of the MB components [264].

From the magnitude of the absorption α and the dispersion of the sound velocity $\Delta v/v$, and also from the formulas of the theory (22.7), (24.1), and (22.2), one can find the values of the relaxation times of the bulk viscosity coefficient τ from two independent measurements. For benzene and carbon tetrachloride, these two values of τ agree with one another; this means (as also in the case of carbon bisulfide) that the formulas of the theory with a single relaxation time quantitatively describe (within the limits of accuracy of the experiment) the propagation of sound in the liquids mentioned.

It is now already evident that the variant of the theory with a single relaxation time is not applicable in all cases [264]. But, on the other hand, liquids obviously exist, in addition to the three named above, to which the simplified variant of the theory is applicable. Methylene chloride clearly belongs to such liquids.

Although previously published measurements in the ultrasonic [452, 453] and hypersonic [33] ranges apparently gave reason to suppose that the dependence of α/f^2 on f cannot be described by Eqs. (24.1), (24.2) with a single relaxation time [36], more accurate measurements of the dispersion of the sound velocity, carried out in [264] with the use of a gas laser, show that the fact is not as had initially been supposed. We do not have at our disposal data on the width of the MB components in methylene chloride, because the ac-

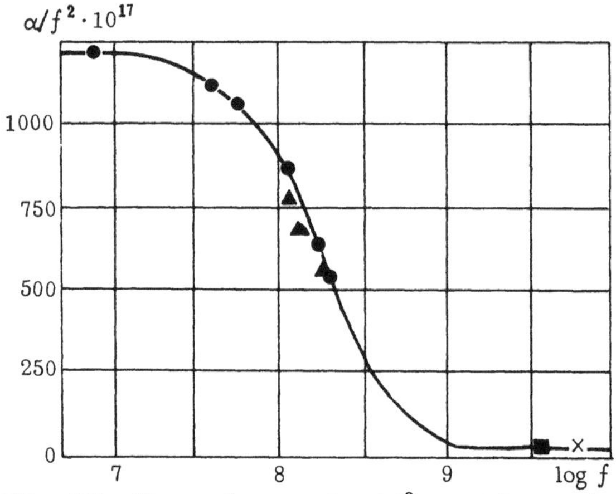

Fig. 73. Dependence of α/f^2 on f in methylene chloride. Circles – ultrasonic measurements [452]; triangles – the same, from [451]; squares – hypersonic measurements with a laser [264]; crosses – previous hypersonic measurements [33].

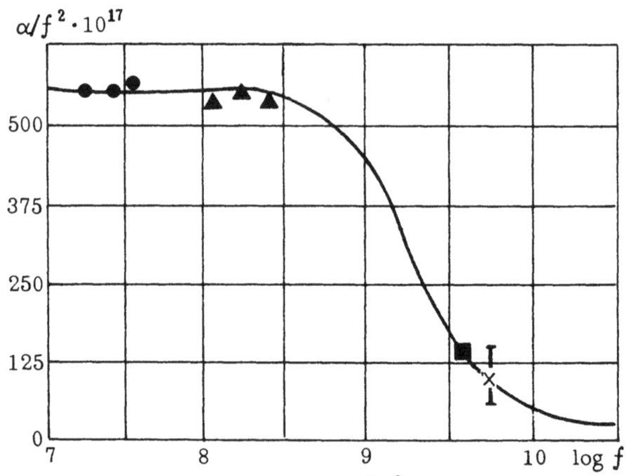

Fig. 74. Dependence of α/f^2 on f in carbon tetrachloride. Circles – ultrasonic measurements [79]; triangles – the same, from [451]; squares – hypersonic measurements with a laser [264]; crosses – previous hypersonic measurements [31].

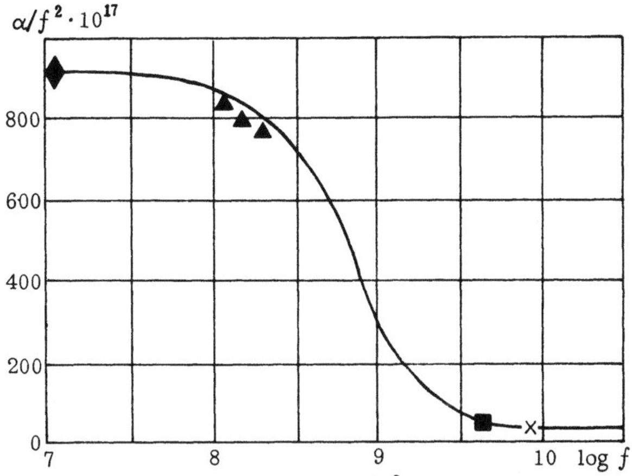

Fig. 75. Dependence of α/f^2 on log f in benzene. Rhombs – ultrasonic measurements [283]; triangles – the same, from [451]; squares – hypersonic measurements with a laser [264]; cresses – previous hypersonic measurements [31].

curacy of our method for such measurements is unsatisfactory to date and, therefore, there are no direct measurements of the absorption coefficient in methylene chloride.

But, from the measurement of the hypersonic velocity and of the ultrasonic measurements of velocity and η', which are more accurate than in [32], α/f^2 can be found from the formulas of the theory with a single relaxation time.

It is shown in Fig. 73 that the experimental points agreed with the general shape of the curve which expresses the requirements of the single τ theory (24.1). Excellent agreement of the simplified theory and experiment allows us to think that the simplified variant of the theory is successful in this case.

It must be admitted that after the measurement of the absorption and dispersion from the fine structure excited by the light of a gas laser, a great deal of unexpected evidence appeared in support of the validity of the simplified variant of the theory for the description of sound propagation in liquids with low shear and comparatively high bulk viscosity coefficients. The curves of the dependence α/f^2 on $\log f$ for benzene and carbon tetrachloride are shown in Figs. 74 and 75.

Table 17 gives some results of the calculation or estimate made from researches already completed [29, 31-33, 36, 53, 246, 264, 446, 447]. In some cases, the numbers in Table 17 are only estimates.

Heasell and Lamb [451] measured the absorption coefficient of sound by a pulse method in 94 organic liquids in the frequency range from 100 to 200 MHz, and discovered that a dependence of α/f^2 on the frequency is observed in some liquids, indicating that the frequency of $2 \cdot 10^8$ Hz is at the very beginning of the relaxation range. These results apply, for example, to isobutyl acetate and some other liquids. For acetic acid and propionic acid, the variation of α/f^2 corresponds to the end of the relaxation region. As has been established earlier, the critical frequency for acetic acid [434] is $f_c = 7.25 \cdot 10^5$ Hz (26°C) while for propionic acid [434] $f_c = 2.02 \cdot 10^6$ Hz (21°C). Special interest for us attached to the relaxation discovered in benzene in [451, 645]. Just this phenomenon was predicted by us [33, 53] after dispersion had been found in benzene, and it was determined that $f_c = 6.5 \cdot 10^8$ Hz. Figure 75 shows the curve of the dependence of α/f^2 on $\log f$ for benzene on the basis of the low frequency and hypersonic values of α/f^2. Although the ultrasonic data refer to the beginning of the relaxation region and the work was carried out with different samples of benzene, we are inclined to consider the results of Heasell and Lamb [451] as evidence in support of the conclusions which follow from the measurements of Fabelinskii and Shustin [31, 53] and also of Mash, Starunov, Tiganov, and Fabelinskii [264]. In the latter work, we found for chloroform $f_c = 26.5 \cdot 10^8$ Hz; this means that the maximum operating frequency used in [451], $f_{max} \sim 2 \cdot 10^8$ Hz, lies before the relaxation region and, consequently, α/f^2 for this liquid should keep its low-frequency value. This discussion applies also to carbon tetrachloride. Actually, the ultrasonic measurements that have been discussed confirm this conclusion for the cases of chloroform and carbon tetrachloride. According to [451], α/f^2 remains unchanged up to frequencies of about $2 \cdot 10^8$ Hz.

Dispersion of the Velocity and Absorption of Sound and the Molecular Theory of Sound Propagation. A systematic molecular theory of sound propagation in a liquid does not exist, as a molecular theory of liquids does not exist at the present time. The necessity of taking into account the complicated play of intermolecular interactions in liquids creates extraordinary difficulties for the theory.

There are no difficulties confronting the molecular theory of gases and, therefore, Kneser [455] succeeded in constructing a molecular theory of sound propagation in gases. It would be attractive to find such a group of liquids which would behave the same as gases relative to dispersion and sound absorption, for which the density, specific heat, etc., would be values which characterize the liquid.

According to the old classification of Pinkerton [456], liquids which exhibit a positive temperature coefficient of absorption are such "gas-like" liquids (Kneser liquids). It is assumed that in Kneser liquids the dispersion and sound absorption are brought about by thermal relaxation with the same mechanism as in gases — the exchange of energy between the internal and external degrees of freedom. This group of liquids is sufficiently broad and varied [283, 424] and includes some of the liquids of Table 17. Among them are benzene, carbon tetrachloride, and carbon bisulfide.

The classification scheme of Pinkerton does not have a universal character, and is not established physically [424], but the many attempts to interpret quantitatively the excess absorption and the dispersion of the sound velocity from the viewpoint of simplified molecular representation are of undoubted interest.

The principal contribution to the development of the molecular theory was made and the experimental results were obtained as a result of the efforts of Kneser [457, 458], Herzfeld [424, 459, 460], Lamb [461-463], Litovitz [424], Nomoto [464-467], their coworkers, and a number of other authors. Only a view of the researches of several authors have been mentioned. Detailed reports of the results can be found, together with bibliographies, in the reviews [422-424].

Here we shall use only the fundamental formulas of the molecular theory. For the absorption coefficient and the dispersion of the sound velocity, we have

$$\frac{\alpha_{\eta'}}{\Omega^2} = \frac{vD\Sigma c_i'}{2v_0^2 c_p (c_V - \Sigma c_i')} \frac{\tau}{1+\Omega^2\tau^2}, \tag{24.1a}$$

$$\left(\frac{v_0}{v}\right)^2 - 1 = -\frac{D\Sigma c_i'}{c_V (c_p - \Sigma c_i')} \frac{\Omega^2\tau^2}{1+\Omega^2\tau^2}. \tag{24.2a}$$

Here we have introduced the notation $D = c_p - c_V$. The actual relaxation time τ' is connected with τ by the relation

$$\tau = \frac{c_V - \Sigma c_i'}{c_V} \tau'. \tag{24.3}$$

τ is a time defined from the frequency f_c (24.2) for which $\Delta v/v$ (24.2) decreases by one half,

$$c_i' = g_i \frac{R_g x^2}{e^x (1 - e^{-x})^2}. \tag{24.4}$$

Here, $x = h\nu_i/R_gT$, where g is the degeneracy factor of the frequency ν_i. Summation in (24.2a) is carried out over all the eigenfrequencies of the molecule. At low sonic frequencies, where one can set $\Omega^2\tau^2 \ll 1$, we get from (24.1a) and (24.2a) that the dispersion is small and

$$\frac{\alpha_{\eta'}}{\Omega^2} = \frac{D\Sigma c_i'}{2v_0 c_p (c_V - \Sigma c_i')} \bar{\tau}. \tag{24.5}$$

From (24.5) one can compute $\bar{\tau}$, which has the meaning of some mean relaxation time. For sound of very high frequency, when $\Omega^2\tau^2 \gg 1$, we get

$$\alpha_{\eta'} = \frac{v_\infty D\Sigma c_i'}{2v_0^2 c_p (c_V - \Sigma c_i')} \frac{1}{\tau}. \tag{24.6}$$

In the case $\Omega^2\tau^2 \gg 1$ in (24.2a), $v \to v_\infty$, and we get the total dispersion of the sound velocity

$$\left(\frac{v_0}{v_\infty}\right)^2 - 1 = \frac{D\Sigma c_i'}{c_V (c_p - \Sigma c_i')}. \tag{24.7}$$

Taking into account that

$$\gamma_0 = \frac{c_p}{c_V}$$

at low frequency, and at high frequency

$$\gamma_\infty = \frac{c_p - \Sigma c_i'}{c_V - \Sigma \overline{c_i'}} = \frac{\overline{c_p}}{\overline{c_V}}, \tag{24.8}$$

after the relaxation region, one can write, in accord with (24.7), that

$$\left(\frac{v_0}{v_\infty}\right)^2 = \frac{\gamma_0}{\gamma_\infty}. \tag{24.9}$$

The formulas (24.1a)-(24.7) written here are known as the Herzfeld formulas.

In the general phenomenological theory of Mandel'shtam and Leontovich [420, 421], it was shown that if one assumes that the only mechanism which produces the absorption $\alpha_{\eta'}$ in a liquid is the exchange of energy between the external and internal degrees of freedom, then the formulas of the general theory should go over into formulas of the Kneser type. In this case, the energy of the sound wave is transformed into the energy of excitation of the molecules of the medium (sound absorption). We shall assume that the internal parameter introduced by Mandel'shtam and Leontovich, $\xi = n_2/n$, where n_2 is the number of molecules excited by the sound wave in 1 cm^3 of volume, and n is the total number of molecules per cm^3. By making such an assumption, formulas are obtained from the phenomenological theory [36, 423] which are identical with the formulas of the molecular theory (24.1a)-(24.7). It is improbable, even in an unassociated liquid, that the absorption $\alpha_{\eta'}$ and the dispersion $\Delta v/v$ are determined only by the same mechanism as in gases. But, as has been shown, it is possible that there is a group of liquids for which the absorption and dispersion are essentially determined by this mechanism.

Herzfeld and Litovitz [424], as an example, cite the values of $\overline{\tau}$, computed from the absorption, and compare them with the τ measured in the gas, and also compare the ratio $\tau/\overline{\tau}$ with the ratio of the densities in the liquid ρ_l and the gas ρ_g. Table 18 gives these data. From a comparison of the last two columns in Table 18, it is clear that the difference in the relaxation time in the liquid and the gas is essentially determined by the density or by the collision number. But it is also evident from this comparison that there are some processes which do not allow us to reduce the entire difference between $\overline{\tau}$ and τ to the difference in the densities of the liquid and the gas.

Table 18. Comparison of $\bar{\tau}$, Computed from the Absorption in a Liquid, with τ, Computed for Vapors (pressure, 1 atm; room temperature) [424]

Substance	Chemical formula	v_0, m/sec	$\frac{\alpha_{\eta'}}{f^2} \cdot 10^{17}$	$\bar{\tau} \cdot 10^{10}$, sec	$\tau \cdot 10^8$, sec	$a = \frac{\tau}{\bar{\tau}}$	$b = \frac{\rho_l}{\rho_g}$	$\frac{a}{b}$
Methyl bromide	CH_3Br	905	304	1.26	7.5	600	460	1.3
Methylene chloride	CH_2Cl_2	1100	1185	5.3	5.8	110	—	—
Chloroform	$CHCl_3$	995	380	1.30	1.35	105	320	0.33
Carbon tetrachloride	CCl_4	930	380	1.30	1.35	105	320	0.31
Ethylene chloride	$(CH_2Cl)_2$	1230	136	0.495	0.5	100	320	0.31
1,1-Dichloroethylene	$(CHCl)_2$	1200	577	2.1	0.6	350	325	1.08
Benzene	C_6H_6	1310	800	2.7	5.0	180	280	0.64
Hexane	C_6H_{14}	1250	36	0.152	0.65	420	190	2.26
Carbon bisulfide	CS_2	1140	5813	28.3	72.7	260	420	0.62
Methyl alcohol	CH_3OH	1123	22	0.13	2.2	1700	630	2.7

Table 19. Molar Heat Capacity, Measured and Computed Values of the Sound Velocity Dispersion

Substance	$t°$, C	Heat capacity, cal/g-mol					$D=c_p-c_V$	$\frac{v_\infty}{v_0}-1$	γ_0	γ_∞	$\left(\frac{\gamma_\infty}{\gamma_0}\right)^{1/2}-1$
		c_p	c_V	c'	$\bar{c}_p$	$\ddot{c}_V$					
Benzene*........	20	32.18	22.18	12.6	19.58	9.58	10	0.11	1.45	2.05	0.19
Carbon bisulfide*...	20	18.7	11.71	3.93	14.24	7.78	6.46	0.9	1.55	1.82	0.9
Carbon tetra-chloride*.......	20	30.82	20.76	11.87	18.95	8.89	10.06	0.13	1.48	2.13	0.20
Methylene chloride*.	20	24.12	15.63	4.17	19.95	11.46	8.49	0.03	1.54	1.74	0.064
Methylene bromide*.	24	25.34	17.0	5.34	20	11.66	8.34	0.13	1.49	1.71	0.07
Chloroform*......	20	28.0	18.47	7.88	20.12	10.59	9.53	0.09	1.51	1.90	0.12
Pyridine†........	20	30.0	23.7	11	21.0	12.7	8.3	0.08	1.35	1.66	0.11
Thiophene†......	20	29.5	21.7	11	18.5	10.7	7.8	0.12	1.36	1.73	0.16
Naphthalene†.....		48.3	35	23	25.3	12	13.3	0.14	1.37	2.01	0.24

*All the data were taken from [36] and [264].

†Data of the second, third, and fourth columns were taken from [447]. The remaining quantities given in the table were computed by the author from these data. v_∞/v_0 is taken from Table 16, where v_∞ is the hypersonic velocity and v_0 the ultrasonic velocity γ_∞/γ_0 is computed from Eq. (24.8); $c' = \Sigma c_i$ is the heat capacity of all the vibrational degrees of freedom.

A comparison of τ found from hypersonic measurements (Table 17) with $\bar{\tau}$ obtained from ultrasonic absorption and molecular theory, for those liquids for which such a comparison is possible (benzene, carbon bisulfide, methylene chloride, and carbon tetrachloride) shows that these quantities are rather close to one another. The existing molecular theory gives excellent agreement in order of magnitude, and this is already a very fine result.

However, much more is required from the primitive molecular theory in some cases. To be precise, Andreae et al. [452, 453] made an attempt at establishing a molecular theory to predict the number of relaxation times, to calculate the values of the dispersion of the sound velocity, etc. It appears to us that such requirements are excessive for the present-day level of development of the theory. This problem has been discussed in detail by Pesin [36], who also showed that the theory expressed in Eqs. (24.1a)-(24.9) is suitable only for estimates of order of magnitude and not for exact calculations.

Table 19 gives the values of the dispersion of the sound velocity, determined experimentally and by calculation. Again one must say that there is excellent agreement of theory with experiment in order of magnitude.

If quantitative agreement of these quantities is required, then one can find it in carbon bisulfide.

In the other cases, the substances shown in Table 19 show, as a rule, less dispersion of the sound velocity than follows from the molecular calculation (see the last column of Table 19). An exception is methylene bromide, but it is entirely possible that the accuracy of the determination of the dispersion in this substance, and also in methylene chloride and in chloroform completed in [36], is strongly over-rated. In [36], the fine structure was excited by light radiation of $\lambda = 4358$ Å from the spectrum of a mercury arc.

However, we do not in any way want to indicate that the same picture will be obtained in all the other cases for the substances shown in Table 19. For an explanation of the observed dispersion, staying within the framework of molecular theory, one takes into account for the calculation of the dispersion not all the eigenfrequencies of the molecule, but only definite groups of frequencies, being guided in the choice of frequencies by certain other considerations. Pesin [36] showed that if among the eigenfrequencies of benzene those are removed which refer to the E type of symmetry (10 frequencies out of 20), c' is computed on the basis of them, and then $(\gamma_0/\gamma_\infty)^{1/2} - 1$, then excellent agreement is obtained with experiment.

For carbon tetrachloride, such a method is not suitable. But here one can select frequencies between which Fermi resonance is possible. In this case, out of five frequencies, it is necessary to select four (eliminating the lowest – 218 cm^{-1}). By grouping the frequencies in such fashion, one again obtains agreement between theory and experiment.

In order that the calculations of the dispersion in carbon bisulfide correspond to experiment, it is necessary to use all the eigenfrequencies of the molecule, including the lowest.

In a number of cases calculations have been made on the relaxation part of the absorption. For example, it has been possible, by taking into account all the vibrational frequencies for liquid CO_2 to describe the relaxing part of the absorption with sufficient accuracy [463, 467].

In the case of sulfur dioxide, for an explanation of the relaxing part of the absorption, one must discard one of the three fundamental frequencies of this molecule in order to make both ends meet [468, 469].

The situation is still worse in cases of strongly associated liquids, such as acetic acid, for example [424]. But these liquids are "non-Kneser," and it is admitted that such an approach is not suitable for them.

In a number of cases, the relaxation absorption is brought about by isomeric rotation [462, 463, 470]. Here we shall only point out this fact, without taking the opportunity to discuss it.

To sum up briefly, it is necessary to turn attention to the fact that in some cases, considered above, it is possible to adjust the calculation so that it satisfies the experiment, but no general method of calculation follows from these considerations.

Therefore, while the molecular theory is in this form, one cannot pretend to a prediction of the exact magnitude of the dispersion, absorption, or relaxation time.

The assumption that the different levels or groups of levels of a molecule can appear in different ways in the process of relaxation of a molecule is natural, but one such assumption still does not make it possible to calculate the dispersion in each specific case beforehand.

Significant progress achieved in ultrasonic measuring technique has already made it possible to determine the absorption coefficient for a large number of liquids up to frequencies of 260 MHz, and to determine the critical frequency f_c [471].

The use of the gas laser as a light source for the study of the fine structure in a liquid has made possible a significant increase in the accuracy of the method and has enabled us to determine experimentally the damping of elastic thermal waves [246, 264, 598, 638, 643, 644]. The joining of the ultrasonic and hypersonic investigations possibly paves the way for a more universal theory.

The development and perfection of the semiphenomenological semimolecular theory, which permits the estimate of the hypersonic velocity [472], and which is based on the Debye theory of the thermal conductivity of a solid, is also of interest [473].

25. Experimental Investigations of the Polarization, Width, and Intensity of the Components of the Fine Structure of the Rayleigh Line

Polarization of the Fine Structure Components. At the very beginning of this book, in Sec. 1, it was pointed out that the adiabatic and isobaric fluctuations of the density do not destroy the isotropy in the volume of fluctuations. Therefore, the light scattered by such fluctuations will be completely polarized. The relaxation theory of light scattering, developed by Leontovich [39] shows, however, that under certain conditions, one can nevertheless expect a little depolarization of the MB components.

The quantitative results of the theory are contained in the formulas (6.37)-(6.39) and show that even in the case $\Omega_L \tau = 10$, $\Delta_u = 0.4$, and $\Omega_T / \Omega_L = 0.75$, the depolarization of the MB components $\rho(\Omega_L) = 4 \cdot 10^{-2}$. When, under the same conditions, $\Omega_L\tau = 1$, $\rho(\Omega_L) \sim 1.5 \cdot 10^{-3}$.

For a large number of liquids studied by different authors, $\Omega_L\tau$ lies within the limits from 0.5 and 0.7 and, consequently, from the viewpoint of existing theory, it is impossible to expect any depolarization of the MB components.

The experimental data described in the literature, referring to this problem, are not in agreement. The first measurement of the polarization of the fine structure components, carried out by Birus [236] showed convincingly that the Rayleigh triplet is completely polarized in toluene. The same result for toluene was obtained even earlier by Raman and Rao.[499]. Qualitative investigations of Gross and Syromyatnikov [475] also showed that depolarization of the MB components does not exist, even in viscous liquids such as n-cresol and phenol. Venkateswaran [257] and Sunanda Bai [258] came to the same conclusions.

In later works, these results, however, were subject to doubt. Rank et al. [243] reported that a significant depolarization of the MB doublet was observed by them, amounting to about 15% in acetone, for example. Later, Venkateswaran [476] reported that, in spite of his old data (the research of [257] was published on this material), there is depolarization of the MB components not noticed by him earlier which the author attempted to explain on the basis of the hole theory of a liquid proposed by Frenkel [151].

Consideration of the researches mentioned above from a methodical side shows that they are not free from inadequacies at various points; these are sometimes very considerable [29, 53, 243]. It is possible that just these methodological difficulties have led to the result that a finite and often significant depolarization of the MB components has been observed in low-viscosity liquids.

The author attempted to make his experimental apparatus free from a whole series of imperfections that are characteristic of the apparatus of his predecessors. In particular, he succeeded in obtaining interference photographs of the fine structure on a Fabry–Perot interferometer with great resolving power and made the measurements simultaneously in the x and z polarizations. The advantage of this method has already been briefly discussed in Sec. 12.

Table 20. Measurement and Estimate of the Depolarization Factor of the Fine Structure

Liquid	Depolarization factor $\rho(\omega) \cdot 10^2$		Authors
	MB components	Central component	
Toluene	Completely polarized	Completely polarized	Raman and Rao
Phenol	No components at room temperature	Depolarized	[499]
Toluene	Completely polarized	Completely polarized	Birus [236]
Benzene		10	Venkateswaran
Tetraline		38	[257]
Phenol		42	
Glycerin		25	
Hydrochloric acid		39 (30°C)	
		24 (75°C)	
Acetic acid		28 (45°C)	
		18 (75°C)	
Butyric acid		14 (120°C)	Sunanda Bai
		20 (45°C)	[258]
Isobutyric acid		11 (120°C)	
Cyclohexane		6 (45°C)	
Carbon bisulfide		21 (45°C)	
Chloroform		7 (45°C)	
Benzene		9 (70°C)	
Carbon bisulfide	Completely polarized with accuracy to within 1%	Completely polarized with accuracy to within 1%	
Benzene			
Toluene			
Acetone			Fabelinskii
Carbon tetrachloride			[29, 53]
Acetic acid		Depolarized	
Triacetin		Depolarized	
Glycerin		Depolarized	
Methylene chloride	Completely polarized	Completely polarized	Pesin and Fabelinskii [33], Pesin [36]
Methylene bromide			
Chloroform			
Benzene	Completely polarized with accuracy to within 6%	Completely polarized with accuracy to within 6%	Cummins and Gammon [646]
Toluene	Completely polarized with accuracy to within 6%	Completely polarized with accuracy to within 6%	

Table 20 (continued)

Liquid	Depolarization factor $\rho(\omega) \cdot 10^2$		Authors
	MB components	Central component	
Carbon bisulfide	Completely polarized with accuracy to within 4%	Completely polarized with accuracy to within 4%	Shapiro, McClintock, Jennings, Barger [641]
Benzene	Completely polarized with accuracy to within 2%	Completely polarized with accuracy to within 6%	

On the photograph of the fine structure of the spectrum of various liquids, shown in Fig. 35, one can plainly see that a continuous uniform background of different intensity in the I_X component is observed in the spectra of carbon bisulfide, benzene, toluene, acetone, carbon tetrachloride, and other liquids with its passage through the anisotropy fluctuations (the wing of the Rayleigh line). The halfwidth of the wing exceeds the region of dispersion of the etalon and is therefore observed as a uniform background. In some liquids, a more or less intense central line is seen in the I_X component.

Table 20 gives the measurements and estimates of the depolarization in the fine structure components. Careful consideration of the I_X component in the spectrum of such liquids as benzene, carbon bisulfide, toluene, acetone, and carbon tetrachloride reveals a very weak Rayleigh triplet which "appears" on the continuous background of anisotropic scattering. The photometric treatment of many photographs of the fine structure carried out in these liquids, and calculation of the aperture, show that the traces of the Rayleigh triplet in the I_X component are brought about not by its depolarization in the scattered light but by the finiteness of the aperture of the incident and scattered light.

This conclusion applies to all three components of the fine structure in benzene, toluene, carbon bisulfide, acetone, and carbon tetrachloride, and to the MB components in acetic acid and triacetin.

Thus the results of our experiments on the determination of the depolarization of the triplet lines lead to the conclusion that all three lines of the triplet are completely polarized under the conditions of the described experiment, with accuracy to within 1%.

The depolarization of the central component of the fine structure, obtained in the work of Venkateswaran [257] and Sunanda Bai [258] (working by the same method and evidently on the same apparatus) is completely understood. Their apparatus possessed a low resolving power and a sufficient region of dispersion. Therefore, the narrowest and most intense part of the wing is added to the central component and depolarization of the central part of the Rayleigh triplet should undoubtedly be discovered. However, it does not entirely follow from this experiment that the part of the scattering associated with the isobaric fluctuations of the density are depolarized. The observed depolarization of the "central component" is a superposition on the central component (isobaric fluctuations) of a more or less narrow wing of anisotropic scattering. This conclusion will be considered again below.

The use of a Ne–He laser for the investigation of the depolarization of the component of the Rayleigh triplet led Cummins and Gammon [646] to the conclusion that all the components of the triplet in benzene are completely polarized (see Table 20). Measurements of the absorption of hypersound have been carried out in the researches [638, 637, 643, 644]. See Table 21.

Thus, it seems to us that the MB components and the central component in low-viscosity mobile liquids can be regarded as completely polarized with excellent photometric accuracy.

In a number of cases, a narrow, intense, and depolarized wing is added onto the central component, and produces an impressive depolarization of the central component.

From the theory of Leontovich [39], it follows that when $\Omega_L \tau$ becomes large, one can expect a measurable depolarization of the MB components. The predicted effect of polarization of the shifted components of the fine structure has not been investigated experimentally up to the present time. Their investigation would be of great interest.

Table 21. Absorption of Hypersound and the Width of the MB Components in Liquids

Substance	Measured* $\delta\nu_{MB} \cdot 10^3$, cm^{-1}	Absorption coefficient of hypersound $\alpha \cdot 10^{-3}$, cm^{-1}		Computed† $\delta\nu_{MB} \cdot 10^3$, cm^{-1}
		Measured*	Calculated†	
Acetone	4 ± 1.6	3		
Water.	6 ± 1.6	4		
	4.3 ± 0.7[a]	3[a]		
Benzene	7	4.5		
Carbon bisulfide . . .	~4	~3	2.6	3.5
Carbon tetrachloride	17	16	26	28.7
Methylene bromide .	4.3 ± 1[c],11[d]	4[c],9.7[d]	13	15
Methylene chloride .			6	8
Chloroform	11	10	41	53
Pyridine			7.7	12.6
Thiophene.			6.3	9.7
Furan			19	25
Quinoline			37.7	62.6
Naphthalene			15.4	24
Acetic acid	25[b]	40	11.3	13.7
Triacetin.	~30	20		
Hydrochloric acid . .	51[b]			
Ethyl acetate	10[b]			
Toluene	13.5[a],7[d]	9[a],4.9[d]		
Nitrobenzene	19	9		

*Scattering was excited by light from a Ne–He laser, λ = 6328 Å.
†Scattering was excited by light from a low-pressure mercury lamp, λ = 4358 Å.
Data of measurements: (a) [637], (b) [602], (c) [638], (d) [641]. The remaining data refer to measurements reported in [53, 246, 264, 643, 644].

Width of the Components of the Fine Structure of the Rayleigh Line. Recently, Mash et al. [246, 264] completed the first experimental determination of the width of the MB components brought about by ultrasonic damping. Some months later, Chiao and Stoicheff [602] published the result of a similar investigation. In the experiment, the 6328 Å line of a Ne–He laser was used as the source of the excitation of the scattered light.

The quantitative data in [246, 264] were obtained for ultrasonic damping in benzene, carbon tetrachloride, and triacetin. Estimates were also obtained for the damping in carbon bisulfide and chloroform. In [602], a significant broadening of the scattering line was discovered in acetic and hydrochloric acids and in ethyl acetate. In a large number of liquids, broadening of the lines was not observed. The amplitude absorption coefficient α is associated with the halfwidth of the MB component by the relation (5.20), which can be rewritten for the halfwidth, expressed in cm^{-1}, as

$$\delta\nu_{MB} = \frac{\alpha v}{\pi c}. \tag{25.1}$$

Consequently, the measurement spectrum of the hypersonic velocity and the halfwidth of the displaced component from the fine structure [by means of (25.1)] make it possible to obtain α.

Until the direct measurement of the absorption became possible, this quantity was determined [29, 30, 53] by measurement of the dispersion of the sound velocity and the bulk viscosity from the formulas of the relaxation theory with a single relaxation time.

Table 21 gives the values of the measured α and those calculated from the dispersion of the sound velocity.

Taking into account that the measurements were carried out at somewhat different hypersonic frequencies (for benzene, for example, for $\lambda = 6328$ Å, $f = 4.9 \cdot 10^9$ Hz, and for $\lambda = 4358$ Å, $f = 6.6 \cdot 10^9$ Hz), it must be admitted that the agreement between the direct measurements and the determinations of the absorption from the dispersion of the sound velocity and the relaxation theory with a single relaxation time are satisfactory for benzene, carbon bisulfide, and carbon tetrachloride [246, 264]. A serious divergence is observed for the cases of chloroform [264] and acetic acid [602]. It is also necessary to explain the very large width (absorption) for hydrochloric acid and ethyl acetate [602]. If we assume the previous measurements of the ultrasonic absorption and dispersion to be valid, and the relaxation theory of sound propagation applicable to these media, then it is difficult to understand the reason for such an appreciable absorption, just as it is difficult to understand why a broadening of the line is not observed in [602] in such liquids as benzene, carbon tetrachloride, and chloroform.

From the intensity distribution in the components of the fine structure for benzene and carbon tetrachloride (Figs. 76a and 76b), it follows that this is a Voigt distribution for each component, close to the Lorentzian distribution.

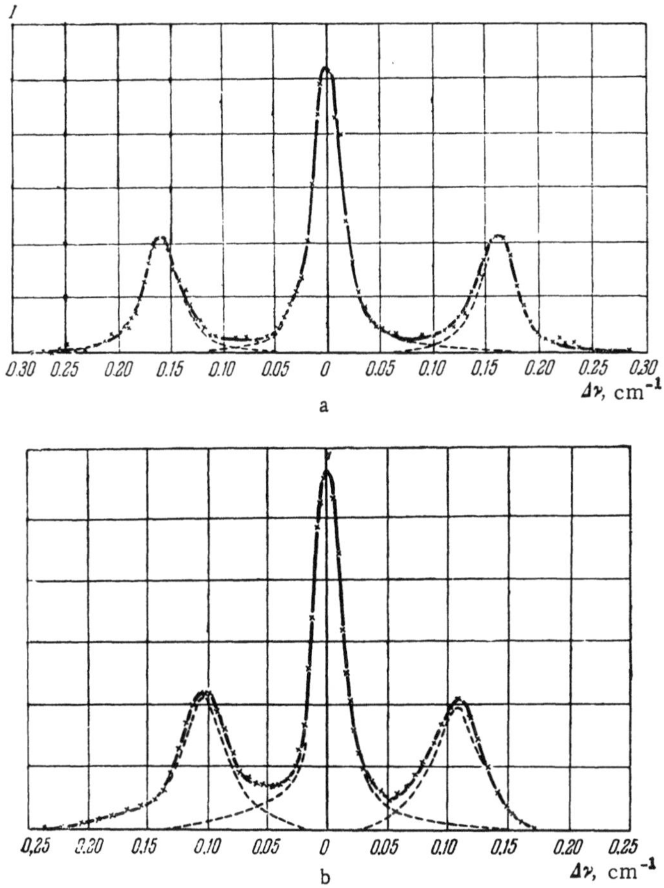

Fig. 76. Intensity distribution in the fine structure of the 6328 Å line of light scattered in the excitation of stimulated emission of Ne−He gas laser. The light is polarized in a plane perpendicular to the plane of scattering (Mash, Starunov, Tiganov, and Fabelinskii [264]). (a) Benzene; (b) carbon tetrachloride. The crosses on the solid curve represent the experimental data. The estimated components of all three components are denoted by the dashed lines.

Table 22. Calculation of the Halfwidth of the Central Component of the Line Structure

Substance	ρ, g/cm^3	$\varkappa \cdot 10^{[illegible]}$, cal/cm-sec-deg	c_p, cal/g-deg	$\chi \cdot 10^4$, cm^2-sec^{-1}	$\Lambda \cdot 10^5$, cm	$\delta\nu_c \cdot 10^4$, cm^{-1}
Benzene	0.879	33.3	0.417	9.1	2.04	6.5
Carbon bisulfide . .	1.262	38.5	0.24	12.6	1.83	3.1
Carbon tetrachloride . .	1.598	28	0.202	8.8	2.10	6.3
Acetic acid	1.049	43	0.483	8.5	2.26	3.75
Acetone	0.792	42.3	0.514	10.5	2.26	4.27
Toluene	0.866	34.5	0.400	9.7	2.02	4.26

Unfortunately, the measurements of the halfwidth of the central component for pure liquids have not yet been carried out. If the processes which characterize the resorption and creation of isobaric fluctuations of the density are actually sufficiently slow (see Secs. 5 and 7), then one can use the thermal constants, measured under static conditions for the estimate of the halfwidth.

For the calculation of the halfwidth of the central component of the fine structure, we rewrite the formula (5.37), expressing $\delta\omega_c$ in cm^{-1},

$$\delta\nu_c = \left(\frac{2\pi}{\Lambda}\right)^2 \frac{\chi}{\pi c} \text{ cm}^{-1}, \tag{25.2}$$

where the coefficient of temperature conductivity $\chi = \varkappa / c_p \rho$. In order to obtain the representation of the order of magnitude of $\delta\nu_c$, we carry out the calculation of this quantity in Table 22 by Eq. (25.2) and the statistical measurements of the thermal capacity c_p and the thermal conductivity $\varkappa$. The comparison of the last columns of Tables 21 and 22 show that the halfwidth of the central components is still narrower than the MB components (see also [326]).

It has already been noted in Sec. 12 that the width of the central component of a solution of polystyrene could be measured by the method of heterodyning the light beams; the width turned out to be equal to 100 Hz [593].

Investigation of the Width of the Central Component in the Approach to the Critical Temperature of a Pure Substance. There is great physical interest in the study of the fine structure of a pure substance as it approaches the critical point, to which we already turned our attention in Sec. 5.

However, experiments of such a nature encounter great difficulties. This is primarily because the MB components shift toward the unshifted line in the approach to the critical point, while they simultaneously increase in width. The experimental investigation of this phenomenon in thermal scattering has not yet been carried out. The shift of the MB components in CO_2 as the critical point is approached was shown by Guberman and Morozov [648]. The first investigations (by the method of light modulation) of the change in the width of the central component as the critical point is approached was carried out in sulfur hexafluoride (SF_6) by Ford and Benedek [640]; later investigations were made in carbon dioxide by Alpert et al. [639] by the method of light heterodyning. This experiment showed that the distance between the shifted and unshifted components rapidly decreased as the critical point was neared (see Sec. 34).

New methods of light modulation [640] and the light heterodyning [593] are better than the usual optical methods in paving the way for solution of this problem (see Sec. 12).*

Ford and Benedek [640] obtained a dependence of the halfwidth of the central line of the fine structure on the temperature $T - T_c$ above and below the transition temperature in SF_6. The density ρ was somewhat lower than critical $[(\rho_c - \rho)/\rho_c = 0.02]$. For this isochore, the minimum value of $\Gamma = \delta_c/2$ [Eqs. (5.36) and (5.37)] amounts to 640p g/sec (~100 Hz). Alpert et al. [639] investigated the dependence of the halfwidth of the central component of the fine structure on the angle of scattering in CO_2 close to the critical point, and also studied the dependence of the halfwidth on the temperature $T - T_p$, where T_p is the temperature of the appearance of the meniscus, which is virtually identical with the critical temperature T_c.

---

*The author is grateful to V. L. Ginzburg, who furnished him with a preprint of the research [639].

It can be seen from Eq. (5.37) that the dependence of $\delta\omega_c$ on $\sin^2(\theta/2)$ is expressed by a straight line whose slope is equal to $2n^2\chi k^2$. In [639], a linear dependence was actually found between $\delta\omega_c$ and $\sin^2(\theta/2)$; from the slope of the line for $T - T_p = 0.002°C$, it was found that the coefficient of temperature conductivity $\chi = 5.7 \pm 1.0 \cdot 10^{-7}$ cm²/sec.

It was further found in [639] that the halfwidth $\delta\omega_c$ is a linear function of $T - T_p$ with slope of 4.6 ± 1 kHz/deg. It follows from these results that the halfwidth of the central line for $T - T_p = 0$ is equal to 92 Hz, including 15 Hz which is the apparatus halfwidth. On the basis of a simple Van der Waals model, c_p was computed in [639] and substituted in (5.37), where $\chi = \varkappa/\rho c_p$; this gives

$$\delta\nu = \frac{16\pi}{\lambda^2}\frac{\varkappa}{\rho}\frac{\sin^2(\theta/2)}{RT}\left[(T - T_c) + \frac{4}{3}T_c\left(\frac{V - V_c}{V_c}\right)^2\right],$$

where V is the specific volume and V_c, the critical volume. Using this last formula and the experimental dependence of $\delta\nu$ on $T - T_p$, it was found that $\chi = 12 \cdot 10^{-7}$ cm² - sec. This simple model describes the results of the experiment surprisingly well.

A detailed and rigorous exposition of the theoretical side of the problem can be found in the review of Mountain [649].

Continuation of the experimental investigation in this difficult region appears extremely important, taking into account what was said in Secs. 2 and 5.

The Relation of the Intensities in the Components of the Fine Structure. The relation of the intensities in the components of the fine structure is determined by the Landau–Placzek formula [140]. The ratio of the integrated intensity of the central component to the integrated intensity of the two MB components is the ratio of the intensity of the light scattered by the isobaric intensity fluctuations to the intensity of the light scattered by the adiabatic density fluctuations.

The expressions were obtained in Sec. 1 for both these intensities (1.53) and (1.54); the ratio of which is also expressed by Eq. (5.40). Although Eq. (5.40) was obtained by a purely thermodynamic derivation, it can be used also for the case in which a significant dispersion is observed. Actually, it was pointed out by Fabelinskii [53] that the light intensity scattered by the adiabatic density fluctuations can be obtained by a nonthermodynamic calculation (7.4). For the calculation of the isobaric fluctuations, the use of thermodynamics is completely admissible, since these fluctuations are characterized by a slow change (halfwidth of the central component $\sim 10^7$ Hz). Rytov [40] obtained a formula that is identical with (7.4) by starting out from a calculation based on correlation statistics [see Eq. (7.23)]. Therefore, the relation (5.40) can also be used for those liquids in which an appreciable dispersion of the sound velocity is observed (see Sec. 24). The Landau–Placzek formula (5.39) differs from the formula (5.40) in two respects.

1. In the derivation of the Landau–Placzek formula (5.39), as is clear from [53, 141] and Eqs. (1.41) and (1.45), neglected the quantity $(\partial \varepsilon/\partial T)^2_\rho \overline{\Delta T^2}$ was neglected. From the ratio of the intensity (1.45) to the intensity (1.41), and taking (1.47) and (1.64) into account, we get the formula (5.39).

2. In the derivation of (5.39), purely thermodynamic relations have been used and, therefore, Eq. (5.39) is inapplicable to the case in which an appreciable dispersion of the sound velocity is observed. Equation (5.40), obtained by us [53, 144, 146], improves Eq. (5.39) in the thermodynamic relation (an accurate thermodynamic derivation) and extends it to the case in which velocity dispersion is present.

With account of the coupling between the adiabatic compressibility and the velocity of sound ($\beta_S = 1/\rho v^2$), Eq. (5.40) can be described in a somewhat different way:

$$\frac{I_c}{2I_{\text{MB}}} = L\frac{\sigma^2 T}{\rho c_p \beta_S} = \frac{L\sigma^2 v^2 T}{c_p}, \qquad (25.3)$$

where $L = [(1/\sigma)(\partial \varepsilon/\partial T)]^2_p / [\rho(\partial\varepsilon/\partial\rho)]^2_S$.

According to the meaning of Eq. (25.3), the velocity v is the velocity of hypersound, measured by the components of the fine structure (Sec. 23) or by some other method, at a frequency which is appropriate for scattering in the liquid under study, and σ and c_p can be taken from static measurements.

One can also find the value of $[(1/\sigma)(\partial \varepsilon/\partial T)]_p$ from the static measurements, while the value of $(\rho\partial\varepsilon/\partial\rho)_S$ must be determined by the dynamic method described in Sec. 13, or, still better, computed from Eq. (1.93), where all the quantities except $(\rho\partial\varepsilon/\partial\rho)^2_S$ are determined by independent measurements at just this frequency which is involved in the light scattering. If we neglect the quantity $(\partial\varepsilon/\partial T)^2_\rho\overline{\Delta T^2}$ in the calculation of the light intensity scattered by the density fluctuations, then, as follows from (1.61) and (1.65),

$$\left(\rho\frac{\partial\varepsilon}{\partial\rho}\right)_T=\left(\rho\frac{\partial\varepsilon}{\partial\rho}\right)_S \text{ and } \left(\frac{1}{\sigma}\frac{\partial\varepsilon}{\partial T}\right)_p=\left(\rho\frac{\partial\varepsilon}{\partial\rho}\right)_T$$

and, consequently, L = 1. By further assuming that there is no dispersion of the sound velocity, we obtain the Landau–Placzek formula (5.39) on the basis of (1.47) and (1.64) in place of (5.40) in (25.3).

The problem of how much L differs from unity for any particular liquid is a problem of the quantitative thermodynamic and nonthermodynamic characteristics of the liquid. Our estimate shows that L varies from 1.7 to 1 for different liquids. It is then clear that it is impossible to assume L = 1 in all cases, although attempts have been mistakenly made to show that L is always equal to unity.*

Table 23 gives the measurements of the ratio of the intensity in the fine structure components that are known to us.

Birus [236] was the first to undertake the experimental proof of the Landau–Placzek relation. His experiment was performed

*In [477], an attempt was made to show that Eq. (25.3) is identical to the Landau–Placzek formula (5.39) in all cases if the relation between the dielectric constant and the density is determined by the Clausius–Mosotti formula or, more correctly, by the Lorentz–Lorenz formula. Here it is assumed that $(1/\rho)(\varepsilon - 1/\varepsilon + 2)$ = const = c, while it follows from the calculations of [477] that c does not depend on the temperature. In other words, the derivative of the dielectric constant with respect to the temperature at constant density $(\partial\varepsilon/\partial T)_\rho$ is neglected. As was shown [145], if we neglect $(\partial\varepsilon/\partial T)_\rho$, then Eqs. (23.3) go over into (5.39). But, again, such a neglect is inadmissible in a number of cases (see Secs. 1 and 5), which evidently remained misunderstood in [477]. These discussions cannot be connected with specific measurements of $I_C/2I_{MB}$.

Table 23. Measurement and Calculation of $I_c/2I_{MB}$ from Data of Different Authors

Liquid	Measured *	Computed		Authors
		Eq. (25.9)†	Eq. (5.39)	
Water	0.36 (0.07)		0.01	Venkateswaran [257]
	0.14		0.01	Rank et al. [243]
	~0	0.0115	0.0068	Fabelinskii [145]
	<0.1			Medhi [478]
	<0.02			Cummins and Gammon [637]
	0.14			Rank et al. [647]
Benzene	0.97 (0.75)		0.32	Venkateswaran [257]
	0.98 ± 0.25	0.8-0.7	0.46	Fabelinskii [145]
	0.72 ± 0.1	0.6	0.46	Mash et al. [264]
	0.84 ± 0.5	[0.66]	0.43	Cummins and Gammon [637]
Acetone	0.79 (0.42)		0.39	Venkateswaran [257]
	0.42		0.39	Rank et al. [243]
	0.59 ± 0.12	0.57	0.40	Fabelinskii [145]
	0.45			Medhi [478]
	0.44 ± 0.01	[0.47]	0.39	Cummins and Gammon [637]
	0.44		0.42	Rank et al. [647]
	0.40 ± 0.05			Tiganov [643]
Carbon tetra-chloride	0.84 (0.49)		0.48	Venkateswaran [257]
	1.10 ± 0.30	0.84	0.46	Fabelinskii [145]
	0.67 ± 0.15	0.84		Mash et al. [264]
	0.72 ± 0.03	[0.62]	0.45	Cummins and Gammon [637]
	0.75			Rank et al. [647]
Ethyl alcohol	0.39 (0.17)		0.18	Venkateswaran [257]
	0.15		0.18	Rank et al. [243]
	0.23 ± 0.01	[0.24]	0.18	Cummins and Gammon [637]
	0.26		0.21	Rank et al. [647]
Carbon bisulfide	0.93 ± 0.18	0.98		Fabelinskii [145]
	0.71 ± 0.04	[0.79]		Cummins and Gammon [637]
	0.77		0.60	Rank et al. [647]
Toluene	0.36		0.40	Birus [236]
	0.45 ± 0.1	0.43	0.36	Mash et al. [264]
	0.42 ± 0.03	[0.41]	0.36	Cummins and Gammon [637]
	0.41		0.34	Rank et al. [647]

*The results corrected in [479] are given in parentheses.

†The results of the calculation in [637] with account of dispersion according to the formula similar to (25.3) are given in brackets.

Table 23 (continued)

Liquid	Measured*	Computed		Authors
		Eq. (25.9)†	Eq. (5.39)	
Ether	0.45 (0.26)		0.34	Venkateswaran [257]
	0.35 ± 0.01	[0.43]	0.37	Cummins and Gammon [637]
Cyclohexane	0.65 (0.38)		0.47	Venkateswaran [257]
Methyl alcohol	0.285 (0.11)		0.21	Sunanda Bai [258]
	0.26 ± 0.01	[0.27]	0.21	Cummins and Gammon [637]
n-Propyl alcohol	0.60		0.34	Sunanda Bai [258]
Isopropyl alcohol	0.45		0.24	
n-Butyl alcohol	0.70		0.3	
Isobutyl alcohol	0.66		0.40	
Chloroform	0.58 ± 0.1	0.58	0.51	Mash et al. [264]
Methylene chloride	0.72 ± 0.1		0.54	
n-Hexane	0.35 ± 0.02 (0.34)		0.28	Cummins and Gammon [637]
Acetic acid	0.41 ± 0.01	0.28	0.21	
Nitrobenzene	0.49 ± 0.06			Starunov, Tiganov, and Fabelinskii [644]

on an instrument with a Michelson echelon. In view of the great complexity of this experiment, which lasted almost two years, only a single liquid was studied qualitatively – toluene. The value he obtained differs from that computed from the formula (5.39). Broader investigations were completed by Venkateswaran [257] and Sunanda Bai [258]. According to their data, the experimental results did not agree with (5.39) in all cases.

Further measurements of the ratio of the intensities in the components of the fine structure of the Rayleigh line were made by Rank et al. [243]. The work was carried out with a Fabry–Perot etalon. To increase the light power and raise the resolving power of the apparatus, Rank and his coworkers used an etalon of thickness 10.5 mm in the investigation of the fine structure. Here the long wavelength component of the fine structure of one order of the interference coincided with the short wavelength component of the next order. The superposition doubled the intensity and the resolving power. However, the accurate coincidence of the orders can be achieved only for one liquid for a given distance between the plates of the interferometer. For other liquids, as the same authors have shown, there will be only partial superposition. The work of the author [145] was carried out on a Fabry–Perot interferometer with the distance between the plates of about 5 mm and with a resolving power of about 10^6. The work of Medhi [478] was also performed with a Fabry–Perot interferometer.

The data of Rank and coworkers [243] and Medhi [478] apply to acetone, water, and ethyl alcohol, and show that there is excellent agreement between the experiment and the results of calculation from (5.39).

In all cases studied by the author [145], the measured quantities are in excellent agreement with those computed by (25.3) and are not in agreement with those computed by (5.39), with the exception of the case of water, for which the measurement gave a value ~ 0, which, in the limits of accuracy of the experiment, agrees with the calculation from (5.39) and (25.3).

We note that Eqs. (25.3) and (5.39) refer to the integrated intensities I_c and I_{MB}. At the same time, the ratios of the maximum blackenings were measured in [257]. So far as the replacing of the integrated measurements by the measurement of maximum intensity is concerned, then, as is shown in Sec. 12, such a substitution is permissible when the width of the exciting line is much greater than the proper width of the scattering line (see Tables 21 and 22). But it seems improper to us to replace the ratio of the intensities by the ratio of the blackenings.

It follows from Sec. 17 that a very serious methodological fault of these researches is that the "background line" is measured arbitrarily "by eye," and this threatens to create a large error in the measured value.

The greatest error is to be expected here in substances with an intense, depolarized continuous scattering spectrum (benzene, carbon bisulfide), and the smallest error in liquids where this form of scattering is weak (water, ethyl alcohol, and to some extent acetone). The sharp delineation of the fine structure components in excitation of the scattering by a laser beam makes it possible to find a more accurate value of the intensity ratio (see Fig. 70).

A method is set forth in Sec. 12 for unambiguous interpretation of the spectrum of the fine structure and the measurement of the integrated intensity of the fine structure components. The measurements of the author, given in Table 23, were carried out by this method.

It is interesting to note that there is no sharp delineation of the central component in the spectrum of the fine structure of water (see Fig. 35). If one carried out measurement of the maximum intensities, while the intensity in the middle between the lateral components is used for the intensity of the central component, then a value of $I_c/2I_{MB} = 0.11$ is obtained, i.e., a value which is close to the data of Rank and differs strongly from the data of Venkateswaran (see Table 23). If we undertake an interpretation of the contours of the lateral components, as was discussed in Sec. 12), then there remains nothing in the fraction of the intensity in the regions between the MB components. That intensity between the shifted components, which was found in the measurement ofthe maximum intensities, is due to the superposition of the "tails" of the MB lines.

In the calculation of $I_c/2I_{MB}$ by Eq. (25.3) for benzene and water, our measurements of $(\rho\partial\varepsilon/\partial\rho)_S$ were taken into account (Table IV, Appendix III) while for the other liquids, this coefficient was determined from the formula for the total intensity of the scattered light (Sec. 1). In the cases of benzene, carbon bisulfide, and carbon tetrachloride, the value of β_S was computed from the velocity of sound measured from the fine structure of the Rayleigh line.

Comparison of the experimental data (Table 23) with the results of calculation from Eqs. (5.39) and (25.3) or from Eq. (II-9) of [637], which is analogous to Eq. (25.3), shows that (25.3) gives much better agreement with the experimental data than Eq. (5.39) in all cases in which these formulas give divergent results. Taking into account the complexity of the procedure of the photometric treatment, the satisfactory agreement with experiment must be recognized while the measurements with the laser are in still better agreement with the calculation according to (25.3). Thus, the results obtained make it possible to speak of a satisfactory agreement between theory and experiment.

We recall that Vaucouleurs [55], from the divergence of his accurate data for the absolute intensity of the scattered light in benzene and ether from Eq. (2.6) on the one hand, and the data of Venkateswaran [257], which do not agree with Eq. (5.39), on the other, drew the conclusion that the theory of Rayleigh scattering underestimates the intensity of the central component. At this point, Vaucouleurs assumes the theory of Rayleigh scattering to be imperfect and proposes to improve it. Rocard [57] took up this assertion and made an attempt to find an additional source of light scattering, by considering the correlation of fluctuations between neighboring volumes which contain only one or two molecules. However, it was shown in Sec. 1 that the contradiction of theory and experiment is due not to the imperfection of the phenomenological theory of light scattering, but to the improper estimate of the values of the parameters entering into Eq. (1.94). The results of [650] can serve as excellent support of what has been said above.

One must point out a special case where the direct measurement of the ratio of the intensities in the components of the fine structure known to disagree with Eq. (25.3). We have in mind the case in which there is a narrow, strongly depolarized wing of the Rayleigh line in the light spectrum scattered in a liquid. Acetic acid, supercooled salol, and several other viscous liquids belong to this group.

In the derivation of the relation (25.3), fluctuations of the anisotropy were not taken into account; therefore this formula cannot pretend to give a description of the phenomenon as a whole.

On the basis of what was said above, one can expect that the ratio of the intensities in such liquids in which there is a low intensity of the wing of the Rayleigh line, will be close to the value computed from the Landau – Placzek formula, with the exception of liquids with a weak wing, but appreciable dispersion of the sound velocity. An example of such an exception is carbon tetrachloride. In water, simple alcohols, and acetone, the wing is of low intensity and dispersion of the sound velocity is absent or very low; therefore, it must be expected that the measured relation will be close to that computed by the Landau – Placzek formula. The difference can be determined only by the value of L (25.3). It can be seen from the measurements of Rank et al. and Medhi (Table 23) that such is the case. However, the data of Venkateswaran and Sunanda Bai differ strongly from the computed. Therefore, Rank [479] made the natural assumption that the data of [257] were distorted by parasitic light of the apparatus, which significantly increased I_c in all cases. On the basis of these considerations, Rank made corrections to the measurements of Venkateswaran. The corrections are shown in parentheses in Table 23.

The fact that the results for benzene, after introduction of the correction, did not approach the value determined from (5.39) (Rank was concerned with this [479]) is natural, inasmuch as dispersion of the sound velocity is observed in benzene (see Sec. 24) and L is larger than unity (see Sec. 5). The corrected quantity obtained by Rank is close to that computed from (25.3).

In the cases of methyl alcohol and ether, the corrections of Rank made the result worse, while, for cyclohexane, the absolute value of the divergence remains the same.

For the case of acetone, the $I_c/2I_{MB}$ cited in [257] and [479] are combined for finding the corrections for other liquids. The result is improved only for water and ethyl alcohol. The coincidence of the correct value for carbon tetrachloride with the value computed from (5.39) produces a perplexity in that dispersion of the sound velocity of about 12%, which makes an increase in the computed quantity by about 25%. Measurements made with a laser in CCl_4 do not agree with the data of Rank (Table 23). New measurements with a laser only confirm the significance of the corrections introduced, which are connected with the dispersion of the sound velocity [637].

All these results and their discussion lead to the thought that the theory of the phenomenon (even for cases in which there exists a dispersion of the sound velocity) is in satisfactory agreement with experiment, while the scatter of the experimental data testify above all to the difficulties of the measurements.

Chapter VII

Scattering of Light in Liquids with a Large Shear Viscosity and in Glasses

26. Intensity and Depolarization of Light Scattered in Silicate Glasses and Viscous Liquids

Scattering of Light in Silicate Glasses. The study of scattered light in silicate glasses, begun by the younger Rayleigh [480], continues to the present day, although the number of researches is not very large. Since Rayleigh, the intensity and depolarization of the light scattered in silicate glasses has been studied by Raman [481], Krishnan [482], Rank and Douglas [483], Levin [484], Maurer [485], and Prudhomme [486]. As a result of these investigations, a broad, various, and reliable collection of experimental material has accumulated.

Raman [481] studied fourteen different types of glasses (crown and flint) with refractive indices from 1.4933 to 1.7782. The intensity of the scattering in these glasses was measured relative to liquid benzene. The measured ratio varied for the different types of glasses from 0.11 to 0.63, while the depolarization factor ranged from 0.045 to 0.295.

The results of Krishnan [482] refer to seventeen different types of glasses with refractive indices from 1.4670 to 1.7850. In these glasses, the intensity of the scattered light, measured relative to liquid ether, varied from 0.8 to 3.3. With account of the fact that $I_{benzene}/I_{ether} = 3.57$, the results lie within the same limits as in the measurements of Raman. The depolarization Δ_u varied in the range from 0.012 to 0.26.

The broad investigations of Levin [484] of the different types of optical glasses for many different melts of each type showed that the intensity of the light scattered by them, measured relative to ether, lay in the limits from 0.7 (for BK-5) to 5.0 (for TF-5). Measurements of the depolarization gave the values 0.022-0.093 for different glasses.

Prudhomme [486] studied thirteen types of glasses with refractive indices in the range from 1.460 to 1.795. The intensity of the scattered light was measured relative to benzene and the obtained results lay in the range from 0.17 ± 0.02 to 2.01 ± 0.05.

Similar results were also obtained in the work of Maurer [485], who studied sixteen types of glasses and found that the intensity relative to benzene varied from 0.08 to 1.03. As a result of these investigations, it was established beyond doubt that the physical characteristic of the silicate glasses was connected not with random circumstances, but with the characteristics of the medium studied.

The value of the intensity of the scattered light in glasses exceeds the calculated intensity of scattering by density fluctuations by approximately an order of magnitude.

The results of the measurement of the scattering intensity in glasses, obtained by different authors and different methods, agree very well with one another. However, it is not possible to speak about measurements of Δ_h and Δ_v for the silicate glasses.

In the researches of Krishnan [482], Rank and Douglas [483], and others on the measurements of the depolarization, conclusions were drawn that the scattering centers are comparable or even larger than the light wavelength.

If this is so, then the scattering indicatrix should have a z asymmetry, essentially different from unity (see Sec. 21). Maurer [485] carried out measurements of z for sixteen glasses and two wavelengths, 5460 Å and 4360 Å. In all cases except two, the result was obtained that z = 1 within the limits of accuracy of the measurements.

Thus, in the majority of cases, the character of the scattering in the glass is very similar to that observed in liquids. However, it is necessary to keep in mind that the temperature treatment of the silicate glass can sometimes produce an inverse change in the character of the scattering. The effect of annealing on the intensity of scattering was observed by Levin and Maurer.

In particular, Levin observed a fall of intensity with increase in temperature in some cases. Vladimirskii [487, 488] had already pointed out the possibility of such a temperature dependence of the intensity of scattering in glass.

For certain sodium–borosilicate glasses, an anomalous dependence of the intensity on the wavelength ($1/\lambda^7$ and even $1/\lambda^8$ in place of $1/\lambda^4$) and departure of the scattering indicatrix from Rayleigh scattering were observed [484, 489, 490].

These phenomena have not yet found a rational explanation. It also remains beyond understanding why the intensity of the light scattered in glass is an order of magnitude larger than what follows from a calculation which takes into account only the density fluctuations, and is larger than is given by a calculation with account of depolarized scattering.

In a number of researches, attempts have been made to explain the observed divergence by different reasons. Sosman [493] and Raman [481] proposed that the increased intensity is connected with the "freezing" of the concentration fluctuations. The idea of freezing of the density fluctuations was held by Mueller [491]. Vladimirskii [487, 488] showed that the freezing of orientation fluctuations is possible; these create an additional source of scattering. The macroinhomogeneities in the glass have also been pointed out both as a reason for the increase in the intensity of the scattered light [487, 488, 491] and for fluctuations in the concentration, which are not usually taken into account in the calculations, and also concentration inhomogeneities. In what follows, we shall again return to the problem of the freezing of fluctuations. It is possible that some of these reasons are actually responsible for the increased light intensity scattered in silicate glasses.

However, the complex composition of the silicate glasses does not allow us to separate the difference sources of scattering unambiguously, and to study each of them apart from the others. Therefore, it is expedient to study the simplest glasses such as, for example, individual chemical substances, which transform easily from the liquid to the solid amorphous state for a small change in temperature.

Processes of transformation from the monomer (liquid) to the polymer (solid amorphous) are themselves of interest, but we shall exclude them from consideration, inasmuch as the transformation from the monomer to the polymer is connected with an important change in the molecule of the substance, which complicates the general picture of the phenomenon.

Intensity and Depolarization of Light Scattered in a Liquid-Glass Transition. The first step in the study of simple liquids which transform to a glass upon decrease in temperature was made by Velichkina [188, 189, 492]. In her researches, the measurement of the intensity and depolarization of scattered light in triacetin,* chloroethyl ether, benzenesulfic acid† (for brevity, β-ether), and salol.‡ In the first two liquids, all the necessary parameters were measured for the calculation of the intensity of the scattered light. The apparatus of Velichkina [188, 189] (see Sec. 11) made it possible to carry out the measurements over a wide range of temperatures.

$CH_2—OCOCH_3$
|
*Triacetin $CH—OCOCH_3$ is a transparent liquid. At 18°C, it has a
|
$CH_2—OCOCH_3$

viscosity 0.27 poise and density 1.16 g/cm^3. At a temperature of -60°C, triacetin transforms into an amorphous state.

† β-Ether 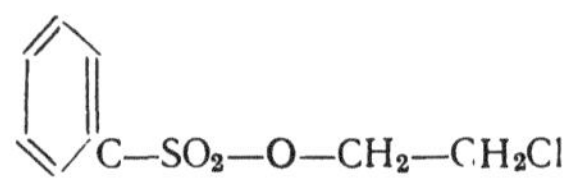 is a transparent liquid; at a temperature of 18°C it has a viscosity of 0.26 poise and density of 1.355 g/cm^3. At a temperature of −70°C, it transforms into a solid amorphous state.

‡ Salol 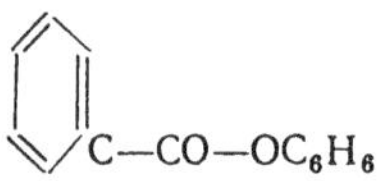 is a crystalline substance. Its melting temperature is 43°C. Pure salol is easily supercooled. The supercooled salol at 18°C has a viscosity of 0.18 poise; at a temperature of about −60°C, it transforms into a solid amorphous state.

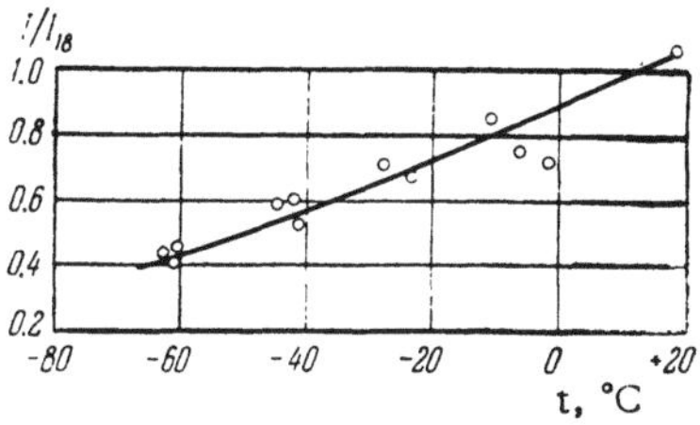

Fig. 77. Temperature dependence of the relative intensity of the scattered light in pentane (Velichkina [188, 189]). Solid curve — calculation. The points are the experimental data.

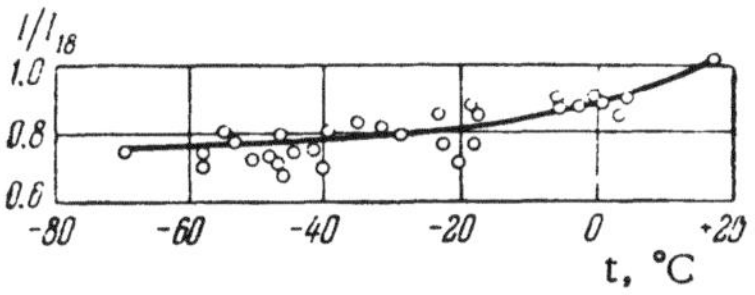

Fig. 78. Temperature dependence of the scattered light in triacetin (Velichkina [188,189]). The curve is drawn through the experimental points.

For testing the operation of the apparatus, and in particular for low temperatures, control measurements of the relative intensity of the scattered light were carried out in pentane over the temperature range from +20 to −70°C. In the region of high temperatures, this liquid had been studied in detail by Rao, and the results are given in Table VI (Appendix III). These results show that the experimental data agree excellently with the Einstein formula (1.94).

Pentane keeps its low viscosity down to the freezing point (−130°C). Figure 77 shows the data of Velichkina [188, 189]. The intensity of the scattered light at 18°C (I_{18}) is chosen as unity. Within the limits of accuracy of the measurement of the intensity (about 10-15%), the theory is in excellent agreement with experiment.

For measurements in triacetin in the temperature range from +20 to −20°C, it was shown that in the limits of accuracy of the experiment the points coincided with the curve constructed from the Einstein formula. The same was true for β-ether. Thus, it was noted that the molecular scattering of the light in a pure medium is confirmed by the measurements.

Experimental data on the temperature dependence of the relative intensity of three studied liquids are given in Table XVI (Appendix III). Figure 78 shows the temperature dependence of I/I_{18} for triacetin. In all cases, the scattering was excited by the 4358 Å line of the mercury spectrum. For the same exciting line, measurements were made of the depolarization factor of the scat-

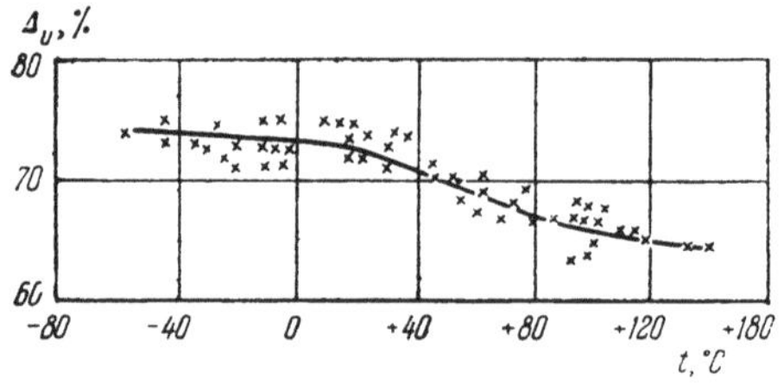

Fig. 79. Temperature dependence of the depolarization of light scattered by salol (Velichkina [189]). The curve is drawn through the experimental points.

tered light in irradiation by natural light. The results of the measurement are shown in Table VIII. The temperature dependence of the depolarization Δ_u in salol is plotted in Fig. 79.

In addition to the measurements in the organic glasses, described above, Velichkina [188, 189] carried out relative measurements of the light intensity scattered in fused quartz. Three samples of different origin were subjected to measurement.

In each specimen, measurements were made at 40-50 points. The difference in the measured intensity from point to point and from specimen to specimen did not exceed 5-10%. This gave support to the idea that molecular light scattering was measured, and not scattering from randomly included impurities.

In addition, the ratio of the intensity of the scattering in triacetin and in β-ether to the scattering of light in liquid benzene was measured for a temperature of 18°C, as well as the intensity of the scattering in fused quartz relative to the intensity of scattering in carbon tetrachloride.

The measurements gave the following ratios: $I_{C_6H_6}/I_{\text{triacetin}} = 4.48$; $I_{C_6H_6}/I_{\beta\text{-ether}} = 1.05$; and $I_{CCl_4}/I_{\text{fused quartz}} = 5.4$.

By using the scattering coefficients for benzene (Tables III and 9) and carbon tetrachloride (Tables III and 10), it is possible to find the scattering coefficients (absolute intensity) for triacetin, β-ether, and fused quartz. On the other hand, on the basis of measurements [188, 189] of the temperature variation of the index

Table 24. Scattering Coefficients at 18°C

Substance	Wave-length λ, Å	$\beta_T \cdot 10^{12}$, bar^{-1}	$\frac{n}{\sigma}\frac{dn}{dT}$	Δ_u	$R_{90} \cdot 10^6$, cm^{-1}	
					Computed	Measured
Triacetin	4358	55.8	0.53	0.32	2.24	1.94
β-Ether	4358	42.3	0.65	0.60	7.7	8.3

Table 25. Coefficient of Scattering in Fused Quartz at 18°C and λ = 4358 Å

T, °K	n	$c_{11} \cdot 10^{-11}$, dyne/cm^2	$c_{44} \cdot 10^{-11}$, dyne/cm^2	p	q	$R_{90} \cdot 10^6$, cm^{-1}	
						Computed*	Measured
291	1.4668	7.88	3.05	0.208	0.073	0.179	1.84

*Computed from Eq. (9.33).

Table 26. Comparison of the Measured Scattering Coefficient in Glass at a Temperature of −70°C and That Computed from the Formulas of Different Theories

Substance	Scattering coefficient $R_{90} \cdot 10^7$, cm^{-1}		
	Measured	Computed (liquid theory)	Computed (crystalline theory)
Triacetin	13.6	7.7	2.7
β-Ether	70	29	6
Fused quartz	18.6	2.3	1.79

of refraction, the density, the expansion coefficient, and the sound velocity (see Sec. 14), the scattering coefficients for triacetin and β-ether are calculated by the formula obtained for the liquid. Using the Neiman constant for fused quartz [494, 495], one can compute the scattering coefficient in fused quartz according to Eq. (9.33).

Tables 24 and 25 give the computed and measured scattering coefficients and the parameters necessary for the calculation of the scattering coefficient in triacetin, β-ether, and fused quartz according to the data of Velichkina [188, 189].

For fused quartz it is seen that the calculated data diverge from the experimental data by an order of magnitude. Comparatively recently, Prudhomme [486] measured the intensity of scattering in fused quartz relative to the scattering intensity in benzene. According to this data, $R_{90} = (0.17 \pm 0.2) \cdot 10^{-5}$ for $\lambda = 5461$ Å; $R_{90} = 0.33 \cdot 10^{-5}$ for $\lambda = 4354$ Å. This quantity is about twice as large as the value measured by Velichkina for fused quartz.

At the present time, there is no theory of molecular scattering of light in a solid amorphous substance. But the theory of scattering in a liquid (Sec. 1) and the theory of scattering in a crystal (Sec. 9) have been brought to a considerable state of completion. Therefore, one must make an attempt to consider the solid amorphous body as a liquid and compute the intensity of the scattering according to Eq. (1.94), which was derived for a liquid, and then consider the amorphous solid as a crystal with an infinite number of axes of symmetry and compute the scattering intensity in the glass according to Eq. (9.33). This very calculation has been carried out by Velichkina [189, 492].

Without giving the calculation in detail, let us write the final results (Table 26). Comparison shows clearly that the agreement of the experimental values of R_{90} for the triacetin and β-ether with the values calculated from the formula from the theory of scattering in a liquid is much better than with the result of the calculation from the formula of the theory of scattering in crystals. At the same time, the experimental value of the scattering coefficient in fused quartz exceeds the data of both calculations by an order of magnitude.

Taking account of the fact that relative quantities, both computed and measured, are always less affected by errors than the absolute quantities, the ratio of the coefficients of scattering R_{90} for β-ether and triacetin is calculated in [189, 492] from the formulas of the theory of scattering in liquids and the formulas of the theory of scattering in a crystal. The following results were obtained: for $R_{\beta\text{-ether}}/R_{\text{triacetin}}$: 2.3 from the theory of crystals,

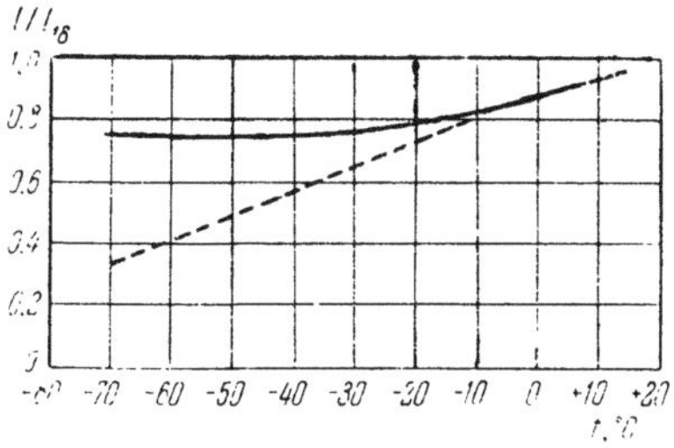

Fig. 80. Temperature dependence of the intensity of scattered light in β - ether (Velichkina [189]). Dashed curve – theoretical; solid line – experimental curve.

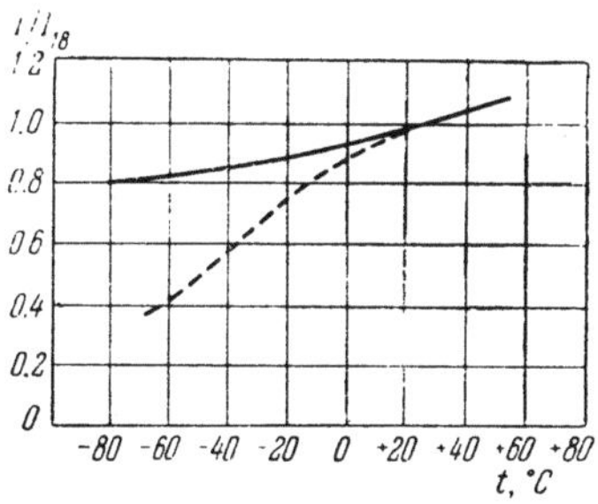

Fig. 81. Temperature dependence of the intensity of the scattered light in triacetin (Velichkina [189]). Dashed curve – theoretical; solid line – experimental curve.

4 from the theory of a liquid, and 5 from experiment. If the absolute values of R_{90} computed from the formula of liquid theory and the measured values in triacetin and in β-ether differ by 100%, then the ratios of the computed/measured R_{90} differ by at most 25%. Figures 80 and 81 show the variation of the relative intensity, both measured and computed by (1.94). The curves of Figs. 80 and 81 give convincing proof that the experimental data agree very well at room temperature with the values computed from (1.94). This serves as a proof of the fact that molecular scattering of light has been observed.

Upon decrease in temperature, the measured and computed values diverge, but the divergence does not exceed 100%.

Unfortunately, such an experiment has not been performed with silicate glasses and fused quartz, and evidently will not soon be done; therefore, the extent to which light scattering bears a molecular character above the glass temperature T_g remains unknown.

Thus, as a result of experiments with organic glasses (although only one set) and fused quartz, one can draw the conclusion that the intensity of scattered light in the case of organic glasses exceeds by about 100% the value computed from the formulas obtained in the theory of scattering in a liquid, and departs very strongly from the intensity computed from the formulas of the theory of scattering in crystals.

The scattering intensity in fused quartz is about an order of magnitude greater than the value computed by both theories. The rational explanation of the observed divergence remains hidden today. At the beginning of this section, we spoke of attempts to explain the divergence of the scattering intensity observed in glasses. The study of organic glasses and fused quartz does not eliminate the necessity of seeking an explanation for the conflicting measured and computed values of the intensity in glasses, although the scale of this divergence now becomes somewhat different. The idea of the "freezing" of the density fluctuations, suggested by Mueller [491], is very tempting, and now [485] supports it. Actually, if one assumes that the scattering in fused quartz is determined by the density fluctuations close to the glass temperature $T_g \sim 2100°K$, then, as was shown by Velichkina [189], the calculation by means of Eq. (1.94) gives $R_{90} = 1.5 \cdot 10^{-6}\ cm^{-1}$, i.e., a value very close to the measured values (see Table 25).

Calculations of such a type for silicate glasses also give a value much closer to the measured value than the calculation using the temperature of the experiment. If this same method of calculation is applied to organic glasses (triacetin and β-ether), then it is seen that the obtained result does not give good agreement with experiment. Thus, the hypothesis of freezing of the density fluctuations does not give a satisfactory result for this very case of organic glasses, which is the simplest, and for which the divergence from theory is minimal (Table 26).

On the other hand, the idea of freezing of the density fluctuations and other fluctuations was first critically considered by Velichkina [189] from the theoretical side.

The density fluctuations, on the freezing of which Mueller spoke, can be represented as consisting of isobaric and adiabatic fluctuations, as was done earlier (see Secs. 1 and 5). The isobaric density fluctuations are dissipated due to the process of thermal conductivity. The time change of the intensity of light scattered by the isobaric fluctuations of density is determined by Eq. (5.35). The time of falloff of the intensity by 1/e (or the dissipation time of the fluctuations) is found from (5.33) and (5.35) to be equal to

$$\delta = \frac{1}{2\chi k^2 \left(2n \sin \frac{\theta}{2}\right)^2} . \tag{26.1}$$

For $\lambda = 4358$ Å, $\theta = 90°$, we get (by assuming that for glass $\chi = 3 \cdot 10^{-3}$ cm^2/sec), $\delta \sim 10^{-9}$ sec. Equation (26.1) was first obtained by Mandel'shtam [24]. It was also shown by him that the dissipation of the isobaric fluctuations takes place quite rapidly. Therefore, it is improbable that such a fluctuation can "freeze" in cooling. The adiabatic density fluctuations are even less likely to freeze in the transition from the liquid to the glassy state. It is true that this circumstance became evident comparatively recently. The adiabatic density fluctuations, as has already been pointed out earlier, cannot remain fixed, but "run" along the body with the speed of hypersound. Modulation of the light by hypersound also leads to the appearance of a sharp fine structure in the spectrum of the light scattering (see Sec. 5). It was demonstrated by the researches of the author and Pesin [34, 35] that the fine structure exists in triacetin and glycerin in the growth of viscosity up to the glass-like state (see Sec. 27). The fine structure was found in fused quartz in the works of Krishnan [225] and Flubacher et al. [244]. It is therefore clear that one cannot even speak of the freezing of adiabatic density fluctuations.

It is still possible to speak of the freezing of the fluctuations of the anisotropy [487]. But if one uses Eq. (1.94) for scattering in liquids with the Cabannes correction, then, as was pointed out in Sec. 1, scattering by the anisotropy fluctuations in this case is taken into account sufficiently completely in the calculation of the temperature dependence of the depolarization factor.

The account of scattering by anisotropy fluctuations in triacetin and β-ether draws the experimental and calculated data close together. In the case of fused quartz, the Cabannes correction is insignificant. Even if one doubles it (taking freezing into account), then this correction changes the result of the calculation insignificantly and consequently does not explain the observed discrepancy.

In the case of silicate glasses, one must consider the possible freezing of the concentration fluctuations.

Actually, by taking into account the fact that the dissipation of the concentration fluctuations obeys an equation of the (5.30) type, we find, on the basis of (5.38), that the dissipation time of the concentration fluctuations is determined by Eq. (26.1), in which the coefficient of temperature conductivity χ must be replaced by the diffusion coefficient D. The problem as to whether the concentration fluctuations "freeze" in the silicate glass is a problem of the specific value of the diffusion coefficient of the particular component of silicate glass or the self-diffusion coefficient for fused quartz and organic glass. We shall not give the numerical data for the diffusion and self-diffusion coefficients in the glass.

For ordinary solutions and electrolytes, $D \sim 10^{-5}$ cm^2/sec and $\delta \sim 10^{-10}$ sec. In the liquid–glass transition, the diffusion coefficient decreases appreciably. If we assume that the diffusion coefficient or self-diffusion coefficient is of the same order as the smallest diffusion coefficient of a solid salt in a solid salt, $D \sim 10^{-8}$ cm^2/sec, then it follows from (26.1) that $\delta \sim 10^{-4}$ sec; in this case, one can scarcely expect that the concentration fluctuations will freeze. But it is possible that the diffusion coefficient is smaller by six or seven orders of magnitude and, hence, one can expect a freezing of the fluctuations in the form of microinhomogeneities.

Once again, however, the possible freezing of the concentration fluctuations cannot explain the observed difference between the measured and calculated R in organic glasses and fused quartz, if there are not foreign contaminants in them.

For an explanation of the increased intensity of scattering in glasses, Vol'kenshtein [496, 497] introduced the interesting hypothesis which follows.

In the glass transition, one can represent the entropy fluctuations ΔS (see Sec. 1) as consisting of two parts: ΔS_1, connected with fluctuations of structure, and ΔS_2 which is identical with the already considered entropy fluctuations ΔS, which are dissipated as the result of the process of thermal conduction (see Sec. 1).

As was shown above, the fluctuations ΔS_2 cannot be frozen, while the fluctuations ΔS_1, in the opinion of Vol'kenshtein, necessarily freeze, inasmuch as the transition to the glass means the cessation of structural changes. The fluctuations ΔS_1 lead to a scattering of the unshifted frequency, and consequently to an increase in the intensity of the central component. Vol'kenshtein [496, 497] computed the ratio of the intensity of the components of the fine structure of the Rayleigh line with account of ΔS_1, but the calculation cannot be carried all the way to numerical results, since such quantities as c_{p1} – the structure heat capacity of the liquid – and other quantities are introduced in them which are not susceptible to measurement or calculation at the present time. However, this hypothesis qualitatively permits the explanation of the increased intensity in glasses generally and the intensity of the central line in particular.

The experimental material considered, and the calculations of the intensity of the scattered light, show that the phenomenon of Rayleigh scattering in an amorphous solid is better explained by the theory of scattering in a liquid than by the theory of scattering in a solid. Such a situation seems natural to us. Actually, in the construction of the theory of the solid, as a consequence of the ordered character of the crystalline structure, it suffices to consider the Debye and Born (acoustical and optical) branches of frequencies in order to consider all the degrees of freedom and thus to construct the complete theory of scattering in the crystal.

If we assume that the amorphous solid is this crystal with an infinite number of symmetry axes (as we have done to obtain the scattering formula for the amorphous solid from the formula obtained for a crystal), then, as the result of the disordered nature of the amorphous solid structure, the anisotropy fluctuations, which appear in the form of a narrow wing, are not taken into account.

The depolarization of the central line of the fine structure in strongly viscous substances indicates the presence of a narrow wing in the spectrum (see Sec. 27).

The successful spectral investigation of the light scattered in a solid amorphous body has only begun. But already the first results (which will be given in the next section) show that many things have already become clear. However, the theory of scattering in an amorphous solid still does not exist and the accumulated material calls insistently for the development of such a theory.

27. Fine Structure of the Line of Scattered Light in Glasses and Liquids with High Viscosity

Some Consequences of the Relaxation Theory of Sound Propagation in Viscous Media. The fundamental formula of relaxation theory, developed by Mandel'shtam and Leontovich [421] (22.1), can be rewritten in the following form:

$$K^* = K_0 + \frac{i\Omega\tau_V \Delta K}{1 + i\Omega\tau_V}, \tag{27.1}$$

where $\Delta K = K_\infty - K_0$, and

$$K_0 = \rho v_0^2, \qquad K_\infty = \rho v_\infty^2 \tag{27.2}$$

are the limiting values of the bulk modulus.

The real part of (27.1) is equal to

$$K(\Omega) = K_0 + \frac{\Delta K \Omega^2 \tau_V^2}{1 + \Omega^2 \tau_V^2}. \tag{27.3}$$

Equations (27.1) and (27.3) refer to the case of the relaxation of the bulk viscosity, which is connected with the moduli K_∞ and K_0 by the relation

$$\eta' = \Delta K \tau_V, \tag{27.4}$$

where τ_V is the relaxation time of the bulk viscosity.

Isakovich [153] extended the theory of Mandel'shtam – Leontovich to the case of relaxation of both the bulk and shear viscosity. This same problem was discussed further by Frenkel [151], Kornfel'd [498], Markham, Beyer, and Lindsay [422], Herzfeld and Litovitz [424], and others. In these and other investigations, results were obtained which are of essential significance in acoustics. The formula of the theory, which takes into account the relation of both viscosity coefficients, and which assumes that the relaxation of each coefficient can be described by a single relaxation time, can be written in the following form:

$$K(\Omega)=K_0+\frac{\Delta K\Omega^2\tau_V^2}{1+\Omega^2\tau_V^2}+\frac{\frac{4}{3}\mu_\infty\Omega^2\tau_S^2}{1+\Omega^2\tau_S^2}. \tag{27.5}$$

Here τ_S is the relaxation time of the shear viscosity and μ_∞ is the shear modulus for infinite frequency. The shear modulus and the shear viscosity are related by an expression analogous to (27.4):

$$\eta=\mu\tau_S. \tag{27.6}$$

It follows from (27.5) and the definition of the bulk modulus (27.2) that one can write for the square of the velocity

$$v^2=\frac{1}{\rho}\left\{K_0+\frac{\Delta K\Omega^2\tau_V^2}{1+\Omega^2\tau_V^2}+\frac{\frac{4}{3}\mu_\infty\Omega^2\tau_S^2}{1+\Omega^2\tau_S^2}\right\}. \tag{27.7}$$

For limiting low frequencies ($\Omega\tau\to 0$), we have, from (27.7),

$$v_0^2=\frac{1}{\rho}K_0, \tag{27.8}$$

and for limiting high frequencies ($\Omega\tau\to\infty$),

$$v_\infty^2=\frac{1}{\rho}\left(K_\infty+\frac{4}{3}\mu_\infty\right). \tag{27.8a}$$

The following expression can be found for the absorption coefficient per unit wavelength [424]:

$$\alpha\Lambda=\pi\,\frac{\dfrac{\Delta K\Omega^2\tau_V}{1+\Omega^2\tau_V^2}+\dfrac{\frac{4}{3}\mu_\infty\Omega^2\tau_S}{1+\Omega^2\tau_S^2}}{K_0+\dfrac{\Delta K\Omega^2\tau_V^2}{1+\Omega^2\tau_V^2}+\dfrac{\frac{4}{3}\mu_\infty\Omega^2\tau_S^2}{1+\Omega^2\tau_S^2}}. \tag{27.9}$$

For low frequency, (27.9) transforms into the expression

$$\alpha\Lambda=\frac{\pi\Omega}{K_0}\left\{\Delta K\tau_V+\frac{4}{3}\mu_\infty\tau_S\right\}=\frac{\pi\Omega}{K_0}\left(\frac{4}{3}\eta+\eta'\right). \tag{27.10}$$

It follows from (27.10) that the ratio of the absorption coefficient measured at low-frequency absorption α to the coefficient α_η, due only to shear viscosity, is equal to

$$\frac{\alpha}{\alpha_\eta}=1+\frac{4\Delta K\tau_V}{3\mu_\infty\tau_S}. \quad (27.11)$$

For high frequency, $\Omega\tau_V \gg 1$ and $\Omega\tau_S \gg 1$, we get, from (27.9),

$$\alpha\Lambda=\pi\frac{\Delta K\frac{1}{\tau_V}+\frac{4}{3}\mu_\infty\frac{1}{\tau_S}}{K_\infty+\frac{4}{3}\mu_\infty}. \quad (27.12)$$

There are data which indicate that τ_S and τ_V are practically the same [424]. Therefore, we set $\tau_V = \tau_S = \tau$. Equations (27.6)-(27.10) are materially simplified and we can write for the square of the velocity

$$v^2=\frac{1}{\rho}\left\{K_0+\frac{\left(\Delta K+\frac{4}{3}\mu_\infty\right)\Omega^2\tau^2}{1+\Omega^2\tau^2}\right\}. \quad (27.13)$$

Taking into account (27.12), (27.13), and the definitions of the moduli K_0 and K_∞, we obtain

$$\frac{v^2-v_0^2}{v_\infty^2-v_0^2}=\frac{\Omega^2\tau^2}{1+\Omega^2\tau^2}, \quad (27.14)$$

while the formula (27.11) goes over into

$$\frac{\alpha}{\alpha_\eta}=1+\frac{\frac{4}{3}\rho\left(v_\infty^2-v_0^2\right)}{\mu_\infty}. \quad (27.15)$$

The relations set down above characterize the propagation of longitudinal acoustic waves. From (27.10), for example, it follows that the absorption increases with increase in viscosity. For high viscosities and in the amorphous state, a transverse elastic wave can also be propagated in addition to the longitudinal wave. Change of the density of the medium does not take place in the propagation of a transverse wave. The velocity of propagation of the transverse elastic wave is determined from the relation

$$v_T^2=\frac{\mu}{\rho}. \quad (27.16)$$

The absorption per unit wavelength is expressed in first approximation by the following formula [151]:

$$\alpha_T \Lambda = \frac{\pi}{\Omega_T \tau} = \frac{\pi \mu}{\Omega_T \eta}, \qquad (27.17)$$

whence it follows that the damping coefficient of the transverse wave α decreases with increase in η. It also follows from (27.17) that in liquids with low viscosity ($\Omega_T \tau \ll 1$), α_T is so large that one can only speak about the existence of a transverse wave in a purely speculative fashion. For high viscosities or in a solid amorphous body, $\Omega_T \tau \gg 1$, and for $\Lambda \sim 10^{-5}$ cm, $\alpha_T \Lambda < 1$, and we can say that a sharp transverse MB doublet should be observed in such a medium.

Taking into account that $v_T < v$, we find that the transverse doublet will be internal, i.e., much closer to the unshifted frequency than the longitudinal doublet. Thus, in liquids with a large shear viscosity, and in solid amorphous bodies, one should expect not two MB components, as in low viscous materials, but four.

Experimental Observation of the Fine Structure of the Rayleigh Line in Glasses and in Liquids with High Viscosity. Immediately after the discovery of the fine structure in crystalline quartz, Gross [27] attempted to discover the fine structure in silicate glass also, but failed to find it.

As a consequence of these experiments, a number of experiments were undertaken which were directed toward the discovery of the fine structure in silicate glasses and viscous liquids.

In the works of Ramm [235], Raman and Rao [499], Venkateswaran [172], Velichkina [188, 189], and Rank and Douglas [483], the MB components in glass were not observed.

On the other hand, all these authors were of one mind, that a rather narrow and very intense central line of the fine structure is observed in the spectrum of light scattered by glasses. Evidently this fact, that all the attempts to find the MB components in glasses led to no positive result, stimulated Rank and Douglas [483] to make the pessimistic remark that no shifted lines of the fine structure can be observed in principle in silicate glass. Their discussion ran about as follows: glass is a liquid with maximum viscosity.

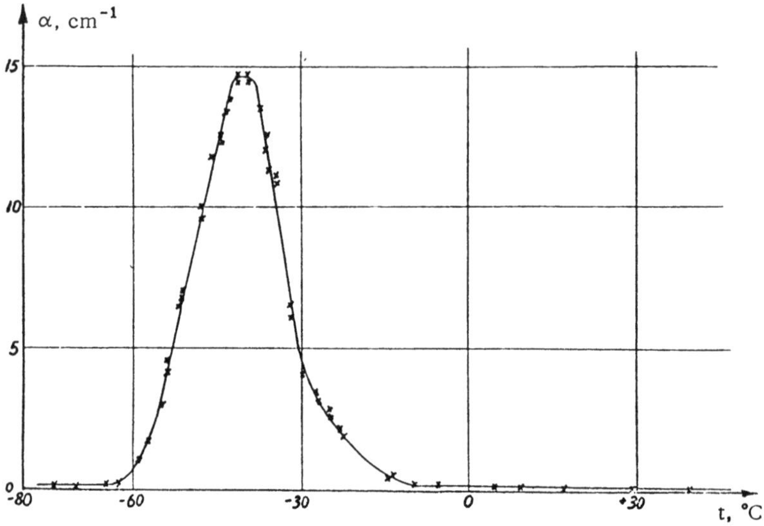

Fig. 82. Temperature dependence of the sound absorption coefficient in triacetin. Sound frequency $f = 3$ MHz (Krivokhizha and Fabelinskii [296]).

The absorption coefficient $\alpha \propto \eta$ [see (5.21)], while the halfwidth of the MB line is proportional to α [see (5.20)]. Therefore, it must be expected that the shifted component of the fine structure in glass is so broadened and moved toward the center that it will not be seen as a discrete line.

Obviously, such a discussion cannot be accepted as valid. If this were the case, then such media as the silicate glasses and glasses generally would not be good conductors of sound. Actually, if Eq. (5.21) were valid for a glass with a static viscosity of $\sim 10^{18}$ poise, then even for a frequency $f \sim 1$ Hz, the absorption coefficient would be $\alpha \sim 10^3$ cm^{-1}, which means a decrease of the intensity of sound by a factor e in a thickness of glass of $\sim 5\ \mu$.

But in reality, glasses are excellent conductors of sound, even at high frequencies. This is so because of the relaxation process, as a consequence of which viscosities of 10^{18} poise do not determine the process of sound propagation. The absorption and velocity of sound, as explained at the beginning of the present section, are determined by the elastic moduli, while the absolute values of the absorption are much less than in a liquid with the measured viscosity. Thanks to the relaxation process, the absorption coefficient of sound α, measured as a function of the temperature or viscosity, increases with increase in viscosity up to a definite maximum, after which the absorption coefficient falls off with continued growth in the viscosity to its original value, or even lower. The curve of the dependence of α on temperature for rosin, obtained by Mikhailov and Gurevich [537], the data obtained by Krivokhizha and Fabelinskii [296] for triacetin and glycerin, and also the data of Herzfeld and Litovitz [424], serve as examples. The form of the dependence of α on t in Fig. 82 is characteristic for a relaxing medium. Consequently, a clear fine structure of the Rayleigh line should be observed in glasses, in accord with the relaxation representation. However, the experiments of the different authors, completed on different apparatus, indicate that the shifted components are not observed.

In the initial period of investigation of the fine structure in viscous liquids and in glasses, the situation was in a definite sense the opposite to what arose in the initial period of study of low-viscosity liquids. Actually, in liquids with small η and with large bulk viscosity η', a clear fine structure was observed, although the hydrodynamic theory did not allow for it. The creation of the relaxation theory was necessary in order to explain the presence of the fine structure.

In the investigation of liquids with a large shear viscosity, and of glasses, the already existing relaxation theory allowed us to expect a fine structure, but it was not observed experimentally.

Later, Venkateswaran [172] carried out an investigation of the fine structure in a number of viscous liquids, including glycerin and castor oil. By using the line spectrum of zinc (4810, 4622, and 4722 Å) for exciting the scattering, he succeeded in observing the MB components in glycerin up to a viscosity of 120.4 poise, and for castor oil, up to 6.04 poise. In these viscosities, the shifted components were very weak in intensity. Upon decrease in the viscosity, the intensity of the components increased. The decrease in the intensity of the MB components with increase in viscosity shows that the effect of relaxation has not yet set in. If this were actually so, then the criterion for the existence of a clear fine structure could be calculated from the classical expression for α [see (5.21)]. Here, it seems that for both glycerin and castor oil, $\alpha\Lambda \gg 1$ and, consequently, the fine structure should not be observed. The fact that it was observed undoubtedly indicated that the relaxation had already set in. And if this is so, then the fine structure ought to become clearer with an increase in the viscosity rather than for a decrease in the viscosity. The results of Venkateswaran apparently contradict what ought to be expected. But this only appears to be so at first glance. In the following, we shall attempt to give a possible explanation of the seeming contradiction. The first report that clear MB components were found in fused quartz was made by Krishnan [225]. His work was carried out on a spectrograph with a camera objective with a focal length of 3 m, and with a resolving power of 30,000 in the region of the line ~2536.5 Å. The scattered light was excited by this line. A cell with mercury vapor at room temperature was placed in the path of the scattered light en route to the slit of the spectrograph. The lines of unshifted frequency were completely absorbed (see Sec. 12). Weak lines were observed in the mercury vapor at a temperature of the quartz of 30°C, and sufficiently intense MB lines were seen at a temperature of 300°C. The shift of the components amounted to $\Delta\nu = \pm 1.5\ \mathrm{cm}^{-1}$, while calculation from the elastic constant gave $\Delta\nu = \pm 1.65\ \mathrm{cm}^{-1}$. Although these results of Krishnan appeared entirely convincing to us, criticism of this and the other researches of Krishnan on the part of Sirkar and Kastha [232, 233] cast some doubt on the credibility of the results obtained.

The critics of Krishnan emphasized that the components of the fine structure are absorbed in the resonant filter, and the result of scattering in the vapors of the resonant filter is observed. They drew the conclusion that the use of the resonant filter is generally not admissible for the study of the fine structure. Other investigations are known in which the resonant filter has led to a material distortion of the results [229, 220].

Therefore, the experiments made with viscous liquids and fused quartz did not give a clear answer to the question as to the presence or absence of fine structure in them.

The reason for such experimental results was not immediately understood, as follows from what has been said above. One of the reasons for the absence of the fine structure in the case considered could be found in the following. From the representa-

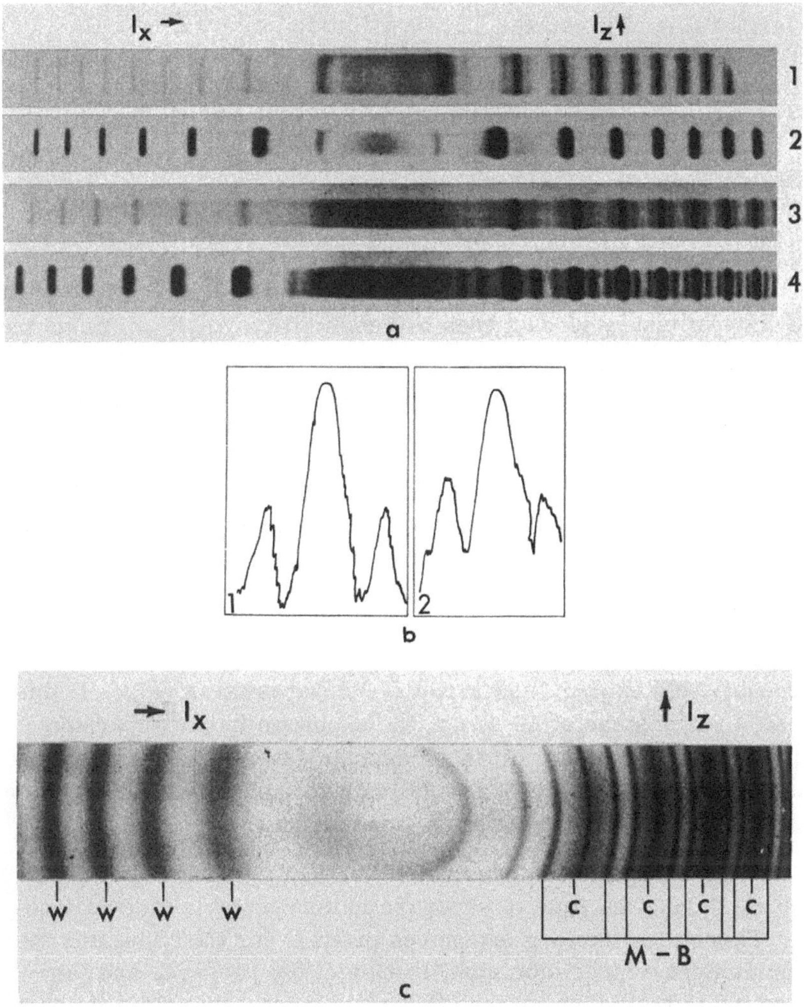

Fig. 83. (a) Spectra. The fine structure of the line of Rayleigh scattering of light in triacetin and glycerin for different temperatures. 1 – Triacetin, 72°C; 2 – triacetin, −15°C; 3 – triacetin, −60°C; 4 – glycerin, −50°C; (b) Microphotographs of spectrum of fine structure in glycerin at −27°C (1) and in triacetin at −40°C (2) (Pesin and Fabelinskii [34, 35]); (c) Spectrum of fine structure in nitrobenzene at room temperature, excited by light of λ = 6328 Å from a Ne–He gas laser. MB – Mandel'shtam–Brillouin components; C – central component; W – diffusion wing of Rayleigh line (Starunov, Tiganov, and Fabelinskii [644]).

tions of relaxation theory, one can assume that in the case in which a large value of η relaxes, $\Delta v/v$ will also be large, i.e., the liquid will behave as a solid at high frequencies. We now assume that $\Delta v/v \approx 1$; this means that the velocity of hypersound is doubled in the transition from small to large viscosities. If this were so, then it would lead to very substantial consequences, chiefly of an experimental character. Actually, the integrated scattering coefficient for both MB components, when observed at the angle θ = 90°, is expressed by Eq. (18.13). If we recall that R_{90} is proportional to β_S, while β_S is inversely proportional to the square of the velocity of hypersound, it then becomes clear that R_{90} depends materially on the hypersonic velocity. If it is assumed that a viscous medium behaves as an amorphous solid and the velocity of sound is, roughly speaking, doubled for frequencies about 10^{10} Hz, then β_S and, together with it R_{90}, decreases by a factor of four. Taking into account that R_{90} (18.13) depends linearly on the absolute temperature, we see that the difference in the intensity is still further increased; consequently, one can assume that the intensity of the MB components in triacetin and glycerin falls off by a factor of four or five in comparison with the intensity of the shifted components in these media at room temperature and low viscosity. Probably this explains the result of Venkateswaran [172]. If this is so, then, instead of 10-12 hr, the exposure in our apparatus ought to take 50-60 hr. In those experiments where the exposure lasted for days [172, 257], now it ought to last for weeks; in other words, the experiment becomes superdifficult.

The change in temperature by 0.5°C or in pressure by 3-4 mm Hg during the time of exposure is unacceptable (see Appendix II). Therefore, the long exposures produce for the apparatus the requirement of precision stabilization of the pressure and temperature in the volume where the interference apparatus is located.

Therefore, in order to discover the shifted lines of the fine structure, one must either take the necessary steps for thermostatting and barostatting the apparatus, or one must increase the light intensity used by an order of magnitude.

Fabelinskii and Pesin [34-36] went by the second path. The light intensity of their apparatus (see Fig. 32) was increased by almost an order of magnitude, and the long-sought results were quickly obtained. Figure 83 shows the spectra of the fine structure of light scattered in triacetin and glycerin at different temperatures; the microphotograms of the fine structure of triacetin were made at −40°C and of glycerin at −27°C.

The spectra and the microphotograms show the completely delineated components of the fine structure, which exist throughout the temperature range down to the glassy state. The velocity of the hypersound in triacetin was measured in the temperature range from +72° to −60°C for a change in viscosity from 0.02 to $2.5 \cdot 10^6$ poise.

In glycerin, the velocity was measured in the temperature range from +150 to −70°C; here the viscosity changed from 0.05 to 10^8 poise.

With an increase in the viscosity by about 10 orders in both substances, the clear fine structure of the Rayleigh line was first observed. The greater the static viscosity here, the sharper were the displaced components, although their intensity decreased.

The velocity of hypersound, determined from the shift of the MB components, is shown in Tables 27 and 28. The accuracy of the determination varied from 5 to 8% of the measured value, while for high viscosities the error is less than for low viscosities. Tables 27 and 28 also give the values of the viscosity.*

The velocity of the hypersound given in Tables 27 and 28 were determined by Fabelinskii and Pesin [34, 35], while the velocity of ultrasound in triacetin was found by Velichkina [188] and also by Krivokhizha and Fabelinskii [296] at frequencies of 1 and 3 MHz. The velocities of the ultrasound in glycerin at the different temperatures are taken from the measurements of Schulz [501].

* The viscosity of the glycerin was measured by Tammann and Hesse [500]. Velichkina [189] measured the temperature dependence of the viscosity in triacetin. For regions of temperatures where the viscosity of these substances has not been measured, the values of the viscosity could be obtained by extrapolation of the experimental curves, or from the formula [500]: $\log\eta = -A + (B/T - T_\infty)$, where for glycerin the constants $A = 4.76$, $B = 950$, and $T_\infty = 132$, while for triacetin, $A = 4.14$, $B = 340$, and $T_\infty = 187$.

Table 27. Temperature Dependence of the Velocity of Hypersound and Ultrasound and Viscosity in Triacetin

t, °C	η, poise	$\Delta\nu$, cm^{-1}	Velocity v, m/sec	
			Hypersound	Ultrasound
+72	0.02	0.19	1246	1214
+20	0.23	0.25	1616	1404
0	0.70	0.32	2065	1480
−15	4.68	0.37	2386	1538
−40	$1.84 \cdot 10^3$	0.40	2536	1800
−60	$2.5 \cdot 10^6$	0.43	2718	2366

Table 28. Temperature Dependence of the Velocities of Hypersound and Ultrasound in Glycerin

t, °C	η, poise	$\Delta\nu$, cm^{-1}	Velocity v, m/sec	
			Hypersound	Ultrasound
+150	0.05	0.25	1580	1670
+ 50	1	0.27	1690	1845
+ 45	1.8	0.38	2420	1855
+ 35	3.2	0.45	2800	1870
+ 22	9	0.46	2900	1900
− 27	$3.9 \cdot 10^3$	0.53	3290	2310
− 45	$2 \cdot 10^5$	0.57	3520	3300
− 70	$1 \cdot 10^8$	0.59	3665	3675

Figures 84 and 85 give the temperature dependence of the velocity of hypersound and ultrasound in triacetin and glycerin,* respectively. As is clear from Fig. 85, in the region 150-50°C, the velocity of hypersound is somewhat less than the velocity of ultrasound. This difference must be ascribed to the errors (about 6-8%) of measurement of the hypersonic velocity at high temperature. Such a difference in the velocities does not exist for triacetin (Fig. 84). Rank et al. [650] measured the velocity of hypersound in several viscous media at different temperatures. A glance at Figs. 84 and 85 is sufficient for one to see that the changes of the

*In the sample of glycerin which was used by Litovitz, 5% water content was measured.

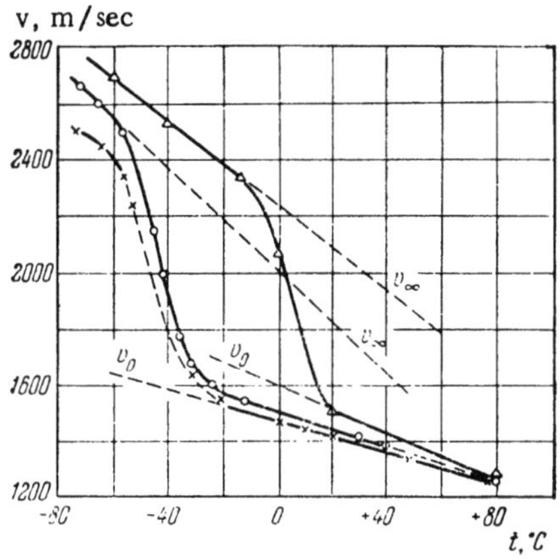

Fig. 84. Temperature dependence of the velocity of hypersound and ultrasound in triacetin. Triangles – velocity of hypersound (Pesin and Fabelinskii [34, 35]); circles – velocity of ultrasound (f = 3 MHz) (Krivokhizha and Fabelinskii [296]); crosses – velocity of ultrasound (f = 1 MHz)(Velichkina [189]).

viscosity with temperature have a relaxational character. Figure 86 gives $(v^2 - v_0^2)/(v_\infty^2 - v_0^2)$, as a function of τ. It is assumed that τ is connected with the viscosity and temperature by the relation

$$\tau = B\frac{\eta}{T}, \tag{27.18}$$

where B is a constant.

Figure 86 shows that the experimental data are not described quantitatively by the formulas of relaxation theory if it is assumed that (27.18) is satisfied. It is well known that the elastic properties of a medium depend strongly on the rate of the reaction on it. If one carries out the measurement of the dependence of the velocity of sound on the temperature under static conditions* in tri-

*By static conditions we mean the measurement of the velocity of sound at such a low frequency that $\Omega_0 \ll 1/\tau$, or the period of the sound vibration $T \gg \tau$.

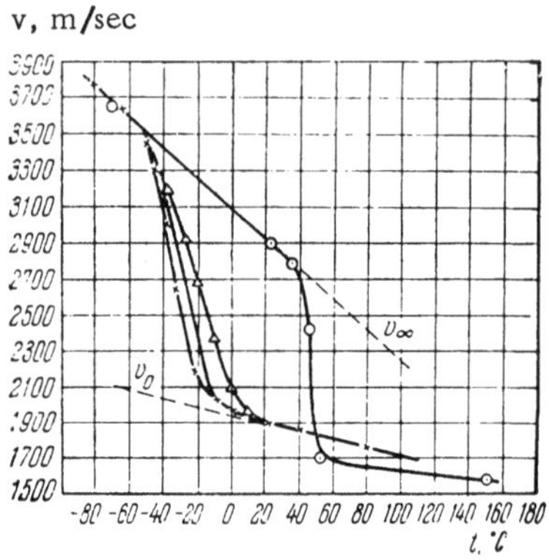

Fig. 85. Temperature dependence of velocity of hypersound and ultrasound in glycerin. Circles – velocity of hypersound (f = 10^4 MHz) (Pesin and Fabelinskii [34, 35]); triangles – velocity of ultrasound (f = 30 MHz) (Litovitz [424, 502, 503]); crosses – velocity of ultrasound (f = 0.5 MHz) (Schulz [501]).

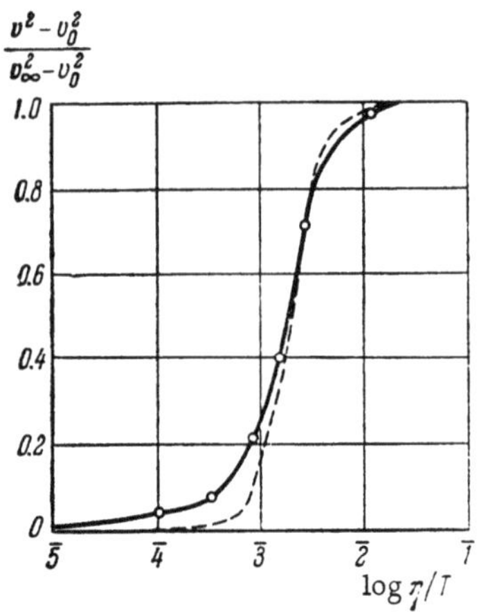

Fig. 86. Dependence of $(v^2 - v_0^2)/(v_\infty^2 - v_0^2)$ on the logarithm of relaxation time τ [$\tau = B(\eta/T)$] in triacetin. The dashed curve corresponds to the theory for a single relaxation time τ.

Table 29. Estimate of the Dispersion of the Velocity and the Temperature Dependence of the Bulk Modulus

t, °C	ρ, g-cm^{-3}	$\frac{v_\infty - v_0}{v_0} \cdot 10^2$	$(K_\infty + \frac{4}{3}\mu) \cdot 10^{-10}$, dyne-cm^{-2}	$K_0 \cdot 10^{-10}$, dyne-cm^{-2}	$(\Delta K + \frac{4}{3}\mu_\infty) \cdot 10^{-10}$ dyne-cm^{-2}
			Triacetin		
40	1.129	40	4.4	2.0	2.4
30	1.142	50	4.8	2.15	2.6
20	1.156	51	5.2	2.3	2.9
10	1.169	52	5.5	2.5	3.0
0	1.182	52	6.1	2.6	3.48
−10	1.195	53	6.6	2.8	3.8
−20	1.208	54	7.1	3.07	4.1
−30	1.222	55	7.6	3.16	4.4
−40	1.235	56	8.2	3.36	4.8
			Glycerin		
40	1.249	47	9.4	4.3	5.1
30	1.255	50	10	4.4	5.6
20	1.261	53	11	4.5	6.1
10	1.268	56	11	4.7	6.7
0	1.275	60	12	4.8	7.3
−10	1.282	62	13	5.0	7.8
−20	1.288	65	14	5.0	8.6
−30	1.295	67	14	5.2	9.2
−40	1.301	70	15	5.3	9.9

acetin, glycerin, or in a similar liquid, then the velocity increases as the temperature falls, and at the glass temperature of the medium T_g, the velocity becomes equal to the velocity in the solid – glass. The viscosity in the region of T_g reaches a value $\sim 10^8$–10^{14} poise. For a study of the temperature dependence of the hypersonic and ultrasonic velocities, it is clear that the transition to the velocity which is characteristic for the solid sets in much earlier (as a consequence of the relaxation process) than under static measurements. As is seen from Figs. 84 and 85, such a transition for hypersound takes place in triacetin at a temperature ~ -15°C, when the viscosity is equal to 5 poise, and for glycerin at a temperature $\sim +40$°C, when the viscosity amounts to only 2 poise. Thus, a liquid of moderate viscosity behaves as a solid for elastic frequencies $\sim 10^9$ Hz. The same conclusion was also made for other forms of a fast reaction in the liquid [498].

We turn our attention to the temperature dependence of the limiting velocities v_∞ and v_0. It is characteristic that the tem-

Table 30. Temperature Dependence of the Shear Modulus μ_∞ and ΔK (glycerin)

t, °C	η, poise	$\mu_\infty \cdot 10^{-10}$, dyne-cm^{-2}	$\Delta K = (K_\infty - K_0) 10^{-10}$, dyne-cm^{-2}
—42	$1.09 \cdot 10^5$	3.52	4.36
—30	$8.89 \cdot 10^3$	3.3	4.0
—10	$6.16 \cdot 10^2$	2.9	3.6

Table 31. Velocity of Longitudinal and Transverse Hypersonic Waves in Fused Quartz at a Temperature of 45°C (Exciting line $\lambda = 2537$ Å; angle of scattering $\theta = 90 \pm 8°$)

	$\Delta\nu_L$, cm^{-1}	$\Delta\nu_T$, cm^{-1}	Velocity of hypersonic waves $\cdot 10^{-5}$, cm/sec	
			Longitudinal	Transverse
Specimen A	1.678 ± 0.02	1.041 ± 0.07	5.993 ± 0.08	3.72 ± 0.3
Specimen B	1.636 ± 0.01	1.111 ± 0.05	5.845 ± 0.04	3.97 ± 0.2

perature coefficients for v_∞ and v_0 are practically the same for ultrasound and hypersound, but they differ significantly from one another. Thus, for glycerin, $(1/v_0)(dv_0/dt) = 1 \cdot 10^{-3}$ deg^{-1}, and $(1/v_\infty)(dv_\infty/dt) = 2.7 \cdot 10^{-3}$ deg^{-1}. For triacetin, these quantities are equal to $1.98 \cdot 10^{-3}$ and $3 \cdot 10^{-3}$ deg^{-1}, respectively.

On the basis of hypersonic measurements [34, 35], Pesin [36] estimated the value of the dispersion of the sound velocity and the bulk modulus in triacetin and glycerin (Table 29). In experiments on the fine structure in glycerin and triacetin, the velocity of longitudinal hypersound has been determined. In order to obtain a representation of the value of μ_∞, we write a table of values of μ_∞ taken from [424]. From Eqs. (27.6) and (27.18) we get, for the shear modulus,

$$\mu_\infty = \frac{T}{B}. \tag{27.19}$$

Such a formula is possibly suitable for an estimate of the order of magnitude of μ_∞ but, as Frenkel [151] has noted, such a relation is devoid of physical meaning. Table 30 shows that μ decreases with increase in temperature, while (27.19) predicts an erroneous

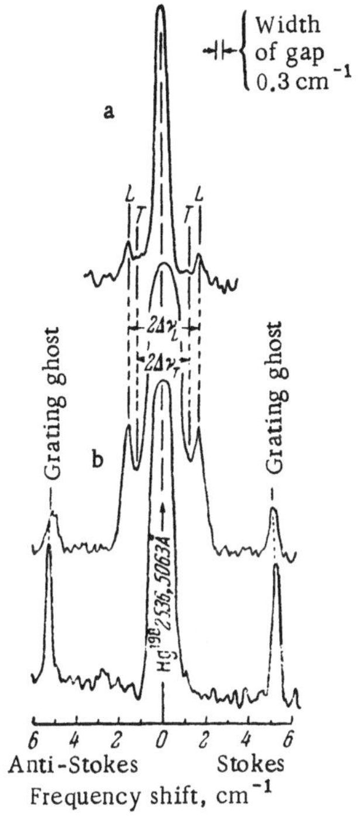

Fig. 87. Microphotograph of the spectrum of fine structure in two specimens (A and B) of fused quartz and of the exciting line (Flubacher et al. [244]).

linear dependence of the shear modulus on the temperature. In our investigations of the fine structure for high viscosities and the glassy state, the MB components produced by transverse waves, which ought to exist in the amorphous solid, were not observed. Estimate of the intensity of the "transverse" components of the fine structure, carried out by Pesin [36] shows that they are weaker by an order of magnitude than the "longitudinal" components, while $\Delta\nu_T$ varies from 0.01 cm^{-1} for +70°C to 0.27 cm^{-1} for −42°C. If one also takes into account the significant width of the apparatus function (the scattering was excited by the λ = 4358 Å mercury arc), it then becomes clear why one could not observe it. The application for excitation of a limitingly narrow line of the stimulated emission of a gas laser probably makes it possible to observe all five components which should be observed in the amorphous solid.

By using a lamp operating on an isotope of mercury, Hg^{198}, Flubacher et al. [244] observed all five components of the fine structure in fused quartz. A microphotograph of the fine structure of the two samples A and B and the spectrum of the exciting line are shown in Fig. 87.

The velocities of longitudinal and transverse hypersonic waves are given in Table 31. By comparing the data of Table 31 with the data of Krishnan for the shift $\Delta\nu_{MB}$ of the MB doublet, the conclusion can be reached that Krishnan was the first to observe the distinct MB components in fused quartz.

However, the work of Flubacher et al. [244] finally settled every doubt as to when the fine structure of the scattering line could be observed in fused quartz. The observations of the fine structure in glycerin and triacetin for huge viscosities and in fused quartz qualitatively confirm the forecast of relaxation theory. However, the quantitative comparison of the results of the experiment with the relaxation theory, which takes into account only a single relaxation time, shows that there is no agreement between theory and experiment. The curve of the dependence of α on τ from the experimental data has a more gently sloping path than the calculated curve of Fig. 86. This relates to the velocity and absorption in the ultrasonic region and to the velocity in the hypersonic region of frequencies. In correspondence with the ideas of Frenkel [151], it was proposed to describe curves of Fig. 86 and the similar curves for other substances by a spectrum of relaxation times. In the description of the experiments, it has been possible to introduce different distribution functions for the relaxation times for the description of the variation of absorption and for the velocity of the longitudinal and transverse vibrations [424]. Of course, the considered phenomena can be described by formulas with infinitely many parameters.

It is quite possible that the propagation of sound or hypersound in very high-viscosity media cannot be described within the framework of existing relaxation theory.

Actually, Krivokhizha and Fabelinskii [296] showed that all the various experimental material pertaining to the measurement of sound propagation in viscous media can be described by formulas of the type $y = A/(1 - A) = \Omega\tau$, where A is equal to the left side of (27.14) and Ω and τ have the same meaning as before. The formulas of relaxation theory require that $y = \Omega^2\tau^2$, which is in contradiction with experiment.

In this region, evidently, experiment carries much farther than theory. The phenomena of scattering, propagation, and absorption of sound and hypersound described in this chapter require an adequate theoretical description which cannot yet be made because of the absence of a satisfactory theory.

Note Added in Proof

Operating on the conclusions of experimental investigations [269], Isakovich and Chaban [632, 633] have developed a nonlocal theory of sound propagation in viscous media. The formulas of the new theory contain a single relaxation time and give a good description of the complex laws of the phenomenon (see Figs. 84-86).

Chapter VIII

Investigation of the Spectral Composition of the Depolarized Scattering of Light (Wing of the Rayleigh Line) in Liquids with Various Viscosities

28. Wing of the Rayleigh Line in Low-Viscosity Liquids at Room Temperature

Remarks on the Nature of the Wing. The first spectral studies of scattered light in liquids revealed that the line of Rayleigh scattering was appreciably broadened in comparison with the exciting line [37, 38]. By the Rayleigh line is meant all three components of the fine structure, which were not resolved in the spectrographic investigation. The observed broadening of the line of scattered light, which is characteristic in varying degree for all liquids, has received the name "wing of the Rayleigh line." The wing immediately butts against the unshifted line on both sides. The intensity of the wing falls off monotonically in both directions from the Rayleigh line, extending perhaps 100-160 cm^{-1} and even beyond. The intensity of the wing on the long wave side is somewhat greater than on the short wave side [144, 228, 510].

The first experiments showed that the wing has a significant depolarization. The wing was interpreted by Raman and Krishnan [38] as the spectrum of combination (Raman) scattering, arising as the result of hindered rotation of anisotropic molecules in the liquid. Later, a similar point of view was developed and experimentally tested by Bhagavantam [509, 511].

Cabannes and Rocard [324] gave a different explanation of the origin of the wing of the Rayleigh line. In their opinion, the broadening of the Rayleigh line is connected with collisions of molecules in the liquid. In other words, the mechanism of broadening of the scattered line in the liquid, according to Cabannes and Rocard, is similar to the Lorentz broadening of the radiation line as the result of collisions of molecules in a gas. As was shown in Sec. 17, Ginzburg [141] took up the hypothesis of Cabannes and Rocard, and showed that the collisions in the liquid should not have an effect on the broadening of the Rayleigh line; therefore, the reason given in [324] should not lead to the observed broadening of the Rayleigh line.

A new hypothesis on the nature of the wing of the Rayleigh line, put forward by Gross [28] found later development in subsequent researches [512-514]. According to the first view of Gross, the lines of low frequency, discovered by Gross and Vuks [513] in crystals, join the wing when the crystal melts and becomes a liquid. The lines of low frequencies are explained by the vibrations of the crystalline lattice. The hypothesis of Gross assumes that vibrations which are characteristic of the crystalline lattice are preserved in the liquid. In other words, in a liquid, fragments of the crystalline lattice exist in a deformed condition. Gross and his coworkers attempted a whole series of experiments on the study of the low-frequency spectrum and the wing of the Rayleigh line; this led the authors to the conclusion that Gross' hypothesis [28] is correct and that [515] can preserve the "total set of facts relating to wings."

A special investigation was undertaken by Venkateswaran [516] of fourteen organic crystals at different temperatures, including benzene, coumarin, phenol, naphthalene, and benzophenone in liquid and crystalline states; salol in liquid and glassy states; as well as catechin and diphenyl. On the basis of his experimental results, Venkateswaran rejected the vibration hypothesis of the origin of the wing of the Rayleigh line advanced by Gross. In coumarin and naphthalene, the lines of low frequency correspond to 110 and 120 cm^{-1}, while in the liquid state of these substances, the wing extends at most to 70-80 cm^{-1}. In other cases, lines of maximal intensity are observed in the spectrum of the crystalline lattice in the region 50-60 cm^{-1}, while a maximum of the intensity in the wing is not observed in the spectrum of the liquid in this region.

Venkateswaran also showed that it is not possible to make a satisfactory explanation of the wing in a liquid, which is observed up to the boiling temperature, from the viewpoint of a quasicrystalline hypothesis. He also found it difficult to explain (from this viewpoint) the fact that the depolarization in the wing is equal to $^6/_7$ and does not depend on the frequency while, according to the data of Gross and Vuks [513], the depolarization of the different lines of low frequency varies in the limits from 0 to 1. Venkateswaran did not observe the wing of the Rayleigh line in the spectrum of the scattered light of glassy salol. This experimental result of Venkateswaran is in contradiction with the results of the experiments of Gross and Raskin [515]. Venkateswaran discovered a clearly marked band in liquid benzophenone, diphenyl, and diphenyl ether in place of the corresponding lines in the solid substances, of width $\sim 100\ cm^{-1}$. The band continues to exist at high temperature and in dilute solution. The observed phenomenon is interpreted as the valence deformation vibration of two benzene rings, and it is emphasized that the vibration of the lattice cannot be related to this phenomenon.

In Sec. 6, it was observed in a brief note that the most rational point of view as to the nature of the wing of the Rayleigh line was, in our opinion, given by Landau and Placzek [140] in their work devoted to the basic question of the relation of the intensity in the components of the fine structure.

According to Landau and Placzek, the wing of the Rayleigh line is explained by relaxational phenomena, while the width of the wing is, roughly speaking, determined by the reciprocal of the size of the relaxation time of the dipole moment. The anisotropy fluctuations and the relaxational phenomena which are produced by repeated Brownian motion, were considered phenomenologically by Leontovich [39]. They are discussed in Sec. 6 of this book. Gross [517] assumed that the wing of the Rayleigh line, directly abutting the unshifted line and having the width 15-20 cm^{-1}, is produced by relaxation phenomena (background of the Rayleigh line), while the remaining part of the wing is considered by him (as before) as the result of the spreading of the line of low frequency in the wing when the crystal transforms into a liquid.

After the experimental and theoretical investigations of the wing mentioned above, and, in particular, after the work of Weiler

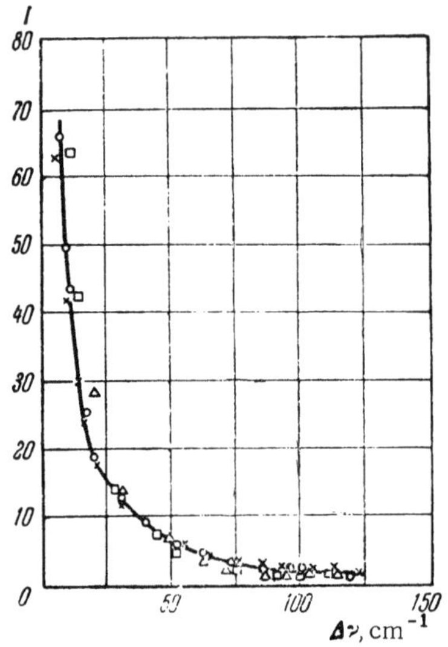

Fig. 88. Distribution of the intensity in the wing of the Rayleigh line in benzene at room temperature (Starunov [505]). The different symbols correspond to data in different experiments.

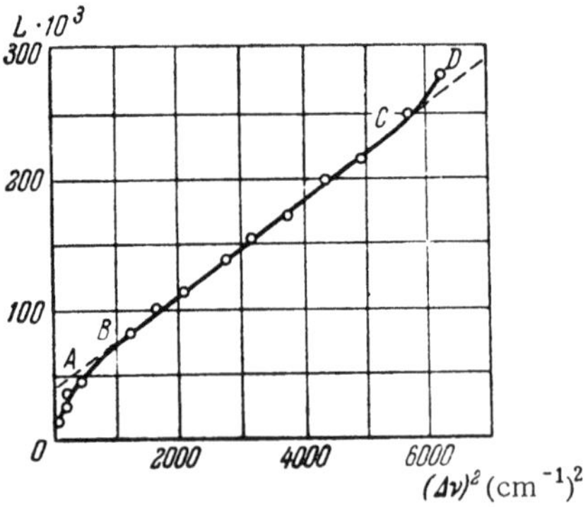

Fig. 89. Dependence of $L = 1/I$ on $(\Delta\nu)^2$ in carbon bisulfide (Fabelinskii [53]).

[518], Rousset [72], Bapagya [519], and Fabelinskii [52, 53, 73, 144, 230, 231, 510], it was made clear that the rotational hypothesis of the origin of the wing in the liquid is not confirmed and should be rejected. This applies to such liquids as benzene, carbon bisulfide, carbon tetrachloride, cyclohexane, acetic acid, chloroform and other ordinary liquids. The existence of investigations in which "maxima" have thus far been discovered in the intensity distribution in the wing of the Rayleigh line in benzene, carbon bisulfide, and other ordinary liquids [520-522] is explained by the large error in the treatment of the observed results.* So far as the free rotation of the molecules in liquids and their effect on the spectrum of the scattered light are concerned, they can be considered only in the special cases of liquid oxygen, nitrogen, methane [523], and ammonia solution [50, 524]. These special cases will not be considered here.

One can now show that the intensity distribution in the wing of the Rayleigh line is described by the existing relaxation theories. Also deserving attention is the theory of the wing which starts out from the statistical quantitative description of the possible forms of rotational Brownian motion of the anisotropic molecule in a liquid, briefly presented in Sec. 8.

From the viewpoint of this theory which was developed by Starunov [41, 505], the part of the wing adjoining the unshifted line and extending 15-25 cm^{-1} from it is explained by the rotational diffusion, while the remaining part of the wing is described by rotational oscillations. The formulas obtained by Starunov (see Sec. 8) have the same structure as the formulas of the phenomenological relaxation theory described in Secs. 6 and 7.

*The error in the treatment of the observations consists first of all of the assumption in [520, 521], and also in [522], that the observed intensity distribution is the sum of the ordinates of the distribution of the apparatus function and the scattering function and, therefore, for finding the intensity distribution in the scattered light from the visible intensity distribution, the contour of the exciting line is to be subtracted. In actuality, the apparatus function and the scattering function are connected by the integral equation of Sec. 19.

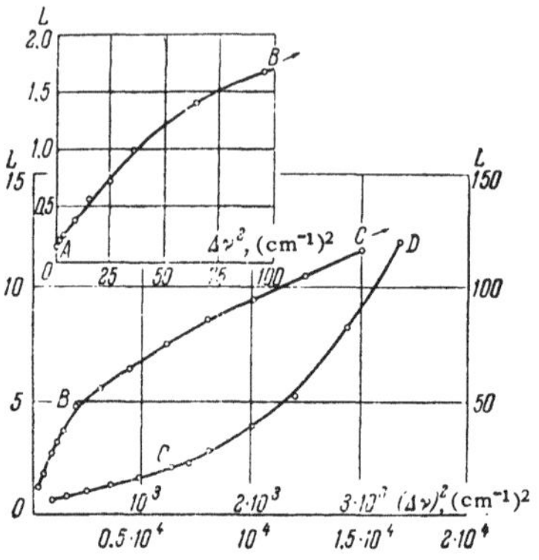

Fig. 90. Dependence of L on $(\Delta\nu)^2$ in benzene (Starunov [505]). The left scale of L and the upper numbers of the frequency scale refer to the curve BC; the right scale of L and the lower figures on the frequency scale refer to the curve CD.

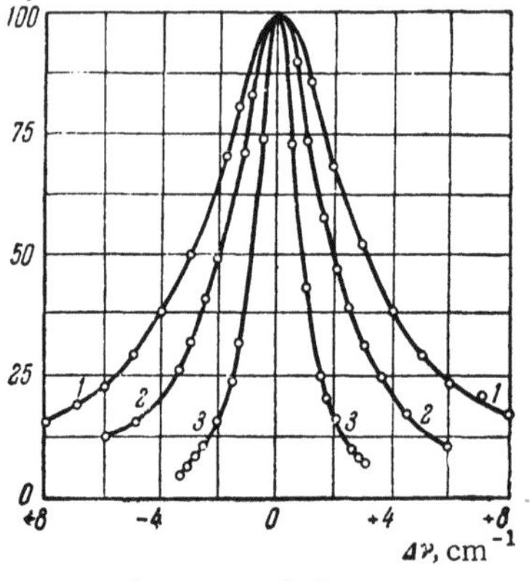

Fig. 91. Distribution of the intensity (arbitrary units) in the narrow part of the wing of the Rayleigh line (Fabelinskii [33]). 1 – Carbon bisulfide; 2 – benzene; 3 – acetic acid.

Determination of the Relaxation Time of the Anisotropy. It has repeatedly been emphasized by the authors of both theoretical and experimental investigations that one can scarcely describe the phenomenon of anisotropy fluctuations by a single relaxation time in such a complicated medium as a liquid. But all the phenomenological theories to date are limited to only one or, in the best case, two relaxation times. It is very surprising that practically all the observed cases can be described by two relaxation times and the inertial parameter a. The short relaxation time of the anisotropy corresponds to the region of applicability of Eqs. (6.32), (6.41), and (7.27) of the phenomenological theory. The intensity distribution in the wing of the Rayleigh line in benzene is shown in Fig. 88. For the determination of the relaxation time τ it is convenient to plot the experimental material on a graph in the coordinates $1/I = L(\Delta\nu)$ and $(\Delta\nu)^2$.

We have Figs. 89 and 90 as examples of such curves. By using Eqs. (6.35) and (6.36) and the linear parts of BC in Figs. 89 and 90, it is easy to find the relaxation time of the anisotropy. On the other hand, a connection is given in the theory of Leontovich [39] between the relaxation time of the anisotropy and the coefficient of double refraction for a flux M, by means of the relation (6.28). From this relation, we find an expression for the relaxation time of the anisotropy τ:

$$\tau = \frac{M^2}{\eta\left(\rho\frac{\partial\varepsilon}{\partial\rho}\right)_T \beta_T}\left[\frac{1}{2\Delta_u} - \frac{7}{12}\right]. \tag{28.1}$$

In the region AB of curves of Figs. 89 and 90, a very long relaxation time exists, a time, as was pointed out, determined by the mechanism of rotational diffusion. The distribution of the intensity in this region is evidently determined by Eq. (8.12); and, consequently, τ is determined also by Eq. (6.36).

As follows from the theory, the relaxation time in this region is equal to $\frac{1}{3}\tau_D$, where τ_D is the relaxation time of the dipole moment, determined by Eq. (6.12).

It must be emphasized that the investigation of the part of the wing in the range 0-5, 0-15 cm^{-1}, and by means of ordinary spectrographs with small linear dispersion (ISP-51) and by a photographic technique is unacceptable first, because the nearby

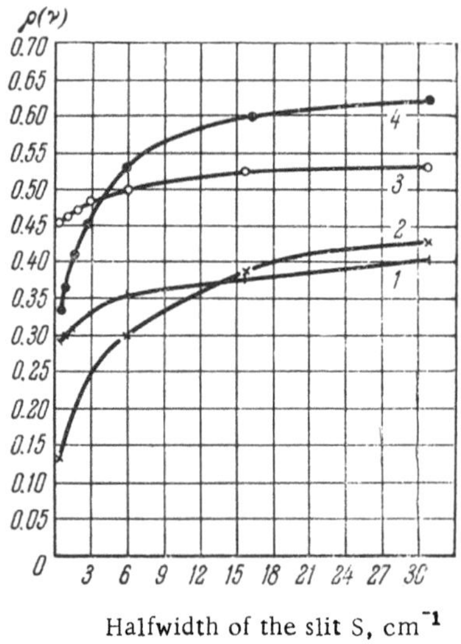

Fig. 92. Dependence of the depolarization factor on the width of the slit of the spectrograph (Saxena [528] and Sunanda Bai [529]). 1 – Acetic acid; 2 – benzene; 3 – tetralin; 4 – carbon bisulfide.

Fig. 93. Interference spectrum of light scattered in liquid salol at different temperatures. (1) +20°C; (2) +100°C; (3) +170°C.

intense Rayleigh line gives extraordinarily large scattering in the photoemulsion and the halo makes the photographs unsuitable for analysis. For this reason, the study of the wing in all detailed investigations in which the work was carried out on a spectrograph began at a distance of 5-10 cm^{-1} from the Rayleigh line. In the

Table 32. Small Relaxation Times of the Anisotropy in Liquids [53, 73]

Substance	Relaxation time of anisotropy $\tau \cdot 10^{12}$, sec		Depolarization of scattered light, $\Delta_u \cdot 10^2$	Viscosity $\eta \cdot 10^3$, poise	Dipole moment $\mu \cdot 10^{18}$
	Scattering of light [Eq. (6.12)] from portion BC (Figs. 89, 90)	Double refraction in flux M [Eq. (28.1)]			
Acetone	0.11	–	23	3.3	2.8
Benzene	0.24	0.24	43	6.5	0
Water.	0.12	–	8	10	1.84
Xylol	0.11	0.47	63	6.4	0.52
Methyl alcohol.	0.14	–	7	5	1.68
Nitrobenzene . .	–	0.17	79	20	3.90
Chloroform . . .	0.12	–	11	5.7	1.05
Carbon bisulfide	0.21	–	68	3.7	0
Acetic acid . . .	0.15	–		12	
Chlorobenzene .	–	0.83	57	8	1.55

photoelectric method of investigation, one can come somewhat closer to the Rayleigh line than in the photographic method of study of the wing. Therefore, a special method was developed by Fabelinskii [53, 144] and by Shustin [234] for the investigation of the wing on a Fabry–Perot interferometer with a large dispersion. The description of the method is given in Sec. 12 and the spectrograms are shown in Fig. 34.

Figure 91 shows the intensity distribution in a narrow part of the wing (the diffusion part), measured by Fabelinskii [53] in carbon bisulfide, benzene, and acetic acid. The broadest wing is observed for carbon bisulfide ($\delta\nu = 2.2$ cm^{-1}), then benzene ($\delta\nu = 1.6$ cm^{-1}) and acetic acid ($\delta\nu = 0.3$ cm^{-1}). In the latter case, the wing is so narrow that it is distinctly seen on an interferogram obtained with a Fabry–Perot etalon with a thickness of the separating ring of 5 mm (region of dispersion 1 cm^{-1}). The same picture is observed in triacetin, salol, benzophenone, and glycerin.

An interferogram of the scattering spectrum in liquid salol is shown in Fig. 93 at different temperatures by way of example.

The use of lasers for the study of the spectrum of the diffusion (narrow) wing allows us now to investigate this phenomenon in more detail than previously [644]. Figure 83b shows the spectrum

Table 33. Large Relaxation Times of the Anisotropy in Liquids

Substance	Measured relaxation times of anisotropy $\tau \cdot 10^{12}$,sec		Computed $\tau \cdot 10^{12}$, sec	Viscosity $\eta \cdot 10^{3}$, poise
	Fabelinskii [53]	Starunov [41, 505]		
Carbon bisulfide . .	2.4	2.3	2.8, 2.1	3.7
Chloroform	–	5.9	5.8	5.7
Toluene	5.3	4.1	8.6	5.9
Benzene	3.3	3.8	7.5, 5.3	6.5
Acetic acid	17.5	14	13	12

of light scattered in nitrobenzene in two polarizations. The half-width of the wing is about 0.1 cm^{-1}.

Table 32 gives the relaxation times determined from the intensity distribution in the part of the wing BC (Fig. 89) and by Eq. (28.1). The values of the relaxation times found by these two methods lead to results which agree with one another, as is predicted by the theory [39] (see Sec. 6).

The determination of the relaxation time from the narrow diffusion wing is given in Table 33 for several liquids. The values of τ in the second column are taken from photographic measurements of the intensity [53] and in the third column the τ are from photoelectric measurements of the intensity [41, 505]. In the fourth column of Table 33 are the values of τ computed from (6.12); the differences of the values for the same liquid are explained by the difference in the data for the dimensions of the molecules used in the calculation. The calculated and measured values of τ agree well with one another. The logic of the separation of the wing of the Rayleigh line into two parts is based on experiment and theory, and the excellent quantitative agreement between the comparable values of the relaxation time of the anisotropy permit us to believe that the correct interpretation of the wing of the Rayleigh line has been found (as a whole) for ordinary liquids. Of course, one must recall that the wing of the Rayleigh line takes up an appreciable portion of the spectrum, abutting the Rayleigh scattering line and, therefore, it is entirely possible that radiation can be superimposed on this same portion of the spectrum without having any relation to the wing. An example is the band found by Venkateswaran [516] in the region of 100 cm^{-1}. Other examples are obviously possible. This means that the wing of the Rayleigh line should be studied in liquids where the phenomenon is not complicated by side effects.

Superposition of the Wing of the Rayleigh Line in the Region of the Fine Structure Components. Direct measurements of the intensity distribution in the narrow part of the wing of the Rayleigh line (Fig. 91) show that in different liquids, different fractions of the intensity of the wing are superimposed on the intensity in the region of the fine structure in ordinary liquids, which occupies the interval ± 0.5 cm^{-1}. From the investigations of the intensity distribution in the wing, the data of the experiments of Rousset [72], Saxena [528], and Sunanda Bai [529] become completely understandable. They carried out measurements of the depolarization factor of the scattered light as a function of the slit width (Fig. 92) on spectrographs with a large linear dispersion and resolving power. It is clear from their results that a decrease in the width of the slit essentially decreases the depolarization factor of the Rayleigh scattering for substances with a broad wing. For substances which are characterized by a narrow wing, such as acetic acid, the change in the width of the slit has an unimportant effect on the change of the depolarization factor up to the very small widths of the slit.

The fundamental value of these measurements lies in the fact that they make it possible to determine the fraction of depolarization scattering superimposed on the central line of the Rayleigh triplet with significant accuracy. In the experiments of Sunanda Bai [529], the very narrow slit was equal to 0.29 cm^{-1}, which encloses the region of the central line but, for example, for benzene does not include the MB components completely.

We designate the intensity of the central line of the triplet and the portion adjoining it, which is embraced by the smallest slit width, by I_0 and the intensity of the portion of the wing determined by the same slit width by i', where $i' = i'_x + i'_z$. The intensity of the remaining part of the wing is denoted by $i'' = i''_x + i''_z$. The complete integrated intensity of the wing $i = i' + i''$. By definition, the depolarization of the scattered light in an infinitely wide slit is

Table 34. Fraction of the Intensity of the Wing Superimposed on the Intensity of the Central Component

Liquid	Δ_u	$\rho\ (\Delta\nu)$	$\frac{i}{I_0}=\frac{13\Delta}{6-7\Delta}$	$\frac{i'}{i}=\frac{\rho(6-7\Delta)}{\Delta_u(6-7\Delta)}$	η
Benzene	0.44	0.12	1.92	0.153	0.00649
Carbon bisulfide	0.63	0.30	5.15	0.195	0.00367
Toluene	0.49	0.22	2.48	0.33	—
Acetone	0.20	0.06	0.56	0.247	0.0033
Ethyl ether.	0.081	0.026	0.20	0.30	0.00234
Carbon tetrachloride	0.052	0.013	0.12	0.24	0.00962
Acetic acid	0.40	0.28	1.63	0.552	0.012192
Formic acid.	0.53	0.44	3.16	0.64	0.01804
Glycerin	0.33	0.30	1.16	0.87	8.3
Benzophenone.	0.75	0.66	13.0	0.86	—
Salol	0.75	0.75	13.0	1.0	0.20
Nitrobenzene	0.62	0.54	4.9	0.44	0.02

$$\Delta_u=\frac{i''_x+i'_x}{I_0+i'_z+i''_z}. \tag{28.2}$$

The depolarization of the wing is

$$\rho_u=\frac{i'_x}{i'_z}=\frac{i''_x}{i''_z}=\frac{6}{7}, \tag{28.3}$$

and, finally, the depolarization factor for a narrow slit is

$$\rho\ (\Delta\nu)=\frac{i'_x}{I_0+i'_z}. \tag{28.4}$$

By combining (28.2), (28.3), and (28.4), we find that the fraction of the integrated intensity of the wing superimposed on the region of the fine structure is

$$\frac{i'}{i}=\frac{\rho\ (\Delta\nu)\ [6-7\Delta_u]}{\Delta_u\ [6-7\rho\ (\Delta\nu)]}. \tag{28.5}$$

The ratio of the integrated intensity of the wing i to the intensity of the Rayleigh line is determined by (1.91).

Table 34 gives the results of the calculation according to (1.94) and (28.5) (the researches of Sunanda Bai and Saxena) for certain liquids.

From Table 34 one can see the division of liquids, roughly speaking, into two groups. The first group is characterized on our spectrograms by a wide wing; a distinct depolarization line or band is missing from the I_x component for the etalon width used; this means that the halfwidth of the wing close to the unshifted line in every case exceeds 0.5 cm^{-1}. For these liquids, the fraction of the intensity of the wing reaching the components of the fine structure lies in the range 0.15 to 0.35 of the total intensity of the wing.

In the other group of liquids, the fraction of the wing intensity in the narrow portions (~0.3 cm^{-1}) is more than half of the total intensity. In these liquids, a narrow depolarized line is seen on the interferogram. The opposite conclusion is also valid: in liquids for which more than half the total intensity of the wing appears in the region of the unshifted line (~0.2 cm^{-1}), a narrow depolarized line will be observed in the spectrum.

Comparison of the results of the depolarization measurement for a narrow slit with depolarization measurements of the central line of the Rayleigh triplet, made by Venkateswaran [257], shows that there is quantitative agreement within the limits of error of both experiments (see Tables 20 and 34). This is one more argument in support of the assumption that Venkateswaran and Sunanda Bai actually measured the depolarization of the central line, brought about by the isobaric density fluctuations, together with the part of the wing superimposed on this line. Consequently, it is impossible to draw conclusions on the depolarization of the light of the central line of the Rayleigh triplet from their measurements. This is explained, as we have already shown above, by the insufficient resolving power of their apparatus. Some depolarization of the MB components is explained in the same way, as Venkateswaran has already pointed out [476].

29. Temperature Investigation of the Wing of the Rayleigh Line and Relaxation of the Viscosity

The most powerful test of the relaxation theory is the comparison of its conclusions with the results of the study of the intensity distribution in the wing of the Rayleigh line as a function of the viscosity and the temperature. The temperature experiments also test all the other hypotheses of the origin of the wing.

For simplicity, it is desirable to study the phenomenon as a function of the change of a single parameter, in order not to bring extra complications into an already sufficiently complex problem. Therefore, liquids were chosen in our experiments for which small variations of temperature lead to appreciable change in the viscosity [52, 53, 73, 144]. In the choice of substances for investigation, we began with the requirement that the substance be chemically distinct, transparent in the visible portion of the spectrum, not crystallize in the transition to high viscosities, capable of "optical cleaning" by multiple distillation with heating to the boiling point, as well as other known requirements. Moreover, the chosen liquid must consist of strongly anisotropic molecules in order that one can count on finding the intense wing of the Rayleigh line in them.

In our investigation, salol, triacetin, benzophenone, and glycerin were chosen. The molecule of salol is characterized by a significant anisotropy. The depolarization factor of the scattered liquid in liquid salol is equal to $\Delta_u = 0.75$. But to "force" the salol to undergo transition from the liquid state (above 42°C) to the supercooled and even to the amorphous state is not simple, and one must resort to different tactics. Triacetin and glycerin do not crystallize in the transition to the solid state. However, their anisotropy is less than that of salol. The depolarization factor of the scattered light for triacetin $\Delta_u = 0.38$, while for glycerin $\Delta_u = 0.33$. Benzophenone is characterized by a large depolarization factor of the scattered light $\Delta_u = 0.75$, but it also has a large tendency to crystallization.

The farther part of the wing of the scattering line in salol (from 5-10 to 50-60 cm^{-1}) was studied by us by the spectrographic method for temperatures from +120 to −45 and −60° (the glassy state) [53, 144, 510]. We succeeded in making photometric photographs, taken in the temperature range from +120 to 0°C. The wing of the Rayleigh line for temperatures below 0°C became so narrow that it could not be separated from the small background of the exciting light source that is always present. The scattering spectrum at −45°C differed practically not at all from the spectrum of the incident light; it was not possible from the measurement of these photographs to make a judgment on the intensity distribution

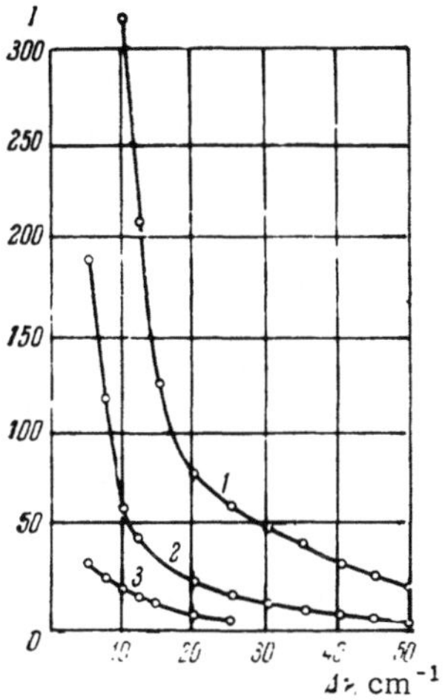

Fig. 94. Distribution of the intensity (arbitrary units) in the wing of the Rayleigh line in salol for different temperatures (Fabelinskii [53]). (1) +120°C; (2) +20°C; (3) 0°C.

in the wing in every case, but it was clearly evident that the intensity of the wing is insignificant under these conditions or is equal to zero within the limits of photometric error.

Figure 94 shows the intensity distribution in the wing of the scattering line in salol for three different temperatures.

The kinetics of the change in the intensity in the wing can be seen from the drawing. As the temperature decreases, the viscosity of the salol increases sharply, changing by a factor of 50 in a temperature decrease from +120 to 0°C, while the change in the absolute temperature amounted to only 40%. In the transition from 120°C to the temperature of the glassy state of salol (−45°C), the temperature changes by about 65%, while the viscosity passes through a value exceeding its value at room temperature by many orders of magnitude. Therefore, there is every reason for supposing that the change in the character of the intensity distribution in the wing of the scattering line in salol is determined practically entirely by the change in the viscosity.

The character of the effect of viscosity growth on the intensity distribution is found in qualitative agreement with the conclusions of relaxation theory. Qualitative observations were made earlier on the change of the wing of the scattering line in salol as a function of the viscosity by Venkateswaran [516] and Raman and Venkateswaran [276]. The authors of these researches came to the conclusion that upon increase of viscosity the wing contracts toward the unshifted line and finally, when the salol becomes glass, the wing is not seen at all, in agreement with our observations.

Gross and Raskin [515] repeated the qualitative experiment with salol over a still greater range of viscosity and did not find any essential change in the wing with change in viscosity. Evidently, our quantitative measurements make it possible to understand the reason for the contradiction. As one can see from Fig. 94 and Table 34, the principal part of the wing of the scattering line in salol is contained in the region immediately adjoining the unshifted line even at room temperature (according to the data of Table 34, the entire wing lies only in the region of the central line, which is, of course, exaggerated). Therefore, one must look for a change in the energy distribution of every weak radiation, which is not simple, the more so that our experiments, although they clearly point to a definite change in the width of the wing, also indicate that this change is not so large as the change in the viscosity. While the viscosity changes by a factor of 50, the shift of the curves $L(\Delta\omega)$ as a function of $(\Delta\omega)^2$ amounts only to a factor of 3-4, which is not easily noted in the qualitative experiment. For further increase in the viscosity, it is impossible to detect a change in the width of the wing.

The temperature dependence of the intensity of the portion of the wing next to the unshifted line of the Rayleigh triplet has been further studied [53, 230]. The investigation was carried out on an apparatus with a Fabry–Perot interferometer according to the arrangement of Fig. 34. The effect of the Rayleigh triplet was practically removed by a rigid limiting of the aperture of the incident light and by the use of a polarization system. The I_x and I_z components are spatially distributed over the spectrogram. Salol, benzophenone, triacetin, and glycerin were studied. In the photographs of the scattering spectra of these substances (Figs. 35 and 93), the kinetics of the change in the intensity of the narrow portion of the wing (I_X component) and of the lines of the fine structure (I_Z components) are clearly seen.

In all cases, without exception, the narrow peak in the I_X component is broadened and the intensity of the continuous background increases (broad portion of the wing) with increase in temperature. The quantitative investigation of the intensity distribution in the I_X component for changes in the viscosity and temperature is made very difficult, chiefly by reason of the appreciable superposition of the continuous background of the neighboring orders and the small difference in the widths of the narrow portions of the wing and the exciting line. Therefore, the errors in the determination of the width of the intensity distribution in the wing or the relaxation time are very great.

However, quantitative experiments on the narrow portion of the wing in triacetin for different temperatures have been completed by Ablekov and Fabelinskii [539].

For the determination of the true halfwidth of the I_X component, a method has been used which was suggested by Ablekov [221]. In Figs. 95a and 95b, the change of the true intensity distribution in the narrow portion of the curve is shown as a function of the temperature. If it is assumed that the intensity distribution is determined only by processes of modulation of the light brought about by rotational diffusion of anisotropic molecules [41, 505], then the halfwidth of the distribution and the relaxation time of the anisotropy are connected by the simple relation

$$\tau = \frac{1}{2\pi\delta\nu}. \qquad (29.1)$$

In such a case, one can determine that (in the limits of change of the temperature shown in Fig. 95b), τ changes from $2 \cdot 10^{-11}$ sec to $6 \cdot 10^{-11}$ sec.

The temperature dependence of the halfwidth is shown in Fig. 96.

Figure 97 gives the data for the temperature dependence of the depolarization factor of the scattered light in the maximum of the scattering line ρ_0 and at a distance of 1.25 cm^{-1} from the maximum $\rho_{1.25}$; for comparison, data are given in the same graph for the total depolarization factor, obtained by Velichkina [188, 492].

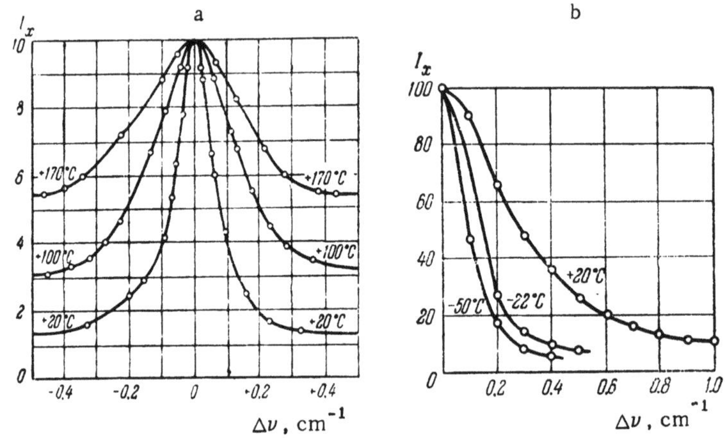

Fig. 95. Distribution of the intensity in the portion of the wing adjoining the unshifted line (I_x component) for different temperatures (a) in salol (Fabelinskii [53]) and (b) in triacetin (Ablekov and Fabelinskii [539]).

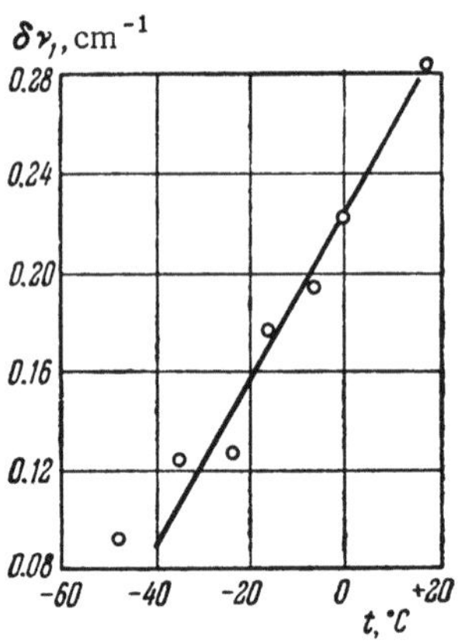

Fig. 96. Temperature dependence of the halfwidth of the near portion of the wing in triacetin (Ablekov and Fabelinskii [539]). The line is drawn through the experimental points.

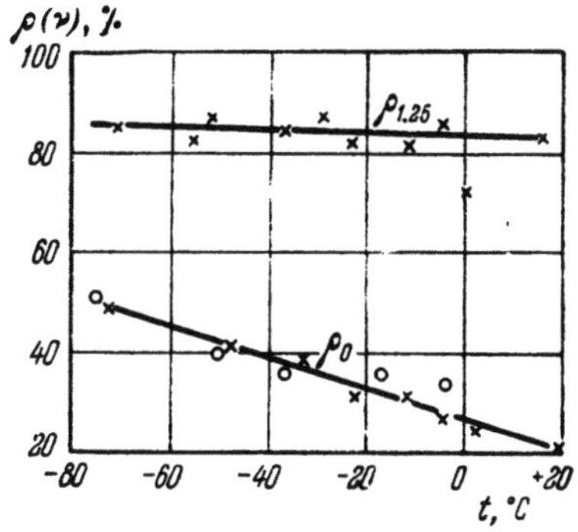

Fig. 97. Temperature dependence of depolarization factor of scattered light in triacetin (Ablekov, Fabelinskii [539]). ρ_0—depolarization in maximum of distribution; $\rho_{1.25}$—depolarization at distance of 1.25 cm^{-1} from maximum of distribution; circles—experimental data [189] for $\lambda = 4358$ Å, obtained without a detailed spectral decomposition of the scattered light.

For a frequency of 1.25 cm^{-1}, at low temperatures, the intensity of the wing is negligible. More distant portions of the wing can be regarded as practically absent. The depolarization factor $\rho_{1.25}$ is virtually independent of the temperature and is equal to $^6/_7$, which corresponds to the depolarization factor of the light scattered by fluctuations of the anisotropy $\Delta\varepsilon'_{ik}$ for excitation by natural light. The depolarization of the wing in benzene was measured with a laser source of excitation; according to the data of [641, 646], $\rho_u = 0.74$ and 0.78, respectively. According to the data of [641], $\rho_u = 0.82$ for carbon bisulfide, 0.70 for toluene.

There arises the serious question of the reason for the divergence of the experimental data [641, 646] and the theory [39,40, 47, 50]. The quantity ρ_0 depends differently on the temperature; it increases strongly for decrease in temperature, which is explained by the interplay of two factors: by a narrowing of the depolarized wing and by a decrease in the light intensity scattered by the density fluctuations. At −70°C, the intensity of the broad part of the wing is negligible and practically all the scattered light is concentrated in the narrow line; therefore, upon increase in temperature, the effect of the lessening of the intensity at the maximum is great.

Qualitatively, the variation of the intensity and the width of the depolarized narrow wing is in accord with the qualitative observations of Raman and Rao [499] for phenol and the data of Gross and Syromyatnikov [475] for phenol and n-cresol. For decrease in the viscosity in the substances mentioned, the intensity of the central depolarized line was reduced and an intense depolarized background appeared in this vicinity (the broad part of the wing).

On the basis of the experimental data given here, one can reach the conclusion that the wing of the Rayleigh line (the depolarized scattering) does not disappear in the transition from the liquid state to the solid amorphous state as it does in the transition from the liquid to the solid crystalline state; it is only narrowed, and its halfwidth decreased. The total depolarization of the scattered light in the transition to the amorphous solid state increases [188, 189], which means an increase in the fraction of the depolarized scattering in the total flux of scattered light. The relaxation times found from light scattering for low-viscosity liquids are compared with the relaxation times which follow from (28.1) and from (6.12) and (8.10). Excellent agreement of the values is seen from Table 33.

For liquids with huge viscosity there are no data for the determination of τ from (28.1), but one can only repeat that τ changes much more slowly than the viscosity in both portions of the wing.

The comparison of τ determined for the narrow portion of the wing (for large viscosity) with that calculated from (6.12) shows that there is no agreement even in order of magnitude.

Equation (6.12), obtained from the hydrodynamic consideration of the motion of a sphere of radius a in a viscous liquid, is, strictly speaking, inapplicable to molecules [532]. However, a number of investigations [77, 152] show that for low-viscosity liquids (6.12) reproduces the experimental data. On the other hand, it has already been noted in the literature that a significant difference is found for high viscosities between the experimental data and the results of the calculation according to (6.12) for relaxation times of the dipole moment [530].

For clarification of the problem of the suitability of the relation (6.12) for the determination of the relaxation time of the dipole moment in viscous liquids a method was developed by Mash, Mayants, and Fabelinskii [531] for the measurement of the dielectric constant and the tangent of the loss angle in dipole liquids in the microwave frequency range. The work was carried out at a line of wavelength 3.22 cm. The dielectric constant and $\operatorname{tg}\delta$ in glycerin were measured in the temperature range from 25 to 160°C.

By using the well-known approximate expression [152]:

$$\operatorname{tg}\delta = \omega\tau_D \tag{29.2}$$

and the data of our research [531], we obtained $\tau_D = 5.1 \cdot 10^{-12}$ sec for 25°C. By assuming that the viscosity of glycerin is 9 poise at 25°C, and taking the size of the molecule of glycerin to be equal to 3 Å, we find $\tau_D = 1.26 \cdot 10^{-8}$ sec on the basis of Eq. (6.12). In the case we investigated, the measured τ_D is about $2 \cdot 10^3$ times smaller than that computed by Eq. (6.12). With decrease of viscosity, the measured and computed values come together, so that there is satisfactory agreement for such substances as the alcohols, according to Debye [152] and Debye and Sack [77]. The divergence increases between the measured and calculated τ with increase in the viscosity. Thus, the phenomenon of light scattering is not the only one in which a quantitative difference is observed between the measured and computed τ. In the case of light scattering, the divergence is so great that one cannot speak of quantitative agreement of theory with experiment, but one quickly reaches the conclusion that the static viscosity by no means determined the process of light scattering.

The static viscosity, as we have seen, does not determine the process of the dispersion of electromagnetic waves in dipole viscosity liquids either.

It must be noted that Eq. (29.2) cannot correctly describe the variation of the tangent of the loss angle with frequency or with temperature, inasmuch as the dielectric constant (ε' and ε'') is not described by a single relaxation time, but should rather be described by a whole spectrum of relaxation times [533-536]. However, the spectrum of times is not so broad that the times can be separated by many orders.

In the simplified picture of the phenomenon, attention is turned to the fact that the divergence between τ obtained from

(29.2) and experiment, and computed from (6.12) amounts to four orders of magnitude for a viscosity of 9 poise. For increase in viscosity, the difference will be still greater.

It was shown earlier that the strong increase in the viscosity for the transition from the liquid to the glassy state involves a strong damping of the sound. The coefficient of sound absorption passes through a maximum and decreases, becoming small for the glassy state [296, 537]. The static viscosity increases monotonically in the glass transition, reaching the huge value of 10^{16}-10^{18} poise [538, 538a], for which, of course, there can be no sound propagation. At the same time, it is well known that the sound in glass, for example, propagates with small losses. Consequently, the viscosity which exists for the process of sound propagation is not large.

The given example clearly shows that the static and dynamic values of the viscosity can differ by many orders of magnitude.

In the phenomenon of dispersion of electromagnetic waves in polar liquids and even more in light scattering, we are dealing with very fast processes. Therefore, it is important to use those values of the viscosity which characterize this phenomenon.

In the problem of the relaxation time of the dipole moment in a viscous medium, one can seek a solution by different paths.

Fabelinskii made the assumption [53, 230] that the static viscosity, substituted in (6.12) for the calculation, does not determine the phenomenon, and, therefore, the result of the calculation diverges strongly from the measured figure. As in the case of sound absorption, the viscosity ceases to be the Newtonian constant coefficient, and becomes a function of frequency. At high frequencies, even a small viscosity begins to depend on the frequency, and a large viscosity becomes a function of the frequency at much lower frequencies. We shall assume that the frequency dependence has the form

$$\eta = \frac{\eta_0}{1+\omega^2\tau_S^2}. \tag{29.3}$$

Here η_0 is the static viscosity and τ_S is the relaxation time of the viscosity, which can be estimated from (22.7), by substituting η' for η. The estimate for the glycerin at 20°C gives $\tau_S \sim 2.5 \cdot 10^{-10}$

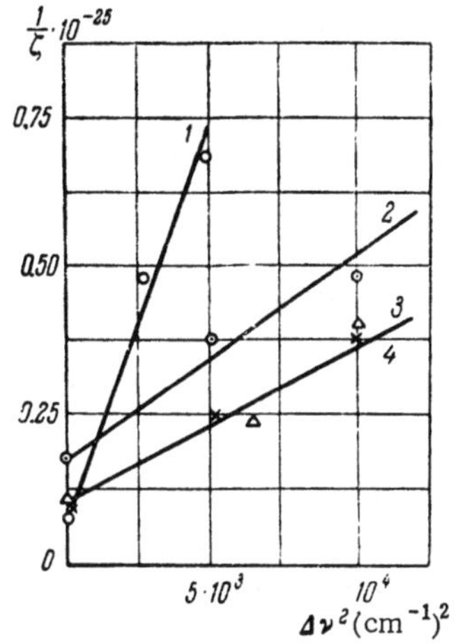

Fig. 98. Dependence of the reciprocal of the internal friction on the square of the frequency (Starunov [505]). 1 – Chloroform (circles); 2 – carbon bisulfide (circles with center dot); 3 – benzene (crosses); 4 – toluene (triangles).

sec. From (29.3) and (6.12), we obtain the result that $\tau_D \sim 5 \cdot 10^{-11}$ sec, and this is already a much better agreement with the τ_D found from experiment and (29.2) than was obtained above.

In the wing of the Rayleigh line, there are also much faster processes of the modulation of the scattered light wave than the process of reorientation in the experiment with dispersion of electromagnetic waves; therefore, the relaxation of the first viscosity coefficient ought to be taken into account in the calculation of the relaxation time of the anisotropy. The region of frequencies of the wing of the Rayleigh line extends to $\sim 10^{14}$ Hz. The different portions of this region extend up to, into, and beyond the relaxation frequency of the shear viscosity coefficient even in low-viscosity liquids [53]. Starting from the already discussed mechanisms of creation of the different parts of the wing, Starunov [505], by using the portion of the wing AB of Figs. 89 and 90 and the formula (8.12), and also the portions BC and CD and Eq. (8.16), found the value of the internal friction $\zeta \sim \eta$ for these three parts. Here it was assumed that in the small parts which extend to 20–40 cm^{-1}, the elastic constant and the viscosity do not depend on the frequency inside the region BCD. For a test of the character of the dependence of ζ on the frequency [505], the dependences of $1/\zeta$ on $(\Delta\nu)^2$ were constructed (beyond the frequency taken for the mean frequency of each part). It is interesting that this dependence was linear for the four cases studied (Fig. 98). Inasmuch as $\zeta \sim \eta$, the linear dependence (Fig. 98) means that the law of the dependence of the viscosity or the frequency (29.3) or (8.17) has been correctly chosen.

From the slope of the curves of Fig. 98, estimates are obtained for the relaxation time of the internal friction or viscosity; $\tau_\zeta = 7.4 \cdot 10^{-14}$ sec for carbon bisulfide, $\tau_\zeta = 8.5 \cdot 10^{-14}$ sec for benzene and toluene, and $\tau_\zeta = 2.4 \cdot 10^{-13}$ sec for chloroform.

Great interest attaches to the study of the far portion of the wing on Figs. 89 and 90, denoted by CD.

In this region of high frequencies, even the low-viscosity liquid behaves as a solid, inasmuch as $\omega_{(CD)} > 1/\tau_\zeta$ in this region. If this is so, then one can assume that the elastic constants of the liquid at such frequencies are close to their limiting values and, therefore, close to the values of these elastic constants in the corresponding crystal. By determining μ_∞ from the portion of the wing CD, Starunov [505] calculated ν_{max} from (8.19).

The very fine agreement of the calculated ν_{max} in the region of the low crystalline frequencies measured testifies to the usefulness of the theory. The presence of only a single frequency ν_{max} is connected with the fact that the molecule is characterized by a single moment of inertia in the simplified model [505].

It must be emphasized that the mechanism considered here of the generation of the wing has nothing in common with the assumption that crystals or crystallites are found in the liquid [512-515, 517] even in a deformed state, or that there is a local orientation order in the liquid.

From what has been said, one can conclude that the new development of the theory of the wing of the Rayleigh line makes it possible to set up a series of experiments from which one can expect a much clearer answer to the problem of the nature of the wing of the Rayleigh line.

30. The Relaxation Time of the Anisotropy and the Internal Time of the Kerr Effect. A New Method of Determining the Relaxation of the Anisotropy

Comparison of the Relaxation Time of the Anisotropy Found from the Scattering of Light and from the Inertia of the Kerr Effect. Immediately after Leontovich [39] had developed the relaxation theory and Fabelinskii had determined the relaxation time of the anisotropy τ from the formulas of the theory and the measurements of the intensity distribution in the wing [73], it became clear that there exists a huge divergence between the values of τ obtained from scattering and those obtained from the measurement of the inertia of the Kerr effect. The latest quantitative measurements of the inertia of the Kerr effect were then, and remain still today, the measurements of Hanle and Mäerks [525]. For convenience in discussing the results of the measurement of the inertia time of the Kerr effect in liquids, we shall refer briefly to the principle of the method [525]. This method consists of the following. A traveling ultrasonic wave is stroboscoped by light modulated by a Kerr cell, which is fed from the same generator as the piezoquartz irradiated by the sound wave. Photographs of the traveling waves in such an experiment show a system of equidistant light and dark bands, the location of which depends on the relation between the phase of the sound wave and the phase of the light modulation.

If the sound wave is photographed, by first modulating the light by the Kerr cell with one liquid (standard) and then by a Kerr cell filled with another liquid (test sample), two sets of bands are obtained in the photograph; these are generally shifted with respect to one another. By assuming that the modulation of the light passing through the Kerr cell takes place according to a sinusoidal law, while the molecular orientation in the field is "dissipated" (after the field is turned on) according to an exponential law, we can write that

$$\operatorname{tg}\varphi=\omega_{M}\tau;\quad \varphi=2\pi\frac{\Delta l}{\lambda}, \tag{30.1}$$

where Δl is the relative band shift mentioned above and ω_M is the modulation frequency. For small φ, we have

$$\tau=\frac{2\pi\Delta l}{\lambda\omega_{M}}.$$

Here τ is the relaxation time of the anisotropy of the studied liquid, measured relative to the standard. Since Δl measures the shift of one system of bands relative to the other, while the initial phase difference (between the sound wave and the electric field intensity) can be assumed to be constant and the same for both condensers, then τ is the difference between the relaxation time desired and that of the standard liquid. Carbon bisulfide was used as the standard liquid.

In the experiments of [525], pure liquids were studied for which the viscosity lay within the limits from 0.01 to 0.9 poise, and also solutions of carbon bisulfide in paraffin and mineral oils, and a solution of nitrobenzene in mineral oil. In all the pure liquids, the relaxation time was found to be about $2 \cdot 10^{-9}$ sec. In solutions of carbon bisulfide in mineral oils, $\tau = 0$ with accuracy to within $5 \cdot 10^{-10}$ sec, while in solutions of nitrobenzene in mineral oil, $\tau = 7 \cdot 10^{-9}$ sec. The divergence between the data from the inertia of the Kerr effect [525] and the data for the same physical quantity given in Tables 32 and 33 is so great that there is no hope of reconciling them. But in this case, the difference need not be viewed as a direct contradiction, since the polarization of the central component of the lines of the Rayleigh triplet has not been studied in detail. Actually, the relaxation time of 10^{-13}-10^{-12} sec, found from scattering (Tables 32 and 33) is inaccessible to measurement by means of the Kerr effect by the very essence of the experimental method of measurement of the inertia of this effect. On the other hand, if the times $\tau \sim 10^{-9}$ sec are real, they cannot be observed in the spectrographic method of investigation of the wing of the Rayleigh line. Actually, the processes of the relaxation of anisotropy with $\tau \sim 2 \cdot 10^{-9}$ sec modulate the scattered light with frequency $f \sim 10^8$ Hz. Consequently, the halfwidth of the spectral distribution of the light intensity scattered by fluctuations of the anisotropy, which changes with a frequency of 10^8 Hz, will be $3 \cdot 10^{-3}$ cm^{-1}. The spectrographic investigation studies the wing, extending 5-10 cm^{-1} from the unchanged frequency. The halfwidth of $2 \cdot 10^{-3}$ cm^{-1} is 5-10 times greater than the actual halfwidth of the central component; however, taking into account that the width of the incident line in our experiments is 0.17 cm^{-1}, one must expect a significant depolarization of the visible central component of the Rayleigh triplet.

Depolarization of the central components could be expected in such liquids as benzene, carbon bisulfide, toluene, etc. But the experiment of Hanle and Mäerks [52] directly reveals it only in such substances as nitrobenzene, its solutions in oil and o-dichlorobenzene, for which the fine structure of the Rayleigh line and its polarization were not studied. Chloroform, for which the inertia time $\tau = 3.8 \cdot 10^{-9}$ sec was obtained, did not reveal any noticeable depolarization of the central component of the fine structure (Fig. 35). In such liquids as acetic acid and a number of others, a very narrow depolarized central line is revealed (Figs. 35 and 83).

In what measure this is connected with the long inertia times of the Kerr effect requires special consideration.

The results of Hanle and Mäerks [525] were subject to criticism and were rejected by Marinin [526]. However, the clear experimental proof of the fact that the inertia of the Kerr effect is characterized by smaller relaxation times than 10^{-9} sec was first obtained in the research of Simkin, Naberezhnykh, and Lukin [527]. In the work of these authors, modulation of light is realized at a frequency of the modulating electric field $\sim 10^{10}$ Hz in a waveguide Kerr cell, filled with nitrobenzene. In order that the light which passes through the Kerr cell contain a variable component, oscillating with a frequency of $2 \cdot 10^{10}$ Hz, it is necessary that the relaxation time of the anisotropic molecules that fill the Kerr element be less than $8 \cdot 10^{-12}$ sec. If it turned out that there were molecules or, rather groups of molecules which are characterized by the time $\sim 2 \cdot 10^{-9}$ sec, then such formations could be tracked from the change of the electric modulating field and, therefore, their contribution to the intensity of the modulated light should be vanishingly small. It follows from what has been said that the quantitative results of the experiments of [525] are in error.

But the final conclusion can be drawn only after the depolarization of the central component of the Rayleigh line has been carefully measured for those substances for which such large relaxation times of the anisotropy were obtained in [525].

Especially interesting results can also be expected from the study of the fine structure of the near part of the wing by means of the narrow line of the gas laser as the source of the exciting light.

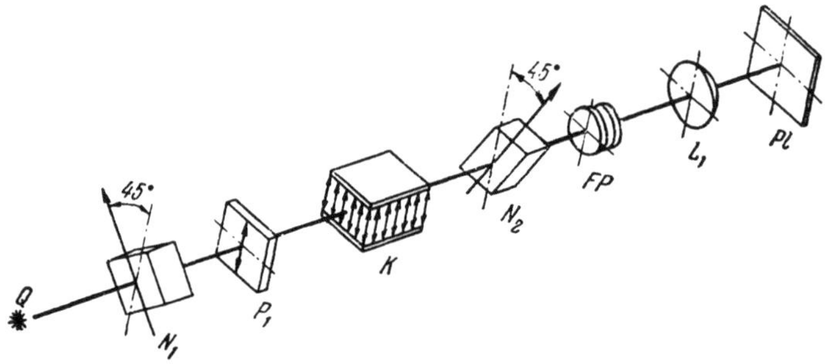

Fig. 99. Arrangement of the different elements of an optical system for the observation of the components of the spectral lines produced by the amplitude modulation of light. Q – Light source (laser); N_1, N_2 – polarization prisms; P_1 – doubly refracting plate, producing a "nonelectric path difference"; K – resonator or waveguide Kerr cell; FP – Fabry–Perot interferometer; L_1 – camera objective; Pl – photographic plate.

To reproduce the experiments of [525] with much greater accuracy is difficult and now it is scarcely necessary.

We shall see still one more new way of making a quantitative measurement of the relaxation time of the anisotropy, which will be explained below.

New Method of Determination of the Relaxation Time of the Anisotropy. The measurement of the relaxation time of the anisotropy or the inertia of the Kerr effect, based on the measurement of the phase shift (30.1), is accompanied by great experimental errors. It seems to us that more results and greater accuracy can be obtained if a different principle is used for the determination of the relaxation time, a principle which was proposed by the author [540]. In this new method, it is proposed to determine the relaxation time of the anisotropy from the relation of the intensities of the components of the spectral line which appear as the result of modulation of light in an ultrahigh-frequency Kerr cell.

It is known that a beam of monochromatic light, which passes through a substance consisting of anisotropic molecules and placed in a Kerr element (condenser, section of a waveguide, resonator) is modulated in phase and amplitude [541] if a variable electric field is applied to the Kerr element.

The Kerr element has been frequently used as a light modulator at low, high, and ultrahigh frequencies of the modulating electric field [527, 541-548].

Of special interest is the study of a different method of modulation of light and the treatment of the different light modulators today when the problem of information transfer by means of the laser ray is very pertinent. From the literature cited above, it is seen that significant successes have already been achieved in this region. In this section, we shall consider the very simple case of the amplitude of the light modulation by a Kerr cell, and show how one can determine the relaxation time from the intensity of the components of the spectral line which is generated as the result of the modulation.

The system of the various elements used is shown in Fig. 99. The optimal amplitude modulation of the light exists for that case in which the principal directions of the polarization devices make an angle of 90° with each other and an angle of 45° with respect to the direction of the electric field. This case is shown in the figure.*

The field of the light wave passing through such a polarizer, Kerr cell, and analyzer, will be expressed in the following fashion:

$$E_a = E_p \sin\frac{k}{2}\left\{\int\limits_{-L/2}^{+L/2}(n_p - n_s)\,dx\right\}\exp i\left[\omega t - \frac{k}{2}\int\limits_{-L/2}^{+L/2}(n_p + n_s)\,dx \pm \frac{\pi}{2}\right], \tag{30.2}$$

where

$$n_p = n + \frac{2}{3}\lambda B\mathscr{E}^2(t), \quad n_s = n - \frac{1}{3}\lambda B\mathscr{E}^2(t). \tag{30.3}$$

*Another case of optimal amplitude modulation is the parallel arrangement of the principal directions of the polarizers, turned by 45° with respect to the field.

Here n_p and n_s are the optical indices of refraction; the electric vector of the light is parallel and perpendicular to the direction of the electric field $\mathscr{E}$, n is the index of refraction without the field, and B is the Kerr constant.

If the frequency of the electric field ω_M is much greater than $1/\tau$ (τ is the relaxation time of the anisotropy), then the molecules cannot follow the change in the field and there will be no component in (30.2) which depends on the frequency of the modulating field, but only a constant component; in the spectrum of the transmitted light, only the incident light frequency will be present.

In the case when $1/\tau \gg \omega_M$, all the components of the modulation spectrum will be present in the transmitted light, with a maximum intensity. In the intermediate case $1/\tau \sim \omega_M$, the intensity of the components of the amplitude spectrum of the modulation will depend on ω_M or, for fixed ω_M, will be determined by the value of τ. The latter case is of the greatest significance to us.

For a quantitative description of the phenomenon, we assume that $y_i \equiv n_i - n$, and $(n_i - n)$ is equal either to $n_p - n$ or to $n_s - n$ at a certain point x of the Kerr cell (the x axis coincides with the direction of propagation of the light) and is described by the following equation:

$$\frac{dy_i}{dt} + \frac{1}{\tau_i} y_i = \frac{1}{\tau_i} b_i \lambda B \mathscr{E}^2 (t). \tag{30.4}$$

Here the index i is either p or s, and

$$b_p = \frac{2}{3}, \quad b_s = -\frac{1}{3}.$$

If $dy_i/dt = 0$, then $y_i = b_i \lambda B \mathscr{E}^2(t)$ and we get (30.3), while, if we turn on the field $\mathscr{E}$, and then instantly turn it off, i.e., we set $\mathscr{E} = 0$ in (30.4), then $y_i = y_i^0 e^{-t/\tau}$; consequently, the artificial anisotropy arising under the action of the field will be dissipated by an exponential law.

For simplicity, we assume that $\mathscr{E}(t)$ does not contain a constant component, and that it is expressed by the harmonic function

$$\mathscr{E}(t) = \mathscr{E}_0 \cos \omega_M t. \tag{30.5}$$

Solving (30.4) for the condition (30.5), we find

$$y_i=\frac{1}{2}\,b_iB\lambda\mathscr{E}_0^2\left\{1+\frac{\cos(2\omega_M t+\varphi)}{\left[1+(2\omega_M)^2\tau_i^2\right]^{1/2}}\right\}. \qquad (30.6)$$

Here φ is the phase shift between the electric field vector and the light wave vector emerging from the Kerr cell while, similar to (30.1), tg $\varphi=2\omega_M\tau_i$.

In an inhomogeneous modulating field, for the determination of $\Psi\equiv\int_{-L/2}^{+L/2}(n_i-n)dx$ it is necessary to integrate (30.6) in the direction of the light ray in the limits from −L/2 to +L/2 if the origin of the coordinates is set at the center of the cell. It must also be assumed that t = (n/c)x in (30.6) and that the phase of the electric field at the moment of passage of the light through the origin of the coordinates is equal to $\omega_M t$. In this case,

$$\Psi=\frac{1}{2}\,b_iL\lambda B\mathscr{E}_0^2\left[1+A_i\frac{\sin kL}{kL}\cos(2\omega_M t+\varphi)\right], \qquad (30.7)$$

where

$$A_i=\left[1+(2\omega_M)^2\tau_i^2\right]^{-1/2} \text{ and } k=\frac{2\pi}{\lambda}.$$

Choosing the length of the Kerr cell such that L = $\lambda_M/4$ (where λ_M is the wavelength of the field in the medium), we get

$$\Psi=\frac{1}{2}\,Lb_i\lambda B\mathscr{E}_0^2\left[1+\frac{2}{\pi}A_i\cos(2\omega_M t+\varphi)\right]. \qquad (30.8)$$

Taking (30.3) and (30.8) into account, and substituting their values in (30.2), we obtain

$$E_a=E_p\frac{1}{2i}\exp i\left(\omega_M t-\frac{4}{3}a-kLn\pm\frac{\pi}{2}\right)\left\{\exp\left[ia\frac{4}{3\pi}A_s\times\right.\right.$$

$$\left.\left.\times\cos(2\omega_M t+\varphi)\right]-\exp\left[ia\frac{8}{3\pi}A_p\cos(2\omega_M t+\varphi)\right]\right\}, \qquad (30.9)$$

where $a=(\pi/2)LB\mathscr{E}_0^2$.

Expanding E_a from (30.9) in a series of Bessel functions, and calculating the intensity of the m-th component with frequency $2m\omega_M$ (m is an integer), we get

$$I_m = E_a^* E_a = \frac{1}{4} I_p \left\{ J_m^2 \left(\frac{4}{3\pi} A_s \right) + J_m^2 \left(\frac{8}{3\pi} a A_p \right) - \right.$$

$$\left. -2(-1)^m J_m \left(\frac{4}{3\pi} A_s \right) J_m \left(\frac{8}{3\pi} a A_p \right) \cos 2a \right\}. \qquad (30.10)$$

The value of A_i changes from zero to unity and for small a the relation of the intensity $I_{\pm 1}$ of the light with frequency $\omega \pm 2\omega_M$ to the intensity of the light with unchanged frequency I_0 will be equal to (if we assume that $A_p = A_s$, i.e., $\tau_p = \tau_s = \tau$)

$$\frac{I_{\pm 1}}{I_0} = \frac{1}{\pi^2 (1 + 4\omega_M^2 \tau^2)}. \qquad (30.11)$$

From Eq. (30.11) we find that

$$\tau^2 = \frac{1 - \pi^2 \left(\frac{I_{\pm 1}}{I_0} \right)}{4\pi^2 \omega_M^2 \left(\frac{I_{\pm 1}}{I_0} \right)}. \qquad (30.12)$$

It is then clear that the relaxation time is determined by the ratio of the intensities of the first shifted component and the central component and the frequency of the modulating electric field. Both these quantities can be measured with considerable accuracy. If the path length of the light in the Kerr cell is equal to $L = \lambda_m/2$, then, as is clear from (30.7), the part of Ψ which depends on $2\omega_M$ vanishes and there will be no modulated light at all. If, in addition to the variable field on the Kerr cell, a constant electric field is impressed in proper fashion, then, in place of (30.5), we must now set

$$\mathscr{E}(t) = \mathscr{E}_0 + \mathscr{E}_1 \cos \omega_M t. \qquad (30.13)$$

In this case, odd satellites appear as the result of the modulation with ordinary frequency ω_M, which will have a maximum intensity for an optical path length in the cell equal to $\lambda_m/2$. These satellites will be absent when the path length is λ_m. For such a length, the even satellites will also be absent. We note that τ can be found even without the simplifying assumptions made in the derivation of (30.11).

Connes et al. [543] have already observed the picture of the discrete splitting of the frequency of light transmitted through a Kerr cell filled with nitrobenzene, for a modulation frequency of 10^7 Hz. The splitting of the light frequency for the modulation as a consequence of the Pockels effect in a crystal of KDP at a frequency of 10^{10} Hz also observed by Kaminow [546]. Evidently, one can hope that the method given above for the determination of τ will rapidly find practical application.

Chapter IX

Molecular Scattering of Light in Crystals

31. Measurement of the Absolute and Relative Intensity, Polarization, and Frequency of the MB Components in Crystals

The measurement of the intensity of scattered light in crystals presents special difficulties. It is now well known that the intensity of the light of molecular scattering, for example in a quartz crystal at a temperature of 273°K, is seven times greater than in gases under normal conditions.

Impurities and other imperfections of the crystal frequency create such a powerful flux of parasitic light that the light of molecular scattering "sinks" completely in it. Therefore, it was at first difficult to establish whether the molecular scattering of light exists in an ideal crystal.

Although one can eliminate the parasitic light in gases and liquids by cleaning the substance and by appropriate choice of the shape of the vessel, such a way out does not exist for crystals.

Only after Landsberg [17, 146] succeeded in finding an objective criterion (see Sec. 11), which made it possible to separate the parasitic light from the light of molecular scattering, was it possible to proceed to the study of the latter [551].

It follows from the theory of molecular light scattering in crystals that the intensity of the molecular scattering is proportional to the absolute temperature. A small additional temperature dependence, associated with the temperature coefficient of the elastic and elasto-optical constants, gives a change in the intensity of the scattered light of about 1% when the temperature changes by 200°C [146] (quartz). Therefore, this effect can be neglected.

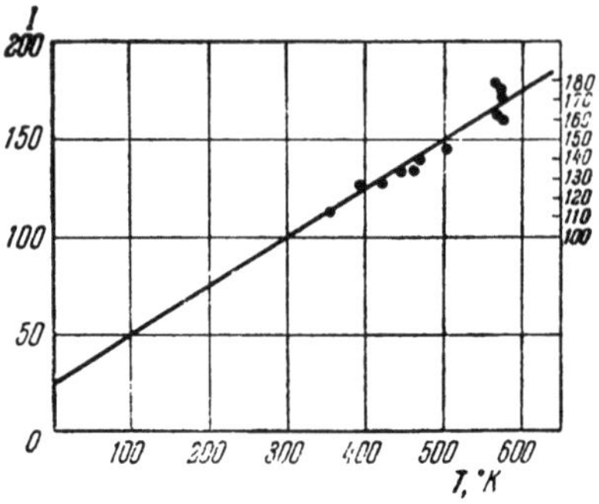

Fig. 100. Temperature dependence of the light scattered by quartz (Landsberg [146]). The intensity for a temperature of 300°K is taken to be 100. The points on the graph correspond to the results of twelve different sets of measurements.

By arranging the experimental data for the temperature dependence of the light intensity scattered in the crystal, and extrapolating this dependence to absolute zero temperature, one can determine the fraction of the intensity corresponding to parasitic scattering.

In the best crystals, the difference between the intensity of the total and parasitic scattering gives the intensity of the molecular scattering. The results of Landsberg [146], which refer to a good crystal, are shown in Fig. 100.

However, there are crystals for which not all the scattered light that is proportional to the absolute temperature is due to the molecularly scattered light.

It has been shown by Born [553], and later by Fürth [554], that the light scattered by dislocations and certain other imperfections of crystals will also be proportional to the absolute temperature. In subsequent experiments, results were obtained from light scattering which are important for the study of the different forms of crystal imperfections. We shall refer to this briefly below; here, it is essential for us to emphasize that there are specimens of crystals in which one can study molecular scattering.

Table 35. Fine Structure of the Lines of Light Scattered by the Crystal NaCl in the Case in Which the Light Is Incident and Observed Along the Corresponding Edges of the Cube

Frequency shift \ Scattering coefficient	R_{zz}	R_{zx}	R_{yz}	R_{yx}
$\Omega_{I}/\omega = 1.78 \cdot 10^{-5}$ (transverse wave polarized perpendicular to the plane of scattering)	0	4.6	4.6	0
$\Omega_{II}/\omega = 2.06 \cdot 10^{-5}$ (transverse wave polarized in the plane of scattering)	0	0	0	0
$\Omega_{III}/\omega = 3.26 \cdot 10^{-5}$ (longitudinal wave)	745	0	0	2.7

In the period associated with the quantitative investigation of the molecular light scattering that was begun by Landsberg, it was possible to find only a small number of sufficiently good crystals, i.e., such in which the scattered light is chiefly determined by the light of molecular scattering.

In first rank among such good crystals, one must list quartz, rocksalt, and diamond. Therefore, we shall specify the general formulas of Sec. 9 for the cases of rocksalt, diamond, and quartz.

Calculation of the Intensity, Polarization, and Frequencies of the MB Components for Rocksalt. For the calculation of the quantities listed in the title of this section, one must assemble the numerical values of the adiabatic elastic, elasto-optical constants, and also the density and index of refraction. The values of these constants [20] are:

$$c_{11} = 46.9 \cdot 10^{10}\ \text{dyne/cm}^2, \quad p_{11} = 0.137,$$
$$c_{12} = 12.85 \cdot 10^{10}\ \text{dyne/cm}^2, \quad p_{12} = 0.178,$$
$$c_{44} = 12.6 \cdot 10^{10}\ \text{dyne/cm}^2, \quad p_{44} = -0.0108.$$

Table 36. Fine Structure of the Lines of Light Scattered by the Crystal NaCl in the Case When the Light Is Incident Along the Diagonal of the Boundary of the Cube and Is Observed Along Another Diagonal of the Cube

Scattering coefficient / Frequency shift	$R_{\perp\perp}$	$R_{\perp\parallel}$	$R_{\parallel\perp}$	$R_{\parallel\parallel}$
$\Omega_{I}/\omega = 1.78 \cdot 10^{-5}$ (transverse wave polarized in the plane of scattering)	0	0	0	0
$\Omega_{II}/\omega = 1.78 \cdot 10^{-5}$ (transverse wave polarized perpendicular to plane of scattering)	0	4.6	4.6	0
$\Omega_{III}/\omega = 3.40 \cdot 10^{-5}$ (longitudinal wave)	676	0	0	9.0

The density $\rho = 2.163$ g/cm^3, the index of refraction n = 1.560.

As was already shown in Sec. 9, the cubic crystal is anisotropic for scattered light. Therefore, two cases are considered.

Case I. Light is incident and scattered along the corresponding edges of the cube.

From the data of Sec. 9, and from the constants of sodium chloride given above, we obtain Tables 35-41, in which the scattering coefficients of the MB components are given in units of $(\pi^2/2\lambda^4)n^8kT \cdot 10^{-16}$ cm^{-1}, and the relative shift of the frequency of the components $\Delta\omega/\omega \equiv \Omega/\omega$. From Eqs. (9.25) and (9.26), we find that, in the irradiation of the crystal by natural light,

$$R_1 = 757 \frac{\pi^2}{2\lambda^4} n^8 kT \cdot 10^{-16} \text{ cm}^{-1}, \qquad (31.1)$$

the depolarization factor is

$$\Delta_{u\,(I)} = 0.01.$$

Table 37. Fine Structure of the Line of Light Scattered in Diamond in the Case in Which the Incident Light Is Directed and the Scattered Light Observed Along the Edge of the Cube

Frequency shift \ Scattering coefficient	R_{zz}	R_{zx}	R_{yz}	R_{yx}
$\Omega_I/\omega = 1.36 \cdot 10^{-4}$ (transverse wave polarized perpendicular to plane of scattering)	0	16.7	16.7	0
$\Omega_{II}/\omega = 1.10 \cdot 10^{-4}$ (transverse wave polarized in the plane of scattering)	0	0	0	0
$\Omega_{III}/\omega = 2.19 \cdot 10^{-4}$ (longitudinal wave)	7.4	0	0	13.1

Case II. Light is incident and scattered along the diagonal to the edge of the cube.

From Table 5b and the numerical values of the constants, we get Table 36. The indices on the scattering coefficients have the same meaning as in Table 5a. For irradiation of the crystal by natural light, we get, from (9.27) and (9.28),

$$R_2 = 694 \frac{\pi^2}{2\lambda^4} n^8 kT \cdot 10^{-16} \text{ cm}^{-1}, \tag{31.2}$$

and the depolarization factor

$$\Delta_{u\,(II)} = 0.02.$$

Comparing both cases of different orientation of the crystal for NaCl, we get

$$\frac{R_1}{R_2} = 1.09.$$

It would be interesting to test this theoretical conclusion experimentally, but such has not yet been done.

The results obtained for NaCl are characteristic for most crystals.

The chief contribution to the intensity of the scattered light is obtained from the diffraction of light by longitudinal or quasi-longitudinal waves. In the longitudinal wave, the chief contribution is obtained for the case in which the incident and scattered light are polarized perpendicular to the plane of scattering (Tables 35, 36, 38-41).

Here the depolarization of the scattered line in the excitation by natural light is much less than unity. An interesting exception is the crystal of diamond. In this crystal, the principal contribution to the flux of scattered light is made by the transverse wave.

The depolarization of the scattered light in the excitation by natural light is much greater than in the previous case.

In addition, it must be noted that the splitting of the MB components in diamond is almost an order of magnitude greater than in ordinary crystals. As an example, only the case in which the incident and scattered light are propagated along the edge of the cube is calculated (Case I).

The constants for diamond obtained by Chandrasekharan [224] are used for the calculation (Table 37),

$$p_{11} = -0.31, \quad p_{12} = 0.09, \quad p_{44} = -0.12$$

and the elastic constants

$$c_{11} = 95 \cdot 10^{11} \quad \text{dyne/cm}^2,$$
$$c_{12} = 39 \cdot 10^{11} \quad \text{dyne/cm}^2,$$
$$c_{44} = 43 \cdot 10^{11} \quad \text{dyne/cm}^2.$$

The index of refraction $n = 2.624$ and the density $\rho = 3.514$ g/cm^3. In the case considered, the scattering coefficient of the light for excitation by natural light is equal to

$$R_1 = 53.9 \frac{\pi^2}{2\lambda^4} n^8 kT \cdot 10^{-16} \text{ cm}^{-1}. \tag{31.3}$$

The depolarization of the scattered light $\Delta_{u(I)} = 1.24$. Because of the large index of refraction, the scattering coefficient in diamond is 4.5 times larger than the scattering coefficient in NaCl for the same Case I.

Calculation of the Intensity, Polarization, and Frequencies of the MB Components for Quartz. We select a set of coordinates such that the axis of symmetry of third order (i.e., the optic axis) is the z axis, and the axis of symmetry of second order (i.e., the electric axis) is the x axis. For right quartz, we shall use a right-handed set of coordinates and, for left quartz, a left-handed system. In this system of coordinates, the matrices of the elastic and elasto-optical constants will have the following form:

$$\begin{vmatrix} c_{11} & c_{12} & c_{13} & c_{14} & 0 & 0 \\ c_{12} & c_{11} & c_{13} & -c_{14} & 0 & 0 \\ c_{13} & c_{13} & c_{33} & 0 & 0 & 0 \\ c_{14} & -c_{14} & 0 & c_{44} & 0 & 0 \\ 0 & 0 & 0 & 0 & c_{44} & c_{14} \\ 0 & 0 & 0 & 0 & c_{14} & 0.5(c_{11}-c_{12}) \end{vmatrix}$$

$$\begin{vmatrix} p_{11} & p_{12} & p_{13} & p_{14} & 0 & 0 \\ p_{12} & p_{11} & p_{13} & -p_{14} & 0 & 0 \\ p_{31} & p_{31} & p_{33} & 0 & 0 & 0 \\ p_{41} & -p_{41} & 0 & p_{44} & 0 & 0 \\ 0 & 0 & 0 & 0 & p_{44} & p_{41} \\ 0 & 0 & 0 & 0 & p_{14} & 0.5(p_{11}-p_{12}) \end{vmatrix}$$

The coefficients Q_{ij} for this class of crystals are equal to

$$Q_{xx} = c_{11}\varkappa_x^2 + \frac{1}{2}(c_{11}-c_{12})\varkappa_y^2 + c_{44}\varkappa_z^2 + 2c_{14}\varkappa_y\varkappa_z,$$

$$Q_{yy} = \frac{1}{2}(c_{11}-c_{12})\varkappa_x^2 + c_{11}\varkappa_y^2 + c_{44}\varkappa_z^2 - 2c_{14}\varkappa_y\varkappa_z,$$

$$Q_{zz} = c_{44}\varkappa_x^2 + c_{44}\varkappa_y^2 + c_{33}\varkappa_z^2,$$

$$Q_{yz} = c_{14}\varkappa_x^2 - c_{14}\varkappa_y^2 + (c_{13}+c_{44})\varkappa_y\varkappa_z,$$

$$Q_{zx} = (c_{13}+c_{44})\varkappa_z\varkappa_x + 2c_{14}\varkappa_x\varkappa_y,$$

$$Q_{xy} = 2c_{14}\varkappa_z\varkappa_x + \left[c_{12} + \frac{1}{2}(c_{11}-c_{12})\right]\varkappa_x\varkappa_y.$$

The coefficients φ'_{ij} are given by the expression:

$$\varphi'_{xx}=p_{11}\varkappa_x\gamma_x+p_{12}\varkappa_y\gamma_y+p_{13}\varkappa_z\gamma_z+p_{14}(\varkappa_z\gamma_y+\varkappa_y\gamma_z),$$
$$\varphi'_{yy}=p_{12}\varkappa_x\gamma_x+p_{11}\varkappa_y\gamma_y+p_{13}\varkappa_z\gamma_z-p_{14}(\varkappa_z\gamma_y+\varkappa_y\gamma_z),$$
$$\varphi'_{zz}=p_{31}\varkappa_x\gamma_x+p_{31}\varkappa_y\gamma_y+p_{33}\varkappa_z\gamma_z,$$
$$\varphi'_{yz}=p_{41}\varkappa_x\gamma_x-p_{41}\varkappa_y\gamma_y+p_{44}(\varkappa_z\gamma_y+\varkappa_y\gamma_z),$$
$$\varphi'_{zx}=p_{44}(\varkappa_x\gamma_z+\varkappa_z\gamma_x)+p_{41}(\varkappa_y\gamma_x+\varkappa_x\gamma_y),$$
$$\varphi'_{xy}=p_{14}(\varkappa_x\gamma_z+\varkappa_z\gamma_y)+0.5(p_{11}-p_{12})(\varkappa_y\gamma_x+\varkappa_x\gamma_y).$$

The adiabatic values of the elastic constants are equal to [552]

$$c_{11}=85.4\cdot10^{10}\ \text{dyne/cm}^2,\quad c_{12}=7.3\cdot10^{10}\ \text{dyne/cm}^2,$$
$$c_{33}=105.6\cdot10^{10}\ \text{dyne/cm}^2,\quad c_{13}=14.4\cdot10^{10}\ \text{dyne/cm}^2,$$
$$c_{44}=57.1\cdot10^{10}\ \text{dyne/cm}^2,\quad c_{14}=16.8\cdot10^{10}\ \text{dyne/cm}^2.$$

The adiabatic values of the elasto-optical constants are equal to [20, 160]:

$$p_{11}=0.137,\quad p_{13}=0.258,\quad p_{14}=-0.029,\quad p_{33}=0.0973,$$
$$p_{12}=0.249,\quad p_{31}=0.257,\quad p_{41}=-0.042,\quad p_{44}=-0.0685.$$

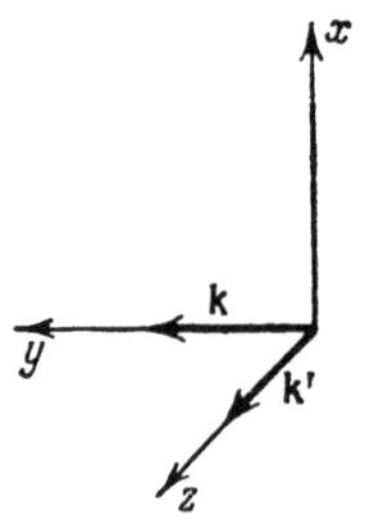

Fig. 101. Calculation of coefficient of light scattering in quartz. The x axis is perpendicular to plane of scattering. Incident light is directed along the +y axis; the scattered light is observed along the z axis.

We consider three cases of different orientation of the quartz relative to the incident beam and the direction of observation.

Case I. The x axis is perpendicular to the plane of scattering (Fig. 101). The incident light is propagated along the y axis; the scattered light is observed along the z axis. Thus, we have the following wave vectors: for the incident wave $\mathbf{k}(0, k, 0)$, for the scattered wave $\mathbf{k}'(0,0,k')$, for the elastic wave $\mathbf{q} = \mathbf{k}' - \mathbf{k}$; $\mathbf{q}(0, -k, k')$. $q^2 = k^2 + (k')^2$,

$$\varkappa_x = 0, \quad \varkappa_y = -\frac{n'_j}{\sqrt{(n'_j)^2 + (n'_i)^2}},$$

$$\varkappa_z = \frac{n'_i}{\sqrt{(n'_j)^2 + (n'_i)^2}}.$$

Since the indices of refraction n_X and n_Z for quartz differ only slightly from one another (thus, for example, for $\lambda = 4358$ Å, $n_X = 1.554$, $n_Z = 1.563$), then we can neglect the difference $n'_i - n'_j$ in the elastic part of the problem. The error in the final result will be negligible and the calculations reduced by a factor of four, because, instead of twelve ultrasonic waves, we now consider only three. Thus, for each four elastic waves that are very similar in direction of propagation and in polarization (and, consequently, also very close in frequency), we substitute a single wave.

Taking the above into account, we obtain for our case

$$\varkappa_x = 0, \quad \varkappa_y = -\frac{\sqrt{2}}{2}, \quad \varkappa_z = \frac{\sqrt{2}}{2}, \quad q^2 = 2k^2;$$

$$Q_{xx} = \frac{1}{2}\left(\frac{c_{11} - c_{12}}{2} + c_{44} - 2c_{14}\right), \quad Q_{yy} = \frac{1}{2}(c_{11} + c_{44}) + c_{14},$$

$$Q_{zz} = \frac{1}{2}(c_{44} + c_{33}), \qquad Q_{yz} = -\frac{1}{2}(c_{14} + c_{13} + c_{44}).$$

$$\left.\begin{aligned} (Q_{xx} - X)\gamma_x &= 0, \\ (Q_{yy} - X)\gamma_y + Q_{yz}\gamma_z &= 0, \\ Q_{yz}\gamma_y + (Q_{zz} - X)\gamma_z &= 0. \end{aligned}\right\}$$

The set of equations for the determination of the frequency and polarization of the elastic waves

We have three solutions:

$$X_{\text{I}} = Q_{xx}, \quad \gamma_x = 1, \quad \gamma_y = 0, \quad \gamma_z = 0 \text{ (transverse wave)}.$$

$$\left.\begin{aligned} X_{\text{II}} &= \frac{1}{2}(Q_{yy} + Q_{zz}) - \frac{1}{2}\sqrt{(Q_{yy} - Q_{zz})^2 + 4Q_{yz}^2}, \\ \gamma_x &= 0, \quad \frac{\gamma_y}{\gamma_z} = -\frac{Q_{yz}}{Q_{yy} - X_{\text{II}}} \end{aligned}\right\} \text{(quasi-transverse wave)}.$$

$$\left.\begin{aligned} X_{\text{III}} &= \frac{1}{2}(Q_{yy} + Q_{zz}) + \frac{1}{2}\sqrt{(Q_{yy} - Q_{zz})^2 + 4Q_{yz}^2}, \\ \gamma_x &= 0, \quad \frac{\gamma_y}{\gamma_z} = -\frac{Q_{yz}}{Q_{yy} - X_{\text{III}}} \end{aligned}\right\} \text{(quasi-longitudinal wave)}.$$

Table 38. Fine Structure of the Line of Light Scattered by the Quartz in the Case in Which the Light Is Incident Along the y Axis and Is Observed Along the z Axis

Frequency shift \ Scattering coefficient	R_{xx}	R_{xy}	R_{zx}	R_{zy}
$\Omega_{I}/\omega = 2.52 \cdot 10^{-5}$ (transverse wave)	0	11.6	11.2	0
$\Omega_{II}/\omega = 2.86 \cdot 10^{-5}$ (quasi-transverse wave)	5.7	0	0	7.0
$\Omega_{III}/\omega = 5.14 \cdot 10^{-5}$ (quasi-longitudinal wave)	618	0	0	63

For quartz, this gives

$$X_{I} = 31.3 \cdot 10^{10}\ \text{dyne/cm}^2, \quad \gamma_x = 1,\ \gamma_y = 0, \quad \gamma_z = 0;$$
$$X_{II} = 40.45 \cdot 10^{10}\ \text{dyne/cm}^2, \quad \gamma_x = 0,\ \gamma_y = 0.680,\ \gamma_z = 0.733;$$
$$X_{III} = 128.95 \cdot 10^{10}\text{dyne/cm}^2, \quad \gamma_x = 0,\ \gamma_y = 0.733,\ \gamma_z = -0.680.$$

The corresponding frequency of the elastic wave is determined from the relation $\Omega/\omega = \sqrt{2X/\rho}\ (n/c)$.

The calculations according to the scheme given in Sec. 9 give the distribution of the scattering coefficient of the light with respect to frequency and to polarization, and are shown in Table 38.

In the calculation of the frequencies of the elastic waves, the mean value of the index of refraction of quartz n = 1.56 was used and the density of quartz $\rho = 2.65\ \text{g/cm}^3$.

For natural incident light R^x (here and below, the superscript denotes the axis perpendicular to the plane of scattering) is equal to

$$R^x = \frac{\pi^2}{2\lambda^4} n^8 kT \cdot 716 \cdot 10^{-16}\ \text{cm}^{-1}. \qquad (31.4)$$

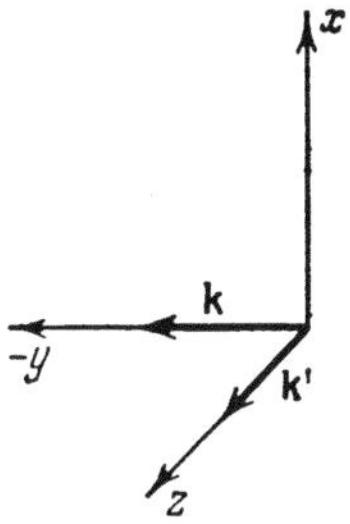

Fig. 102. Calculation of the coefficient of light scattering in quartz. The x axis is perpendicular to plane of scattering. Incident light is directed along −y axis; scattered light is observed along the z axis.

The depolarization of the scattered light is

$$\Delta_{u1}=0.13.$$

For quartz, as for any piezoelectric, the directions +x and −x are not equivalent. Therefore, if for this same direction in space for the incident and scattered light we take not right-handed but left-handed quartz, then this will correspond to the case in which the light is propagated along the negative direction, i.e., along −y, while it is scattered along the positive direction of the z axis. The positive direction of the x axis is perpendicular to the plane of scattering. These two cases are equivalent to the two directions of observation of the scattered light in quartz (both right- and left-handed), making an angle of ±90° with the direction of the incident light (observation to the left and to the right of the incident beam).

Case I'. The x axis is perpendicular to the plane of scattering (Fig. 102). The light is incident along −y; the scattered light is observed along z. In this case, we have

$$\boldsymbol{k}(0,\,-k,\,0),\quad \boldsymbol{k}'(0,\,0,\,k),\quad \boldsymbol{q}=\boldsymbol{k}'-\boldsymbol{k},\quad \boldsymbol{q}(0,\,k,\,k),$$

$$\varkappa_x=0,\quad \varkappa_y=\frac{\sqrt{2}}{2},\quad \varkappa_z=\frac{\sqrt{2}}{2};$$

$$Q_{xx}=\frac{1}{2}\left(\frac{c_{11}-c_{12}}{2}+c_{44}+2c_{14}\right),$$

$$Q_{yy}=\frac{1}{2}(c_{11}+c_{44})-c_{14},$$

$$Q_{zz}=\frac{1}{2}(c_{44}+c_{33}),$$

$$Q_{yz}=\frac{1}{2}(-c_{14}+c_{13}+c_{44}),$$

$$Q_{zx}=Q_{xy}=0.$$

Table 39. The Fine Structure of the Line of Light Scattered by Quartz in the Case in Which the Light Is Incident Along the Negative Direction of the y Axis and Is Observed Along the Positive Direction of the z Axis

Frequency shift \ Scattering coefficient	R_{xx}	R_{xy}	R_{zx}	R_{zy}
$\Omega_{I}/\omega = 3.64 \cdot 10^{-5}$ (transverse wave)	0	56	94.0	0
$\Omega_{II}/\omega = 2.75 \cdot 10^{-5}$ (quasi-transverse wave)	58	0	0	2.5
$\Omega_{III}/\omega = 4.48 \cdot 10^{-5}$ quasi-longitudinal wave)	492	0	0	26

The solutions of the corresponding elastic problem will be

$$\left.\begin{array}{l} X_{I} = 64.9 \cdot 10^{10} \text{ dyne/cm}^2, \\ \gamma_x = 1, \ \gamma_y = 0, \ \gamma_z = 0 \end{array}\right\} \text{(transverse wave)},$$

$$\left.\begin{array}{l} X_{II} = 37.4 \cdot 10^{10} \text{ dyne/cm}^2, \\ \gamma_x = 0, \ \gamma_y = -0.848, \ \gamma_z = 0.528 \end{array}\right\} \text{(quasi-transverse wave)},$$

$$\left.\begin{array}{l} X_{III} = 98.4 \cdot 10^{10} \text{ dyne/cm}^2, \\ \gamma_x = 0, \ \gamma_y = 0.528, \ \gamma_z = 0.848 \end{array}\right\} \text{(quasi-longitudinal wave)}.$$

The frequency and polarization distribution of the light scattering coefficient for quartz is given in Table 39.

For natural incident light:

$$R^{x'} = \frac{\pi^2}{2\lambda^4} n^8 kT \cdot 728 \cdot 10^{-16} \text{ cm}^{-1}. \qquad (31.5)$$

The depolarization of the scattered light is equal to $\Delta_u' = 0.13$. As is seen, the total intensity and depolarization differ little in these two cases, while the frequency and polarization distributions of the intensity differ significantly. The reason for such a difference is that the elastic waves which are propagated along the two bisectors of the angles between the y and z axes in quartz have a different frequency and a different polarization for the same value of the wave vector **q** (because of the polarity of the x axis).

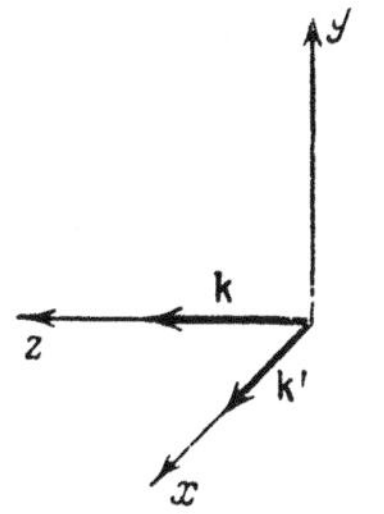

Fig. 103. Calculation of light scattering coefficient in quartz. The y axis is perpendicular to the plane of scattering. Incident light is directed along the z axis; scattered light is observed along the x axis.

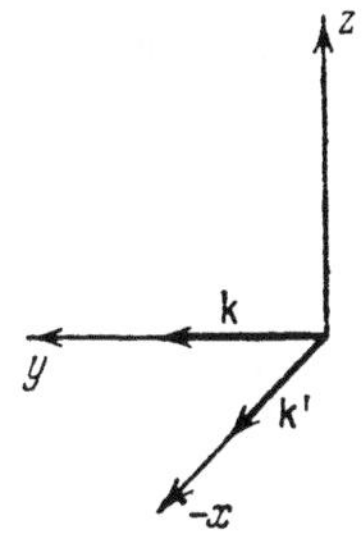

Fig. 104. Calculation of light scattering coefficient in quartz. The z axis is perpendicular to the plane of scattering. Incident light is directed along the y axis; scattered light is observed along −x.

Case II. The y axis is perpendicular to the plane of scattering (Fig. 103). Light is incident along the z axis; the scattered light is observed along the x axis. In this case, we have

$$\boldsymbol{k}(0,\ 0,\ k),\quad \boldsymbol{k}'(k,\ 0,\ 0),\quad \boldsymbol{q}=\boldsymbol{k}'-\boldsymbol{k},\quad \boldsymbol{q}(k,\ 0,-k),\quad q^2=2k^2,$$

$$\varkappa_x=\frac{\sqrt{2}}{2},\quad \varkappa_y=0,\ \varkappa_z=-\frac{\sqrt{2}}{2},$$

$$Q_{xx}=\frac{1}{2}(c_{11}+c_{44}),\ Q_{yy}=\frac{1}{2}\left(\frac{c_{11}-c_{12}}{2}+c_{44}\right),\ Q_{zz}=\frac{1}{2}(c_{44}+c_{33}),$$

$$Q_{yz}=\frac{1}{2}c_{14},\ Q_{zx}=-\frac{1}{2}(c_{13}+c_{44}),\ Q_{xy}=-c_{14}.$$

The corresponding characteristic equation has the form

$$X^3-20.07\cdot 10^{11}X^2+115.04\cdot 10^{22}X-199.4\cdot 10^{33}=0,$$

and the solution of the elastic problem will be

$$\left.\begin{aligned} X_{\mathrm{I}} &= 34.79\cdot 10^{10}\ \text{dyne/cm}^2,\\ \gamma_x &= 0.683,\quad \gamma_y=0.601,\ \gamma_z=0.415 \end{aligned}\right\}\ \text{(quasi-transverse wave)},$$

$$\left.\begin{aligned} X_{\mathrm{II}} &= 49.06\cdot 10^{10}\ \text{dyne/cm}^2,\\ \gamma_x &= 0.323,\ \gamma_y=-0.763,\ \gamma_z=0.559 \end{aligned}\right\}\ \text{(quasi-transverse wave)},$$

$$\left.\begin{aligned} X_{\mathrm{III}} &= 116.85\cdot 10^{10}\ \text{dyne/cm}^2\\ \gamma_x &= 0.653,\ \gamma_y=0.247,\ \gamma_z=0.716 \end{aligned}\right\}\ \text{(quasi-longitudinal wave)}.$$

Table 40. Fine Structure of the Line of Light Scattered by Quartz in the Case in Which the Light Is Incident Along the z Axis and Is Observed Along the x Axis

Frequency shift \ Scattering coefficient	R_{yy}	R_{yz}	R_{xy}	R_{xz}
$\Omega_{I}/\omega = 2.66\cdot 10^{-5}$ (quasi-transverse wave)	30.0	2.2	9.6	0.7
$\Omega_{II}/\omega = 3.17\cdot 10^{-5}$ (quasi-transverse wave)	17.7	44.2	13.1	2.6
$\Omega_{III}/\omega = 4.88\cdot 10^{-5}$ (quasi-longitudinal wave)	539	4.2	12.3	46.4

The distribution of the scattering coefficient is shown in Table 40. The scattering coefficient for natural incident light is

$$R^{y} = \frac{\pi^2}{2\lambda^4} n^8 kT \cdot 722 \cdot 10^{-16} \text{ cm}^{-1}. \tag{31.6}$$

The depolarization of the scattered light is equal to

$$\Delta_u = 0.17.$$

It is not difficult to prove that left-handed quartz gives the same result as right-handed quartz in this case (i.e., the elastic waves which are propagated along the two bisectors of the angles between the x and z axes have different frequencies but the same polarization).

Case III. The z axis is perpendicular to the plane of scattering (Fig. 104). The light is incident along the y axis, the scattered light is observed along the negative direction of the x axis, i.e., along −x). In this case, we have

$$\boldsymbol{k}(0,\ k,\ 0),\quad \boldsymbol{k}'(-k,\ 0,\ 0),$$

$$\boldsymbol{q} = \boldsymbol{k}' - \boldsymbol{k},\quad \boldsymbol{q}(-k, -k,\ 0),\quad q^2 = 2k^2,$$

$$\varkappa_x = -\frac{\sqrt{2}}{2},\quad \varkappa_y = -\frac{\sqrt{2}}{2},\quad \varkappa_z = 0,$$

Table 41. Fine Structure of the Line of Light Scattered by Quartz in the Case in Which the Light Is Incident Along the y Axis and Is Observed Along the Negative Direction of the x Axis

Frequency shift \ Scattering coefficient	R_{zz}	R_{zy}	R_{xz}	R_{xy}
$\Omega_{\rm I}/\omega = 2.57 \cdot 10^{-5}$ (quasi-transverse wave)	23.1	5.8	10.5	0.4
$\Omega_{\rm II}/\omega = 3.46 \cdot 10^{-5}$ (quasi-transverse wave)	144.0	59.3	9.8	0.2
$\Omega_{\rm III}/\omega = 4.29 \cdot 10^{-5}$ (quasi-longitudinal wave)	630.0	5.0	35.7	39.2

$$Q_{xx} = \frac{1}{4}(3c_{11} - c_{12}),\ Q_{yy} = Q_{xx},\ Q_{zz} = c_{44},$$

$$Q_{yz} = 0,\ \ Q_{zx} = c_{14},\ \ Q_{xy} = \frac{1}{4}(c_{11} + c_{12}).$$

The characteristic equation has the form

$$X^3 - 18.15 \cdot 10^{11} X^2 + 101.5 \cdot 10^{22} X - 172.6 \cdot 10^{33} = 0.$$

The corresponding solutions are

$$\left.\begin{aligned} X_{\rm I} &= 32.55 \cdot 10^{10}\ \text{dyne/cm}^2, \\ \gamma_x &= -0.693,\ \gamma_y = 0.543,\ \gamma_z = 0.474 \end{aligned}\right\}\ \text{(quasi-transverse wave)},$$

$$\left.\begin{aligned} X_{\rm II} &= 58.87 \cdot 10^{10}\ \text{dyne/cm}^2, \\ \gamma_x &= 0.085,\ \gamma_y = -0.592,\ \gamma_z = 0.806 \end{aligned}\right\}\ \text{(quasi-transverse wave)},$$

$$\left.\begin{aligned} X_{\rm III} &= 90.07 \cdot 10^{10}\ \text{dyne/cm}^2, \\ \gamma_x &= 0.716,\ \gamma_y = 0.596,\ \ \gamma_z = 0.364 \end{aligned}\right\}\ \text{(quasi-longitudinal wave)}.$$

The distribution of the scattering coefficient is shown in Table 41. The total scattering coefficient of the light for natural incident light is

$$R^z = \frac{\pi^2}{2\lambda^4} n^8 kT \cdot 963 \cdot 10^{-16}. \tag{31.7}$$

The depolarization of the scattered light is

$$\Delta_u = 0.13.$$

Left-handed quartz gives the same result as right-handed in this case.

Absolute and Relative Intensity. The experimental investigation of the molecular scattering of light in crystals was devoted principally to the determination of the absolute value of the intensity of the scattered light, the depolarization of the scattered light, the effect of the anisotropy of the scattered light, and, finally, the fine structure of the Rayleigh scattering line.

The first determination of the absolute value of the intensity of the scattered light in a solid was made by Landsberg and Vul'fson [549] by measurement of the ratio of the scattering coefficient in quartz to the scattering coefficient in gaseous carbon dioxide (CO_2).

As a result of this extraordinarily difficult experiment, it was determined that

$$\frac{R_{Qu}}{R_{(CO_2)}} = 3.68. \tag{31.8}$$

From Tables 6 and 7, it follows that $R_{CO_2} = 4.74 \cdot 10^{-8}\ \text{cm}^{-1}$ for $\lambda = 4358$ Å.

In the experiment of [549], the orientation of the quartz was not accurately known; therefore, in the comparison with the theory, it is reasonable to use the mean value of the scattering coefficient calculated from Eqs. (31.5)-(31.7). This mean value is equal to

$$\overline{R} = \frac{\pi^2}{2\lambda^4} n^8 kT \cdot 804 \cdot 10^{-16}\ \text{cm}^{-1} \tag{31.9}$$

for $\lambda = 4358$ Å, $\overline{R}_{Qu} = 1.55 \cdot 10^{-7}\ \text{cm}^{-7}$.

The coefficient of molecular scattering in quartz was determined from the temperature variation of the intensity. Here the light of Raman scattering also depends on the temperature; this light, from the estimate of Motulevich [20], amounts on the average to 30% of the total scattering, which depends linearly on the temperature.

Therefore, for scattering by statistical fluctuations of the density, the relation (31.8) should be corrected; the result obtained is the following:

$$\frac{R_{Qu}}{R_{CO_2}} = 2.58.$$

It then follows that the experimentally found value $R_{Qu} = 1.2 \cdot 10^{-7}$ cm^{-1} is in good agreement with theory (31.9).

Recently, the absolute value of the scattering coefficient has been found in quartz and NaCl in the works of Tourel and Chapelle [557-559] by comparison with the scattering in benzene. For the quartz crystal, when R_{zz} was measured (the optic axis is perpendicular to the plane of scattering, the incident and scattered light are polarized along the z axis), we obtain

$$\left(\frac{R_{zz}}{R_{C_6H_6}}\right) = 7.1 \cdot 10^{-3}. \qquad (31.10)$$

For comparison of the obtained result with the theoretical calculation, it is also necessary here to take into account the contribution of Raman scattering. For the given orientation of the quartz, the fraction of Raman scattering, according to the calculation of [20], amounts to 26% of the total molecular scattering. Therefore, taking Raman scattering into account in (31.10), we obtain

$$\left(\frac{R_{zz}}{R_{C_6H_6}}\right)_{\exp} = 5.6 \cdot 10^{-3}. \qquad (31.11)$$

The scattering coefficient in benzene (see Sec. 18) is equal to $R_{C_6H_6} = 32 \cdot 10^{-6}$ cm^{-1}, then the experimental value $R_{zz} = 1.79 \cdot 10^{-7}$ cm^{-1}, while the theoretical value of this quantity is $R_{zz} = 1.85 \cdot 10^{-7}$ cm^{-1}. Here the experiment is in very good agreement with theory.

For sodium chloride, the scattering coefficient was measured in [557] relative to benzene. For orientation of the sodium chloride corresponding to Case I (Table 35), the following ratio is obtained:

$$\left(\frac{R_{zz}}{R_{C_6H_6}}\right)_{\exp} = 0.43 \cdot 10^{-2}.$$

In sodium chloride, the lines of Raman scattering of first order are missing; therefore, the corresponding correction should not be applied.

Again taking $R_{C_6H_6} = 32 \cdot 10^{-6}$ cm^{-1}, we get

$$R_{zz} = 1.38 \cdot 10^{-7} \text{ cm}^{-1}.$$

The theoretical value (see Table 35) is:

$$R_{zz} = 1.44 \cdot 10^{-7}\ \text{cm}^{-1}.$$

Even in this case, the experiment agrees well with theory. The relations of the intensity of light scattering in sodium chloride to the intensity of light scattered in quartz were obtained by Landsberg and Mandel'shtam [550]. The ratio obtained by them was

$$\frac{R_{NaCl}}{R_{Qu}} = 0.69.$$

This same experiment was repeated by Motulevich [20] for quartz so oriented that the x axis was perpendicular to the plane of scattering. The ratio obtained by him was

$$\frac{R_{NaCl}}{R_{Qu}} = 0.65.$$

If we apply the correction for Raman scattering to quartz, then the experimental value will be equal to

$$\left(\frac{R_{NaCl}}{R^*_{Qu}}\right)_{exp} = 1.00.$$

The theory for the orientation theory chosen gives the value

$$\left(\frac{R_{NaCl}}{R^*_{Qu}}\right)_{theor} = 1.07.$$

The agreement of experiment with theory here also is good.

Anisotropy of the Molecular Scattering of Light. Calculation of the intensity of the scattered light for different orientations of the crystal [Eqs. (31.1), (31.2), and (31.4)-(31.7)] shows that the intensity of the scattered light should depend on the orientation of the crystal. This effect was first discovered experimentally in crystals of quartz by Vul'fson and Lombert [21] and studied in detail by Motulevich [20]. The experiment of [20] gave the following values for the two samples (after corrections for Raman light scattering):

$$R^z : R^y : R^x = 1.26 : 1 : 0.98.$$

The theory [Eqs. (31.4)-(31.7)] gives the following ratios:

$$R^z : R^y : R^x = 1.315 : 1 : 1.$$

In this case also, the agreement of theory with experiment is completely satisfactory.

Polarization of Scattered Light. The polarization of scattered light was studied experimentally in detail by Chapelle and Tourel [558] in a quartz crystal. The crystal was so oriented that the optic axis was perpendicular to the plane of scattering. Only for such an orientation is the effect of the rotation of the plane of polarization in the quartz eliminated.

The results of the measurement are given in Table 42, where the value of R_{ZZ} is taken to be 100. In this same table are listed the values taken from Table 41. The agreement of experiment and theory in this case is satisfactory. The measured depolarization factor in this case is equal to

$$\Delta_u = \frac{R_{xy} + R_{zy}}{R_{zz} + R_{xz}} = 0.125.$$

The theoretical value of this quantity is equal to 0.13, which agrees well with experiment.

Imperfections of the Crystalline Structure. In this paragraph, we shall give the basic experimental results of the investigation of the molecular scattering of light and their comparison with the theoretical calculation.

The theory with which such a comparison is carried out was set forth in Sec. 9, and takes into account only the scattering of light by thermal fluctuations.

As was noted in the beginning of the section, the scattering of light takes place at dislocations and other microinhomogeneities. The character of such scattering differs from the character of molecular scattering from thermal fluctuations in a number of ways.

In particular, it was shown that repeated heating and cooling of a crystal specimen with a large number of microdefects is ini-

Table 42. Coefficient of Scattering in Quartz for Light of Different Polarizations

Scattering coefficient	Experimental value	Theoretical value
R_{zz}	100	100
R_{zy}	7.3	8.8
R_{xz}	6.8	7.0
R_{xy}	6.0	5.0

tially irreversible. Only after several series of heatings does the intensity become reversible; at room temperature the reversible value of the intensity is much smaller than it was originally at the same temperature and is proportional to λ^{-4} [563]. Several days after the reversing is stopped, the intensity returns to its original value. This phenomenon was studied on several crystals and in a number of researches [560-563].

As a result of the experimental investigation of the light scattered in crystals, Humphreys-Owen [555, 556] showed that in his samples of crystals of sodium chloride and quartz, along with the molecular scattering of light produced by the thermal fluctuations, there is a part of the light scattering brought about by the microimperfections of the crystals. In the study of the dependence of the intensity of the total scattering on λ^{-4} over a wide range of wavelengths, a curve is obtained with a maximum, while for molecular scattering by thermal fluctuations, such a dependence should be a straight line, passing through the origin.

The dimensions of the scattering imperfections of the crystals can be found from the experimental data [555, 556] and theoretical considerations [553, 554, 564]. They amount in the different cases to 1500-3000 Å. Theimer and Plint [564] developed a theory of scattering by an almost ideal crystal and, using experimental data, a part of which was recalled above, drew conclusions on the shape and dimensions of the imperfections in the crystals.

Investigation of the Fine Structure of the Rayleigh Line in Crystals. In different places in this book (see, for example, the Introduction), it has already been noted that the first fine structure of the Rayleigh line was predicted by Mandel'shtam [24] and Brillouin [26], and was discovered in the spectrum of light scattered by crystalline quartz [6, 27].

In the first investigations, agreement was established between the general observed characteristics of the phenomenon and the conclusions of the theory.

To date, the fine structure of the Rayleigh line has been found in several crystals. Here there is no possibility of considering all the interesting investigations carried out in this area; therefore, we limit ourselves to putting down only the fundamental results of the most thorough investigations, carried out by Krishnan [565], Krishnan and Chandrasekharan [223], and Chandrasekharan [224] in the study of the fine structure of the Rayleigh line in diamond and quartz.

One sample of diamond of size $10 \times 6 \times 1.3$ mm with a finely polished front face and lateral surface was studied by Krishnan and Chandrasekharan. The length, width, and thickness of the crystal are parallel to the [011], [111], and [211] directions, respectively.* The exposure in the different cases lasted from two to five days. The central line (exciting light) was absorbed by a resonance filter (see Sec. 12).

The reproduction of the photographs for several orientations of the crystal is given in Fig. 105c, d, e.

Here one can see that the internal MB(T) transverse components of MB in diamond are more intense than the longitudinal MB(L) components.

This same conclusion follows from the theory (see Table 37). Table 43 gives the results of the measurement of the splitting of the fine-structure components for various orientations of the diamond crystal. We call attention to the fact that the calculated values of $\Delta\nu$ for the longitudinal components are somewhat larger than the observed values, while, for the transverse components, they are somewhat less than the measured values. Taking into ac-

* The photography of the spectra of the fine structure was performed on a quartz spectrograph (Hilger) with a focal distance of the camera objective of 3 m, resolving power ~300,000, and dispersion 14.1 cm^{-1}/mm in the region of $\lambda = 2534.8$ Å. This line also served for excitation of the scattering although the crystal was of slightly yellowish hue.

Table 43. Splitting of the Fine-Structure Components in a Crystal of Diamond and the Relative Intensity of the Transverse and Longitudinal Components for Different Orientations of the Crystal

No.	Direction of the incident light	Direction of the scattered light	Angle of scattering, $\theta°$	Nature of components	Shift of components: Chandrasekharan	Shift of components: Krishnan	Calculated $\Delta\nu$, cm^{-1}	Ratio of the intensity of the transverse and longitudinal components	Ratio of the intensity of the 1332 cm^{-1} line and the sum of the MB components
1	[211]	[111]	90	Longitudinal	8.56	—	8.66	2.3	12
	[211]	[111]	90	Transverse	5.63	—	5.39 4.38		
2	[211]	[111]	90	Longitudinal	8.08	8.2	8.11	3.3	8
	[211]	[111]	90	Transverse	5.88	6.7	5.36 5.29		
3	[211]	[011]	90	Longitudinal	8.56	9.0	8.68	1.8	7
	[211]	[011]	90	Transverse	5.58	5.5	5.02 4.78		
4	[111]	[011]	90	Longitudinal	—	8.7	8.43	—	—
	[111]	[011]	90	Transverse	5.59	5.6	5.12	—	—
5	[111]	[111]	180	Longitudinal	—	—	12.47	—	—
	[111]	[111]	180	Transverse	7.0	—	6.69	—	—

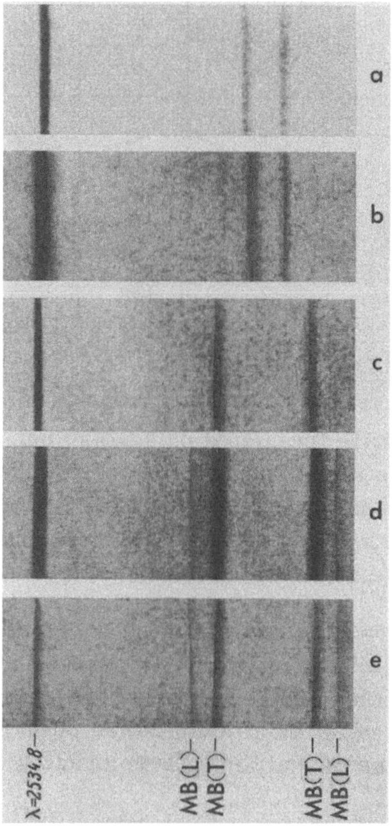

Fig. 105. Reproduction of the photographs of the fine structure of the Rayleigh line in crystals of diamond and quartz (Chandrasekharan [224]). (a) Fine structure in quartz ($\theta = 180°$); (b) components of the fine structure of quartz and diamond, taken on one plate ($\theta = 90°$); (c, d, e) fine structure in a diamond crystal for different positions of it ($\theta = 90°$). MB(L) – components produced by the longitudinal hypersonic wave; MB(T) – components produced by a transverse hypersonic wave.

Table 44. Splitting of the Fine-Structure Components in a Quartz Crystal

	Exciting light directed along the axis	Scattered light directed along the axis	Scattering angle, θ°	Measured Δν, cm⁻¹	Calculated Δν, cm⁻¹
1	$-y$	z	90	1.90	1.80
2	y	z	90	2.07	2.07
3	x	z	90	1.96	1.97
4	x	y	90	1.75	1.73
5	x	x	180	2.36	2.38
6	y	y	180	2.22	2.49
7	z	z	180	2.51	2.64

count that the crystal is irradiated by a broad beam of light, which led to splitting of the fine-structure components, and also to appreciable error in the determination of the elastic and elasto-optical constants, the agreement between the observed and calculated values of $\Delta\nu$ must be regarded as satisfactory.

Figure 106 shows the change in the intensity of the fine-structure components as a function of the temperature. It is seen from this figure that the lines are very intense for a temperature of 570°K and are quite weak for a temperature of 85°K (liquid air), although the line of Raman scattering here practically does not change its intensity.

From the results given for $\Delta\nu$ in Table 43, one can conclude [see (12.2)] that the velocity of longitudinal sound in a diamond reaches ~ 18,000 m/sec, while the velocity of transverse sound ~11,000 m/sec.

In the study of the fine structure of the spectrum of the light scattered by quartz [223], the crystal was chosen in the shape of a parallelepiped of dimensions 6.5 × 4.5 × 3.5 cm, with crystallographic axes perpendicular to the boundaries of the crystal. The arrangement is the same as in [224], which was used for the study of the fine structure in diamond.

One pair of MB components was observed in the scattered

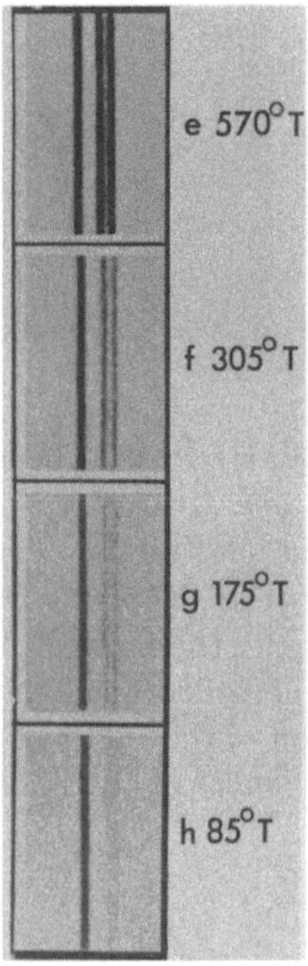

Fig. 106. Temperature dependence of the intensity of the fine-structure components of the Rayleigh line in diamond (Krishnan and Chandrasekharan [223]).

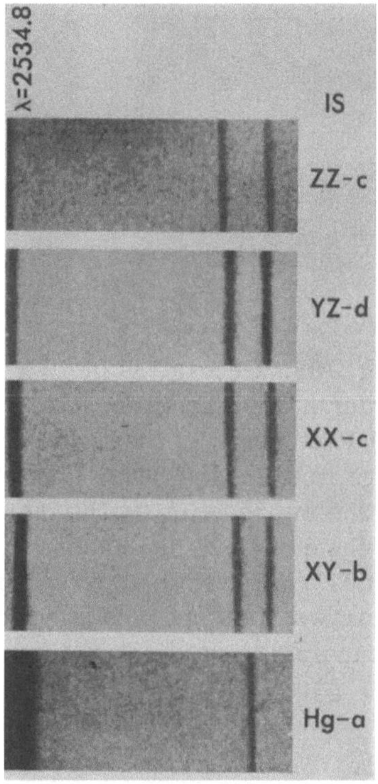

Fig. 107. Fine structure of the Rayleigh line in a quartz crystal for different orientations (Krishnan and Chandrasekharan [223]).

light, due to the longitudinal hypersonic wave. The results of the measurement of the frequencies are given in Table 44. The last two columns of the table show the excellent agreement between the measured and computed value of $\Delta\nu$. Figure 107 shows the fine structure of a quartz crystal and the directions of observation of the scattered light. The problem of the sound in the quartz crystal and their corresponding fine structure was studied in detail by Vul'fson [566].

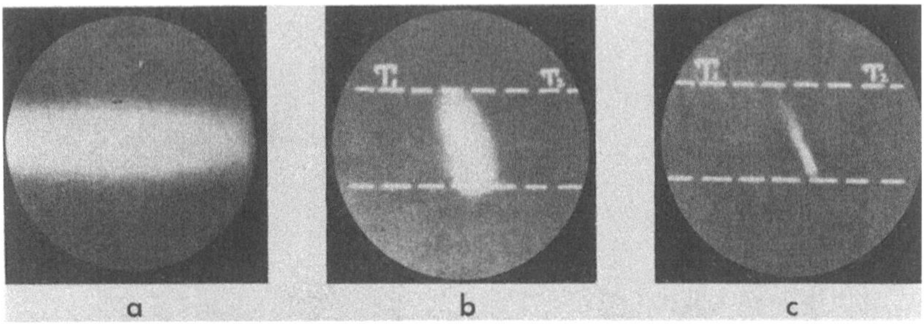

Fig. 108. Generation of "fog" in crystalline quartz in the second-order phase transition (Yakovlev, Velichkina, and Mikheeva [22]). (a) Photograph of the beam of incident light outside the temperature range of the phase transition (1-hr exposure); (b) photograph of the intersection by the incident light beam with the band of "fog" in the quartz (1-sec exposure, horizontal temperature gradient $\approx 0.03°$/mm); (c) photograph of the intersection of the incident light beam with the band of "fog" in the quartz (1-sec exposure, horizontal temperature gradient $\approx 1°$/mm).

32. Molecular Scattering of Light in a Quartz Crystal in a Second-Order Phase Transition

In the exposition of the theory of molecular light scattering in second-order phase transitions in Sec. 2 it was pointed out that a phenomenon reminiscent of critical opalescence was discovered in crystalline quartz by Yakovlev, Velichkina, and Mikheeva [22].

In this section, we shall describe the principal features of this phenomenon and its quantitative characteristics, which were studied in [567-569]. In the explanation, use is made of the theory of light scattering in second-order phase transitions, developed by Ginzburg [23, 105, 106] (see Sec. 2).

The phenomenon of which we shall be speaking could have been discovered almost immediately after the discovery of molecular light scattering in crystals, i.e., almost forty years ago. But, as frequently happens, it was not observed. The theory of light scattering in second-order phase transitions developed by Ginzburg [23, 105, 106] predicted the phenomenon before it was discovered experimentally.

Independently of the theory, searches for the phenomenon were undertaken in all physical considerations [22, 567, 568] and only in the most interesting final stage of the investigation could the discovered phenomenon and the theory be compared. The phenomenon described above occurs in quartz in a narrow temperature range of about 0.1°C and, therefore, special attention was paid to the thermal side of the experiment, the details of which were given in Sec. 11.

A quartz crystal cut in a definite shape, with dimensions 20 × 20 × 40 mm, was placed in a special oven (see Fig. 21), in which the temperature could be changed slowly and smoothly* up to temperatures above the $\alpha \rightleftharpoons \beta$ transition (in crystalline quartz, this temperature is about 573°C) and in which cooling of the crystal could also be carried out.

All efforts were directed to the end that the temperature gradient be as small as possible inside the crystal (see Sec. 11). But it was not possible to avoid gradients everywhere, considering that there were three windows in the oven for entrance and exit of the light and for observation of the scattered light at an angle of 90° to the direction of the incident beam.

The structure of the oven was such that an isothermal layer of crystal was placed almost perpendicular to the direction of the incident light. Therefore, in the gradual heating, the temperature, corresponding to the transition, "entered" the crystal at one end and, moving through the crystal, "went out" through the opposite end. In the cooling, the whole picture could be observed in reverse order. The width of the transition region was larger for a small horizontal gradient and less for a large gradient. The phenomenon was studied on three samples of quartz belonging to two deposits. The effect was observed for different crystallographic directions for different states of polarization of the incident and scattered light.

The scattered light could be observed visually or recorded by a photoelectric photometer, or photographed. In the visual observation, one could see how a band of strong light scattering ap-

---

* The entire cycle of heating and cooling lasted 72 hr. The rate of change of temperature near the temperature transitions was 0.3°/hr.

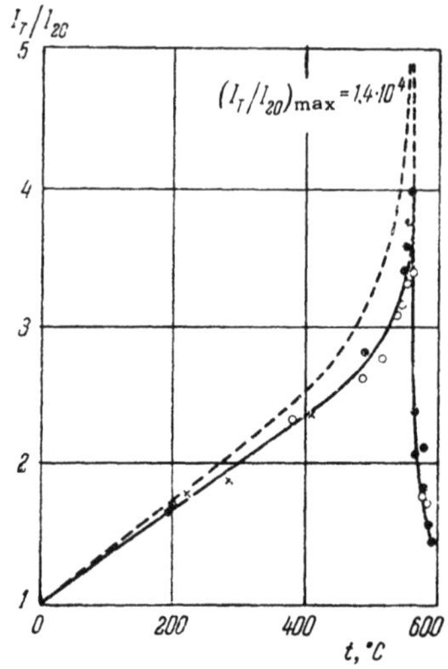

Fig. 109. Temperature dependence of the scattered light intensity on the temperature in a quartz crystal (Yakovlev, Velichkina, and Mikheeva [22]). ● – Heating; ○ – cooling; × – repeated experiment; dashed line – theoretical curve (for the range of temperatures 573-600°C, it is entirely the same as the experimental curve).

peared during the approach of the transition temperature from the heated end of the crystal. This band moved to the opposite end. In cooling, the band of "fog" moved in reverse fashion. If desired, it could be stopped for a long interval by temperature stabilization.

Figures 108b and c show the bands of "fog" for different temperature gradients, while Fig. 108a gives the case for a uniformly scattering quartz crystal. The boundaries of the beam of primary light are shown by the dashed lines.

Photoelectric measurements for the same case as in Fig. 108 gave the results shown in Fig. 109.

The linear dependence of the intensity on temperature is maintained up to a temperature somewhat above 500°C agreeing with the measurements of Motulevich to within 4% [20].

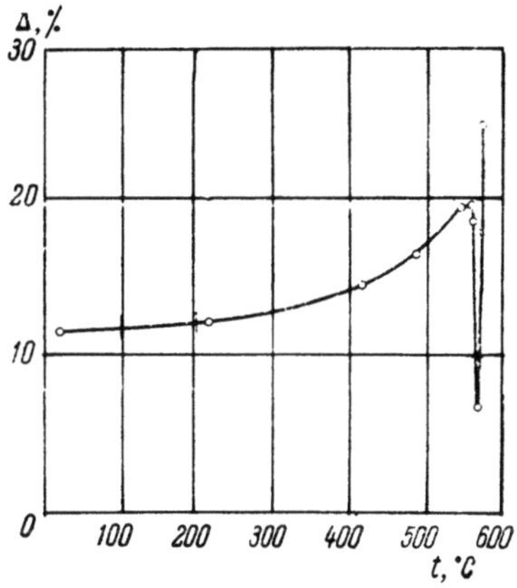

Fig. 110. Depolarization of the scattered light as a function of temperature (Yakovlev, Velichkina, and Mikheeva [22]).

For varying temperatures, the increase in the intensity becomes steeper, and a "fog" appears in a region ~ 0.1°C around the transition temperature, and the intensity becomes $1.4 \cdot 10^4$ times greater than at room temperature.

It was shown by a special experiment that Rayleigh's law ($I \sim \lambda^{-4}$) is satisfied with sufficient accuracy over the entire range of change in the intensity, including the transition region, and, therefore, one could draw the conclusion that the scattering in the "fog" takes place from objects smaller than the wavelength of the visible light.

The depolarization of the scattered light (Fig. 110) at the point of the phase change was shown to be equal to 0.06, while at room temperature it is 0.12. As is clear from Fig. 110, the decrease in the depolarization takes place in this same narrow temperature interval.

It has been established experimentally that Rayleigh's law is satisfied in the region of the phase transition; this means that in (2.16), $f^* \sim 0$; otherwise the dependence on the wavelength would have been different.

Therefore, R_1 in (2.16) can be written in the following fashion for $\theta = 90°$:

$$R_{1,90} = \frac{\pi^2}{2\lambda^4}\left(\frac{\partial \varepsilon}{\partial (\eta^2)}\right)^2_{\rho, T} \frac{kT}{(C^2 - 2AD)^{1/2}} \cdot \tag{32.1}$$

For a second-order phase transition, far from the critical Curie temperature, when $\eta_0^2 = -A/C$ [51], Eq. (32.1) now transforms into [105]:

$$R_{1,90} = \frac{\pi^2}{2\lambda^4}\left[\frac{\partial \varepsilon}{\partial (\eta^2)}\right]^2_{\rho, T} \frac{kT}{C} \cdot \tag{32.2}$$

Quantities entering into (32.1) and (32.2) can be determined experimentally if we take into account the connection of AC and DC with the specific heat and the value of η_0^2 near the transition temperature [51, 105], while the quantity $[\partial \varepsilon/\partial(\eta^2)]^2_{\rho,T}$ can be determined from the measurement of ε as a function of temperature up to the transition temperature. Starting out from this, one can make estimates of the value of (32.2), as was done in [105].

For the second-order transition, far from the critical Curie point [51], the jump in the specific heat is associated with the parameter (2.15) by the relation

$$\Delta C = \frac{T_c}{C}(A')^2, \tag{32.3}$$

where

$$A' = \left(\frac{\partial A}{\partial T}\right)_{T=T_c} = \frac{A}{T - T_c} \cdot \tag{32.4}$$

From (32.2) and (32.4) we find

$$\frac{1}{C} = \frac{T_c \eta_0^4}{\Delta C (T_c - T)^2}, \tag{32.5}$$

$$\left[\frac{\partial \varepsilon}{\partial (\eta^2)}\right]_{\rho, T} = \frac{\varepsilon_2 - \varepsilon_1}{\eta_0^2} = \frac{2n\Delta n}{\eta_0^2} \cdot$$

Here Δn is the difference in the indices of refraction at the transition point, where $\eta_0^2 = 0$, and at the point with the given value of η_0^2.

Substituting the given values in (32.2), we get

$$R_{1,90}=\frac{\pi^2}{2\lambda^4}\,kT\,\frac{4n^2(\Delta n)^2\,T_c}{\Delta C\,(T_c-T)^2}\,. \tag{32.6}$$

Inasmuch as we have neglected the anisotropy in the theory set forth in Sec. 2, we shall equate (32.6) with R_{90}, which corresponds to the scattering by density fluctuations [Sec. 1, Eq. (1.94) for $\Delta = 0$], then

$$\frac{R_{1,90}}{R_{90}}=\frac{4n^2\,(\Delta n)^2\left(\frac{T_c}{T}\right)}{\Delta C\,(T_c-T)^2\left(\rho\frac{\partial\varepsilon}{\partial\rho}\right)_T^2\beta_T}\,. \tag{32.7}$$

Using the actual numerical data that determine (32.7) and that were already used in [22, 105], namely $T_C = 846°K$, $n = 1.56$, $\Delta n = 1.2 \cdot 10^{-2}$ in the range $T_C - T = 0.1°$, $\Delta C = 0.4$ cal/deg = $4.2 \cdot 10^7$ erg/deg $\cdot$ cm^3, $\rho = 2.51$ g/cm^3, $(\rho\partial\varepsilon/\partial\rho) \sim 1.6$, and $\beta_T = 3.3 \cdot 10^{-12}$ cm^2/dyne, and substituting them in (32.7), we get

$$\frac{R_{1,90}}{R_{90}} \sim 1.1 \cdot 10^4,$$

which is in excellent agreement with the results of experiment.

Thus, one can draw the conclusion that the theory of light scattering developed by Ginzburg and his coworkers quantitatively describes the observed phenomenon in quartz.

However, for such a method of estimating the intensity, when the experimentally found values of Δn are substituted in Eq. (32.6), it is impossible to establish which scattering mechanism, $\Delta\varepsilon(\eta^2)$ or $\Delta\varepsilon(\rho, T)$ [see (2.12)] plays the principal role in increasing the intensity.

Further investigation of phase changes in a solid by the light scattering method were completed by Shustin [677], who found a strong increase in the intensity of the light scattering in the phase transition in a single crystal of ammonium chloride (NH_4Cl) at a temperature of $-30°C$.

The phase transition in ammonium chloride at a temperature of $-30°C$ is evidently a first-order phase transition, but with a small heat of transition.

From the set of experimental investigations of phase transitions in quartz and ammonium chloride, one can draw the conclusion that the increase in the light scattering is characteristic not only for phase transitions of the second kind, but it also characterizes phase transitions of the first kind, close to the critical Curie point [677].

The investigation of the spectrum of thermal scattering in a phase transition of the second kind in triglycine sulfate was carried out by Gammon and Cummins [678].

It was shown that $\Delta\nu$ of the MB components ($\Delta\nu$ is proportional to the hypersonic velocity v) in a narrow region in the vicinity of the Curie temperature T undergoes a characteristic change for the particular components of the longitudinal and transverse hypersonic waves. In the latter case, a large change in $\Delta\nu$ and a broadening of the component is observed.

These experimental results can be described by the theory of Landau and Khalatnikov [679], which, for the coupling of the sound velocity, frequency, and relaxation time [Eq. (27.14)] gives an expression for the relaxation time of the form $\tau = A/(T_c - T)$.*

Investigation of the spectrum of thermal scattering of light in the region of a phase transition allows us to find the velocity and absorption of hypersound simultaneously with great accuracy, as well as the dependence of these qualities on the temperature.

This same stimulated MB scattering is used for a study of phase transitions. See Sec. 34.

33. Light Scattering in a Nonuniformly Heated Body and the Damping of Hypersonic Waves

The intensity of the light scattered by adiabatic density fluctuations, as was described in Secs. 5, 7, 9, and in a number of other places in this book, is closely connected with the propagation and intensity of the thermal elastic waves.

---

*For reasons that are not yet clear, there is a divergence in order of magnitude between the determination of A from this experiment [678] and the experiment with ultrasound [680].

In Chapter VI, Secs. 24 and 25, we estimated the damping of the thermal elastic wave of hypersonic frequency in a liquid on the basis of direct measurements and measurements of the velocity dispersion of sound waves.

There is no basis for expecting dispersion of the velocity of sound in crystals up to frequencies $\sim 10^{10}$ Hz, and direct measurements in this range have only now become possible; therefore, such estimates of damping which have been used successfully for liquids have not yet been made for solids.

Techniques of measurement of absorption in crystals have now been developed for frequencies $\sim 10^{10}$ Hz by an acoustic method [436, 575, 576]. But even before these important successes in the technology of ultrasonics, Mandel'shtam [570] proposed in principle a new method for the measurement of the damping of sound in solids and liquids, which is of great scientific interest, and which has not lost its practical value even today.

The scattering of light by adiabatic density fluctuations or by thermal elastic waves in a solid, in which a constant temperature gradient has been established, will depend on the extent to which elastic waves, which determine the light scattering, are damped in this body. If we choose the dimensions of the crystal so that there is practically no damping, while, on the other hand, a large temperature gradient is set up in this crystal, then, in spite of the arbitrary temperature gradient, the intensity of the scattering from any place in such a crystal will be the same. The damping of the thermal waves can become appreciable upon increase in the dimensions of the crystal, and the scattering of the light from different volumes of such a crystal in the presence of a temperature gradient will no longer be the same. Therefore, Mandel'shtam [570] immediately noted that the effect predicted by him depends on the shape and dimensions of the crystal. The less the damping of the elastic wave, the more will the light intensity scattered at a given point of a nonuniformly heated crystal differ from the intensity of the scattering by a uniformly heated crystal at the same temperature. Consequently, the intensity of the scattering at a given point in a nonuniformly heated crystal depends not only on the temperature at this point, but also on the temperature distribution throughout the crystal. The experimental investigations of this phenomenon were undertaken by Landsberg and Shubin [571], while the quantitative theory of the effect has been given by Leontovich [572, 573].

In the exposition of the relevant quantitative considerations, we shall confine ourselves to the work of [573].

The frequency of the elastic waves, which are important for the scattering of light, as has been pointed out in detail in Sec. 5, lies within the limits from 0 at $\theta = 0°$ to $3 \cdot 10^{10}$ Hz at $\theta = 180°$, $n = 1.5$, $v = 5 \cdot 10^5$ cm/sec, and $\lambda = 5000$ Å. There is a very small fraction of internal energy of the crystal in this frequency range.

The problem of the energy distribution over the spectrum of elastic vibrations was solved by Leontovich [573] in the same way as the optical problem of the distribution of the light intensity in an irradiating and absorbing medium was solved.

The elastic thermal waves in a given range of frequencies, which are propagated in all directions, are considered as noncoherent beams of different directions. The exchange of energy between the irradiated and the very-high-frequency spectrum of the elastic waves is taken into account by the introduction of absorption and radiation.

We shall denote by $Ad\Omega do$ the flux of energy which the elastic waves transport, contained in the frequency interval from Ω to $\Omega + d\Omega$, through a unit area perpendicular to the direction of the flux $\mathbf{s}_1$ and lying inside a solid angle do. For waves of the considered frequency, the "radiation capacity" of the crystal is a function of the temperature at the given point, and does not depend on the spatial derivatives of T, which one would have to take into account for the very-high-frequency spectrum of elastic waves.

In the case of a stationary process, A is a function of the coordinates and the direction $\mathbf{s}_1$ and satisfies the equation

$$\frac{\partial A}{\partial S} = -2\alpha A + u, \tag{33.1}$$

where S is the length of the body in the direction $\mathbf{s}_1$.

In thermodynamic equilibrium, when the state of the crystal is the same at different points, $\partial A/\partial S = 0$, and $u/2\alpha = A$. Consequently, A is isotropic and is proportional to the energy of the elastic waves in the interval considered; this energy is in turn proportional to the absolute temperature: $u = 2\alpha CT$ (C is a constant). Substituting this value of u in (33.1), we get

$$\frac{\partial A}{\partial S} = -2\alpha (A - CT). \qquad (33.2)$$

The case was considered by Leontovich [573] in which the crystal is infinite in the x and y directions and finite in the z direction. A constant temperature gradient is directed along the z axis: $T = T(z) = T_0 + Gz$, where the temperature gradient is $G = (T_l - T_0)/l$. Here l is the thickness of the crystalline plate and T_l and T_0 are the absolute temperatures of its faces. By virtue of the symmetry of the problem, A is a function only of z and the angle ϑ (angle between the direction $\mathbf{s}_1$ and the z axis).

Then (33.2) is written

$$\cos\vartheta \frac{\partial A}{\partial z} + 2\alpha A = 2\alpha CT = 2\alpha C (T_0 + Gz). \qquad (33.3)$$

The general solution of (33.3) is the following:

$$A(z, \vartheta) = CT(z) - CG\frac{\cos\vartheta}{2\alpha} + \left[A(0, \vartheta) - CT_0 + CG\frac{\cos\vartheta}{2\alpha}\right] e^{-\frac{2\alpha z}{\cos\vartheta}}, \qquad (33.4)$$

$$0 < \vartheta < \frac{\pi}{2}$$

and

$$A(z, \vartheta) = CT(z) - CG\frac{\cos\vartheta}{2\alpha} + \left[A(l, \vartheta) - CT_l + CG\frac{\cos\vartheta}{2\alpha}\right] e^{-\frac{2\alpha(l-z)}{\cos\vartheta}}, \qquad (33.5)$$

$$\frac{\pi}{2} < \vartheta < \pi.$$

For the determination of the boundary values of A, which have an important effect on the next derivation, it is assumed that the crystal is surrounded by a medium which strongly absorbs the waves of the considered frequency range. In this medium, the intensity of the elastic waves of the studied range has an equilibrium value or, in other words, is determined by the temperature at the given point. The intensity of the elastic waves propagating inside the crystal at the boundaries will consist of the intensity of the reflected wave and the intensity of the wave entering the crystal from the surrounding medium.

By finding $A(0,\vartheta)$ and $A(l, \pi - \vartheta)$ and substituting their values in (33.4) and (33.5), replacing $\pi - \vartheta$ here by ϑ, we get

$$A(z,\vartheta) = C\left\{T - G\frac{\cos\vartheta}{2\alpha}\left[1 - \frac{(1+r)\,e^{-\frac{2\alpha z}{\cos\vartheta}}}{1+re^{-2\alpha l/\cos\vartheta}}\right]\right\}, \tag{33.6}$$

$$0 < \vartheta < \frac{\pi}{2},$$

$$A(z,\vartheta) = C\left\{T - G\frac{\cos\vartheta}{2\alpha}\left[1 - \frac{(1+r)\,e^{\frac{2\alpha(l-z)}{\cos\vartheta}}}{1+re^{2\alpha l/\cos\vartheta}}\right]\right\}, \tag{33.7}$$

$$\frac{\pi}{2} < \vartheta < \pi.$$

In the latter expressions, the reflection coefficient at the boundary of the crystal and the surrounding medium is denoted by r.

The intensity of the light scattered by the adiabatic density fluctuations is proportional to the sum of the intensity of two elastic waves, whose wavelength Λ is determined by (1.27) and their direction corresponds to the Bragg condition (1.19). The Bragg condition is satisfied for two mutually opposite directions of propagation of the elastic waves under consideration.

If the angle between the direction of the temperature gradient, the z axis, and the direction of propagation of one of the elastic waves is denoted by ϑ_B and the other by ϑ'_B, then the intensity of the scattered light will be

$$I = \frac{1}{2}\left[A(z, \vartheta_B) + A(z, \vartheta'_B)\right]. \tag{33.8}$$

The expression (33.8) is written with accuracy up to a factor which depends on the scattering angle, but it does not depend on the temperature distribution. We substitute the values of (33.4) and (33.5) in (33.8), and take into account that (for $\vartheta_B > 0$)

$$\cos\vartheta'_B = -\cos\vartheta_B. \tag{33.9}$$

We get an expression for the intensity of the scattered light

$$I = C\left\{T + \frac{G\cos\vartheta_B}{2\alpha}\,\frac{(1+r)\exp\left(-\frac{2\alpha l}{2\cos\vartheta_B}\right)}{1+r\exp\left(-\frac{2\alpha l}{\cos\vartheta_B}\right)}\,\mathrm{sh}\left[\frac{2\alpha\left(\frac{l}{2}-z\right)}{\cos\vartheta_B}\right]\right\}. \tag{33.10}$$

If we express the angle ϑ_B in terms of the scattering angle θ and use the Bragg condition (1.19) and (1.26) for the case in which the beam of incident light is directed parallel to the temperature gradient, while the temperature gradient is directed toward increase of T ($T_l > T_0$), then the intensity of the scattered light is written in the following fashion:

$$I(z,\ \theta)=C\left\{T+\frac{G\sin\theta/2}{2\alpha}\,\frac{(1+r)\exp\left(-\dfrac{2\alpha l}{2\sin\theta/2}\right)}{1+r\exp\left(-\dfrac{2\alpha l}{\sin\theta/2}\right)}\,\mathrm{sh}\left[\frac{2\alpha\left(\dfrac{l}{2}-z\right)}{\sin\theta/2}\right]\right\} \qquad (33.11)$$

Equation (33.11) is introduced, as has already been said, under the assumption that the crystal is infinite in the x and y directions and finite in the z direction.

The formula, obtained by Leontovich, can be applied to a real plate if its thickness is much less than the linear dimensions of the cross section. In the case in which this condition is not satisfied, the observed effect will be less than for an unbounded plate.

As follows from (33.11), the greatest effect is to be expected at the ends of the plate, $z_1 = 0$ and $z_2 = l$. In the middle of the plate ($z = l/2$), the intensity of the light in a nonuniformly heated crystal does not depend on the temperature gradient. The intensity of the scattered light will also not depend on the temperature gradient if the light which produces the scattering is directed along the perpendicular to the temperature gradient along x and the observation is carried out along the y axis. In this case, the scattered light is determined by the intensity of the elastic waves propagating parallel to the (xy) plane, while the intensity of these waves is determined by the temperature at the given point and does not depend on the temperature gradient. This is seen from (33.5), where it is necessary to set $\vartheta = \pi/2$ in this case. Finally, if the crystal is unbounded not only in x and y but also in z, then, as is clear from (33.10), by setting $l = \infty$ and $z = l/2$, we see that the dependence of the intensity on the gradient falls off if the gradient is constant (the temperature depends on the coordinates according to a linear law). If there is a parabolic dependence of the temperature on the coordinates, then, as Leontovich has pointed out [573],

$$I=C\left[T+\frac{\cos^2\vartheta_B}{4\alpha^2}\frac{\partial^2 T}{\partial z^2}\right]. \qquad (33.12)$$

As was shown above, the scattering of light in a nonuniformly heated crystal has been investigated experimentally by Landsberg and Shubin [571]. In their research, a slice of dimensions 3.9 × 3.4 × 3.1 cm in the shape of a parallelepiped was cut from a large block of crystalline quartz. The optic axis of the crystal is parallel to the vertical edge. The temperature on the lower and upper boundaries was measured by thermocouples, the ends of which were placed in a narrow slot between the crystal and plates cut from this same crystal and set into optical contact. The basic experimental difficulty of the work is that there is a very great danger of cracking the crystal in the production of large temperature gradients. To decrease this danger, it is necessary to create a temperature distribution according to such a law which would guarantee the minimum internal stress in the crystal.

Kirpichev [574] has shown that the necessary and sufficient condition which the temperature distribution must satisfy in order that no internal stresses arise in the free isotropic body is that the temperature at different points of the body would be a linear function of the cartesian coordinates. For the case of a parallelepiped, the temperature should change according to the law $T = a + bz$, remaining constant within the limits of any cross section.

By a special calculation, Leontovich showed that the theory of Kirpichev could also be extended to crystals. The small temperature dependence of the thermal conductivity of quartz made difficult the establishment of the needed temperature distribution law, but, by keeping the temperature constant on the upper and lower faces of the crystal, it was possible to approximate it.

In the experiments of Landsberg and Shubin, the excitation was carried out with white light. The light was directed up from below, entering the crystal through the hot end, and leaving through the cold end. The observation of the scattered light was made at an angle of $\theta = 90°$ to the incident light and the temperature gradient. Calculations of the intensity were carried out at three points (the measurement was made from the lower face): (1) at a distance of 11 mm (low point), (2) at a distance of 21 mm (middle), and, (3) at a distance of 31 mm (upper).

The measured intensity I contained the light scattered by the impurities and other obstacles, I_f, the light of Raman scattering

I_k, and the light of Rayleigh scattering I_R. The effect under study refers only to this part I_R, which is connected with the adiabatic density fluctuations.

The intensity of the light scattered by isobaric density fluctuations is very small in crystals and, therefore, it is assumed that I_R is determined only by the adiabatic fluctuations.

By denoting the light intensity which refers to uniform heating of the crystal by $\tilde{I}$, we get, in first approximation, that $I_f = \tilde{I}_f$; $I_k = \tilde{I}_k$, and then

$$I - \tilde{I} = I_R - \tilde{I}_R .$$

In the experiments of Landsberg and Shubin [571], repeated series of measurements were carried out with a large number of measurements in each series. From the result of an experiment by them, the conclusion was drawn that, in the limits of accuracy of measurement (1%), the intensity of the lower, middle, and upper parts of the nonuniform heated crystal do not differ from the intensity of the uniformly heated crystal.

On the basis of this experimental result and Eq. (33.11), one can estimate the lower boundary for α. Setting l = 3.9 cm, $\sin\theta/2 = 0.7$, and for z = 1.1 cm, T_1 = 450°K, G = 65 deg/cm (lower point) and for z_3 = 3.1, T_3 = 602°K and G_3 = 88 deg/cm (upper point), and assuming, on the basis of the experiment, that $(\tilde{I} - I)/\tilde{I} \leq 0.01$, we find from (33.11) that $\alpha > 0.75\ \text{cm}^{-1}$.

On the other hand, inasmuch as the fine structure is observed for $\alpha\Lambda \ll 1$, then, taking $\Lambda \sim 2 \cdot 10^{-5}$ cm, we obtain the result that $\alpha < 5 \cdot 10^4\ \text{cm}^{-1}$. Thus, estimates of the upper and lower values of α are very widely separated from one another. We shall speak below of direct measurements of α at frequencies of 10^{10} Hz.

It now remains to speak of another interesting possibility of the experimental investigation of the effect of the temperature gradient on the light scattering, as suggested by Vladimirskii [488, 578].

The possibility is based on the fact that, as is clear from Eqs. (33.6)-(33.9), the intensity of the light scattered by the thermal wave of one direction differs from the light intensity scattered by the wave of the opposite direction. Consequently, the Stokes and anti-Stokes MB components are of different intensity.

One can calculate the value of such a difference in the intensity. Setting z and $l - z$ in Eqs. (33.6) and (33.7) – the distances of the point considered from the faces of the crystal – equal to infinity, and taking (33.9) into account, we get

$$I_{\text{st}} = C\left[T + \frac{G \cos \vartheta_B}{2\alpha}\right], \qquad (33.13)$$

$$I_{\text{ast}} = C\left[T - \frac{G \cos \vartheta_B}{2\alpha}\right]. \qquad (33.14)$$

For a uniformly heated body, $I_{st} = I_{ast} = I = CT$. From the latter formulas, we find I_{st}

$$\frac{I_{\text{st}} - I_{\text{ast}}}{I} = \frac{G \cos \vartheta_B}{\alpha T}. \qquad (33.15)$$

In the real case of absorption in a quartz crystal at $2 \cdot 10^9$ Hz for $\alpha = 2.44\ \text{cm}^{-1}$, the left side of (33.15) can be replaced by ~0.1 for G = 100 deg/cm, scattering angle $\theta = 10°$, and T = 290°K, under the condition that the primary light is directed parallel to the temperature gradient. In this method, as was shown in [488], there is an advantage in comparison with the method already tested experimentally [571], but there is a series of inadequacies connected with the extremely great difficulty of the experiment. The basic advantage of the method suggested by Vladimirskii is that the observation can be carried out in any one region of the crystal, far from its faces, and does not require an accurate knowledge of the temperature gradient.

For an estimate of the perspectives of the observation of the effect of the temperature gradient on the Rayleigh scattering, we return to the measurements of the absorption coefficient.

At the present time, there are direct measurements of the absorption in crystalline quartz and some other crystals at frequencies up to 10^9 Hz. The absorption measurements of Lamb, Redwood, and Shteinshleifer [575] in the frequency range from $2 \cdot 10^8$ Hz to 10^{10} Hz for crystalline quartz and in the frequency range from $8 \cdot 10^7$ to $\sim 4 \cdot 10^8$ Hz for fused quartz have shown that right up to the highest frequencies obtained, the proportionality of

Table 45. Absorption of Hypersound in Crystalline Quartz at a Frequency of 10^9 Hz

Cut of the crystal parallel to which the sound wave is directed	Temp., °K	Velocity $v \cdot 10^{-5}$, cm/sec	Absorption* α, cm^{-1}	Character of hypersonic wave	Author
x	293	–	0.61	Longitudinal	Lamb, Redwood, and Shteinshleifer [575]
x	140	5.6	0.42	Longitudinal	Bömmel and Dransfeld [576]
z	140	6.3	0.32		
AC	140	3.32	0.8	Transverse	
BC	140	5.04	0.18		

*The absorption coefficient α in [575] is given in dB/cm, in [576] in dB/in. The values of α given in column four of the table have been recalculated by the author.

the absorption coefficient to the square of the sound frequency is preserved. The values of the absorption coefficient of sound in crystalline quartz measured at a frequency of 10^9 Hz are given in Table 45. By extrapolating the value of α for 10^9 Hz and 293°K by a square law to a frequency of $2 \cdot 10^9$ Hz, we find $\alpha = 2.44$ cm^{-1}. The same such extrapolation for the data of Bömmel and Dransfeld [576] gives a value of α lying in the limits from 3.2 cm^{-1} for AC cut to 0.72 cm^{-1} for BC cut transverse sound wave, and from 1.68 cm^{-1} for x cut to 1.28 cm^{-1} for z cut.

If it is taken into consideration that the absorption can change from sample to sample of quartz because of different dislocation and other imperfections of the crystals [577], then one must come to the conclusion that the absorption of the hypersound from room temperature (293°K) to 140°K remains practically unchanged or changes very slightly. A very significant result of the work of Bömmel and Dransfeld [576] is that the absorption coefficient continues to remain constant in the range of temperatures from 140°K to 40-60°K, while it falls catastrophically beyond 40-60°K, reaching a value that is immeasurably small at 10-15°K.* This effect

*The part of α which does not depend on the temperature is not given in the graphs of [576] and is a quantity much smaller than 0.04 cm^{-1}.

was observed in [576] for the longitudinal wave (different for the x and z cuts) and for the transverse wave (different for the AC and BC cuts).

At the temperature of liquid hydrogen (20°K), the absorption coefficient for a longitudinal wave in a quartz crystal at a frequency of $2.4 \cdot 10^{10}$ Hz (a rough estimate from the graph of [700]) is $\alpha = 0.04\ \text{cm}^{-1}$.

An attempt to calculate the absorption in a solid by assuming that all the absorption is brought about by the thermal conductivity for a number of crystals [577], including x-cut quartz [576], shows that at a frequency of 10^{10} Hz, this mechanism gives a very small value of α which does not explain the observed phenomenon.

For transverse waves, the absorption mechanism connected with the thermal conductivity generally does not take place.

An entirely new absorption mechanism, not connected with the thermal conductivity, was proposed in the theory of Akhiezer [579]. Later, the absorption theory was generalized by Pomeranchuk [580].

This new mechanism is as follows. The sound wave passing through the crystal disrupts the equilibrium Planck distribution of phonons. The reestablishment of the equilibrium distribution of the phonon gas involves an increase in the entropy, which in turn produces absorption of the sound wave.

A group of phonons of definite direction and polarization taken out of the equilibrium state will approach equilibrium, exchanging energy with another group of phonons. The finite exchange time of the energy is the relaxation time, introduced by Bömmel and Dransfeld [576], which allowed them to obtain a relaxation formula whose structure is the same as (22.3). This estimate of the value of the absorption coefficient is in excellent agreement with experiment. The temperature variation of the absorption coefficient corresponding to the experiment is also obtained. From the given data for sound absorption in a quartz crystal at a temperature of 20°K and for a scattering angle of $\theta = 90°$, it follows that if a gradient of 2 deg/cm is created in this temperature range, then $(I_{st} - I_{ast})/I \approx 2$. In a sapphire crystal, under these same conditions, the absorption is still smaller [659] and the effect under consideration must be larger than for quartz. At the much lower helium temperatures, the absorption is so greatly reduced that a gradient of a fraction of a degree suffices in order that a significant effect be observed in the crystals studied [658, 659, 700, 701].

The experiment becomes more difficult at low temperatures, but the effect is so large that it can be established with certainty in a qualitative experiment. We have spoken earlier of the advantage of such a type of experiment. With the appearance of gas lasers, the existence of such an experiment became possible for crystalline quartz.

This experiment is also possible for the excitation of scattering of light from an ordinary low-pressure mercury lamp, but only in a diamond crystal, where the distance between the transverse components of the fine structure is large. To date, these experiments remain unperformed.

Chapter X

Stimulated Molecular Scattering of Light

In different parts of this book, it has been said that with the introduction of lasers into laboratory practice, the perspective opened up for new investigations which were in earlier times extremely difficult or even completely impossible.

The low-power Ne—He gas laser made it possible to excite the fine structure of the Rayleigh line by means of the extraordinarily sharp line $\lambda = 6328$ Å of stimulated emission.

The velocity of hypersound has been determined with great accuracy, from the position of the shifted components of the fine structure, while the absorption of hypersound in liquids was found from the width of the shifted components (see Secs. 23 and 24).The width of the unshifted line has already been used for the determination of the dimensions of macromolecules in solution and can be used for the measurement of the hypersonic velocity and absorption but, as has already been pointed out (see Sec. 12), this problem could be solved at least in principle, with the help of gas discharge tubes, operating on a single mercury isotope, or on another nearby element.

The application of ruby lasers, which emit light of great intensity, to the study of the interaction of light with matter makes it possible to observe and investigate phenomena connected with the nonlinearity of this interaction, the experimental study of which was previously impossible. A new region of science arose from the study of the interaction of the intense radiation of the ruby laser with matter — nonlinear optics, which is less than four years old.

Reviews of the first successes in this new region of physical optics have been made by Akhmanov and Khokhlov [607] and by Bloembergen [608].

In this chapter, we shall give the fundamental experimental results and the simplified classical theory of the new nonlinear optical phenomena in stimulated molecular scattering of light: (1) the stimulated MB scattering (SMBS), and, (2) the stimulated scattering of light of the wing of the Rayleigh line (SSWRL).

The new nonlinear phenomena differ from one another in radical fashion, and each of them creates a new direction of investigation of physical optics.

The stimulated combination scattering of light (the Raman effect), which is not considered here, was discovered even earlier.

Before going further, we shall make one note on terminology.

The scattering of light by optical inhomogeneities, which arises as the result of the thermal fluctuations of the density and anisotropy, has in recent times been called the "spontaneous" scattering of light. Now, this scattering is induced by the exciting light and does not take place of itself (spontaneously). Therefore, we shall call such a scattering in what follows thermal or ordinary light scattering, reserving for the term "stimulated scattering" that meaning which is now given to it, although even here, greater precision would be desirable.

By stimulated molecular scattering of light is meant both forms of stimulated scattering noted above.

One of the essential characteristics of stimulated molecular scattering of light is that the intensity of this scattering increases nonlinearly with increase in the linear dimensions of the scattering volume in the direction of observation of the scattered light.

On the other hand, the intensity of the thermal scattering by fluctuations, whose dimensions are much less than the wavelength of the exciting light, increases linearly with increase in the linear dimensions of the scattering volume in the direction of observation of the scattered light.

If the nature of the stimulated scattering reduces to the parametric interaction of waves, then the nonlinear increase in the in-

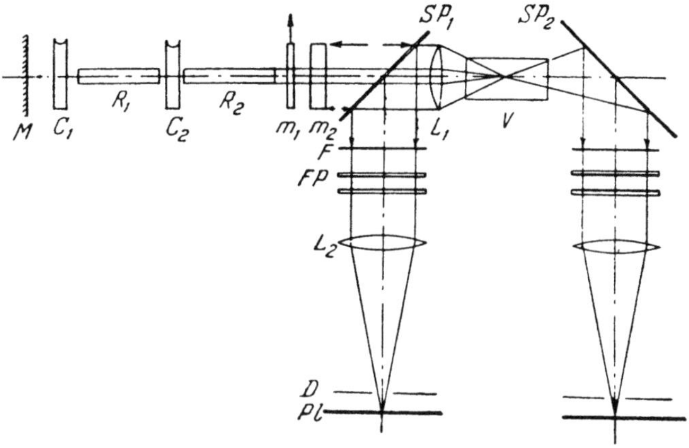

Fig. 111. Apparatus for excitation and observation of stimulated MB scattering and scattering of the wing of the Rayleigh line (Mash et al. [630, 631]). M – Mirror (R ~ 100%); C_1, C_2 – vessels with a solution of cryptocyanin in methyl alcohol (concentration $1.8 \cdot 10^{-6}$ M); R_1, R_2 – ruby rods of length 120 mm, diameter 12 mm; m_1,m_2 – plane parallel plates, mode separators; SP_1, SP_2 – half-silvered mirrors; L_1 – lens (f = 3 cm); L_2 – objective (f = 120 cm); V – test cell; F – light filter; FP – Fabry–Perot interferometer; D – diaphragm; Pl – photographic plate.

tensity of the scattered light begins after the intensity of the exciting line reaches a certain threshold value. The expressions and values of the thresholds will be discussed below.

The stimulated MB scattering in crystals of quartz and sapphire was discovered by Chiao, Townes, and Stoicheff [595]. This same phenomenon was then discovered in certain liquids by Brewer and Rieckhoff [601, 609] in amorphous solids and supercooled liquids by Mash, Morozov, Starunov, Tiganov, and Fabelinskii [630], and in gases by these same authors [615], by Hagenlocker and Rado [651], and also by Rank, Wiggins, Wick, and Eastman [652].

The stimulated scattering of light of the wing of the Rayleigh line was discovered by Mash, Morozov, Starunov, and Fabelinskii [631].

Table 46. Measured and Computed Frequencies of the Shift of the Components in Stimulated MB Scattering

	Scattering angle	Shift $\Delta\nu$, cm^{-1}	
		Measured	Computed from the elastic constants [612]
Quartz [595]	180°	0.99	0.97
	180°	0.85	0.88
	90°	0.73	0.67
Sapphire [595]	180°	2.07	2.01
Fused quartz [630]	180°	0.81	0.84
Crown K-8 [630]	180°	0.86	

In Sec. 34 we shall discuss the fundamental results of experimental researches on stimulated MB scattering and the elementary simplified theory of this phenomenon. In Sec. 35 we shall discuss the basic results relative to stimulated scattering of the light of the wing of the Rayleigh line (SSWRL) in liquids.

34. Stimulated Mandel'shtam – Brillouin Scattering

Observation of Stimulated MB Scattering in Crystals and Amorphous Solids. Excitation and observation of the spectrum of stimulated MB scattering and stimulated scattering of the light of the wing of the Rayleigh line are carried out in the same types of apparatus. As an example, a schematic diagram is shown in Fig. 111 of the apparatus with which the author and his coworkers [603, 631, 653] investigated both types of stimulated scattering in various media.

The power of the apparatus of Fig. 111 varied from 100 to 300 MW with a pulse duration of about 10^{-8} sec. The lens L_1 had various focal lengths from 3 to 10 cm. The scattered light could be observed at scattering angles of 0° and 180°.

In [595, 630, 653], only the Stokes components of stimulated MB scattering were observed. In [654], when a generator of 500-MW power was used for the stimulated MB scattering in the quartz crystal, five components were observed; there is a suspicion that two correspond to Stokes and three to anti-Stokes shifts, although there is no full certainty that the identification was made correctly.

Multiple lines of stimulated MB scattering, located on the long wavelength side, were observed in crystals, glasses, liquids, and gases. Garmire and Townes [599], who observed multiple equidistant components of stimulated MB scattering in liquids, interpreted them as the result of repeated stimulated scattering.

In the phenomenon of repeated stimulated scattering of light, light previously scattered in the medium of Fig. 111 passes through the excited ruby, is amplified in it, returns to medium V, and can be shown to be so intense that it itself produces stimulated MB scattering, etc. Thus, several equidistant components appear in the spectrum of Fig. 112. In all these cases, the components of the stimulated MB scattering are comparable in intensity with the exciting line.

Giuliano [671] used an electron-optical transformer of the picture, achieving very rapid development of the picture of the spectrum in time [667] and showed in direct experiment that the second Stokes line of the stimulated MB scattering in quartz is later than the first. Thus, the presence of the successive scattering can be regarded as a fact.

The shifted components of stimulated MB scattering in quartz, sapphire, fused quartz, and glass were measured and their values compared with the computed values. The results are shown

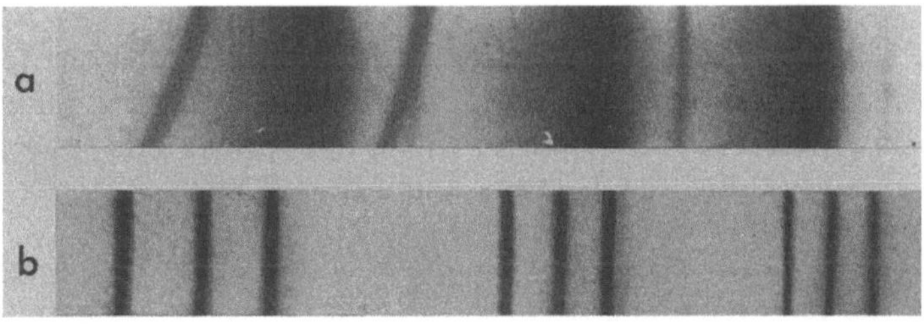

Fig. 112. Spectrum of the stimulated MB scattering in carbon bisulfide and in fused quartz (Mash et al. [630]). (a) Carbon bisulfide; dispersion range of the interferometer 8.33 cm^{-1}; 17 components of stimulated light scattering are shown in the photograph; (b) fused quartz; dispersion range of the interferometer 5 cm^{-1}; two Stokes components of the stimulated scattered light are seen in the photograph.

in Table 46. It is well known that two generators are created simultaneously in the process of stimulated MB scattering. These are a light generator at the shifted frequency (the MB component) and a generator of hypersound or ultrasound, the frequencies of which depend on the scattering angle and some other characteristics, of which we shall speak below.

In principle, the frequency of the sound which arises as the result of stimulated MB scattering changes from zero at an angle of $\theta = 0°$ to a value of about 10^{11} Hz in such crystals as diamond and sapphire at $\theta = 180°$. The lowest frequencies of acoustic vibration which appear in stimulated MB scattering were obtained in the work of Hsu and Kovage [655], where an anisotropic body – crystalline quartz – was used as the nonlinear element. In their work, sound of frequency $7.5 \cdot 10^7$ Hz was obtained and detected by a piezo receiver.

Interactions can arise in an anisotropic medium between the waves of the incident and scattered light, for which the index of refraction is different (ordinary and extraordinary waves).

In this case, instead of (1.26) and (5.8), we must write for the frequency of the sound wave

$$\Omega = \omega\,(v/c)\{(n_1 - n_2)^2 + n_1 n_2[2\sin(\theta/2)]^2\}^{1/2} \tag{34.1}$$

n_1 and n_2 are the refractive indices of the incident and scattered light, respectively, and the remaining notation is the same as before. In an isotropic medium or along the optical axis, $n_1 = n_2$ with great accuracy, and (34.1) transforms into (5.8).

In a crystal, n_1, n_2, and v depend on the direction of propagation of the incident and scattered light and on their states of polarization. For propagation of longitudinal sound waves along the x axis of quartz, it was found in [655] that if the incident light is regarded as the ordinary ray, and the scattered light as the extraordinary ray, then the frequency of the sound for $\theta = 0$ will be equal to $7.5 \cdot 10^7$ Hz, while for $\theta = 180°$, it will be $2.56 \cdot 10^{10}$ Hz.

The minimal frequency can be made still lower by bringing the ray of incident light closer to the optic axis of the crystal.

It is quite remarkable that in the experiments of [655] a sharp threshold was established for the generation of sound, which indicates the generation of sound as the result of stimulated MB scattering while, on the other hand, the threshold for the appearance of low-frequency sound is lower than the threshold of destruction of the crystal. The authors see the explanation of this in the small absorption of ultrasound of frequency of about 10^8 Hz in crystals of quartz, even at room temperature.

However, the problem of the destruction of transparent dielectrics under the action of a powerful laser radiation has not yet been explained with sufficient completeness. In all the cases known to us, every time that stimulated MB scattering arises in the observation of backward scattering, destruction of the sample results. Undoubtedly, the destruction takes place as the result of the absorption of a significant fraction of the energy or power of the light pulse.

Various points of view exist as to the mechanism of such absorption. Giuliano [629] considered the phonons which are produced by the stimulated MB scattering, and which are strongly absorbed, as the reason for the destruction of glass in the focused beam of a gigantic light pulse. He showed that the observed destruction cannot be brought about by the explosive wave which arises in a plasma, by light pressure, or by optical phonons which arise as the result of stimulated Raman light scattering.

Whiteman and Wilson [698] considered that the phonons which appear in stimulated MB scattering were the reason for the destruction of a white natural diamond, subjected to the action of the focused gigantic pulse.

In the wide-ranging investigations of Ashkinadze, Vladimirov, et al. [699], crystals, glass and plastics were subjected to destruction in focused gigantic and ordinary pulses of a ruby laser and of a neodymium glass laser. The authors came to the conclusion that the reason for the destruction in all cases are the phonons which arise as the result of stimulated MB scattering. Brewer and Rieckhoff [601] suggest that the destruction in the laser beam is connected with an explosive wave which arises as a result of the powerful absorption in the plasma.

Brewer [668] also showed that destruction takes place in such crystals as LiF, NaCl, and MgO, under the action of the focused gigantic impulse, but stimulated MB scattering does not appear.

Belikova and Sviridenkov [656] destroyed a sample of ruby by the light from a ruby laser, and came to the conclusion that the reason for the destruction is a resonant two-photon absorption of light in the ruby. Harper [657] focused the pulse of a ruby laser without modulated Q inside samples of glass, and concluded that the energy of the acoustic waves was too small to explain the observed destruction.

In the opinion of Harper, sound waves transform the glass in a small volume into a superheated liquid, and this liquid absorbs a great deal of energy from the incident light, producing the observed destruction. In the experiment of [657], continuous radiation was observed with an effective temperature of 5000°C. In the experiments of Belyaev et al. [662], the destruction of crystal samples was observed, which is accompanied by triboluminescent radiation. We cannot discuss this important problem in detail, but shall only point out the obvious fact that the mechanisms of destruction and radiation can be different under different conditions. We shall assume for the present that the phonons arising in stimulated MB scattering are certainly of importance in the destruction of a transparent dielectric by a focused gigantic light pulse from a laser.

If this is so, it appears natural to avoid the destruction by means of the greatest possible lowering of the acoustic losses in the irradiated material.

Bömmel and Dransfeld [576], Nava et al. [658], Chernets and Gonopol'skii [659], have shown experimentally that the absorption of the sound in a single crystal of quartz decreases enormously with temperature, and becomes vanishingly small at liquid helium temperatures. Therefore, Krivokhizha, Mash, Morozov, Starunov, and Fabelinskii [653] undertook the investigation of stimulated MB scattering in single crystals of quartz in the temperature range from room temperature down to 2.1°K.

We expected that stimulated MB scattering would appear at a temperature of 2.1 or 4.2°K, but that the quartz crystal would not be destroyed.

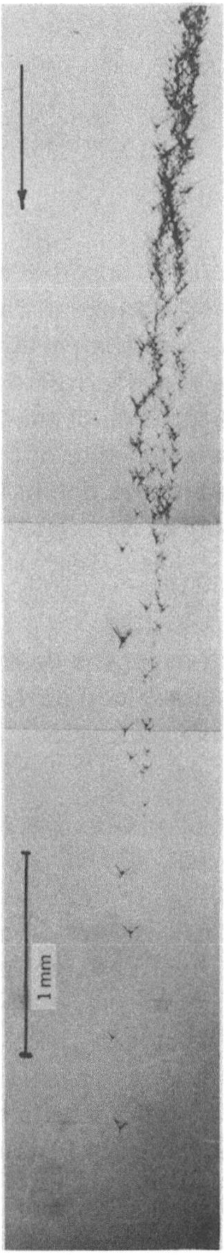

Fig. 113. Microphotographs of the path of destruction in quartz at liquid helium temperatures. The arrow indicates the direction of propagation of the incident light (Krivokhizha et al. [653]).

A gigantic pulse of power ~250 MW is focused inside a single crystal of Brazilian quartz by a lens with a focus of 5 cm. Destruction arose in the quartz crystal under these conditions, both when the quartz was at room temperature and when it was at helium temperature. However, the form and dimensions of the destruction at low temperature were different than at room temperature.

Figure 113 shows a microphotograph of the portion of the destruction in the temperature of liquid helium. The total extent of the region of destruction amounts to 0.8-1.5 cm.

By the marks of the destruction, one can see a division of the beam into several filaments, but thus far there is not sufficient experimental material to determine whether these filaments are explained by the mechanism suggested by Bespalov and Talanov [660] or whether this is the result of the inhomogeneity of the original beam, as suggested by Raizer [661].

Thus, the quartz was destroyed at a temperature of ~2°K, although the absorption coefficient of the sound is a few orders smaller than that at room temperature. A completely different picture is observed at a temperature of 80°K (liquid nitro-

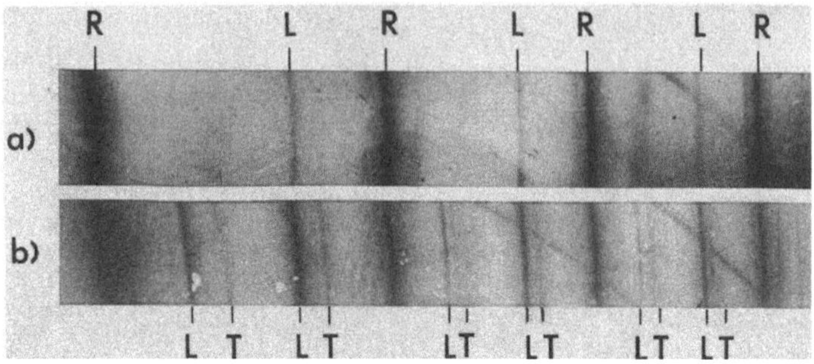

Fig. 114. Spectrum of stimulated MB scattering in a polycrystal of quartz at different temperatures. The dispersion range of the Fabry–Perot interferometer is 2.5 cm^{-1}. (a) At a temperature of 293°K; (b) at 80°K; θ = 180°. The electric field of the light wave is directed along the x axis, the direction of propagation of the incident light is along the optic axis. R is the emission line of the ruby, and L and T are the stimulated MB components due to the longitudinal and transverse waves, respectively (Krivokhizha et al. [653]).

gen) under otherwise equal conditions. In this case, the destruction is not observed anywhere, although the same place of the crystal is subjected to the multiple action of the focused gigantic pulse of laser light.

At a temperature of 80°K, there is no destruction, but at the same time there is stimulated MB scattering not only for longitudinal, but also for transverse waves (Fig. 114). In both cases, the shift of the MB component in the longitudinal wave at 80 and 300°K is equal to 0.93 cm^{-1} when the incident light is directed along the optic axis, and the scattering angle θ = 180°. For transverse waves under these same conditions, when the vector of the electric field of the light wave is directed along the x axis, the shift amounts to 0.65 cm^{-1}. The spectrum of stimulated scattering in quartz at liquid helium temperatures is needed in the subsequent investigation; its origin is not clear at the present time [653].

We emphasize again that in the experiments of [653] a phonon generator operating at 80°K was obtained at a frequency of about $3 \cdot 10^{10}$ Hz and a power of about 50 kW, without destruction of the sample. Such a generator allows us to introduce hypersound in other media, including nontransparent ones.

The fact that the destruction of the quartz crystal stops at 80°K, while it is destroyed under the same conditions at higher and at lower temperatures, can be explained at least qualitatively.

One of the possible variants of such an explanation is that, as the temperature is lowered and the absorption decreases, the nonlinearity of the sound wave increases [576, 658, 650]. The growth of the nonlinearity leads to an increase in absorption and the distortion of the shape of the sound wave of low amplitude. The quantity which characterizes the degree of nonlinearity of the sound wave is analogous to the Reynolds number [663], where, if Re $\ll$ 1, then the absorption dominates over the nonlinearity, and distortion of the wave is virtually absent. If Re $\gg$ 1, then the nonlinearity dominates, the sound wave goes over into a periodic shock wave with a thickness $\delta x \sim \Lambda/\mathrm{Re}$. Using the expression for the absorption coefficient of quartz obtained in [576], [658], Polyakova [664] found that the parameter Re is given by

$$\mathrm{Re} = \epsilon p(1 + \Omega^2\tau^2)/GcT\Omega\tau\,,$$

where ε is a nonlinear parameter of about 4-5, Ω and p are the frequency and pressure amplitude of the sound wave, C is the heat capacity per unit volume, and G is the Grüneisen constant. Correspondingly, τ is the time for establishment of the equilibrium state of the phonon gas.

A strong dependence of C and τ on the temperature leads to a temperature dependence also for the parameter Re.

The absorption coefficient of nonlinear sound α is connected with the sound absorption coefficient at low amplitude α_0 for the condition Re $\ll$ 1 by the simple relation

$$\alpha = (2/\pi)\alpha_0\,\mathrm{Re}\,.$$

Graphs constructed for α and Re as a function of temperature [664] for the experimental conditions of [653] show that for the region ~100°K, Re ~ 1, and α is effectively determined by the sound absorption at low amplitude. Upon further decrease in the temperature, the parameter Re increases strongly and α practically does not change. At 25°K, Re ~ Λ/a (where a is the lattice constant), the sound wave at this and much lower temperatures becomes a periodic shock wave with front thickness $\delta x \approx a$.

Polyakova [664] turned her attention to the fact that in the front of such a shock wave there are appreciable forces and elastic shear stresses which can be regarded as one of the reasons for the destruction of the crystal at low temperatures.

The very low threshold of stimulated MB scattering in triglycinsulfate and especially in Rochelle salt, was discovered by De Martini [666]. In Rochelle salt, it was ~1 MW, while in triglycin-sulfate it was twice as large. In this experiment, destruction of the crystal was not observed close to the thresholds of stimulated MB scattering in Rochelle salt which, in the opinion of the author of [666], can be only partially explained by the dependence of the threshold of stimulated MB scattering on the elastic constants of the crystal, and partially connected with the large electrostrictive constant of these crystals.

A sharp change is observed in the threshold of stimulated MB scattering and in the effect of an external electric field on the intensity of stimulated MB scattering in Rochelle salt close to the upper Curie point. The application of stimulated MB scattering to the study of second-order phase transitions or phase transitions in general is of undoubted interest, although evidently much broader information on phase transitions can be had from the study of the spectrum of the thermal scattering of light, which has already been applied to these ends [678].

In our experiment [630], the lines of stimulated MB scattering were observed in fused quartz and in a number of silicate glasses (Fig. 112 and Table 46).

In the silicate glasses in this experiment, so far as we know, the MB components were observed first. This result again demonstrates (see Sec. 27) that the pessimism relative to the possibility of observing (in principle) the fine structure in glasses [483] has no foundation.

In the recent experiment of Shapiro, Gammon, and Cummins [675], the fine structure in silicate glasses has also been discovered in the thermal scattering of light.

It must be pointed out that the thermal and stimulated light scattering form an effective method for the study of the acoustic properties of supercooled liquids [630], which are crystallized by the introduction of a sound beam in them by the usual ultrasonic method.

Stimulated Mandel'shtam – Brillouin Scattering in Liquids. As a result of the first studies of stimulated MB scattering in crystals, researches have appeared in which the same phenomenon has been observed in liquids.

The Stokes and anti-Stokes components of the stimulated MB scattering were observed by Brewer and Rieckhoff [601, 609] in light scattered by benzene and water. In this same work, up to three Stokes components and two anti-Stokes components were observed (benzene). In the opinion of the authors of [601], the additional components refer to harmonics connected with the character of the nonlinearity of the stimulated light scattering.

In the work of Garmire and Townes [599], harmonics were not observed, but an effect was observed of successive stimulated scattering.

Wiggins et al. [672], using a photodiode, recorded the successive appearance of lines of repeated stimulated light scattering in certain liquids. It was established that the distance between the light generator and the scattering medium is connected with the time of appearance of the shifted component by a linear law, in which extrapolation to zero distance gives zero time. This means that the stimulated MB scattering arises within several nanoseconds or even within a small fraction of a nanosecond.

In the experiments described in [599, 609], it was observed that up to nine equidistant lines appear in successive scattering, but in the work of [630] on the scattering in carbon bisulfide, 17 components were observed. The measurement of the hypersonic velocity is given in Table 16. Inasmuch as the width of the fluorescence band of the ruby R amounts to 11 cm^{-1}, while the shift of the MB component in liquids amounts to $\sim 0.2\ cm^{-1}$, then up to 50 lines of successive stimulated scattering can be observed in principle. To date, only the smaller number of components mentioned above have been observed. It is possible that this is explained by insufficient intensity of the exciting light, but it is possible that

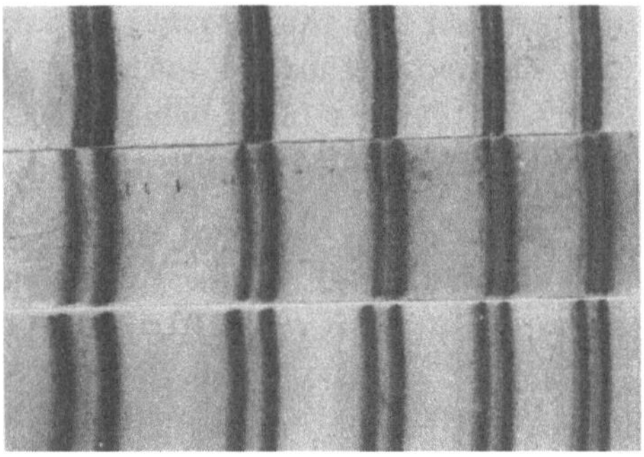

Fig. 115. The spectrum of stimulated MB scattering in nitrobenzene for different temperatures (Mash et al. [631]). (a) 170°C; (b) 65°C; (c) 20°C.

there is another reason. This question should be subject to further analysis. In Fig. 115, the spectrum of stimulated MB scattering in nitrobenzene is given for different temperatures. The results obtained of experiments on successive scattering allow us to hope that the hypersonic velocity can be determined with great accuracy from a large number of sharp, equidistant lines. Optimistic estimates [599] suggest an increase in the present accuracy of measurement of the hypersonic velocity by two orders of magnitude. However, such a high accuracy of determination of the hypersonic velocity is scarcely realizable because of the uncontrolled heating arising in the focal region of the laser beam [630].

Another important reason limiting the accuracy of determination of the hypersonic velocity follows from the work of Brewer [673], who has rightly shown that the quantum generator cannot be regarded as a passive amplifier for repeated scattering, but as an amplifier which consists of an assembly of modes. The amplification takes place only at the mode frequencies.

By using a sensitive method of light beats, Brewer demonstrated that the heating of the scattering medium (toluene) does not change the beat frequency and, consequently, the location of the shifted component, up to the point where the change in velocity reaches such a value which changes the frequency to the next mode. Then the beat frequency changes by a jump.

There is also another view of the possible reason for the observed decrease in the velocity. This reason was pointed out by Chaban [665], who established the fact that such an effect can be expected every time that there is an anisotropy of the amplification.

Stimulated MB scattering was observed in the excitation of the pulsed light of a ruby laser without modulated Q [599], but the gigantic pulse is preferable because of its better spectral qualities. In this latter case, mode separators are used for the excitation of stimulated MB scattering in liquids which separate the spectral parts of widths of 0.04 cm^{-1} [599] and 0.03 cm^{-1} [630], or even 0.01 cm^{-1} [609].

The stimulated MB scattering in carbon bisulfide was excited by Takuma and Jennings [610] with an off-axis resonator. The angle of the axis of the resonator with the direction of the incident light (angle of scattering) amounted to ~2.5°. These same authors [611] determined the hypersonic velocity at a frequency of 123 MHz by the method of heterodyning the light scattered in carbon bisulfide at an angle of ~2.5° with the light that excites the scattering (see Table 16). The experimental determination of the threshold intensity of the stimulated MB scattering shows that it varies exceedingly for different substances [599]. Thus, the threshold for stimulated scattering in carbon bisulfide, for example, amounts to 30 MW/cm^2, and in benzene and other liquids to 1200 MW/cm^2. The threshold in salol and triacetin is much higher [630].

In such liquids as carbon bisulfide, nitrobenzene, toluene, benzene, and acetone, the stimulated MB scattering was observed, together with stimulated Raman scattering, which has a much lower threshold in these liquids. On the other hand, for water, carbon-tetrachloride, and methanol, there was no observed stimulated Raman scattering for the intensity used. For the power levels given above, rugged glass containers with liquids remained undamaged and were suitable for re-use [599, 601].

Further important results of experiments on stimulated MB scattering in liquids were given by Brewer [674], who discovered experimentally the second harmonic of hypersound excited as a result of the nonlinearity of the fundamental hypersonic wave in stimulated MB scattering. This was the first result which showed experimentally that the hypersonic waves that arise in the process of stimulated MB scattering possess a nonlinear character.

Stimulated Mandel'shtam–Brillouin Scattering in Gases. The problem of the possibility of the existence of a fine structure of the Rayleigh line in gases was considered in detail by the author of this book [335] on the basis of the gas kinetic theory and was developed in Sec. 17.

From the considerations described, it follows that in the observation of scattered light at an angle $\theta = 90°$ in exciting light $\lambda = 4358$ Å, the discrete components of the fine structure should be observed in hydrogen (H_2), nitrogen (N_2), oxygen (O_2), and carbon dioxide (CO_2) at pressures above 20-25 atm. This conclusion is in contradiction with the experimental investigation of [330] (see Sec. 17).

Until the introduction into laboratory practice of laser sources of light, experiments on the observation of the scattering spectrum in gases required exposures of many days' duration (in [330] it took two weeks); it is possible, therefore, that the experiments of [330] were not repeated.

After the appearance of powerful sources of excitation, stimulated MB scattering was found almost immediately in compressed gases by various groups of physicists independently of one another.

With the apparatus of Fig. 111, and using light pulses with output powers of ~250 MW, Mash et al. [615] observed stimulated MB scattering in gaseous hydrogen, oxygen, nitrogen (Fig. 116) at pressures from 75 to 150 atm. Only the Stokes components were observed in hydrogen; in oxygen, weak anti-Stokes components of stimulated MB scattering also appeared.

In the region of the critical point of CO_2, stimulated MB scattering was observed by Guberman and Morozov [648], who established a strong shift of the components of stimulated MB scattering toward the central line as the scattering substance approached its critical point. In a broad range of pressures up to 1000 atm, stimulated MB scattering was observed by Hagenlocker and Rado [651] in nitrogen and in argon. This same phenomenon was observed by Rank et al. [652] in nitrogen, methane, and carbon dioxide at elevated pressures.

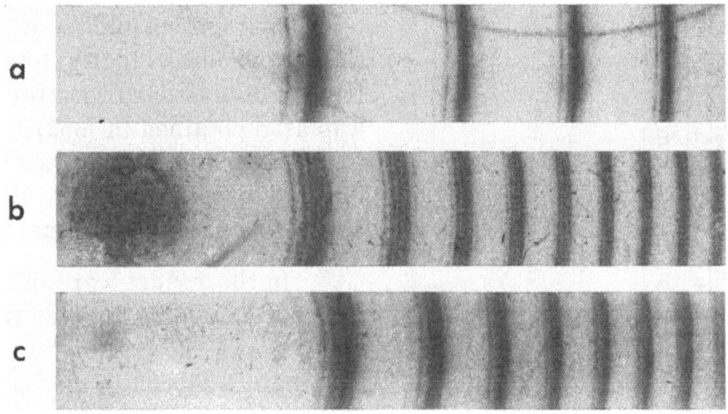

Fig. 116. Spectra of stimulated MB scattering in compressed gases. (a) Hydrogen at a pressure of 95 atm. The dispersion range of the Fabry—Perot etalon is 1 cm^{-1}. (b) Oxygen, pressure 150 atm. (c) Nitrogen, pressure 125 atm. In cases (a) and (b) the dispersion range of the interferometer is 0.333 cm^{-1}; $\theta = 180°$. (Mash, Morozov, Starunov, and Fabelinskii [615].)

It was easy to find the hypersonic velocity from the position of the distinct lines of stimulated MB scattering in gases [615] (Fig. 116). Table 47 gives the corresponding measurements, which are compared with the velocity of ultrasound.

Two circumstances require attention:

1. The gas kinetic calculation [335] from Eqs. (17.11) and (17.12) gives the result that $\alpha\Lambda > 1$ for helium at pressures ~100 atm, and therefore a discrete fine structure cannot be expected and actually has not been observed under the conditions of the experiment [615] (Table 47).

2. It should be expected that the adiabatic value of the velocity, determined from the shift of the components of stimulated MB scattering in gases, will in every case be no less than the velocity from ultrasonic measurements.

However, experiment has shown otherwise. For H_2 and N_2, values of the hypersonic velocity are obtained close to the isothermal velocities of ultrasound.

Table 47. Velocity of Hypersound in Compressed Gases, Determined from Stimulated MB Scattering

Gas	Pressure p, atm	Index of refraction, n	Hypersonic frequency $f \cdot 10^{-9}$, Hz	Hypersonic wavelength $\Lambda \cdot 10^{5}$, cm	$\alpha\Lambda$	Measured in stimulated MB scattering: $\Delta\nu \cdot 10^{2}$, cm^{-1}	Measured in stimulated MB scattering: Velocity of hypersound, m/sec	Ultrasonic measurements: Adiabatic, m/cm	Ultrasonic measurements: Isothermal, m/sec
N_2	125	1.035	0.84	3.3	0.06	2.8 ± 0.1	280 ± 10	352	297
O_2	150	1.038	0.99	3.3	0.06	3.3 ± 0.3	330 ± 30	331	280
H_2	95	1.012	3.3	3.4	0.14	11 ± 1	1130 ± 100	1334	1127
He	140	1.005	2.6 (comp.)	3.5 (comp.)	1.7	—	—	1008	783

The isothermal value of the hypersonic velocity, found from stimulated MB scattering, was also obtained in [652]. No other determinations of the velocity from the stimulated MB scattering are known to us.*

In the recent work of Greytak and Benedek [670], the fine structure of the Rayleigh line has been observed in the thermal scattering of light in argon, xenon, nitrogen, carbon dioxide (CO_2), and methane (CH_4) at practically atmospheric pressure and for excitation of the scattering by light of the line $\lambda = 6328$ Å of a Ne–He gas laser at a scattering angle $\theta = 10.6°$.

In all cases, the values found for the velocity agree with the adiabatic values of the velocities of ultrasound for these gases. The isothermal values of the hypersonic velocity observed by us [615] and in [652] have not found a unique explanation.

It is possible that the reason for this lies in the physical processes which call for further investigation, and it is possible also that this result is connected with the technique of

*In [651] graphs are given from which it is difficult to establish what value of the velocity was observed.

measurement or with the peculiarities of stimulated MB scattering in repeated scattering, on which we have already spoken above [673].

The investigation of the spectrum of stimulated and thermal scattering in gases has continued, but even now the investigations mentioned here remove (in principle) the indeterminacy of the question of the existence of fine structure in gases, which existed somewhat more than a year ago.

Further important results have been obtained by Mink, Hagenlocker, and Rado [676]; these have revealed stimulated light scattering in deuterium at pure rotational frequencies.

Elements of the Approximate Classical Theory of Stimulated Mandel'shtam–Brillouin Scattering of Light. We shall first consider the problem from a qualitative point of view, and then proceed to the approximate classical theory.

1. Qualitative Explanation of the Phenomenon. The physical reason for the stimulated MB scattering is that at high intensities of the incident light, the effect of the incident and scattered light on the character of the motion of the medium begins to play a significant role (an elastic wave arises of high intensity) while in thermal scattering, this effect was insignificant and the motion of the medium was completely described by the Debye thermal waves. If an electric field acts on the body, then an electrostriction appears in such a body which produces a change in the volume of the body. The relative change of volume of the body is [47]

$$\frac{\Delta V}{V}=\frac{1}{8\pi}\beta_S\left(\rho\frac{\partial\varepsilon}{\partial\rho}\right)_S E^2. \qquad (34.2)$$

If the rate of change of $\Delta V/V$ does not exceed 10^{13} Hz, then the adiabatic characteristics of the medium entering into (34.2) should be used.

Using the determination of β_S (see the notation in Sec. 1), we can rewrite (34.2) in the form

$$|\Delta p|=p=\frac{1}{8\pi}\left(\rho\frac{\partial\varepsilon}{\partial\rho}\right)_S E^2, \qquad (34.2a)$$

where $\Delta p = p$ is the excess pressure produced by electrostriction. If we take the intensity of the field in an unfocused solar ray equal to $E_\odot = 10$ V/cm and let this intensity be 10^4 times greater in the focused ray, then $p_\odot \sim 5 \cdot 10^{-3}$ kg/cm^2, while the electric field intensity in the focused beam of the laser reaches the enormous value of 10^9 V/cm [613]. In this case, the pressure $p_{las} = 5 \cdot 10^5$ kg/cm^2. From these very approximate estimates, it follows that the focused (and even the unfocused) rays of the laser are capable of producing an enormous pressure in the medium and, therefore, such nonlinear effects as electrostriction, while not playing a role under ordinary conditions, become phenomena of outstanding importance here. The interaction between the light wave of the laser, the elastic wave, or pressure wave produced by electrostriction, and the polarization wave that is produced in turn by the pressure wave can be explained by the phenomenon of stimulated light scattering.

For a complete classical consideration of this problem, we need to write Maxwell's equations and the equation of elasticity theory for a solid, or the equations of hydrodynamics for liquids with account of the nonlinearity of the medium.

The common solution of these equations should describe the phenomenon of stimulated MB scattering and other nonlinear effects. We proceed to such considerations; initially, in order to make clear the physical nature of stimulated scattering, we consider the previously known rough approximation, when there is in the medium the field of the incident wave $\mathbf{E}_0 \cos(\omega t - \mathbf{kr})$ (gigantic pulse of the ruby laser) and – as a result of thermal scattering – the field of the Stokes satellite $\mathbf{E}_1 \cos[(\omega - \Omega)t - \mathbf{k}_1'\mathbf{r}]$ and the field of the anti-Stokes satellite $\mathbf{E}_2 \cos[(\omega + \Omega)t - \mathbf{k}_2'\mathbf{r}]$. We recall that the field of these satellites arose as the result of the scattering of light at an angle satisfying the Bragg condition (1.19) and as the result of the modulation of the scattered light by a thermal wave with frequency Ω and wave vector $\mathbf{q}$. We shall assume that $\mathbf{E}$ in Eq. (34.2a) consists of the sum of the field of the incident light and the fields of the Stokes and anti-Stokes components [599, 614]. After squaring the three terms, and after some elementary trigonometric transformations, we obtain the result that p consists of the high-frequency components of frequencies $\sim\omega$ and 2ω, and the hypersonic frequency components $\sim\Omega$ and 2Ω. The sound (pressure) of frequency $\sim\omega$ and 2ω cannot propagate. We shall not consider here the role of these components. The hypersonic components can be written as

$$p = \frac{1}{8\pi}\left(\rho\frac{\partial\varepsilon}{\partial\rho}\right)_S \{E_0E_1\cos[\Omega t - (k - k_1')r] + E_0E_2\cos[\Omega t + (k - k_2')r] + + E_1E_2\cos[2\Omega t + (k_1 - k_2')r]\}. \tag{34.3}$$

If the incident light is propagated along the direction of its wave vector **k** (isotropic medium) and observation of the scattered light is made in the opposite direction ($\theta = 180°$), then the first component gives the wave of sound or pressure coinciding exactly with the original thermal hypersonic wave, since $\mathbf{k} - \mathbf{k}_1' = \mathbf{q}$ and the sound wave is propagated in the same direction as **k** ($|\mathbf{k}| > |\mathbf{k}_1|$). The second term gives a wave of this same frequency and phase, since $|\mathbf{k}_2| > |\mathbf{k}|$, and, consequently, $\mathbf{k} - \mathbf{k}_2 = \mathbf{q}$. A sound wave of double frequency is propagated in this same direction, since the difference $\mathbf{k} - \mathbf{k}_2$ is negative. Thus, although the field of the scattered waves originally arose as the result of two thermal waves of a single frequency Ω and opposite directions of propagation $\pm\mathbf{q}$, the wave of the same frequency and phase as the thermal wave which produces the Stokes component is amplified as a result of electrostriction alone. The anti-Stokes component of stimulated scattering cannot appear in this approximation.

So far as the Stokes component is concerned, it is clearly amplified, because the wave produced by electrostriction is added to the thermal wave, being identical with it in frequency and phase. And although $\mathbf{E}_1$ is initially small, because of the enormous value of $\mathbf{E}_0$, the product $\mathbf{E}_0\mathbf{E}_1$ is sufficient that the Stokes component increases the intensity, and this in turn leads to a growth in p, etc. This is also the process of parametric amplification or parametric resonance. In this process, "pumping" of energy from the light wave of the ruby laser to the light of the Stokes component and into hypersound takes place. The scattering in the reverse direction ($\theta = 180°$) as the result of parametric resonance for sufficient intensity of the incident light can immediately produce two quantum generators – a light generator at frequency ($\omega - \Omega$) and a hypersonic generator at a frequency Ω (see below). In the experiments of [595], the power of the hypersonic wave in a crystal amounted to kilowatts and it was several times larger in the experiments of [630].

2. Approximate Classical Theory of Stimulated Mandel'-shtam–Brillouin Scattering. When the author wrote this section of the book for the Russian edition, he could rely only on the first reviews of nonlinear optics [607, 608] and several original papers, devoted principally to the theory of stimulated Raman light scattering [618, 619].

The then still-unpublished work of Akhmanov and Chin Dong-A [616, 617] was devoted to the theory of stimulated MB scattering and was based on the already developed radio physical theory of the parametric interaction of waves in a nonlinear medium.

Our exposition of the simplified classical theory of stimulated MB scattering is based on the fundamental remarks in [616, 617] and has an illustrational character, dealing with the representation of the existence of the process of stimulated MB scattering.

After publication of the book, various theoretical investigations of stimulated MB scattering appeared [681-686] which should be taken into account in the more complete theoretical study of the question.

As has already been pointed out above, it is necessary for the description of stimulated MB scattering to solve simultaneously the set of Maxwell equations and those of hydrodynamics* with account of the nonlinearity of the medium, produced by the high intensity of the exciting light. After elimination of **H** from the first two Maxwell equations (1.6), we get

$$\frac{\partial^2 \boldsymbol{D}}{\partial t^2} + c^2 \operatorname{rot} \operatorname{rot} \boldsymbol{E} = 0. \tag{34.4}$$

In Eqs. (1.6), the electric displacement **D** was determined by the dielectric constant ε which fluctuates because of the thermal motion, but did not depend on the field **E**, or through the polarization **P**, which depended only on the first power of the electric field in-

---

* Here we shall consider the simple case of stimulated scattering in an isotropic liquid medium. The case of stimulated light scattering in a solid is also considered in the researches mentioned above.

tensity **E**. In the nonlinear medium, the dielectric constant becomes a function of the field or, what amounts to the same thing, the polarization depends on the second, third, and higher powers of the field [see Eq. (19.4)].

Here, we shall consider the stimulated MB scattering only in first order; therefore, one does not have to write the polarization in the form (19.4), but it suffices to consider the weak nonlinearity. In other words, we shall assume that, because of the electrostriction, an additional dielectric constant $\Delta\varepsilon(p)$ arises, which is a function of the strictive pressure and the corresponding nonlinear contribution of the polarization P^{NL}. In this case, for the electric induction, we can write the expression

$$\boldsymbol{D}=[\varepsilon^*+\Delta\varepsilon(p)]\boldsymbol{E}=\boldsymbol{E}+4\pi(\boldsymbol{P}^L+\boldsymbol{P}^{NL}). \tag{34.5}$$

The complex optical dielectric constant is $\varepsilon^* = \varepsilon' - i\varepsilon''$, where

$$\left.\begin{aligned}\varepsilon' &= n^2\left[1-\left(\frac{k_\omega}{k_i}\right)^2\right],\\ \varepsilon'' &= \frac{2n^2}{|k_i|}k_\omega.\end{aligned}\right\} \tag{34.6}$$

Here k_ω is the amplitude of the optical damping coefficient and k_i is the wave number of the light in the medium.

It follows from Eq. (34.5) that

$$\left.\begin{aligned}\boldsymbol{P}^L &= \frac{\varepsilon^*-1}{4\pi}\boldsymbol{E},\\ \boldsymbol{P}^{NL} &= \frac{\Delta\varepsilon(p)}{4\pi}\boldsymbol{E}=\frac{1}{4\pi}\left(\rho\frac{\partial\varepsilon}{\partial\rho}\right)_S\beta_S p\boldsymbol{E}.\end{aligned}\right\} \tag{34.7}$$

After substitution of (34.5) in (34.4) we get

$$\varepsilon^*\frac{\partial^2\boldsymbol{E}}{\partial t^2}+c^2\,\mathrm{rot}\,\mathrm{rot}\,\boldsymbol{E}=-Y\beta_S\frac{\partial^2}{\partial t^2}(p\boldsymbol{E}). \tag{34.8}$$

The abbreviation $Y = (\rho\,\partial\varepsilon/\partial\rho)_S$ is substituted in Eq. (34.8). The propagation of elastic waves in a dissipative medium is described by Eq. (5.10), in which we now have p in place of Φ, and on the right-hand side of Eq. (5.10) there ought to be a divergence of the ponderomotive force, which arises in a dielectric placed in an electric field. Then the equation for p will have the form:

$$\frac{\partial^2 p}{\partial t^2} - v^2\nabla^2 p - \Gamma\nabla^2\frac{\partial p}{\partial t} = -v^2 \operatorname{div} \boldsymbol{F}. \tag{34.9}$$

Γ is connected with the kinematic viscosity by Eq. (5.10) and

$$\boldsymbol{F} = \frac{1}{8\pi} Y \operatorname{grad} |E|^2. \tag{34.10}$$

Substituting (34.10) in (34.9) we get, finally,

$$\frac{\partial^2 p}{\partial t^2} - v^2\nabla^2 p - \Gamma\nabla^2\frac{\partial p}{\partial t} = -\frac{Yv^2}{8\pi}\nabla^2 E^2. \tag{34.11}$$

Equations (34.8) and (34.11) represent a set of desired nonlinear equations which, in the chosen approximation, describe the interaction of the optical and elastic waves. For an elucidation of the character of this interaction, one must find the solution of the set of Eqs. (34.8) and (34.11). However, to find the exact solution of these equations at once is not possible. Before seeking the approximate solution of (34.8), we take into consideration that the MB scattering is polarized in the same plane as the incident light. Let the direction of the electric vector of the light wave be parallel to the z axis. The direction of propagation of the incident and scattered light lies in the x,y plane. We then get, in place of Eq. (34.8),

$$\varepsilon^* \frac{\partial^2 E_z}{\partial t^2} - c^2\left(\frac{\partial^2 E_z}{\partial x^2} + \frac{\partial^2 E_z}{\partial y^2}\right) = -Y\beta_S \frac{\partial^2}{\partial t^2}(pE_z). \tag{34.12}$$

The method of the small parameter of Van der Pohl, the mathematical basis of which is given by Mandel'shtam and Papaleskii [620], is an adequate approximate method of solution of the set of equations (34.11) and (34.12). This method and its application are given in sufficient detail in the book of Andronov, Vitt, and Khaikin [621] and in the already mentioned book of Akhmanov and Khokhlov [607]. The nonlinear terms in the set of equations under consideration are small in comparison with the linear. This means that the spatial change of p and E per unit wavelength is small.

The small nondimensional parameter of Van der Pohl μ in this case can be characterized by the quantities $Y\beta_S(p_0/\varepsilon') = \mu \ll 1$, $\Delta\varepsilon(p)/\varepsilon' = \mu \ll 1$. The optical losses are also assumed to be small: $\mu = \varepsilon''/\varepsilon' \ll 1$, and the acoustic losses $\alpha\Lambda = \mu \ll 1$, etc. The smaller the value of μ, the closer will be the approximate solution of the set of equations (34.11), (34.12) to their exact solution. It is further assumed that the nondimensional parameter μ has the same order of smallness in the cases mentioned.

We shall seek the solution of the set of equations by assuming that there are four waves present in the medium:

1. The incident wave (frequency ω and wave vector $\mathbf{k}$)
2. The sound or hypersonic wave (frequency Ω and wave vector $\mathbf{q}$)
3. The scattered light wave (frequency $\omega_1 = \omega - \Omega$ and wave vector $\mathbf{k}_1'$) – the Stokes component
4. The scattered light wave (frequency $\omega_2 = \omega + \Omega$ and wave vector $\mathbf{k}_2'$) – the anti-Stokes component.

In this case, we shall seek a solution in the form of the superposition of these waves with slowly changing amplitude*

$$E = \sum_{n=0}^{2} E_n(\mu r)\, e^{i(\omega_n t - k_n r)} + \text{complex conjugate}, \tag{34.13}$$

$$p = p(\mu r)\, e^{i(\Omega t - qr)} + \text{complex conjugate} \tag{34.14}$$

(in what follows, the index 0 on $\mathbf{E}$ will be omitted).

In the solutions, written in the form (34.13) and (34.14), the time dependence of the amplitude is not considered and, consequently, the time derivatives will be omitted below in the equations for $E_n(n, r)$ and $P(n, r)$.

Therefore, the stationary solutions obtained below can be applied only to a limited region of space and time of interaction of the waves.

---

* One could have made this solution more accurate in the zero approximation by the addition of terms of order μ, but in setting up the solution in such a description, terms of order μ give quantities of order μ^2, which are also thrown away. Therefore, these correction terms are not considered in (34.13) and (34.14).

The solutions which take transition processes into account were obtained by Kroll [685]. Substituting (34.13) in (34.12), and (34.13) in (34.11), and equating the coefficients of the corresponding exponents, we obtain a set of so-called abridged equations, which, with accuracy up to μ^2, are identical with the initial set of equations (34.11) and (34.12):

$$\left.\begin{aligned} &(\boldsymbol{k}\nabla)E+\frac{\varepsilon''\omega^2}{2c^2}E+i\,\frac{Y\beta_S\omega^2}{2c^2}(pE_1e^{-i\Delta_1r}+p^*E_2e^{i\Delta_2r})=0;\\ &(\boldsymbol{k}'_1\nabla)E_1+\frac{\varepsilon''\omega_1^2}{2c^2}E_1+i\,\frac{Y\beta_S\omega_1^2}{2c^2}Ep^*e^{i\Delta_1r}=0;\\ &(\boldsymbol{k}'_2\nabla)E_2+\frac{\varepsilon''\omega_2^2}{2c^2}E_2+i\,\frac{Y\beta_S\omega_2^2}{2c^2}Epe^{i\Delta_2r}=0;\\ &(\boldsymbol{q}\nabla)p+\frac{\Gamma q^2\Omega}{2v^2\left(1+i\frac{\Gamma\Omega}{v^2}\right)}+\frac{iY}{16\pi\left(1+i\,\frac{\Gamma\Omega}{v^2}\right)}[EE_1^*(\boldsymbol{q}-\boldsymbol{\Delta}_1)^2e^{i\Delta_1r}+\\ &\qquad+E^*E_2(\boldsymbol{q}-\boldsymbol{\Delta}_2)e^{i\Delta_2r}]=0.\end{aligned}\right\}\quad(34.15)$$

In Eqs. (34.15) the following notation has been introduced:

$$\left.\begin{aligned}\boldsymbol{\Delta}_1&=\boldsymbol{q}-\boldsymbol{k}+\boldsymbol{k}'_1,\\ \boldsymbol{\Delta}_2&=\boldsymbol{q}+\boldsymbol{k}-\boldsymbol{k}'_2.\end{aligned}\right\}\quad(34.16)$$

Here $\boldsymbol{\Delta}_1$ and $\boldsymbol{\Delta}_2$ take into account a certain "disorder" or departure from the Bragg condition (1.19). Therefore, Eqs. (34.15) can be applied for investigation of the parametric amplification of sound by transfer into the sound wave of the energy of the light wave or two light waves of different frequencies.

Experiments on the direct amplification of sound from the energy of the light wave and generation of sound from the nonlinear interaction of two light waves of different frequency have been carried out by Korpel, Adler, and Alpiner [622] in a liquid and by Caddes, Quate, and Wilkinson [687] in a solid.

The most effective interaction of the waves takes place when $\boldsymbol{\Delta}_1=\boldsymbol{\Delta}_2=0$, i.e., when the Bragg condition is fulfilled exactly.

The Bragg condition is automatically satisfied for scattered light. Therefore, one can set $\boldsymbol{\Delta}_1$ and $\boldsymbol{\Delta}_2$ equal to zero in Eqs. (34.15). In addition, one must take into account that Eqs. (34.11) are valid when $\Gamma\Omega/v^2=\alpha\Lambda/\pi\ll 1$ (see Sec. 5), and that some effective interaction of the Stokes and anti-Stokes light waves with the same elastic wave is possible only for small angles of scattering. For angles for which such an interaction is unimportant, the set of Eqs. (34.15) splits into two independent sets of equations of the following form:

$$\left.\begin{aligned}&\frac{\partial E}{\partial x}+d_k E+ig_k pE_1=0,\\&\frac{\partial E_1}{\partial x}+d_{k_1}E_1+ig_{k_1}p^*E=0,\\&\frac{\partial p}{\partial x}+d_q p+ig_q EE_1^*=0;\end{aligned}\right\}\qquad(34.17)$$

$$\left.\begin{aligned}&\frac{\partial E}{\partial x}+d_k E+ig_k p^*E_2=0,\\&\frac{\partial E_2}{\partial x}+d_{k_2}E_2+ig_{k_2}pE=0,\\&\frac{\partial p}{\partial x}+d_q p+ig_q E^*E_2=0.\end{aligned}\right\}\qquad(34.18)$$

Taking (34.6) into account, one can write the contractions introduced in (34.17) and (34.18) in the following fashion:

$$\left.\begin{aligned}&d_{k_i}=\frac{nk_\omega}{\cos(\widehat{\boldsymbol{k}_i\boldsymbol{x}})},\quad g_{k_i}=\frac{Y\beta_S|k_i|}{2\cos(\widehat{\boldsymbol{k}_i\boldsymbol{x}})},\\&d_q=\frac{\alpha}{\cos(\widehat{\boldsymbol{q}\boldsymbol{x}})},\quad g_q=\frac{Y|\boldsymbol{q}|}{16\pi\cos(\widehat{\boldsymbol{q}\boldsymbol{x}})}.\end{aligned}\right\}\qquad(34.19)$$

<u>Stimulated Mandel'shtam–Brillouin Scattering near the Threshold Value of the Intensity of the Exciting Light.</u> It is of interest to elucidate how the intensity of the scattered light and the intensity of the elastic wave behave immediately after the threshold of stimulated scattering has been reached, when the intensities of the Stokes and anti-Stokes components are still much less than the intensity of the incident light and, consequently, it can be assumed that the field intensity of the incident wave does not depend on the coordinates ($\partial E/\partial x = 0$). For most graphic results, we assume initially that the optical and acoustical losses are absent: $d_k = d_q = 0$.* Under

* Later we shall return to the case in which the losses and $\partial E/\partial x \neq 0$.

the simplifying assumptions that have been made, we consider the following two cases for Stokes and anti-Stokes components separately.

1. The projections of the wave vectors of the MB components and the elastic wave on the direction of the wave vector of the incident light have the same direction (traveling wave). This means that the scattering angle lies in the limits $0° \leq \theta \leq 90°$.

2. The projection of the wave vectors of the MB components and the elastic wave on the direction of the wave vector of the incident light have opposite directions (reverse wave). This means that the scattering angle lies in the limits $90° \leq \theta \leq 180°$.

We now consider the first case for the Stokes component.

Case 1. Under the assumptions made, Eqs. (34.17) reduce to two equations, which can be written in the following form:

$$\left.\begin{aligned} &\frac{\partial E_1}{\partial x} + i g_{k_1} E p^* = 0, \\ &\frac{\partial p^*}{\partial x} - i g_q E_1 E^* = 0. \end{aligned}\right\} \tag{34.20}$$

(Here $g_k > 0$ and $g_q > 0$.) The solution of (34.20) for the boundary conditions $\mathbf{E}_1(0) = 0$ and $p(0) = p_0$ will be

$$\left.\begin{aligned} &|E_1(x)|^2 = \frac{g_k}{g_q} p_0^2 \operatorname{sh}^2(E\sqrt{g_{k_1} g_q}\, x), \\ &|p(x)|^2 = p_0^2 \operatorname{ch}^2(E\sqrt{g_{k_1} g_q}\, x). \end{aligned}\right\} \tag{34.21}$$

The solution (34.21) shows that the intensities of the Stokes MB components and of the sonic or hypersonic waves increase exponentially in the direction of propagation (Fig. 117a), i.e., spatial instability appears. In order that generation take place (temporal instability), it is necessary to place the medium possessing the properties (34.21) in a resonator.

Case 2. For the Stokes component under these same assumptions, and for the boundary conditions $E_1(l) = 0$, $p(0) = p_0$, we get

$$\left.\begin{aligned} &|E_1(x)|^2 = \frac{|g_{k_1}|}{g_q} |p_0|^2 \frac{\sin^2\left[\sqrt{|g_{k_1}| g_q |E|^2}\,(x-l)\right]}{\cos^2\left[\sqrt{|g_{k1}| g_q |E|^2}\, l\right]}, \\ &|p(x)|^2 = |p_0|^2 \frac{\cos^2\left[\sqrt{|g_{k_1}| g_q |E|^2}\,(x-l)\right]}{\cos^2\left[\sqrt{|g_{k_1}| g_q |E|^2}\, l\right]}. \end{aligned}\right\} \tag{34.22}$$

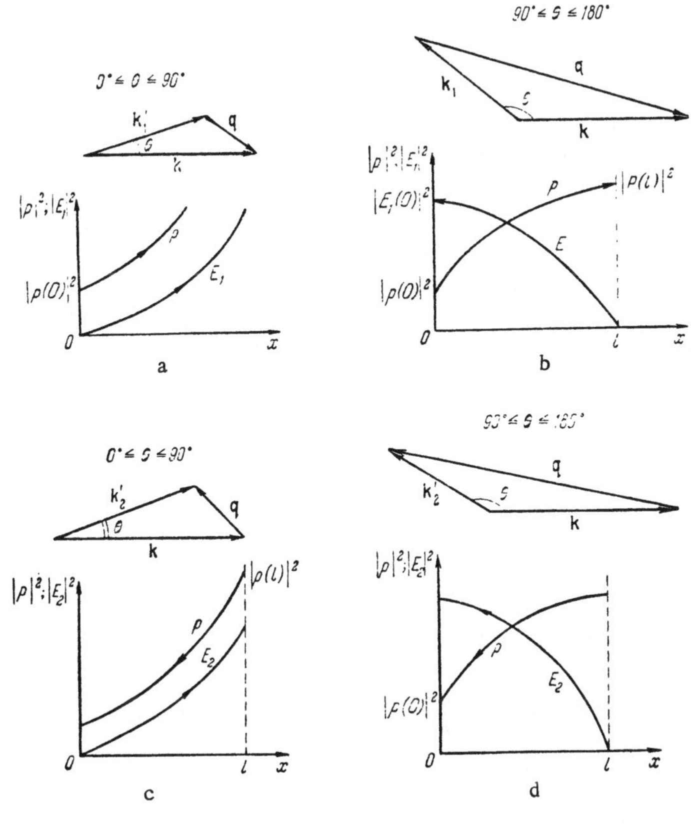

Fig. 117. Character of the intensity change of a light wave and of the pressure in a sound wave in various cases (Akhmanov and Chin Dong-A [616, 617]). (a) Stokes component; the scattering angle lies in the first-fourth quadrants; (b) Stokes component; the scattering angle lies in the second-third quadrants; (c) anti-Stokes component; the scattering angle lies in the first-fourth quadrants; (d) anti-Stokes component; the scattering angle lies in the second-third quadrants.

The dependence of $|E_1(x)|^2$ and $|p(x)|^2$ on the coordinate x is shown graphically in Fig. 117b. It is very important to emphasize that, for a given value of l, equal to

$$l=\frac{\pi}{2\sqrt{g_{k_1}g_qE^2}}, \tag{34.23}$$

the system becomes unstable in time, the intensities of the Stokes MB components and the corresponding hypersonic waves increase unboundedly within the framework of the given approximations and, therefore, generation of light is possible at the frequency $\omega - \Omega$ and of hypersound at the frequency Ω without use of the resonator. For the anti-Stokes MB component in Case 1, the solution of the set of equations (34.18), under the same assumptions and for boundary conditions $p(l) = p_l$ and $E(0) = 0$, will be (Fig. 117c)

$$|E_2(x)|^2=p_l^2\frac{|g_q|}{g_{k_2}}\frac{\mathrm{sh}^2\left(\sqrt{g_{k_2}|g_q|E^2}\,x\right)}{\mathrm{ch}^2\left(\sqrt{g_{k_2}g_qE^2}\,l\right)}, \tag{34.24}$$

$$|p(x)|^2=p_l^2\frac{\mathrm{ch}^2\left(\sqrt{g_{k_2}|g_q|E^2}\,x\right)}{\mathrm{ch}^2\left(\sqrt{g_{k_2}|g_q|E^2}\,l\right)}. \tag{34.25}$$

For the anti-Stokes component in Case 2, under the same assumptions and for the boundary conditions $E_2(l) = 0$, $p(l) = p_l$, the solutions are

$$|E_2(x)|^2=p_l^2\frac{g_{k_2}}{g_q}\sin^2\left[\sqrt{|g_{k_2}||g_q|E^2}\,(x-l)\right], \tag{34.26}$$

$$|p(x)|^2=p_l^2\cos^2\left[\sqrt{|g_{k_2}||g_q|E^2}\,(x-l)\right]. \tag{34.27}$$

The graphic representations of (34.26) and (34.27) are shown in Fig. 117c. It is interesting to note that in the observation of light scattered forward, the intensity of the Stokes component can be much stronger than that of the anti-Stokes. Actually, from (34.22), (34.25), and (34.27), one can form the ratio

$$\left|\frac{E_1(x)}{E_2(x)}\right|^2=\frac{1}{\cos^2\left[\sqrt{|g_{k1}||g_q|E^2}\,l\right]}. \tag{34.28}$$

The ratio (34.28) can become very great if l approaches the value expressed by Eq. (34.23). We also note that the intensity of the anti-Stokes component of (34.26) and (34.27) always remains finite in contrast with the intensity of the Stokes component (34.22).

The neglect of optical and acoustical losses leads to the result that in the case of nonlinear interaction of the waves, stimulated scattering can arise for an arbitrarily low intensity of the incident light. Actually, this is not the case. Real optical and acoustical losses do not allow stimulated scattering to develop at an arbitrary intensity of the incident light. The stimulated scattering arises only when the intensity of the incident light exceeds the given threshold value.

For an estimate of the threshold value of the intensity of the Stokes component, we find the solution of the set of Eqs. (34.17) by assuming that $d_k \neq 0$, $d_q \neq 0$, and that, as before, $\partial E/\partial x = 0$. For Case 1, $0° \leq \theta \leq 90°$ (traveling wave), the solution of the set of equations (34.17) will be

$$E_1(x) = e^{-dx}(C_1 e^{ax} + C_2 e^{-ax}),$$

$$p(x) = e^{-dx}(B_1 e^{ax} + B_2 e^{-ax}). \qquad (34.29)$$

Here C_1, C_2, B_1, and B_2 are constants, $d = (d_{k_1} + d_q)/2$, and $a = \frac{1}{2}[(d_{k_1} - d_q)^2 + 4g_{k_1}g_{k_q}E^2]^{1/2}$.

In the solution (34.29), the exponential growth of E(x) and p(x) will exist only when

$$a - d > 0. \qquad (34.30)$$

Substituting the values of a and d obtained for the threshold values of the square of the field intensity in this last inequality, we get

$$E^2_{\text{thres}} > \frac{d_{k_1} d_q}{g_{k_1} g_q} = \frac{32\pi n k_{\omega}\alpha}{\left(\rho \frac{\partial \varepsilon}{\partial \rho}\right)^2_S |k_1||q|\beta_S}. \qquad (34.31)$$

For Case 2, $90° \leq \theta \leq 180°$ (reverse wave), proceeding as in the previous case, we find the condition for which $E_1(x)$ and p(x) oscillate [see Eq. (34.22)]. From this condition, we find the threshold value

$$E^2_{\text{thres}} > \frac{(d_{k_1} + d_q)^2}{4g_{k_1}g_q}. \qquad (34.32)$$

For the scattering angle $\theta = 180°$, we get from (34.32)

$$E^2_{1\,\mathrm{thres}} > \frac{8\pi(\alpha + nk_\omega)^2}{\left(\rho\frac{\partial\varepsilon}{\partial\rho}\right)^2_S |q||k|\beta_S}. \tag{34.33}$$

If we assume that $d_q > d_k$, then the threshold in Case 2 is much higher than in Case 1.

For an estimate of the threshold of the anti-Stokes component, it is necessary to proceed in the same way as for the estimate of the threshold of the Stokes component, only to solve the set of equations (34.18). The result is obtained that the expression for the threshold value of the square of the intensity for the anti-Stokes component $E^2_{2\mathrm{thres}}$ for Case 1 is identical in form with the expression (34.31), and it is only necessary to replace d_{k_1} and g_{k_1} by d_{k_2} and g_{k_2}. For Case 2, the threshold of the anti-Stokes component will be expressed by the same formula as (34.32) with the corresponding substitution entering into it [616, 617]. The expressions that are given for the threshold values of the square of the intensity of the light field (34.31) and (34.32) are similar to the corresponding expressions given in [595].

In the solution of the set of equations (34.17) and (34.18), it was assumed that up to the boundary of the volume in which the stimulated scattering arises, the intensity of the MB components is not equal to zero, but is determined by the thermal scattering. The value of the intensity of the components of the thermal scattering depends on the nature of the scattering material, the intensity of the incident light, and on the scattering volume or its linear dimensions up to the boundary at which the stimulated scattering arises. If we take into account that the intensity of the MB components is initially different from zero, then the value of the threshold is not changed. Account of the thermal scattering is reflected only in the value of the intensity of the stimulated scattering.

Up to the present time, the problem of the intensity change of the stimulated scattering has been solved under the assumption that the intensity of the Stokes and anti-Stokes components is small in comparison with the intensity of the incident light ($\partial E/\partial x = 0$). This assumption made it possible to simplify the set of equations (34.17), to find their solution, and to trace the character of the intensity change of the lines of stimulated scattering in the different cases. This stage of the solution of the problem is important because the behavior of the intensities of the components of the fine structure of the scattered light and the acoustic waves immediately after the threshold determines, to a great extent, the further fate of the system. It is difficult to show whether the initial stages of the development of the phenomenon of stimulated scattering can be observed in a real experiment because, in this phase, the intensity of the shifted components is very small. Until now, the phenomenon has been observed when the intensity of the lines of stimulated scattering was already comparable with the intensity of the incident line. Consequently, it is impossible to assume that $\partial E/\partial x = 0$ in this case, and for the description of the experiment it is necessary to return to the set of equations (34.17) and (34.18) by assuming that $\partial E/\partial x \neq 0$.

Large Intensity of the Mandel'shtam–Brillouin Components in Stimulated Light Scattering. The problem of the interaction of waves of the incident and scattered light in an elastic wave when the intensity of the scattered wave is comparable with the intensity of the incident light (for Case 1), $0 \leq \theta \leq 90°$, reduces to the problem of parametric amplification of a traveling wave.

Pogorelova and Khokhlov [623] developed the nonlinear theory of parametric amplification in traveling waves with account of damping of the system, while Jurkys et al. [624] and Tienkin [625] constructed a theory of parametric amplification without account of damping. Akhmanov and Khokhlov [607] transferred these theoretical radiophysical representations to the description of parametric effects in optics.

The solution of the interaction of waves for $\partial E/\partial x \neq 0$ in (34.17) and (34.18) can be obtained comparatively simply in analytic form if we neglect the damping of the system in Eqs. (34.17) and (34.18). Under this assumption, we write the equation and solve them similarly to [624, 607]. We shall then show how the changes arise due to the damping, as was done in [623].

For account of strong interaction between waves, one must write the equation with account of the amplitudes and phases of these waves. By setting $d_{k_1} = d_q = 0$ in (34.17), and replacing E by $Ee^{i\varphi_k}$, E_1 by $E_1e^{i\varphi_{k_1}}$, and p by $pe^{i\varphi_q}$, we find the abbreviated equations for the actual amplitudes and phases of the incident light of the Stokes component and the elastic acoustic wave:

$$\left.\begin{aligned}&\frac{\partial E}{\partial x}-g_k pE_1\sin\Psi=0,\\&\frac{\partial E}{\partial x}+g_{k_1}pE\sin\Psi=0,\\&\frac{\partial p}{\partial x}+g_q E_1 E\sin\Psi=0,\\&\frac{\partial\Psi}{\partial x}+\left(g_q\frac{E_1E}{p}+g_{k_1}\frac{pE}{E_1}-g_k\frac{pE_1}{E}\right)\cos\Psi=0,\end{aligned}\right\}\tag{34.34}$$

where $\Psi = \varphi_k - \varphi_{k_1} - \varphi_q$.

We shall assume that the value of Ψ was established and does not change appreciably; therefore, $\partial\Psi/\partial x = 0$, $\cos\Psi = 0$, while $\sin\Psi = 1$. This question was considered in detail in [607]. Then (34.34) reduces to a system consisting of the first three equations in which $\sin\Psi = 1$. We find the solution of this set of equations for the following boundary conditions ($x = 0$):

$$\left.\begin{aligned}&E(0)=E_0,\ E_1(0)=0,\ p(0)=p_0,\\&\Psi(0)=\Psi_0=\frac{\pi}{2}.\end{aligned}\right\}\tag{34.35}$$

For the boundary conditions (34.35), the solution of the set of equations (34.34) can be obtained in terms of the elliptic Jacobi functions

$$E=E_0\,\mathrm{sn}\left(\mathcal{K}+\frac{\alpha_0}{k}x\right),\tag{34.36}$$

or

$$E=E_0\,\mathrm{cd}\left(\frac{\alpha_0}{k}x\right),\tag{34.36a}$$

$$E_1=\sqrt{\frac{g_{k_1}}{g_k}}E_0\,\mathrm{cn}\left(\mathcal{K}+\frac{\alpha_0}{k}x\right),\tag{34.37}$$

or

$$E_1=-\sqrt{\frac{g_{k_1}}{g_q}}p_0k\,\mathrm{sd}\left(\frac{\alpha_0}{k}x\right),\tag{34.37a}$$

$$p(x)=\sqrt{\frac{g_q}{g_k}}E_0\frac{1}{k}\mathrm{dn}\left(\mathcal{K}+\frac{\alpha_0}{k}x\right),\tag{34.38}$$

or

$$p(x) = p_0 \,\mathrm{nd}\left(\frac{\alpha_0}{k}x\right). \qquad (34.38a)$$

Here we have introduced the notation: $\mathcal{K}$ is the complete elliptic integral of the first kind:

$$\mathcal{K} = \int_0^1 \frac{dy}{[(1-y^2)(1-k^2y^2)]^{1/2}},$$

$$k = \left(1 + \frac{g_k E_1^2}{g_{k_1} E_0^2}\right)^{-1/2}, \quad \alpha_0 = E_0 \sqrt{g_k g_q}.$$

All three of Eqs. (34.36)-(34.38) are periodic functions of the coordinates.

From the expressions for E(x), E_1(x), and p(x), one can form expressions which describe the intensities (energy flux), which change along the x axis. Such intensities will be proportional to the squares of (34.36a), (34.37a), and (34.38a) [624]:

$$I(x) \sim |E(x)|^2 = I_0 \,\mathrm{cd}^2\left(\frac{\alpha_0}{k}x\right), \qquad (34.39)$$

$$I_1(x) \sim |E_1(x)|^2 = \frac{g_{k_1}}{g_k} p_0^2 k^2 \mathrm{sd}^2\left(\frac{\alpha_0}{k}x\right), \qquad (34.40)$$

$$S_q(x) \sim |p(x)|_2 = S_{q0} \,\mathrm{nd}_2\left(\frac{\alpha_0}{k}x\right). \qquad (34.41)$$

The dependence of the intensities (34.39)-(34.41) on the x coordinate is shown graphically in Fig. 118. The graph shows that as the incident light moves along the x axis inside the medium, a transfer of energy takes place from the incident light to the light of the Stokes MB component and to the elastic wave.* Furthermore, the

---

*The graphs of Fig. 118 are taken from [624], in which the parametric amplifier is described; we have changed the notation: pumping for incident light, signal for the sound wave, and the idler wave for the wave of the MB component. Inasmuch as the graph is shown as an illustration, the lack of scale correspondence is of no importance.

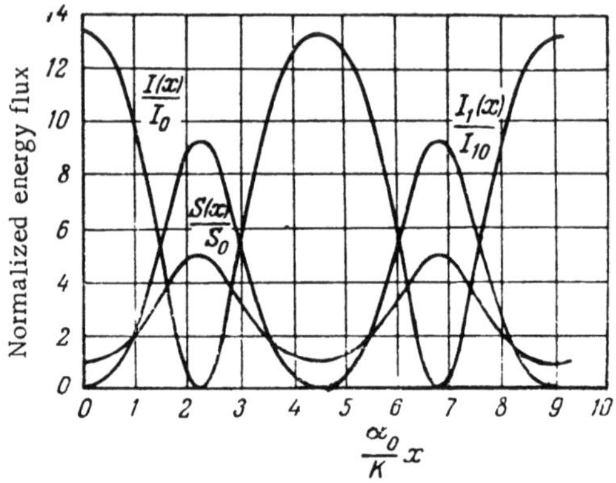

Fig. 118. Spatial change of $I(x)/I_0$; $I_1(x)/I_{10}$, and $S(x)/S_0$ (the intensity of the incident light, the Stokes MB component, and the sound wave) for $k^2 = 0.8$.

MB component and the elastic wave transfer their energy to the incident light, etc. In other words, the interactions of the waves under consideration have the character of spatial beats.

The spatial period of these beats is

$$\Lambda_0 = \frac{2\mathcal{K}k}{\alpha_0}.$$

If $(\omega/\Omega)(p_0^2/I_0) \ll 1$ and $k \to 1$ (i.e., the intensities of the MB components and the sound wave are small), then the elliptic functions (34.39)-(34.41) transform to the hyperbolic:

$$|E(x)|^2 = I_0 = \text{const},$$
$$|E_1(x)|^2 = p_0^2 \frac{g_k}{g_q} \operatorname{sh}^2(\alpha_0 x),$$
$$|p(x)|^2 = p_0^2 \operatorname{ch}^2(\alpha_0 x),$$

and we obtain expressions that are identical with those already obtained earlier (34.21) for $I_1 \ll I_0$ and $S_{q0} \ll I_0$. If there are no optical and acoustical losses, then the law of energy conservation should be satisfied, i.e., one can write the following equation from (34.39)-(34.41):

$$I+I_1+S_q=I_0+S_{q0}, \tag{34.42}$$

which should be satisfied for all values of x. When the energy of the incident wave is transformed into the energy of the Stokes MB component, and the energy of the sound wave, then these latter have a maximum amplification and an energy distribution following that of Manley and Rowe [626]. From (34.41), the maximum amplification can be estimated

$$S_q(x)_{\max}=S_{q0}+\frac{\Omega}{\omega}\frac{v}{c}I_0, \tag{34.43}$$

where Ω, ω are the frequencies of the sound and the exciting light. For $\Omega \sim 10^{10}$, $\omega \sim 10^{14}$ sec^{-1}, and $v/c \sim 10^{-5}$, $S(x)_{\max}$ can be increased to $\sim 10^{-9} I_0$.

In the case of the MB component, the frequency $\omega_1 \sim \omega$, and the intensity of the component can be of the order of the intensity of the incident light ($I_{1\max} \approx I_0$).

Case 2: forward light scattering, when the scattering angle lies in the range $90° \le \theta \le 180°$ (reverse wave), is especially interesting because, under the assumption $\partial E/\partial x = 0$, solution of the set of equations (34.17) showed that instability arises for definite conditions, and I_1 and p^2 can increase indefinitely.

The solution of the set of equations (34.34) under the assumption that $\partial E/\partial x \neq 0$, $d_k = d_q = 0$, and for the boundary conditions

$$E(0)=E_0,\quad p(0)=p_0,\quad E_1(l)=0 \quad \text{and} \quad \Psi(0)=\Psi_0=\frac{\pi}{2}$$

was obtained by Akhmanov and Chin Dong-A [616, 617].

The solution here can be written in the following fashion:

$$\left.\begin{aligned} I(x) &\sim |E(x)|^2 = \left(E_0^2+\frac{g_k}{g_q}p_0^2\right)\left\{1-\frac{\operatorname{sn}^2[\mathcal{K}+\xi(x-l)]}{1+\beta_1}\right\},\\ I_1(x) &\sim |E_1(x)|^2 = \frac{g_{k_1}}{g_q}p_0^2\frac{1+\beta_2}{1+\beta_1}\operatorname{cn}^2[\mathcal{K}+\xi(x-l)],\\ p^2(x) &= p_0{}^2\frac{1+\beta_2}{1+\beta_1}\operatorname{sn}^2[\mathcal{K}+\xi(x-l)],\end{aligned}\right\} \tag{34.44}$$

where $\mathcal{K}$ is the complete elliptic integral with the parameter

$$k=(1+\beta_1)^{-1/2}, \tag{34.45}$$

and the abbreviations

$$\xi=\left[g_{k_1}(g_q E_0^2+g_k p_0^2)\right]^{1/2},$$
$$\beta_1=\frac{g_q E_0^2}{g_k p_0^2},\quad \beta_2=\frac{g_q E^2}{g_k p_0}.$$

are introduced. By comparing Eqs. (34.44) with Eqs. (34.39)-(34.41), one can see that even in this case (reverse wave), the interaction of the waves has the character of spatial beats with a period determined above [see p. 517, where k should be determined by the value of (34.45)].

It follows from the solutions of (34.44) that, as a result of the interaction of the waves, the intensity of the Stokes component cannot increase without limit under any conditions. The intensity of the Stokes component can only approach the value of the intensity of the incident light, while the maximum value of the intensity of the elastic wave, as also in the previous case, can be estimated from a relation of the form (34.43) [616, 617] and, consequently, the maximum will be $\sim(\Omega/\omega)I_0$.

In the real case, all three waves in both cases considered are damped to a greater or lesser extent. The effect of damping on the interaction of the three waves in Case 1 (traveling wave), as has already been pointed out, was considered by Pogorelova and Khokhlov [623]; it was found that the picture of spatial beats (Fig. 118) will be preserved; however, with increase of x, the amplitude of the beats will fall off. In the case when the damping is strong, then only a single maximum will be observed in the x dependence of the intensity of each of the three waves.

For the consideration of the problem of stimulated MB scattering, it was assumed that the scattering medium is irradiated by a plane wave while, in the real experiment, the laser light is focused inside the medium to achieve intensities exceeding the threshold and, consequently, the medium is irradiated by spherical waves. But such is not the case if we do not take diffraction into consideration in the formation of the image. The diffraction theory of the spatial distribution of the intensity in the region of a geometric focus of an ideal lens was developed by Linfoot and Wolf and treated in detail in [391]. According to this theory, the maximum intensity will be on the optic axis at a point corresponding to the geometric focus:

$$I_{max} = I_0 \frac{\pi^2 a^4}{16\lambda^2 f^2 n^2}. \tag{34.46}$$

Here, a is the lens diameter, λ is the wavelength of light in the medium, and f is the focal length of the lens in free space.

The isophotes in the immediate vicinity of the focus have the shape of ellipsoids of revolution. Therefore, one can select a small cylinder for which one can assume that the light wave is plane. The dimensions of this cylinder, as follows from the meridional cross section of the intensity distribution in the region of the geometric focus [391], is the following: the length of the cylinder $l'' \approx 8\lambda(f/a)^2$ and diameter $d'' \approx 2\lambda(f/a)$, volume of the cylinder $V'' \approx 8\pi\lambda^3(f/a)^4$. Appreciable stimulated MB scattering takes place in this cylinder.

In the real case, because of the imperfection of the lens, the size of the cylinder will be much greater and the intensity of the light inside the cylinder will be much less than that computed from the formula (34.46).

A structure of the light beam of the laser, incident on the collecting lens, can also appear in the intensity in the focal region. An approximate estimate, according to (34.31) (traveling wave), of the threshold intensity for excitation of stimulated MB scattering shows that for benzene, for example [616, 617], for irradiation by a gigantic pulse of a ruby laser ($\lambda \sim 7000$ Å), $I_{thres} \approx 10$ MW/cm^2. For the case of the reverse wave, the estimate of threshold from Eq. (34.33) gives a value $I_{thres} \approx 7 \cdot 10^5$ MW/cm^2.

The experimental investigations of Brewer [668] on the dependence of the intensity of the components of stimulated MB scattering on the pulse intensity and the optical path length of the unfocused light of a gigantic pulse (parallel beam) in a test cell (n-hexane) shows that the experimental result can be described by the Kroll theory [685] for appropriate choice of the Pockel constant and acoustic absorption (optical absorption is neglected). As the optical path length of the ray increases, the intensity of the component of stimulated MB scattering gradually tends toward a certain limiting value, which depends on the initial intensity of the incident light.

However, these experiments do not eliminate the threshold for stimulated MB scattering for a very short optical path.

35. Stimulated Scattering of Light of the Wing of the Rayleigh Line

Qualitative Explanation of the Phenomenon. Observation of the Phenomenon and Its Basic Characteristics. The depolarized light spectrum of thermal scattering by time-dependent orientation fluctuations or, in other words, the wing of the Rayleigh line, and its nature, were discussed in detail in Chapters II and VIII of this book.

The statistical thermal fluctuations of the orientation of the axes of anisotropic molecules and their change with time change the conditions of light propagation in a medium where such fluctuations exist.

On the other hand, there should also be the reciprocal effect of the field of the light wave on the orientation of anisotropic molecules, and on the character of motion of the medium.

The orienting force **F**, due to the presence of the electric field, acts on an individual nonpolar molecule with an axis of symmetry and is proportional to the square of the electric field intensity and inversely proportional to the energy of thermal motion kT, which tends to disorient the anisotropic molecules [151]:

$$\mathbf{F} \propto \frac{b_1 - b_2}{kT}\mathbf{E}^2, \tag{35.1}$$

where b_1, $b_2 = b_3$ are the principal polarizabilities of the molecule and **E** includes the sum of all electric fields of the incident and scattered light.

The field intensity in the usual (pre-laser) light sources used for excitation of scattering was so small that the force corresponding to this field (35.1) could not change the character of the motion of the molecules of the medium to any extent.

However, if the gigantic pulse of a ruby laser is used for the excitation of scattering, then a very appreciable orienting force (35.1) arises which is capable of altering the character of the motion of the molecules of the medium, of creating a preferential direction in them, and thus producing an anisotropic medium.

In the expression for the orienting force (35.1), a constant component, a high-frequency component, and a low-frequency component will all be present.

The constant component will orient the molecule in the same way as the Kerr effect when a constant field is applied. This part of the force **F** can play a very important role in the formulation of a self-focusing channel.

The high-frequency components (the frequency of which is of the order of the light frequency or greater) show no effect on the orientation of the molecules, inasmuch as the relaxation time of the anisotropy τ is much greater than the period of the high-frequency light vibrations.

The low-frequency components will produce a time-varying orientation of the molecules, in which the process will be the more effective the closer the frequency to $1/\tau$. These components create a time-varying anisotropy and, consequently, a time-dependent index of refraction, which modulates the scattered light and changes the character of the propagation of the light wave.

As a result of such a process, a parametric interaction arises between the incident light, the scattered light, and the "wave" anisotropy.

Such an interaction leads to the appearance of a new phenomenon — the stimulated scattering of the wing of the Rayleigh line (SSWRL).*

---

* This phenomenon is sometimes called stimulated inelastic Rayleigh scattering. In the scattered light spectrum, the central component of the fine structure is called by some the Rayleigh line (in principle, stimulated scattering can also be observed from this line, although this has not yet happened), which can have no relation to the effect considered. Therefore, we keep the term SSWRL.

As has been shown, the most effective interaction will occur when the variable component of the force (35.1) will change in time with frequency $\Delta\omega = 1/\tau$; therefore, in the stimulated scattering, one should observe a maximum in the halfwidth of the thermal wing of the Rayleigh line.

The new phenomenon qualitatively described here – the scattering of the light of the wing of the Rayleigh line – was discovered by Mash et al. [631].

After a description of the basic experimental results and effects related to the new phenomenon, we shall turn to the approximate classical theory of SSWRL.

The observation of SSWRL was completed on the apparatus whose scheme is shown in Fig. 111. The power of the gigantic light pulse of the ruby laser was about 100 MW and consisted of radiation with a spectral width of about 0.02 cm^{-1}. The lens L_1 has a focal length $f = 3$ cm. The light pulse was focused inside the test cell. The scattered light was observed at a scattering angle equal to 0°. The stimulated light scattering of the wing of the Rayleigh line was discovered in carbon bisulfide and nitrobenzene at room temperature, and in salol at a temperature of 170°C.

In benzene, toluene, acetic acid, and triacetin, SSWRL was not observed in experiments under these same conditions [631]. Figure 119 shows the spectrum of SSWRL in the three liquids already mentioned. The spectrum of SSWRL naturally has a different extent in different materials. The greatest width of the SSWRL was observed in carbon bisulfide (Figs. 119a and 119b), where it amounted to 15 cm^{-1}, while, for a still more powerful ray of the laser (~300 MW), the photographs taken on the spectrograph showed that the extent of the SSWRL spectrum reaches ~30 cm^{-1}. It was not possible to observe the maximum of the intensity in the stimulated wing of the Rayleigh line in carbon bisulfide because of the broad apparatus function of the Fabry–Perot interferometer and the spectrograph. In addition, smeared-out components of the stimulated MB scattering could be found between the predicted maximum and the shifted line.

The SSWRL spectrum in nitrobenzene has a total spread ~1 cm^{-1} (Fig. 119d). In this photograph one can see the projected maximum at a distance of ~0.5 cm^{-1}, but accurate determination

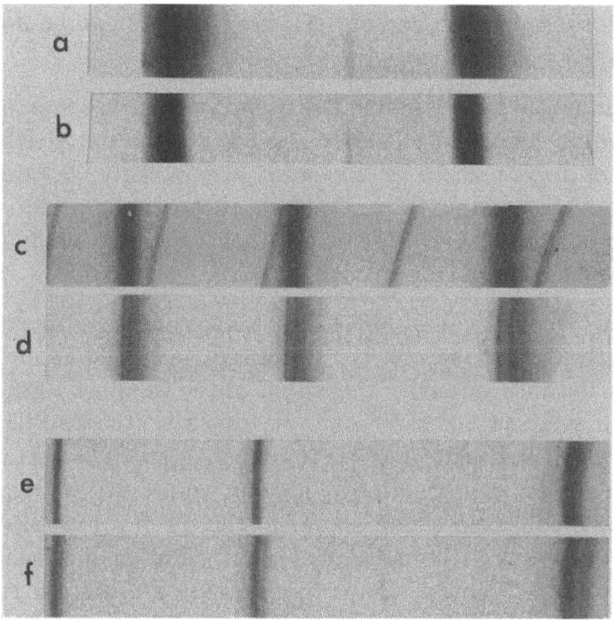

Fig. 119. Spectra of stimulated light scattering of the wing of the Rayleigh line; $\theta = 0°$. (a) Stimulated scattering of the wing of the Rayleigh line (SSWRL) in carbon bisulfide at a temperature of 20°C. (b) The same case for power of the incident light increased by 10–15%. (c) Light of the exciting line $\lambda = 6943$ Å. (d) SSWRL in nitrobenzene at a temperature of 20°C. (e) SSWRL in salol at a temperature of 170°C. In cases (a) and (b), the dispersion range of the Fabry–Perot etalon was 50 cm^{-1}. In cases (c) and (e), the dispersion range was 5 cm^{-1} (Mash, Morozov, Starunov, and Fabelinskii [631]).

of its position is difficult because of the superposition of the MB components.

Photographs of SSWRL in salol were made at a temperature of 170°C. On this photograph (Fig. 119e), a definite band is seen with a maximum at about 0.16 cm^{-1}.

At this temperature the component of stimulated MB scattering in salol has a frequency of 0.11 cm^{-1} but, as can be seen in Fig. 119e, this line did not appear.

It must be noted that in our observations in those cases where the phenomenon of SSWRL appears, stimulated MB scattering did not appear at all or appeared very weakly. Evidently, this effect is explained by the mechanism of the concurrence between the various processes. We believe that a detailed study of this mechanism would be of importance.

The phenomenon of SSWRL was also observed from the long wavelength (Stokes) side of the unshifted line and was characterized by a definite intensity threshold of the incident light, which is estimated at about 10^5 MW/cm^2. In intensity, the SSWRL is comparable with the intensity of the incident light. The anti-Stokes part of the wing was practically absent, which will be explained below in the discussion of the classical theory of the phenomenon. Bloembergen and Lallemand [688] obtained an anomalously large line width for stimulated Raman light scattering as the result of the time-varying index of refraction of the medium, brought about by the temporal change in the orientation of anisotropic molecules under the action of the field.

The same phenomenon of temporal change of the index of refraction brought about by the stimulated, varying orientations of the molecules, plays a fundamental role in the process of SSWRL. In the experiments of Bloembergen and Lallemand, the gigantic pulse of the ruby laser consisted of two identical polarized modes of similar intensity and differing in frequency by 1.6 cm^{-1}.

The lines of stimulated Raman light scattering of cyclohexane brought about by this pulse, focused inside the test cell, remained practically unresolved. But the addition of carbon bisulfide – a liquid consisting of very anisotropic molecules – to the cell with the cyclohexane led to the widening of the lines of cyclohexane, in which the line widths increased as the concentration of carbon bisulfide was raised. The line width reached 85 cm^{-1}. The dependence of the line width on the value of τ was found in a similar, somewhat modified experiment. This quantity changed with change in the viscosity of the "modulating medium." With increase in τ, the width of the modulating line passes through a maximum, as was to be expected, inasmuch as the effectiveness of the "amplitude" of the modulation falls off on both sides of $\Delta\omega = 1/\tau$, in the same way as occurs in SSWRL. The theoretical treatment given in [688] gives the correct qualitative interpretation of the observed phenomenon.

In the work of Lallemand [689], with the same source of excitation (gigantic pulse containing two modes, separated by 1.6 cm), it was shown experimentally that SSWRL determines the asymmetry of the line of stimulated Raman scattering in benzaldehyde, bromobenzene, nitrobenzene, and iodobenzene. Especially clear results were obtained with benzaldehyde where, for a change in temperature of this liquid from +85°C to −25°C, the width of the modulated line narrowed with increase in τ.

This recently discovered circle of phenomena gives reason for hope that in every case the part of the phenomenon which is not understood in stimulated Raman scattering will be explained, and that the SSWRL plays a significant role in the formation of self-focusing in liquid media with anisotropic molecules.

Approximate Classical Theory of Stimulated Scattering of the Light from the Wing of the Rayleigh Line. For a detailed quantitative description of SSWRL one must find the nonlinear kinetic equation for the anisotropy tensor and Maxwell equations, and obtain their simultaneous solution.

This problem has been considered in detail by Starunov. Here we shall put forth only the principal course of the discussions, which lead to a simplified approximate solution of the classical problem.

For the investigation of the problem of the orientation of nonpolar molecules in variable fields, Frenkel [151] found a kinetic equation for the anisotropy tensor. As an anisotropy tensor, we introduce

$$S_{ik} = \overline{\cos\vartheta_i \cos\vartheta_k} - \frac{1}{3}\delta_{ik}; \qquad (35.2)$$

the angles ϑ_i and ϑ_k characterize the location of the molecule relative to a fixed system of coordinates; then S_{ik} (35.2) characterizes the departure of the axes of the molecules from an isotropic distribution. In the absence of a force orienting the molecules, the molecules are distributed randomly, the medium is isotropic, and $S_{ii} = 0$.

If a force F_{ik} arose deflecting the anisotropic molecules from a completely random distribution, then the kinetic equation obeyed by the anisotropy tensor S_{ik} would have the same form as Eq. (30.4):

$$\dot{S}_{ik} + \frac{1}{\tau} S_{ik} = F_{ik}, \tag{35.3}$$

where τ has the meaning of an anisotropy relaxation time.

In the specific case of our problem, when the light of a gigantic pulse is focused in the volume of a liquid, polarized along the z axis (the pulse is propagated along the x axis) one must take into account from (35.2) only

$$S_{zz} \equiv S = \overline{\cos^2\vartheta} - \frac{1}{3} \tag{35.4}$$

and Eq. (35.3) becomes correspondingly simplified. The mean value of $\cos^2\vartheta$ can be computed by taking into account that the energy of the individual molecule (not interacting with its own neighbors) in an electric field E is equal to

$$V = -\frac{1}{2}(b_1 - b_2)(\cos^2\vartheta)E^2 - \frac{1}{2} b_2 E^2. \tag{35.5}$$

If we employ the condition $[(b_1 - b_2)/kT]E^2 \ll 1$ which, for example, for molecules of carbon bisulfide, nitrobenzene, and benzene, is valid up to fields $\sim 10^7$ V/cm, then it is easy to get

$$S = \overline{\cos^2\vartheta} - \frac{1}{3} = \frac{4}{45} \frac{b_1 - b_2}{kT} E^2. \tag{35.6}$$

Taking into consideration that, in the stationary case, $S = \tau F$, we get the desired kinetic equation for the anisotropy tensor:

$$\dot{S} + \frac{1}{\tau} S = \frac{4}{45} \frac{b_1 - b_2}{kT\tau} E^2. \tag{35.7}$$

The set of Maxwell equations (34.4) give

$$\frac{\partial^2 \mathbf{D}}{\partial t^2} + C^2 \operatorname{rot} \operatorname{rot} \mathbf{E} = 0. \tag{35.8}$$

Here

$$\mathbf{D} = \mathbf{E} + 4\pi(\mathbf{P}^{L} + \mathbf{P}^{NL}), \tag{35.9}$$

where the linear part of the polarization is equal to

$$\mathbf{P}^{L} = \frac{\epsilon^{*} - 1}{4\pi}\mathbf{E} \tag{35.10}$$

while the nonlinear part of the polarization is

$$\mathbf{P}^{NL} = (1/4\pi)(\partial\epsilon/\partial S)\,S\cdot\mathbf{E} \tag{35.11}$$

Writing (35.8) for the polarized light wave, and taking into account that $\varepsilon^{*} = \varepsilon' - i\varepsilon''$, just as in (34.6) we get

$$\epsilon'\frac{\partial^2 E}{\partial t^2} - i\epsilon''\frac{\partial^2 E}{\partial t^2} - C^2\left(\frac{\partial^2 E}{\partial x^2} + \frac{\partial^2 E}{\partial y^2}\right) = -\frac{1}{4\pi}\left(\frac{\partial\epsilon}{\partial S}\right)\frac{\partial^2}{\partial t^2}(SE) \tag{35.12}$$

The set of equations (35.7) and (35.12) in simultaneous solution should provide an answer on the fundamental properties of SSWRL. The set of equations named can be solved approximately as in the case of stimulated MB scattering (see Sec. 34) by the small parameter method of Van der Pohl.

We shall assume that the scattering is excited by monochromatic light of frequency ω_0 and amplitude E_0. The scattered light consists of a collection of Stokes and anti-Stokes components. We denote the Stokes component by the index 1 and the anti-Stokes by 2.

We shall seek a solution in the form

$$E = \sum_{e}\sum_{n=0}^{2} E_{ne}\, e^{i(\omega_{n,e}t - \mathbf{k}_{n,e}\mathbf{r})} + \text{complex conjugate} \tag{35.13}$$

$$S = \sum_{e} S_e\, e^{i(\Omega_e t - \mathbf{q}_e\mathbf{r})} + \text{complex conjugate}. \tag{35.14}$$

Neglecting terms $\sim E_{1e_{k_1}}\cdot E_{1e_{k_2}}$ in comparison with terms $\sim E_0 E_{1e_k}$, and taking the laws of conservation of energy and momentum into account, we obtain

$$S_e = \frac{4}{45}\,\frac{b_1 - b_2}{kT}\,W_e(E_0 E^{*}_{1,e} + E^{*}_0 E_{2,e}), \tag{35.15}$$

where

$$W_e = W_{1e} - iW_{2e} = \frac{1}{1 + \Omega_e^2 \tau^2} - i\,\frac{\Omega_e \tau}{1 + \Omega_e^2 \tau^2}.$$

From Eq. (35.12), the conservation laws, and (35.13), we get a system of contracted equations which is greatly simplified if we assume that $\partial E_0/\partial x \sim 0$ or, in other words, we seek a solution at the very beginning of the development of the interaction process of the waves. In this case, we get

$$\begin{aligned} &\partial E_{1e}/\partial x + d'_{k_1} E_{1e} + ig'_1 S_e^* E_0 = 0 \\ &\partial E_{2e}/\partial x + d'_{k_2} E_{2e} + ig'_2 S_e E_0 = 0, \end{aligned} \tag{35.16}$$

where

$$d'_{k_1} = \epsilon''|k_1|/2 \cos(\hat{kx}) \qquad g'_1 = \left(\frac{\partial \epsilon}{\partial S}\right)|k_1|/2 \cos(\hat{k_1 x}). \tag{35.17}$$

Assuming further that the interaction of the Stokes and anti-Stokes components can be neglected, we get from the first equations of (35.16) and (35.15),

$$|E_{1e}(x)|^2 = |E_1(0)|^2 \exp\left\{\left[-d_{k_1} + \frac{4}{45}\frac{(b_1 - b_2)}{kT} g'_1 W_{2e} |E_0|^2\right] x\right\}. \tag{35.18}$$

It follows from (35.17) and (35.15) that if the condition

$$E_0^2 \geq \frac{45}{4}\frac{1}{(\partial\epsilon/\partial S)}\frac{kT}{(b_1 - b_2)}\frac{\epsilon''(1 + \Omega_e^2\tau^2)}{\Omega_e \tau} \tag{35.19}$$

is satisfied, then the Stokes part of the wing will grow exponentially in propagation in a medium in the region of nonlinear interaction. In this approximation, the solution for the anti-Stokes part of the wing is always being damped and, therefore, the effect of stimulated scattering should not be observed for it.

One can see from Eq. (35.19) that the right side of this expression has a maximum for $\tau = 1/\Omega_e$ and, consequently, one should expect a maximum in the spectrum of stimulated scattering at this frequency.

It should be noted that the character of the intensity distribution in the spectrum of depolarized scattering is not so simple as was assumed in the determination of the form of (35.15) (see Secs. 6–8). Therefore, the expression for the threshold of SSWRL would be written more accurately in the following form, where (34.6) has already been taken into account:

$$\left(\frac{E_0^2}{8\pi}\right)_{\text{thres}} = \frac{45}{16\pi^2} \frac{K_\omega (kT) n^2}{(b_1 - b_2)|k_1|(\partial \epsilon / \partial S)\Omega \mathfrak{I}(\Omega)}. \tag{35.20}$$

Here, by $\mathfrak{I}(\Omega)$ (dimensionality, sec) is understood the intensity distribution over the frequency (normalized to unity) in the spectrum of thermally depolarized scattered light.

An estimate of the threshold value of the intensity for SSWRL shows that it is higher than the threshold for stimulated MB scattering.

The exact calculation of the value of (35.20) is difficult. Actually, calculation according to (35.20) gives

$$I_{\text{thres}} = 10^5 \cdot K_\omega \text{MW/cm}^2$$

for the threshold intensity. If one takes $K_\omega \sim 10^{-3}$ cm^{-1}, then $I_{\text{thres}} \sim 100$ MW/cm^2. However, such an estimate is incorrect, since, by K_ω is meant the effective absorption coefficient, which should take into account the losses by stimulated MB scattering, stimulated Raman light scattering, and the expenditure of the energy on other possible processes.

It is quite possible that, by the exact determination of the threshold of SSWRL on the one hand, and the exact value of all losses on the other, it could be shown that the intensity remaining in the fraction of SSWRL is insufficient to excite it. And if SSWRL is not excited, then this process is evidently due in significant measure either to satisfactory concurrence conditions or to the effect of self-focusing.

In the formation of a self-focusing channel, electrostriction and the orientation of anisotropic molecules play a role (Kerr effect).

From the estimates of Zel'dovich and Raizer [695], it follows that the Kerr effect evidently plays a more important role than the electrostriction effect in the formation of a self-focusing channel in liquids with anisotropic molecules.

The emergence of self-focusing leads to a strong increase in the intensity and under these conditions the appearance of SSWRL and all other phenomena characterized by an intensity threshold is made easier. The effect of self-focusing on stimulated Raman scattering was demonstrated in [696, 697].

The results recorded here of the simplified theory enable us to understand why the phenomenon of SSWRL develops principally from the Stokes side, gives expressions for the threshold of the effect, and also enables us to understand why a maximum can be observed in SSWRL.

The elements of the theory developed here refer to the stationary case and are valid for a duration of the interacting waves much greater than the relaxation time of the anisotropy.

The theory of SSWRL does not pretend to more than this. We note that the complicated intensity distribution in the wing of the Rayleigh line in thermal scattering, as discussed in detail in Chapters VII and VIII, shows at least two relaxation times – the anisotropy relaxation time, which corresponds to the process of rotational diffusion, and the smaller time connected with the rotational oscillations of the molecules. Therefore, there should be two maxima, in principle, in SSWRL, but to date only one has been observed, corresponding to the most intense diffusion wing. The development of a more rigorous theory would permit us to obtain more from experiment, but its construction is difficult and the problem is not yet solved.

Appendix I

Calculation of the Fluctuations of Some Thermodynamic Quantities

In his calculation of the intensity of scattered light in liquids, Einstein [14] computed the fluctuations of the density and concentration. After this research, calculations of the fluctuations of thermodynamic quantities were performed by different authors on a number of occasions.

A broad, very general investigation of fluctuations was carried out by Leontovich [16, 83, 135]. He obtained formulas which made it possible to calculate the fluctuations both under ordinary conditions and at the critical temperature, and on the interface between two media.

A detailed exposition of everything relating to this circle of problems can be found in the texts on statistical physics by Leontovich [16] and by Landau and Lifshitz [51]. Here, for the convenience of the reader and for brevity of presentation in the text of this book, we shall sketch the calculation of the fluctuations in the most important cases.

Let the physical quantity x deviate from its mean value $\bar{x}$ by the quantity $x - \bar{x} = \Delta x$, where Δx is the fluctuation of x. Let the probability of finding the value of Δx contained in the interval $d(\Delta x)$ be given by the Gaussian distribution:

$$W(\Delta x)\, d(\Delta x) = \text{const} \cdot \exp\left\{-\frac{(\Delta x)^2}{2\,\overline{|\Delta x|^2}}\right\} d(\Delta x), \tag{I.1}$$

where $\overline{|\Delta x|^2}$ is the mean square of the fluctuations of the physical quantity. In other words, the probability of departure of the thermodynamic quantity x from its equilibrium value x_0 is determined by the Boltzmann principle:

$$W(\Delta x)\,d(\Delta x) = \text{const}\cdot\exp\left\{-\frac{\Delta\Psi V'}{kT}\right\} d(\Delta x), \tag{I.2}$$

where the volume V' is a small part of the volume of the entire thermodynamic system. V' can be the scattering volume V or the volume of the fluctuations V*, and $\Delta\Psi$ can be, in different cases, the change of the free energy, the enthalpy, the thermodynamic or the chemical potential.

The minimum work A_{min} required to bring about a reversible transition of the thermodynamic quantity from its initial value [51] is $A_{min} \equiv \Delta\Psi = \Delta E - T\Delta S + p\Delta V$.

If we consider a small change $\Delta\Psi$, which again describes the fluctuations, then $\Delta\Psi$ can usually be expanded in a Taylor series in any two thermodynamic variables and cut off at the quadratic term in the series.

Let $\Delta\Psi$ be a function of two independent variables, for example, the entropy S and the pressure p; then

$$\Delta\Psi(S,p) = \left(\frac{\partial E}{\partial S}\right)_p \Delta S + \left(\frac{\partial E}{\partial p}\right)_S \Delta p + \frac{1}{2}\left\{\left(\frac{\partial^2 E}{\partial S^2}\right)_p (\Delta S)^2 + \left(\frac{\partial^2 E}{\partial p^2}\right)_S (\Delta p)^2 + \right.$$

$$\left. + 2\frac{\partial^2 E}{\partial S\partial p}(\Delta p\Delta S)\right\} - T\Delta S + p\left(\frac{\partial V}{\partial p}\right)_S \Delta p + p\left(\frac{\partial V}{\partial S}\right)_p \Delta S, \tag{I.3}$$

where the derivatives are taken under equilibrium conditions for p and S. The sum of the coefficients for ΔS and also the sum of the coefficients for Δp vanish [51]. Therefore,

$$\Delta\Psi(S,p) = \frac{1}{2}\left\{\left(\frac{\partial^2 E}{\partial S^2}\right)_p (\Delta S)^2 + \left(\frac{\partial^2 E}{\partial p^2}\right)_S (\Delta p)^2 + 2\frac{\partial^2 E}{\partial S\partial p}\Delta S\Delta p\right\} =$$

$$= \frac{1}{2}\left[\Delta p\Delta\left(\frac{\partial E}{\partial p}\right)_S + \Delta S\Delta\left(\frac{\partial E}{\partial S}\right)_p\right]. \tag{I.4}$$

Taking into account the fact that

$$\left.\begin{aligned} &\left(\frac{\partial E}{\partial p}\right)_S = -p\left(\frac{\partial V}{\partial p}\right)_S, \quad \left(\frac{\partial E}{\partial S}\right)_p = T - p\left(\frac{\partial V}{\partial S}\right)_p, \\ &\left(\frac{\partial V}{\partial S}\right)_p = \left(\frac{\partial T}{\partial p}\right)_S, \quad \Delta T = \frac{T}{c_p}\Delta S + \left(\frac{\partial T}{dp}\right)_S \Delta p, \end{aligned}\right\} \tag{I.5}$$

we get from (I.4) and (I.5):

$$\Delta\Psi(S,p)=\frac{1}{2}\left\{\frac{T}{c_p}(\Delta S)^2-\left(\frac{\partial V}{\partial p}\right)_S(\Delta p)^2\right\}. \tag{I.6}$$

Substituting (I.6) in (I.2), we find

$$W=\text{const}\cdot\exp\left\{-\frac{1}{2}\frac{(\Delta S)^2}{kc_p}V'+\frac{1}{2}V'\left(\frac{\partial V}{\partial p}\right)_S\frac{(\Delta p)^2}{kT}\right\}. \tag{I.7}$$

Equation (I.7) is divided into two parts, where each part depends either on S only or on p only. It then follows that the fluctuations of Δp and ΔS are statistically independent.

Comparing (I.7) with the Gaussian distribution (I.1), we find

$$\overline{(\Delta p)^2}=\frac{kT}{V'\beta_S}, \tag{I.8}$$

$$\overline{(\Delta S)^2}=\frac{1}{V'}kc_p\rho, \tag{I.9}$$

$$\overline{(\Delta p\Delta S)}=0. \tag{I.10}$$

Solving the problem in the same way with the variables V and T, and taking into account that in this case $\Delta\psi = A_{min} = d\Phi$, where Φ is the free energy per unit volume

$$\Phi=E-TS, \tag{I.11}$$

we find

$$\overline{(\Delta V)^2}=\frac{\rho}{V'}kT\beta_T \tag{I.12}$$

or

$$\overline{(\Delta\rho)^2}=\frac{\rho^2}{V'}kT\beta_T, \tag{I.13}$$

$$\overline{(\Delta T)^2}=\frac{1}{V'\rho}\frac{kT^2}{c_p} \tag{I.14}$$

and

$$\overline{(\Delta V\Delta T)}=0. \tag{I.15}$$

In this way, the fluctuations of the concentration in solutions and the fluctuations of other thermodynamic quantities can be calculated.

If $\Delta\Psi$ can be considered as a function of just a single variable x, and all the remaining quantities are constant, then $\Delta\Psi$ is written in the following form:

$$\Delta\Psi=\left(\frac{\partial\Psi}{\partial x}\right)_{x_0}\Delta x+\frac{1}{2!}\left(\frac{\partial^2\Psi}{\partial x^2}\right)_{x_0}(\Delta x)^2+\frac{1}{3!}\left(\frac{\partial^3\Psi}{\partial x^3}\right)_{x_0}(\Delta x)^3+\ldots \tag{I.16}$$

Neglecting the terms of higher order than quadratic, and again assuming that for a closed system in a state of equilibrium $(\partial\Psi/\partial x)_{x_0} = 0$, we get

$$\Delta\Psi(x)\approx\frac{1}{2}\left(\frac{\partial^2\Psi}{\partial x^2}\right)_{x_0}(\Delta x)^2. \tag{I.17}$$

Substituting (I.17) in (I.2), and comparing with (I.1), we obtain a general expression for the fluctuations in the given approximation:

$$\overline{(\Delta x)^2}=\frac{kT}{V'\left(\frac{\partial^2\Psi}{\partial x^2}\right)_{x_0}}. \tag{I.18}$$

If it is necessary to compute the concentration fluctuations, then, by considering (16.51), we have

$$\left(\frac{\partial^2\Psi}{\partial c^2}\right)_{x_0}=\frac{\partial p}{\partial c}, \tag{I.19}$$

where p is the osmotic pressure of the dissolved substance and c is the concentration in g/cm^3. Using the same method, we can show that $\overline{\Delta c\Delta T} = 0$, $\overline{\Delta c\Delta\rho} = 0$, and

$$\overline{\Delta c^2}=c\frac{kT}{V(\partial p/\partial c)}. \tag{I.18a}$$

In the foregoing calculations of the fluctuations, it was assumed that the free energy of a certain volume is an additive function of its separate parts. Such an assumption ceases to be valid in a narrow temperature range close to the critical point. In the latter case, it is necessary to take into account the energy of interaction of the neighboring volumes in which the fluctuations arise. Then the free energy will be a function not only of $(\Delta\rho/\rho)^2$, but also of the gradient of this quantity [16]:

$$\Delta\Psi = \frac{1}{2}\left\{\frac{1}{\beta_T}\left(\frac{\Delta\rho}{\rho}\right)^2 + f^*\left[\nabla\left(\frac{\Delta\rho}{\rho}\right)^2\right]\right\}. \tag{I.20}$$

In the critical region, a calculation of the Fourier component of the density fluctuation is of interest [51]; therefore, we expand $\Delta\rho$ in a Fourier series in the scattering volume:

$$\Delta\rho = \sum \Delta\rho_q e^{iqr}, \text{ where } \Delta\rho_q = \Delta\rho_q^*, \tag{I.21}$$

with the coefficients

$$\Delta\rho_q = \frac{1}{V'}\int \Delta\rho e^{-iqr} dV. \tag{I.22}$$

Substituting (I.22) in (I.20) and carrying out the integration over the volume, we find that

$$\Delta\Psi = \frac{1}{2}\left\{\frac{1}{\beta_T\rho} + f^* q^2\right\} |\Delta\rho_q|^2, \tag{I.23}$$

where q is determined by (I.26). Substituting (I.23) in (I.2) and comparing with (I.1), we find

$$\overline{\Delta\rho_q^2} = \frac{\rho^2 kT}{V[1/\beta_T + (f^*/\lambda^2)(1-\cos\theta)]}, \tag{I.24}$$

where f^* is a constant characterizing the intermolecular interaction.

In the consideration of the molecular fluctuations of the surface of the liquid, the change in the free energy [15, 16, 111] is equal to

$$\Delta\Psi = \frac{\gamma^*}{kT}\left\{\left(\frac{\partial\zeta}{\partial x}\right)^2 + \left(\frac{\partial\zeta}{\partial y}\right)^2 + g\rho\zeta^2\right\}, \tag{I.25}$$

where ζ is the fluctuating elevation above the ideally smooth surface of the liquid, γ^* is the coefficient of surface tension, and g is the acceleration due to gravity. Here we proceed in the same way as was done in the previous case. For the Fourier component of the fluctuating elevation over the smooth surface,

$$\overline{|\zeta_q|^2} = \frac{c_1\left(\frac{q\lambda}{2\pi}\right)^2}{c_2 + \sin^2\theta_0 + 2\sin\theta_0\sin\theta\cos\varphi} \tag{I.26}$$

where

$$c_1 = \frac{kT}{4\pi\gamma^*}, \quad c_2 = \frac{g\rho\lambda^2}{4\pi\gamma^*}, \quad q = \frac{2\pi}{a}$$

(a is an expansion parameter in the Fourier series; the angles θ_0, θ, and φ are clear from Fig. 7).

Appendix II

The Most Important Characteristics of the Fabry–Perot Interferometer, the Conditions of Its Operation, and Some Methods of Interpretation of Interference Spectra

Fundamental Characteristics of the Interferometer

It was recalled in Sec. 12 that all forms of spectral apparatus have been, and still are, required for the investigation of the spectral composition of scattered light. However, special analysis of the possibilities of different spectral instruments, carried out by different authors, and in particular by Zhakino [237], has shown that preference must be given to the Fabry–Perot interferometer above all other types of spectral instruments.

The basic results of the author and his coworkers in the study of the fine structure of the Rayleigh line and its narrow wing were obtained by means of the Fabry–Perot interferometer. This instrument continues to be used effectively today and undoubtedly will be used in the future.

Its applications have now become extremely varied, when it is used both as spectral apparatus and as a resonator for light generators – lasers.

The theory of the Fabry–Perot interferometer has been explained in monographs [239, 474], papers [240, 590], textbooks [110,

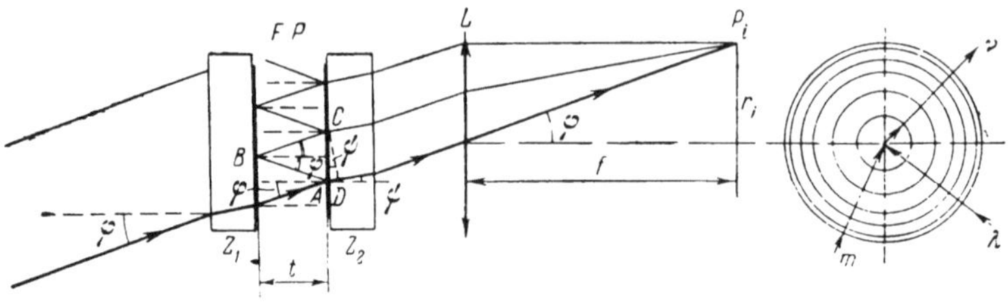

Fig. 120. Basic features of the Fabry–Perot interferometer. Z_1, Z_2 – plates of the interferometer with inner reflecting coatings; the distance between the plates is t. L – objective, in the focal plane of which the interference picture is viewed; f – focal distance of L. r_i – radius of the i-th interference ring. φ – angle of incidence; ψ – angle of refraction. The direction of the wave number ν, the wavelength λ, and the spectral order m are shown by arrows against the background of the interference rings to the right in the drawing.

391], and in other books [209, 210, 238]. It is not possible to enumerate all the sources here. For convenience, the basic characteristics of the Fabry–Perot interferometer that are necessary for work with it will be set forth.

Figure 120 shows a schematic representation of the plates of the Fabry–Perot interferometer Z_1 and Z_2, on which a monochromatic light wave is incident at an angle φ. The rays shown in the figure represent normals to the wave front. The inner sides of the plates Z_1 and Z_2 are coated with a reflecting material. The path difference of two adjacent rays, represented in wavelengths or in orders of interference, is equal to

$$m = \frac{1}{\lambda}(AB + BC - nAD)$$

(n is the index of refraction of the plate). It is clear from Fig. 120 that AB + BC − nAD = 2tn' cos φ, whence it follows that

$$2tn' \cos\varphi = m\lambda \qquad \text{(II.1)}$$

(n' is the index of refraction of the medium between the reflecting surfaces). Furthermore, we shall assume n' = 1. In obtaining Eq.

(II.1), the small phase change of the light which is produced by reflection from the mirror layer has not been taken into account. It has also been assumed that the plates were ideally homogeneous, the surfaces ideally plane and parallel to one another. In addition, the light absorption in the layer t has not been taken into consideration.

If the path difference of the interfering rays is an integral multiple of the wavelength λ, then such waves reinforce one another and give a maximum intensity. If the path difference is an integral odd number of half wavelengths, then the waves cancel one another and yield a minimum intensity.

The objective L with the focal length f, located behind the interferometer, collects all the rays coming out at the angle φ_i (Fig. 120) at the single point P_i. If the azimuth of the incident rays here changes from 0 to 2π, then a light ring appears instead of the point P_i. For a different φ_i corresponding to another, but integral, m, a light ring of different radius r_i appears.

The radius of the ring is a function of the angle φ_i and the focal length of the objective L, where

$$\cos\varphi_i = \frac{f}{(f^2 + r_i^2)^{1/2}} \simeq 1 - \frac{r_i^2}{2f^2} = 1 - \frac{D_i^2}{8f^2}. \tag{II.2}$$

Substituting (II.2) in (II.1), we get

$$m = \frac{2t}{\lambda}\cos\varphi_i = \frac{2t}{\lambda}\left(1 - \frac{r_i^2}{2f^2}\right). \tag{II.3}$$

Differentiating (II.3), it is easy to establish the connection between the change of order m by unity and the change in the radius of the interference ring by Δr_i:

$$\Delta r_i = \frac{\lambda f^2}{2tr_i}. \tag{II.4}$$

With increase in r_i, Δr_i decreases and, consequently, the rings become closer and closer to one another as the distance from the center of the interference picture increases.

The maximum interval of wavelengths $<\Delta\lambda>$ or wave numbers ($<\Delta\nu>$) of the light incident on the interferometer for which the rings of neighboring orders are still not superimposed on one another, is called the dispersion range of the Fabry–Perot interferometer.

If the direction (on the ring) of order m (see Fig. 120) of wavelength λ_1 coincides with the direction of (m + 1)-st order of wavelength λ_2, then it follows from (II.1) that $(m+1)\lambda_2 = m\lambda_1$, whence $\lambda_1 - \lambda_2 = \lambda_1/m$. By definition, this difference of wavelengths is also the dispersion range:

$$\langle \Delta\lambda \rangle = \frac{\lambda}{m}.$$

Using (II.1), we find for the dispersion range

$$\langle \Delta\lambda \rangle = \frac{\lambda^2}{2t}, \tag{II.5}$$

$$\langle \Delta\nu \rangle = \frac{1}{2t}. \tag{II.6}$$

The angular dispersion of the etalon can be obtained from the fundamental relation (II.1):

$$|D_\varphi| = \left|\frac{d\varphi}{d\lambda}\right| = \frac{m}{2t\sin\varphi} = \frac{1}{\lambda\,\mathrm{tg}\,\varphi}. \tag{II.7}$$

The linear dispersion is

$$|D_l| = \left|\frac{dr}{d\lambda}\right| = \frac{mf^2}{2t r_i} \sim \frac{f^2}{\lambda r_i}. \tag{II.8}$$

It is essential that the angular and linear dispersions do not depend on the thickness of the etalon t, but on the angle (II.7) or the radius of the ring r_i (II.8). With decrease in the angle φ or the radius r_i, the dispersion increases, going to infinity for r_i and φ equal to zero.

For the characteristic of the spectral apparatus, we use the quantity $D_l' = 1/D_l$, and express the linear dispersion in Å/mm.

A camera objective $f = 600$ mm is usually employed in the apparatus of Figs. 33 and 36. In this case, for $\lambda = 4358$ Å, at a distance of 5 mm from the center of the interference picture, $D_l \approx 0.06$ Å/mm.

The distribution of the intensity in the interference picture formed in the focal plane of L (Fig. 120) is determined by the following expression [239, 391]:

$$I=\frac{T^2}{(1-r)^2+4r\sin^2\pi m}. \tag{II.9}$$

Here r is the reflection coefficient and T is the transmission coefficient of each plate of the interferometer. It is assumed that both plates have the same r, T, and e (e is the absorption coefficient). These quantities are related by the equation

$$r+T+e=1. \tag{II.10}$$

The intensity of the light wave incident on the interferometer is taken to be unity. For m equal to an integer, Eq. (II.9) gives the intensity of the light at the maximum of the interference picture

$$I_{\max}=\frac{T^2}{(1-r)^2}=\frac{T^2}{(T+e)^2}=\left(1-\frac{e}{1-r}\right)^2. \tag{II.11}$$

When absorption is absent, e = 0,

$$I_{\max}=1.$$

The minimum of the intensity of the interference picture is found from (II.9) for m equal to a half integer:

$$I_{\min}=\frac{T^2}{(1+r)^2}. \tag{II.11a}$$

The contrast interference picture is determined by the factor

$$C=\frac{I_{\max}}{I_{\min}}=\frac{(1+r)^2}{(1-r)^2}. \tag{II.12}$$

Turning to the fundamental relation (II.9), we set $m = m_0 + \delta_m$, where m_0 is an integer and δ_m is a fractional part. Substituting the value of m in (II.9), we get

$$I_{\delta_m}=\frac{I_{\max}}{1+F_1\sin^2\pi\delta_m}, \tag{II.13}$$

where $F_1 = 4r/(1-r)^2$ is called the sharpness of the band. For small values of δ_m we get from (II.13), in first approximation:

$$I_{\delta_m}=\frac{I_{\max}}{1+F_1\pi^2\delta_m^2}. \tag{II.14}$$

The halfwidth of the distribution (II.14) will be equal to

$$\delta_m = \frac{1}{\pi F_1^{1/2}} = \frac{1}{F} = \frac{1-r}{2\pi\sqrt{r}}, \tag{II.15}$$

where $F = \pi F_1^{1/2}$ is called the finesse. The apparatus halfwidth* expressed in fractions of the order will be

$$\delta\nu_a = 2\delta_m = \frac{r-1}{\pi\sqrt{r}}. \tag{II.16}$$

If we use (II.5) and (II.6), we then find

$$\delta\nu_{a_\lambda} = \frac{\lambda^2}{2\pi t}\frac{1-r}{\sqrt{r}}\ \text{Å},$$

$$\delta\nu_{a_\nu} = \frac{1}{2\pi t}\frac{1-r}{\sqrt{r}}\ \text{cm}^{-1}. \tag{II.17}$$

In the next approximation, $\delta\nu_a$, expressed in fractional order, will be

$$\delta\nu_a = 2\delta_m = \frac{1-r}{\pi\sqrt{r}} + \frac{1}{24}\left(\frac{1-r}{\sqrt{r}}\right)^3. \tag{II.18}$$

On the basis of the Rayleigh resolution criterion, the resolving power of the Fabry–Perot etalon is expressed by the relation

$$A_{FP} = \frac{2.98 \cdot m r^{1/2}}{1-r}. \tag{II.19}$$

For a diffraction grating, the resolving power is

$$A_G = mN, \tag{II.20}$$

where N is the number of slits in the grating (number of rays), and m is, as in (II.19), the order of the spectrum. Comparing (II.19) with (II.20), one can introduce the concept of the effective number

*The halfwidth produced by the spectral instrument when it is illuminated by monochromatic light (see Sec. 12).

of rays for the interferometer, equal to

$$N_{\text{eff}} = \frac{2.98}{\pi} F = \frac{2 \cdot 2.98 \cdot \sqrt{r}}{1-r} \approx 0.95F. \tag{II.21}$$

Using (II.1), we get from (II.19), by setting $\cos\varphi \sim 1$,

$$A_{\text{FP}} = \frac{2.98}{\pi\lambda} 2tF = \frac{2.98 \cdot 2t\sqrt{r}}{\lambda(1-r)} = \frac{2.98 \cdot 2t\sqrt{r}\,\nu}{(1-r)}. \tag{II.22}$$

For t = 0.5 cm, r = 0.9, and λ = 4358 Å, A ~ 0.65 · 10^6. The resolving power by (Rayleigh) definition is A = λ / δλ, where δλ is the limit of resolution. Equating the latter expression for A to (II.22), we get

$$\delta\lambda = 0.335 \frac{\lambda^2}{2t} \frac{(1-r)}{\sqrt{r}}, \tag{II.23}$$

or, in wave numbers,

$$\delta\nu = 0.335 \frac{1}{2t} \frac{(1-r)}{\sqrt{r}}. \tag{II.24}$$

It follows from the formulas written above that it is very important that the reflection coefficient of the mirrors of the interferometer be as close to unity as possible, and that the absorption coefficient be as close to zero as possible.

In this ideal case, the formulas show that the resolving power A and the finesse F tend to infinity, while the transmission of the etalon tends to unity.

In actuality, the resolving power is currently limited not by the reflection coefficient, but by the quality of preparation of the plates of the interferometer.

Thanks to the use of multilayer dielectric coatings for the interferometer mirrors, it has been possible to obtain very high values of r for very low absorption [582-589].

The problem of the effect of imperfections in the preparation of the mirror on the contrast or the resolving power has been studied in great detail by Dufour (see Appendix II in [239]).

The value of the contrast C (II.12) can be ~10^4 under modern conditions of high reflection coefficients. The finesse F ~ 300 and

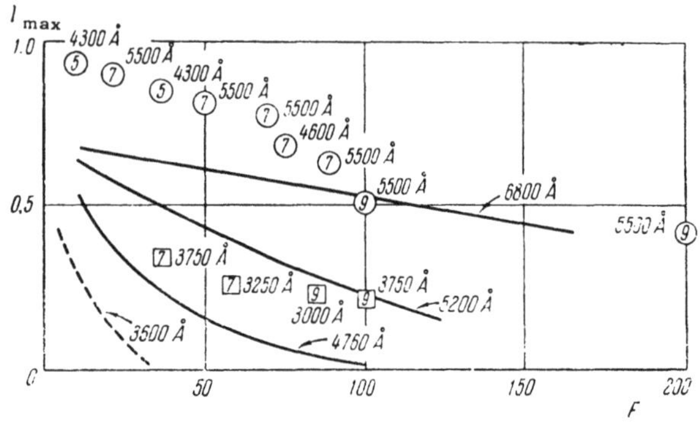

Fig. 121. Dependence of the transmission of light by an interferometer at the maximum of the interference band I_{max} (in arbitrary units) on the finesse F for different reflecting surfaces and wavelengths (it is assumed that the plates are ideally plane and are parallel to one another). The graph was constructed by Born and Wolf [391] from various data. Solid lines – mirror of a bright silver layer [586]; broken line – mirror of aluminum layer [587]; ○ – mirror of layers of ZnS–cryolite; □ – mirror of layer of lead chloride–magnesium fluoride [589]. The number of dielectric layers is shown inside the circles and squares; next to them is given the optimal wavelength of the light in angstroms.

the apparatus halfwidth $\sim 3 \cdot 10^{-3}$ of an order. However, imperfections of the mirror surfaces change all these characteristics for the worse.

Figure 121 represents the data for the maximum transmission I_{max} (II.11) as a function of the band sharpness F. The graph is constructed in [391] for the bright layers of silver and aluminum, and for certain dielectric coatings.

As r approaches unity, F approaches its limiting value, the magnitude of which is determined by the character and dimensions of the imperfections of the surface. In a simple case, in which, in place of a plane, we have a spherical distortion of such a size that the edges differ from the center by $^1/_{100}\,\lambda$, then the limiting value $F \rightarrow F_d = 50$ for a bright silver layer and $\lambda = 5200$ Å, as is shown in [391]. In this case, the intensity at the maximum of the band falls off sharply.

Effect of Temperature and Pressure on the Resolving Power of the Fabry – Perot Etalon

The change of temperature of the interferometer and of the air pressure in the gap between the mirrors Z_1 and Z_2 (Fig. 120), as is clear from Eq. (II.1), influence the value of the angle φ and, consequently, the position of the interference maximum in the focal plane L. A change in temperature is reflected chiefly in the thickness of the etalon t, while an air pressure change has its effect on the index of refraction n'.

If the change of the pressure and temperature during the time of observation is such that the maxima of the interference rings for two barely resolved lines are displaced by an amount equal to the distance between them (limit of resolution), then the picture of the discrete lines will be blurred. Consequently, such changes in temperature and pressure are inadmissible in research on the limit of resolution.

A quantitative connection between the change in temperature ΔT during the time of exposure, which shifts the interference bands by an amount equal to one half the limit of resolution, and the resolving power A, was obtained by Hansen [591] and the corresponding relation for the change in pressure was obtained by Sterin [241]. This dependence can be written for the temperature change:

$$|\Delta T| = \left|\frac{1}{A\beta}\right|, \tag{II.25}$$

where A is the resolving power and β is the coefficient of linear expansion of the material of the ring, which gives the distance between the plates Z_1 and Z_2 (the temperature change of n' is neglected). For a pressure change

$$|\Delta p| = \left|\frac{1}{A\xi}\right|, \tag{II.26}$$

where $\xi = (1/n)(\partial n/\partial p)$. Setting $n - 1 = cp$, where p is the pressure, we get $\xi = (n-1)/np$. For air, $n - 1 = 0.0003$ and $\xi = 4 \cdot 10^{-7}$.

In a Fabry–Perot etalon with a resolving power $A = 1.3 \cdot 10^6$ ($t = 0.5$ cm, $r = 0.95$) and a quartz spacer ($\beta = 0.26 \cdot 10^{-6}$), the shift

Table A1. Tolerance Limits for Change of Atmospheric Pressure for Fabry–Perot Etalons of Different Thickness with Different Coefficients of Reflection of the Mirror (λ = 4358 Å)

Coefficient of reflection r \ Thickness of etalon t, mm	0.3	0.5	0.7	1	3	5	10	15
	Δp, mm Hg							
0.75	86	51	37	25	8	5	2.5	1.6
0.80	66	40	28	20	6	4	2.0	1.3
0.85	48	29	21	15	5	3	1.5	1.0
0.90	32	19	13	9	3	2	1.0	0.6
0.95	15	9	6	4.5	1.5	0.9	0.5	0.3
0.99	3.5	2	1.4	1	0.33	0.2	0.1	0.07

in the interference bands by an amount equal to one half the limit of resolution is produced by a temperature change $\Delta T \sim 3°$. For an invar ring, $\beta = 0.9 \cdot 10^{-6}$, $\Delta T \sim 0.8°$. For the same resolving power, the shift of the ring by half the limit of resolution takes place for a pressure change $\Delta p \sim 1$ mm Hg.

Given in Table A1 are the values of Δp which change the location of the band by one half the limit of resolution for different t and r (Sterin [241]).

From the given calculations and Table A1, it is clear that the use of an isobaric and isothermal chamber for the Fabry–Perot interferometer is extremely important for $A \sim 10^6$.

Interpretation of Interference Spectra

1. Method of Squares of the Ring Diameters. Let the part of the spectrum under investigation consist of several closely placed components, with no superposition of interference orders, and with frequency difference $\Delta\nu$ between the components amounting to a fraction of an order Δm_i. In this case, finding the difference of frequencies between the components reduces to finding the corresponding difference Δm_i, which can be expressed in terms of the measurement parameters of the interference picture.

We rewrite (II.3) for the wave numbers, taking into account that $m = m_i + \Delta m_i$, where Δm_i is the fractional part of an order, we get

$$D_{m_i}^2 = 8f^2\left(1 - \frac{m_i + \Delta m_i}{2t\nu}\right).$$

We can write for the component a

$$D_{a,m}^2 = 8f^2\left(1 - \frac{m + \Delta m_a}{2t\nu}\right),$$

$$D_{a,m+1}^2 = 8f^2\left(1 - \frac{m + 1 + \Delta m_a}{2t\nu}\right),$$

whence

$$\Delta m_a = \frac{D_{a,m}^2}{D_{a,m+1}^2 - D_{a,m}^2} - m.$$

For the component b, we get, by the same method,

$$\Delta m_b = \frac{D_{b,m}^2}{D_{b,m+1}^2 - D_{b,m}^2} - m,$$

the difference is

$$\Delta m_a - \Delta m_b = \frac{D_{a,m}^2 - D_{b,m}^2}{C}, \tag{II.26a}$$

where

$$C = D_{a,m+1}^2 - D_{a,m}^2 = D_{b,m+1}^2 - D_{b,m}^2 = D_{c,m+1}^2 - D_{c,m}^2, \text{ etc.}$$

is the same for the difference of the squares of the diameters of the corresponding rings for all components.

The difference of the wave numbers of the components a and b is

$$\delta\nu_{a,b} = (\Delta m_a - \Delta m_b)\frac{1}{2t}. \tag{II.27}$$

On the basis of (II.26a) and (II.27), we find

$$\delta\nu_{a,b} = \frac{1}{2t}\frac{D_{a,m}^2 - D_{b,m}^2}{C} = \frac{1}{2t}\frac{C_{ab}}{C}. \tag{II.28}$$

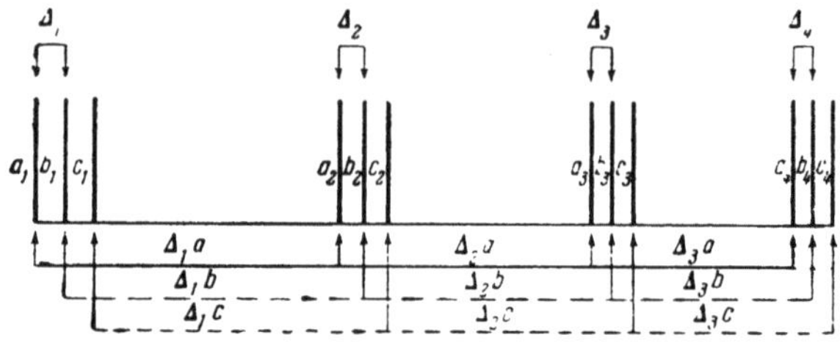

Fig. 122. Interpretation of the interference spectrum in the case of noncentral bands.

To save time in the calculation of the differences of the frequencies of the components, Tolansky [239] proposed that a table in the form of Table A2 be established.

2. Method of Interpretation of the Noncentral Interference Bands. In this case, only a part of the interference orders can be shown in the photograph of the interference picture. These orders are located on one side of the center. It is shown thus, for example, in the x and z components of polarized light when the fine structure of the Rayleigh line is photographed in low-viscosity liquids (Fig. 35). In this case, the method of squares of the diameters is unsuitable.

Table A2

Component	Number of ring						
	1		2		3		4
a	$D^2_{a.1}$	C_{12}	D^2_{a2}	C_{23}	D^2_{a3}	C_{34}	D^2_{a4}
	C_{ab}		C_{ab}		C_{ab}		
b	D^2_{b1}	C_{12}	D^2_{b2}	C_{23}	D^2_{b3}	C_{34}	D^2_{b4}
	C_{bc}		C_{bc}		C_{bc}		C_{bc}
$\pm C$	D^2_{c1}	C_{12}	D^2_{c2}	C_{23}	D^2_{c3}	C_{34}	D^2_{c4}

The interpretation of the noncentral bands is carried out according to the following scheme [239].

Figure 122 shows a schematic of three components a, b, c repeated in four orders. By knowing the mutual position of the components of the fine structure in several orders, one can obtain independent values of Δm, namely:

$$\Delta m_{ab} = \frac{2\Delta_2}{\Delta_{2b} + \Delta_{2a}} = \frac{2\Delta_3}{\Delta_{2b} + \Delta_{3a}}.$$

The difference between the components b and c can be obtained in exactly the same fashion:

$$\Delta m_{bc} = \frac{2\Delta_2'}{\Delta_{2b} + \Delta_{1c}} = \frac{2\Delta_3'}{\Delta_{3b} + \Delta_{2c}}.$$

It is not difficult to extend such a calculation to the large number of components.

Δm is calculated in fractions of orders. For computation in cm^{-1}, we write

$$\Delta\nu = \frac{\Delta m}{2t \cos \varphi}. \qquad (II.29)$$

Without great error, $\cos \varphi$ can be set equal to unity. The error of this method of calculation is not large and it is smaller the further from the center of the interference picture the orders used in the calculation of Δm are located [239].

3. Dispersion Curve Method. In those cases in which it is necessary to study the frequency distribution of the intensity inside a single component, application of the method of the difference of the squares of the diameters and the method of noncentral bands requires an extraordinarily large expenditure of time. Therefore, a simple and sufficiently accurate method of interpretation was proposed and used by the author [53]. It consists of the following: the linear dispersion of the instrument can be expressed by the following formula:

$$D_l = \frac{r}{\nu f^2} \qquad (II.30)$$

The radius of the interference ring r_i in any order can be reliably determined if there are several rings on the photograph. In the case in which there is only half an interference picture in the scattered light, the radius of the ring of any order can be found from the interference rings of the incident light.

A graph of the dependence of the dispersion in the maximum of the interference orders is constructed as a function of the radius of corresponding order [53]. It is quite suitable for the interpretation of the interferogram according to the wavelength or frequency just as the corresponding dispersion curve of a spectrograph.

Still more accurate results are obtained with great simplicity and convenience if we represent analytically the frequency dependence $\Delta\nu$ on the dispersion D_0' and the distance l on the photographic plate. The dependence of D' on the radius of the interference ring can be represented in the form

$$D' = D_0' + \mathrm{tg}\,\beta\,(r_i - r_{i0}), \tag{II.31}$$

where D' is expressed in cm^{-1}/mm, r_i in millimeters, and $\mathrm{tg}\,\beta = \Delta D/\Delta r_i$. It follows from this relation that

$$\Delta\nu = D_0' l \pm \frac{1}{2} l^2\,\mathrm{tg}\,\beta, \tag{II.32}$$

where l is the distance on the photographic plate, measured from the maximum of the order (the dispersion D_0') to the right (+) or left (−) and expressed in millimeters. Thus, in the vicinity of any order, the interferogram can be interpreted with sufficiently high accuracy and with a small expenditure of time. The accuracy of the measurement is determined by the accuracy of measurement of D_0' and l.

D_0' is determined by f^2, ν, and r_i. The focal distance of the objective and the frequency are known with great accuracy. The radius of the ring r_i on the interferogram is measured on a double Abbe comparator. The reproducibility of the measurements lies within the limits of hundredths of millimeters for an absolute value of the radius from 5 to 16 mm, i.e., the relative error of measurement of the radius amounts to a fraction of one percent; l is measured in the process of photometry on a semiautomatic microphotometer with an accuracy up to 0.01 mm, which leads to an error in the frequency measurement of about 0.1%. It is not difficult to increase this accuracy.

We now point out one of the possible methods of finding the true intensity distribution according to the visible experimentally observed contour. This method was proposed by Ablekov [221]. It has already been noted that the visible intensity distribution in the spectrum is connected with the true intensity distribution by an integral equation (12.5). In the case of the Fabry–Perot etalon, the apparatus function (II.13) can be rewritten in the following form:

$$a(\nu)=\frac{I_{\max}}{1+F\sin^2\frac{\nu}{\langle\Delta\nu\rangle}}. \qquad \text{(II.33)}$$

If we now substitute (II.33) in (12.5), then, even when $I(\nu)$ is described by a simple dispersion law, the integral cannot be computed simply. Ablekov [221] got around this difficulty by expanding (II.33) in a series in the Lorentzian distribution of the following form:

$$\frac{I_0}{1+\left(\frac{\nu}{\delta\nu_a}\right)^2}, \qquad \text{(II.34)}$$

which is equivalent to being located on the frequency axis.

We note that the consideration of the Fabry–Perot etalon as a hollow plane resonator, made by Sommerfeld [266], gives a distribution of the characteristic frequencies of the resonator which is identical with the maximum of (II.33).

Use of (II.34) and (12.5) gives for the visible distribution

$$W(\nu)=\sum_{n=-\infty}^{\infty}\int_{-\infty}^{\infty} I(\nu-x)\frac{I_0}{1+\left(\frac{x+n\langle\Delta\nu\rangle}{\delta\nu_a}\right)}dx. \qquad \text{(II.35)}$$

If $I(\nu)$ is the Lorentzian distribution with halfwidth $\delta\nu_x$, then the integral (II.35) also gives the Lorentzian distribution with halfwidth

$$\delta\nu_w=\delta\nu_a+\delta\nu_x.$$

Summation of the integrals over all orders of interference gives

$$W(\nu) = \frac{I_{\max}}{1 + F_{\mathrm{eff}} \sin^2 \pi \frac{\nu}{\langle \Delta\nu \rangle}}, \tag{II.36}$$

where

$$I_{\max} = I_0 \pi \frac{\delta\nu_w}{\langle \Delta\nu \rangle} \operatorname{cth} \pi \frac{\nu}{\langle \Delta\nu \rangle}, \tag{II.37}$$

$$F_{\mathrm{eff}} = \operatorname{cth}^2 \pi \frac{\delta\nu_w}{\langle \Delta\nu \rangle} - 1. \tag{II.38}$$

From the last three relations (II.36)-(II.38), one can find the three quantities $\delta\nu_W$, I_0, and F_{eff} for the experimentally determined I_{max} or I_{min} and $W(\nu)$.

The described methods of interpretation of the interference spectra are completely reliable, but involve a great deal of time. They can be used successfully when the material under study is limited. On the other hand, if the material is voluminous, then it is probably better to follow the method of Harrison [238] and construct a machine which is capable of making the necessary interpretation with great accuracy in a short time.

Appendix III

Table I. Depolarization Factor Δ_u and the Kerr Constant K for Various Gases and Vapors*

Substance	Chemical formula	$\Delta_u \cdot 100$	$K \cdot 10^{15}$ reduced to 760 mm Hg, λ 5460 Å	$K_1 \cdot 10^{15}$	$K_2 \cdot 10^{15}$	Measurement temp., °C
Argon	Ar	$\leqslant 5 \cdot 10^{-4}$ (0.041 ± ± 0.005)	0.04	0.04	0	25 (20)
Helium	He	0.03				
Neon	Ne	0.01				
Krypton	Kr	0.0055				
Xenon	Xe	0.0055 (0.043 ± ± 0.001)				 (20)
Hydrogen	H_2	0.9 (0.93 ± ± 0.05)	0.04	0.04	0	34.6 (20)
Nitrogen	N_2	3.6 (1.05 ± ± 0.01)	0.14	0.14	0	25 (20)
Oxygen	O_2	6.4 (3.09 ± ± 0.02)	0.45	0.45	0	0 (20)
Chlorine	Cl_2	4.3	2.3	2.3	0	24
Cyanogen	$(CN)_2$		4.3	4.3	0	25
Carbon monoxide	CO	3.2				
Nitric oxide	NO	2.7				
Carbon dioxide	CO_2	7.2 (4.13 ± ± 0.021)	1.28	1.28	0	25 (20)
Carbon bisulfide	CS_2	11—14	21.6	21.6	0	56.7
Nitrous oxide	N_2O	10.2	3.08	2.68	0.4	26

*This table is taken from the data of Stuart [116, 413], Landolt–Börnstein [312], Cabannes [303, 314, 315], Vaucouleurs [56], and Bridge and Buckingham [592]. The measurements of [592] are given in parentheses.

Table I (continued)

Substance	Chemical formula	$\Delta_u \cdot 100$	$K \cdot 10^{15}$ reduced to 760 mm Hg, λ 5460 Å	$K_1 \cdot 10^{15}$	$K_2 \cdot 10^{15}$	Measurement temp., °C
Sulfur dioxide	SO_2	3.1—4.1	—9.4	1.7	11.1	18
Hydrogen chloride	HCl	0.7	5.75	0.44	5.31	18
Hydrogen bromide	HBr	0.84				
Hydrogen iodide	HI	1.3				
Water vapor	H_2O	2.0				
Carbonyl sulfide	COS	8.8				
Hydrogen sulfide	H_2S	0.3—0.9	1.59	0.28	1.31	18
Ammonia	NH_3	1.0				
Methane	CH_4	$20 \cdot 10^{-4}$	0.2	0.2	0	18
Ethane	C_2H_6	0.5—1.3	0.6	0.6	0	18
Ethylene	C_2H_4	2.9	1.09	1.09	0	18
Acetylene	C_2H_2	4.5	1.85	1.85	0	25
Cyclopropane	C_3H_6	0.5				
Propane	C_3H_8	0.7—1.5	0.95	0.95	0	18
Propylene	C_3H_6	2.9				
n-Butane	C_4H_{10}	0.9; 1.5	1.4	1.4	0	18
Isobutane	C_4H_{10}	0.5; 1.0				
n-Pentane	C_5H_{12}	1.2	1.06	1.06	0	18
Isopentane	C_5H_{12}	1.2				
Isoamylene β	C_5H_{10}	2.2				
Cyclohexane	C_6H_{12}	0.8—1	0.65	0.65	0	127
Cyclohexene	C_6H_{10}	2.05				
n-Hexane	C_6H_{14}	1.56	2.54	2.54	0	110
n-Heptane	C_7H_{16}	1.4				
Methyl cyclohexane	C_7H_{12}	2.12	2.4	2.9	0	155
n-Octane	C_8H_{18}	1.66	3.1	3.1	0	175
Benzene	C_6H_6	4.2	5.5	5.5	0	113.6
Toluene	C_7H_8	4.2—4,6	9.0	7.6	1.4	137.7
m-Xylol	C_6H_{10}	4.6	9.2	8.6	0.6	177
Ethyl benzene	C_8H_{10}	4.3				
Methyl chloride	CH_3Cl	1.5—2.0	36.5	0.8	35.7	18
Methylene chloride	CH_2Cl_2	2.9	—5.6	1.0	—6.6	83.5
Chloroform	$CHCl_3$	1.6—1.8	—7.7	1.1	—8.8	89.5
Propyl bromide	C_3H_7Br	2.03				
Isopropyl iodide	C_3H_7I	2.13				
Isobutyl chloride	C_4H_9Cl	1.69				
Bromoform	$ClBr_3$	2.9				
Carbon tetrachloride	CCl_4	0.15 (0.5)	<±0.2	<0.2	0	99.4
Ethyl chloride	C_2H_5Cl	1.5	52.2	1.63	53.6	18
Dichloroethane	$C_2H_4Cl_2$	3.27	4.7	2.8	1.9	108.5
Ethyl bromide	C_2H_5Br	2.2; 3.2				
Dibromoethane	$C_2H_4Br_2$	3.54				
Propyl chloride	C_3H_7Cl	1.49				
Isobutyl bromide	C_4H_9Br	1.53				
Chlorobenzene	C_6H_5Cl	4.58	37.2	7.5	29.7	153.7
Bromobenzene	C_6H_5Br	4.83				

Table I (continued)

Substance	Chemical formula	$\Delta_u \cdot 100$	$K \cdot 10^{15}$ reduced to 760 mm Hg, λ 5460 Å	$K_1 \cdot 10^{15}$	$K_2 \cdot 10^{15}$	Measurement temp., °C
Benzyl chloride	C_7H_7Cl	4.40				
Benzal chloride	$C_7H_6Cl_2$	4.13				
Methyl alcohol	CH_4O	1.64	$<\pm 0.4$	0.3	0.1—0.7	98.8
Ethyl alcohol	C_2H_6O	0.88	$<\pm 0.5$	0.3	0.2—0.8	102
Propyl alcohol	C_3H_8O	1.19				
Isopropyl alcohol	C_3H_8O	1.3				
Allyl alcohol	C_3H_6O	2.5				
Butyl alcohol	$C_4H_{10}O$	1.62				
Isobutyl alcohol	$C_4H_{10}O$	0.77				
Trimethylcarbinol	$C_4H_{10}O$	0.69				
Amyl alcohol	$C_5H_{12}O$	1.2				
Isoamyl alcohol	$C_5H_{12}O$	1.03				
Dimethyl ether	C_2H_6O	1.2—1.6	—5.0	0.9	—5.9	18
Diethyl ether	$C_4H_{10}O$	2.5	—3.9	3.7	—7.6	62.7
Di-n-propyl ether	$C_6H_{14}O$	1.4	—2.3	—	—	123.9
Formaldehyde	CH_2O	1.9				
Acetaldehyde	C_2H_4O	2.7				
Acetone	C_3H_6O	1.6—1.7	32.1	1.0	31.1	83.1
Methyl ethyl ketone	C_4H_8O	2.2	16.0	1.8	14.2	104.3
Methyl propyl ketone	$C_5H_{10}O$	2.0	7.8	2.1	5.7	129.7
Diethyl ketone	$C_5H_{10}O$	3.2	5.3	3.2	2.0	134.7
Diisopropyl ketone	$C_7H_{14}O$	1.2				
Formic acid	CH_2O_2	3.8				
Acetic acid	$C_2H_4O_2$	2.8				
Propionic acid	$C_3H_6O_2$	2.6				
Butyric acid	$C_4H_8O_2$	1.8				
Benzoic acid	$C_7H_6O_2$	5.3				
Salicylic acid	$C_7H_6O_3$	5.3				
Methyl formate	$C_2H_4O_2$	2.1				
Ethyl formate	$C_3H_6O_2$	2.1				
Propyl formate	$C_4H_8O_2$	1.6				
Methyl acetate	$C_3H_6O_2$	2.1				
Ethyl acetate	$C_4H_8O_2$	2.0				
Propyl acetate	$C_5H_{10}O_2$	1.4				
Methyl benzoate	$C_8H_8O_2$	5.9				
Anhydrous acetic acid	$C_4H_6O_3$	1.8				
Anhydrous propionic acid	$C_6H_{10}O_3$	2.8				
Methylamine	CH_3NH_2	0.8				
Ethylamine	$C_2H_5NH_2$	1.0				
Nitrobenzene	$C_6H_5NO_2$	5.6	146		139	235.5
o-Nitrotoluene	$C_7H_7NO_2$	4.8				
m-Nitrotoluene	$C_7H_2NO_2$	5.25				
Aniline	$C_6H_5NH_2$	4.9—5.6				
Pyridine	C_5H_6N	4.5	25.7	4.1	21.6	146.3
Quinoline	C_9H_7N	7.9				

Table II. Average and Principal Polarizabilities of Various Molecules, cm^3 [413]

Molecule	Chemical formula	Average polariz. $b \cdot 10^{25}$	$b_1 \cdot 10^{25}$	$b_2 \cdot 10^{25}$	$b_3 \cdot 10^{25}$	Location of the optic axis and the electric moment, symmetry properties
Chlorine	Cl_2	46.1	66.0	36.2	36.2	b_1 – axis of symmetry, elongated
Hydrogen fluoride . . .	HF	24.6	(9.6)	(7.2)	(7.2)	
Hydrogen chloride. . .	HCl	26.3	31.3	23.9	23.9	
Hydrogen bromide. . .	HBr	36.1	42.2	33.1	33.1	
Hydrogen iodide	HI	54.5	65.8	48.9	48.9	
Nitrous oxide	N_2O	30.0	48.6	20.7	20.7	Symmetry axis is unsymmetrically elongated
Carbon monoxide . . .	CO	19.5	26.0	16.25	16.25	b_1 – axis of symmetry
Carbon dioxide.	CO_2	26.5	40.1 40.0	19.7— 19.3	19.7— 19.3	
Hydrogen sulfide. . . .	H_2S	37.8	40.9	34.4	40.1	$\mu = \mu_3$, b_2 perpendicular to the HSH plane, bent
Sulfur dioxide.	SO_2	37.2	54.9	27.2	34.9	$\mu = \mu_3$, b_2 perpendicular to the OSO plane, bent
Carbon bisulfide	CS_2	87.4	151.4	55.4	55.4	b_1 – axis of symmetry, elongated
Ammonia	NH_3	22.6	24.2	21.8	21.8	b_1 – axis of symmetry, $\mu = \mu_1$ pyramid
Cyanogen	$(CN)_2$	50.1	77.6	36.4	36.4	b_1 – axis of symmetry, elongated
Hydrogen cyanide	HCN	25.9	39.2	19.2	19.2	b_1 – axis of symmetry, elongated

Table II (continued)

Molecule	Chemical formula	Average polariz. $b \cdot 10^{25}$	$b_1 \cdot 10^{25}$	$b_2 \cdot 10^{25}$	$b_3 \cdot 10^{25}$	Location of the optic axis and the electric moment, symmetry properties
Tin tetrachloride. . . .	$SnCl_4$	137.7	137.7	137.7	137.7	Right tetrahedron
Methane	CH_4	26.0	26.0	26.0	26.0	b_1 – axis of symmetry, elongated (Methane and Ethane)
Ethane	C_2H_6	44.7	54.8	39.7	39.7	
Ethylene	C_2H_4	42.6	56.1	35.9	35.9	
Acetylene	C_2H_2	33.3	51.2	24.3	24.3	b_1 – axis of symmetry, elongated
Propane.	C_3H_8	62.9	50.1	69.3	69.3	b_1 perpendicular to CCC plane
Cyclohexane	C_6H_{12}	108.7	92.5	116.8	116.8	b_1 perpendicular to the plane of the ring
Benzene	C_6H_6	103.2	123.1	63.5	123.1	b_2 – axis of symmetry
Toluene	C_7H_8	122.6	136.6	74.8	156.4	
n-Xylol	$C_6H_4(CH_3)_2$	142.0	(156.0)	(88.0)	(182.0)	Plane molecule, $\mu = \mu_3$, b_2 perpendicular to plane of molecule (n-, o-, m-Xylol)
o-Xylol	$C_6H_4(CH_3)_2$	141.8	178.3	85.5	161.6	
m-Xylol	$C_6H_4(CH_3)_2$	141.0	(161.3)	(83.0)	(179.6)	
Methyl chloride	CH_3Cl	45.6	54.2	41.4	41.4	b_1 – axis of symmetry, $\mu = \mu_3$ (Methyl chloride and Methyl bromide)
Methyl bromide	CH_3Br	55.5	68.5	49	49	
Chloroform	$CHCl_3$	82.3	66.8	90.1	90.1	b_1 – axis of symmetry, $\mu = \mu_1$
Carbon tetrachloride.	CCl_4	105	105	105	105	Right tetrahedron
Ethyl chloride.	C_2H_6Cl	64	66.0	50.1	75.9	Bent molecule
Chlorobenzene	C_6H_5Cl	122.5	132.4	75.8	159.3	Plane molecule, $\mu = \mu_3$, b_2 perpendicular to plane of molecule

Table II (continued)

Molecule	Chemical formula	Average polariz. $b \cdot 10^{25}$	$b_1 \cdot 10^{25}$	$b_2 \cdot 10^{25}$	$b_3 \cdot 10^{25}$	Location of the optic axis and the electric moment, symmetry properties
n-Dichlorobenzene . .	$C_6H_4Cl_2$	144.7	(124.8)	(88.3)	212.9	Plane molecule, $\mu = \mu_3$, b_2 perpendicular to the plane of the molecule
m-Dichlorobenzene. .	$C_6H_4Cl_2$	142.3	171.4—	89.6—	165.7—	
			176.6	86.6	163.8	
o-Dichlorobenzene . .	$C_6H_4Cl_2$	141.7	143.1—	95.1—	186.8—	
			155.2	84.8	185.0	
Methyl alcohol.	CH_3OH	32.3	40	25.6	31.4	Computed for μ; μ perpendicular to OC $(b_1\mu) = 70°$, bent molecule
Dimethyl ether.	CH_3OCH_3	51.6	63.0	43.1	48.6	$\mu = \mu_3$, b_2 perpendicular to COC plane, bent molecule
Diethyl ether	$C_2H_5OC_2H_5$	87.3	112.6	70.7	78.7	
Di-n-propyl ether . . .	$C_3H_7OC_3H_7$	125.2	136.5	122	117	
Dioxane	$C_4H_8O_2$	94.4	76.0	103.6	103.6	b_1 perpendicular to the plane of the ring
Acetone	CH_3COCH_3	63.3	71.0	48.2	70.8	$\mu = \mu_3$, b_2 perpendicular to COC plane, bent molecule
Methyl ethyl ketone. .	$CH_3COC_2H_5$	81.3	98.3	60.2	85.5	
Diethyl ketone	$C_2H_5COC_2H_5$	99.3	126.4	71.5	101.1	
Methyl propyl ketone.	$CH_3COC_2H_7$	99.3	120.6	76	101.4	
Diisopropyl ketone . .	$(CH_3)CHCOCH(CH_3)_2$	135.3	(162)	(117)	(126.6)	
Nitrobenzene	$C_6H_5NO_2$	129.2	132.5	77.5	77.6	Plane molecule, $\mu = \mu_3$, b_2 perpendicular to the plane of the molecule
Pyridine	C_5H_5N	95	118.8	57.8	108.4	

Table III. Experimental Data for the Absolute Value of the Scattering Coefficient in Various Liquids

Liquid	t, °C	$R_{90} \cdot 10^6$, cm^{-1} λ 4358 Å	$R_{90} \cdot 10^6$, cm^{-1} λ 5461 Å	Year	Authors
Benzene	20	32.8	–	1922	Martin and Lehrman [336]
	15	–	10.7	1927	Cabannes and Daure [207]
		55.1	–	1923	Raman and Rao [338]
	24	34.9	–	1938	Peyrot [179]
	20	29.6	–	1942	Sunanda Bai [54]
	25	48.2	16.4	1950	Doty and Steiner [339]
	25	48.4	–	1950	Brice, Holwera, and Speiser [168]
	25	48.5	16.3	1950	Carr and Zimm [198]
	20	32.3	12.0	1949, 1951	Vaucouleurs [55, 56]
	20	48.4	–	1953	Oth, Oth, and Desreux [340]
	25	–	17.6	1953	Halwer, Nutting, and Brice [341]
	25	28	–	1953	Harrand [342]
	25	47.1	–	1953	Oster [343]
	25	–	15.6	1954	Trap and Hermans [344]
		46.5	16.8	1954	Sedlacek [345]
	25	49.7	17.2	1954	Maron and Lou [346]
		46.3	–	1954	Kremen and Shapiro [428]
		46	–	1954	Kushner [347]
	25	44	15.4	1954	Cantow and Schulz [348]
	20	44.4	–	1953, 1955	Stamm and Button [349]
	20	32		1955	Tourel and Chapelle [557, 558]
	25	48.4	16.6	1956	Trossarelli and Saini [350]
		50	15.7	1956	Carpenter and Krigbaum [351]
	20	47	14.9	1958	Challot and Morlot [311]
	23	45.56	15.8	1960	Coumou [268]
	20	32.8	12.3	1962	Vuks and Rozhdestvenskaya [329]
Water		2	–	1922	Martin and Lehrman [336]
	32	3.08	–	1929	Raman and Rao [338]
		–	1.05	1954	Goring and Napier [352]
		–	0.97	1954	Fessenden and Stein [353]
		2.89	1.05	1955	Kraut and Dandliker [354]

Table III (continued)

Liquid	t, °C	$R_{90} \cdot 10^6$, cm^{-1}		Year	Authors
		λ 4358 Å	λ 5461 Å		
Water (cont.)	25	4.3	–	1956	Trossarelli and Saini [350]
		37*	–	1965	Leite, Moore, Porto, and Ripper [412]
Ether	20	–	9.2	1922	Martin and Lehrman [336]
	20	10.1	4.1	1951	Vaucouleurs [56]
	30	15.5	5.78	1950	Carr and Zimm [198]
Carbon tetrachloride	25	15.40	5.79	1953	Stamm and Button [349]
	25	14.9	5.32	1954	Maron and Lou [346]
	25	14.90	5.35	1956	Trossarelli and Saini [350]
	23	14.8	5.5	1960	Coumou [268]
	20	12.5	4.8	1962	Vuks and Rozhdestvenskaya [329]
		13.7*	–	1965	Leite, Moore, Porto, and Ripper [412]
Carbon bisulfide	25	151	47	1949	Blaker, Badger, and Gilmann [355]
	25	240	72.5	1956	Cantow and Schulz [348]
Toluene	28-25	32.2	11.8	1953	Stamm and Button [349]
	25	60	20.5	1954	Maron and Lou [346]
	25	55,60	19,30	1956	Trossarelli and Saini [350]
		60	21	1956	Carpenter and Krigbaum [351]
	25	49.0	17.2	1956	Cantow and Schulz [348]
		38.7*	–	1965	Leite, Moore, Porto, and Ripper [412]
Acetone	25	14.80	–	1956	Trossarelli and Saini [350]
Chloroform	25	18.60	–	1956	Trossarelli and Saini [350]
n-Octane	23	13.2	4.8	1960	Coumou [268]
n-Decane	23	13.5	4.9	1960	Coumou [268]
n-Hexadecane	23	15.6	5.7	1960	Coumou [268]
Methyl ethyl ketone	20	10.7	4.1	1962	Vuks and Rozhdestvenskaya [329]

*For the wavelength λ 4358 Å, the data from [412] were recalculated according to the formula $R_{90} = (3\alpha_S/16\pi)(6328/4358)^4$.

Table IV. Some Parameters for Calculation of the Scattering Coefficient for 20°C, λ = 4358 Å*

Liquid	ρ, g/cm²	n	$\sigma = \frac{1}{v} \cdot \frac{dv}{dT} \cdot 10^5$, deg^{-1}	$c_p \cdot 10^{-7}$, erg/deg	$\beta_S \cdot 10^{12}$, bar^{-1}	$\beta_T \cdot 10^{12}$, bar^{-1}	$\frac{dn}{dT} \cdot 10^5$, deg^{-1}	$\left(\frac{1}{\sigma}\frac{\partial \varepsilon}{\partial T}\right)_p = 2n\frac{1}{\sigma}\frac{dn}{dT}$	Dynamic		Static	$\Delta_u \cdot 10^2$
									$\left(\rho\frac{\partial \varepsilon}{\partial \rho}\right)_S$	$\left(\rho\frac{\partial \varepsilon}{\partial \rho}\right)_T$	$\left(\rho\frac{\partial \varepsilon}{\partial \rho}\right)_T$	
Water......	0.997	1.3397	20.7	4.18	45.7	45.7	−11	−1.42	0.82	0.82	0.868	8.8
Methyl alcohol	0.791	1.338	119	2.56	100.5	127.1	−40.5	−0.91	0.82	0.83	0.902	6.0
Ethyl alcohol .	0.789	1.369	110.1	2.39	92.8	109.5	−41	−1.02	0.95	0.96	1.041	6.7
Ether	0.713	1.3637	162	2.27	138.7	184.2	−57	−0.96	0.89	0.91	0.989	9.2
Benzene	0.879	1.5220	121	1.70	52.6	93.2	−64	−1.61	1.56	1.58	1.62	9.3
Carbon bisulfide...	1.262	1.67377	118	1.0	49.5	92.0	−81.5	−2.31	2.39	—	2.37	62
Toluene	0.865	1.5130	106	1.68	70.0	88.4	−61	−1.74	1.60	1.645	1.602	48
Acetone	0.791	1.3677	143	2.21	89.0	123.4	−50	−0.96	—		0.99	11
Chloroform ..	1.489	1.4582	126	0.966	45.8	87.1	−61	−1.41			1.41	20
Carbon tetrachloride.	1.595	1.4720	122	0.841	58.0	90.5	−56	−1.35			—	6

*Table is composed of data from [29, 273-276, 279, 312, 359-366].

Table V. Relative Intensity of the Light of Molecular Scattering in Various Liquids

Liquid	Intensity relative to benzene	Intensity relative to ether
Water	0.0546	0.195
Hexane	0.298	1.06
Benzene	1.00	3.57
Toluene	1.12	3.98
Carbon bisulfide	4.15	
Carbon tetrachloride	0.323	
Chloroform	0.357	
Acetone	0.257	
Chlorobenzene	1.46	
Aniline	1.01	
Nitrobenzene	3.45	
Acetophenone	1.79	
o-Xylol	1.29	4.60
m-Xylol	1.30	4.62
n-Xylol	1.61	5.76
Cyclohexane	0.217	0.776
Cyclohexene	0.400	1.43
Cyclohexadiene 1-3	0.633	2.26
Methyl alcohol	0.146	0.520
Ethyl alcohol	0.178	0.635
n-Propyl alcohol	0.180	0.643
Isopropyl alcohol	0.168	0.601
n-Butyl alcohol	0.191	0.681
Isobutyl alcohol	0.182	0.649
n-Amyl alcohol	0.217	0.774
Isoamyl alcohol	0.214	0.769
n-Hexyl alcohol	0.235	0.840
n-Heptyl alcohol	0.266	0.940
n-Octyl alcohol	0.307	1.09
Allyl alcohol	0.274	0.978
Benzyl alcohol	1.02	3.66
Cyclohexanol	0.180	0.642
Acetaldehyde	0.263	0.939
Paraldehyde	0.200	0.715
Diethyl ether	0.280	1.00
Formic acid	0.351	1.25
Acetic acid	0.303	1.08
Propionic acid	0.366	1.31
n-Butyric acid	0.376	1.34
Isobutyric acid	0.293	1.05
n-Valeric acid	0.353	1.26
Isovaleric acid	0.332	1.19

Table VI. Temperature Dependence of the Relative Intensity [371]

t, °C	$\Delta_u \cdot 10^2$ (measured)	$\beta_T \cdot 10^6$, atm^{-1} (measured)	I_t/I_{30} (measured)	I_t/I_{30} (computed)
		Isopentane		
30	5.6	238	1.00	1.00
60	5.6	383	1.53	1.54
80	5.3	474	1.76	1.78
90	5.16	580	2.16	2.15
100	4.8	740	2.49	2.65
110	4.7	907	2.97	3.09
120	4.2	1 090	3.35	3.53
130	3.9	1 430	3.98	4.16
140	2.9	1 956	5.61	4.44
150	2.7	2 626	6.80	6.87
160	2.15	3 884	8.74	8.83
170	1.8	12 400	24.72	26.73
180	1.5	43 970	78.80	90.7
		n-Pentane		
30	7.2	235	1.00	1.00
60	7.2	437	1.76	1.76
80	7.1	606	2.35	2.33
90	6.5	706	2.60	2.61
100	5.76	830	2.97	2.95
110	5.6	970	3.29	3.33
120	5.0	1 120	3.36	3.67
130	4.0	1 320	3.66	4.02
140	3.4	1 550	4.06	4.41
150	2.56	2 130	4.92	5.53
160	2.0	9 880	6.0	7.02
170	1.8	5 030	9.28	10.63
180	1.57	9 060	18.00	19.90
190	1.24	39 800	60.00	62.10
		Ethyl ether		
30	8.0	212	1.00	1.00
50	8.0	258	1.20	1.21
70	7.6	323	1.40	1.42
100	7.3	539	2.06	2.16
110	7.13	696	2.56	2.68
120	7.0	880	2.99	3.22
130	6.1	1 140	3.86	3.77
140	4.9	1 700	4.76	5.29
150	3.3	3 280	8.30	9.30
160	2.5	4 660	10.80	11.80
170	1.5	6 500	13.80	14.30
180	1.24	11 400	19.80	20.80
187	1.03	45 000	60.00	62.50

Table VII. Depolarization Factor of Scattered Light in Liquids (Room temperature) *

Liquid	Chemical formula	$\Delta_u \cdot 10^2$
Inorganic Compounds		
Chlorine	Cl_2	24
Water	H_2O	8.8
Carbon bisulfide	CS_2	62 (66.9± 0.2)
Sulfur dioxide	SO_2	22
Hydrocarbon and Cyclic Compounds		
n-Pentane	C_5H_{12}	7.3
2-Methyl butane	C_5H_{12}	5.0
n-Hexane	C_6H_{14}	8.0
2-Methyl pentane	C_6H_{14}	6.8
n-Heptane	C_7H_{16}	8.7
2-Methyl hexane	C_7H_{16}	7.9
3-Methyl hexane	C_7H_{16}	7.3
2,4-Dimethyl pentane	C_7H_{16}	7.0
3,3-Dimethyl pentane	C_7H_{16}	6.1
2,2-Dimethyl pentane	C_7H_{16}	5.8
2,2,3-Trimethyl butane	C_7H_{16}	5.4
n-Octane	C_8H_{18}	8.8
2,4-Dimethyl hexane	C_8H_{18}	7.8
2,3-Dimethyl hexane	C_8H_{18}	7.7
2-Methyl-3-ethyl pentane	C_8H_{18}	7.5
3-Methyl hexane	C_7H_{16}	7.6
3-Methyl-3-ethyl pentane	C_8H_{18}	7.4
2,5-Dimethyl hexane	C_8H_{18}	7.3
2,2-Dimethyl hexane	C_8H_{18}	7.3
4-Methyl hexane	C_7H_{16}	7.0
3-Ethyl hexane	C_8H_{18}	6.5
3,3-Dimethyl hexane	C_8H_{18}	6.3
2,3,4-Trimethyl pentane	C_8H_{18}	6.3
2,2,4-Trimethyl pentane	C_8H_{18}	5.6
2,2,4-Trimethyl pentane	C_8H_{18}	5.5
Allene	C_3H_4	23.5
Allylene (propyne)	C_3H_4	32.0
Isoamylene β	C_5H_{10}	25.8
Cyclopentane	C_5H_{10}	11.1 (3.6± 0.1)
Ethyl cyclopentane	C_7H_{14}	12.0
Cyclohexane	C_6H_{12}	5.6 (5.9± 0.1)
Methyl cyclohexane	C_7H_{14}	11.1
Ethyl cyclohexane	C_8H_{16}	15.5

*Depolarization of the first twenty-five hydrocarbons was measured in [174]. Other data are taken from [56, 102]. The results of measurement of the depolarization using a gas laser as the light source are shown in parentheses [594].

Table VII (continued)

Liquid	Chemical formula	$\Delta_u \cdot 10^2$
Hydrocarbon and Cyclic Compounds (cont)		
1,2-Dimethyl cyclohexane	C_8H_{16}	10.7
1,3-Dimethyl cyclohexane	C_8H_{16}	10.8
1,4-Dimethyl cyclohexane	C_8H_{16}	11.8
Cycloheptane	C_7H_{14}	11.1
Methyl cycloheptane	C_8H_{16}	17.1
Cyclooctane	C_8H_{16}	13.2 (6.4 ± 0.1)
Methyl cyclooctane	C_9H_{18}	22.2
Cyclopentene	C_5H_{18}	23
Methyl cyclopentene	C_6H_{10}	33.3
Ethyl cyclopentene	C_7H_{12}	32
Cyclopentene oxide	C_5H_8O	37.5
Cyclohexene	C_6H_{10}	23.2
1-Methyl cyclohexene	C_7H_{12}	31
3-Methyl cyclohexene	C_7H_{12}	24.8
1-Ethyl cyclohexene	C_8H_{14}	32.0
1,4-Dimethyl cyclohexene	C_8H_{14}	32.0
1,3,4-Trimethyl cyclohexene	C_9H_{16}	36.1
Cyclohexene oxide	$C_6H_{10}O$	22.4
Cycloheptene	C_7H_{12}	27.7
1-Methyl cycloheptene	C_8H_{14}	32.6
Cyclooctene	C_8H_{14}	33.3
Halide and Nitrogen Compounds		
Dichloromethane	CH_2Cl_2	31.0
Chloroform	$CHCl_3$	20 (20.4 ± 0.2)
Carbon tetrachloride	CCl_4	6 (3.8 ± 0.1)
Methyl bromide	CH_3Br	21.7
Ethyl bromide	C_2H_5Br	25
Ethylene chloride	$C_2H_4Cl_2$	40
Ethylene bromide	$C_2H_4Br_2$	61
Propyl chloride	C_3H_7Cl	16.2
Isopropyl chloride	C_3H_7Cl	16.2
Propyl bromide	C_3H_7Br	24
Isopropyl bromide	C_3H_7Br	24
Isopropyl iodide	C_3H_7I	5
Allyl chloride	C_3H_5Cl	36
Allyl bromide	C_3H_5Br	59
Isobutyl chloride	C_4H_9Cl	17
Amyl chloride	$C_5H_{11}Cl$	16
Amyl bromide	$C_5H_{11}Br$	24
Isobutyl nitrite	C_4H_9ONO	20.4
Isobutyl nitrate	$C_4H_9ONO_2$	39
Amyl nitrate	$C_5H_{11}ONO_2$	41.4

Table VII (continued)

Liquid	Chemical formula	$\Delta_u \cdot 10^2$
Sulfur Compounds		
Dimethyl sulfide	C_2H_6S	12.9
Diethyl sulfide	$C_4H_{10}S$	18.2
Thiophene	C_4H_4S	38.5
Alcohols		
Methyl alcohol	CH_4O	4.9
Ethyl alcohol	C_2H_6O	5.6
n-Propyl alcohol	C_3H_8O	5.8
Isopropyl alcohol	C_3H_8O	3.9
n-Butyl alcohol	$C_4H_{10}O$	9.3
Isobutyl alcohol	$C_4H_{10}O$	5.5
n-Amyl alcohol	$C_5H_{12}O$	8.5
Isoamyl alcohol	$C_5H_{12}O$	6.9
n-Hexyl alcohol	$C_6H_{13}OH$	8.9
n-Heptyl alcohol	$C_7H_{15}OH$	14.4
n-Octyl alcohol	$C_8H_{17}OH$	16.5
Allyl alcohol	C_3H_6O	26.6
Benzyl alcohol	C_7H_8O	58.6
Fatty Acids		
Formic acid	CH_2O_2	52.8
Acetic acid	$C_2H_4O_2$	38.6
Propionic acid	$C_3H_6O_2$	32.9
n-Butyric acid	$C_4H_8O_2$	35.7
Isobutyric acid	$C_4H_8O_2$	20.6
n-Valeric acid	$C_5H_{10}O_2$	28.4
Isovaleric acid	$C_5H_{10}O_2$	26.0
Ethers and Anhydrides		
Diethyl ether	$C_4H_{10}O$	7
Acetic anhydride	$C_4H_6O_3$	43
Propionic anhydride	$C_6H_{10}O_3$	44.0
Methyl ether formic acid	$C_2H_4O_2$	28.1
Ethyl ether formic acid	$C_3H_6O_2$	21.3
Propyl ether formic acid	$C_4H_8O_2$	21
Ethyl ether acetic acid	$C_4H_8O_2$	23.5
Propyl ether acetic acid	$C_5H_{10}O_2$	21.7
Isobutyl ether butyric acid	$C_8H_{16}O_2$	17.3
Ethyl ether lauric acid	$C_4H_{28}O_2$	27.8
Ethyl ether palmitic acid	$C_{18}H_{36}O_2$	26 (30°C)
Ethyl ether stearic acid	$C_{20}H_{40}O_2$	22 (40°C)
Ethyl ether oleic acid	$C_{20}H_{38}O$	42
Aldehydes and Ketones		
Acetaldehyde	C_2H_4O	18.9
Benzaldehyde	C_7H_6O	69

Table VII (continued)

Liquid	Chemical formula	$\Delta u \cdot 10^2$
Aldehydes and Ketones (cont.)		
Salicylaldehyde	$C_7H_6O_2$	73
Acetone	C_3H_6O	23.6
Methyl ethyl ketone.	C_4H_8O	16.6
Diethyl ketone	$C_5H_{10}O$	18
Methyl propyl ketone.	$C_5H_{10}O$	19.6
Acetophenone.	C_8H_8O	68.3
Benzophenone.	$C_{13}H_{10}O$	61.9
n-Metaxylbenzophenone.	$C_{14}H_{12}O$	60.8
Benzene and Its Derivatives		
Benzene	C_6H_6	43 (43.8 ± 0.2)
Toluene	C_7H_8	48 (52.8 ± 0.2)
Ethylbenzene	C_8H_{10}	54.0
o-Xylol	C_8H_{10}	49.7
m-Xylol	C_8H_{10}	50.6
n-Xylol	C_8H_{10}	56.4
Propenylbenzene.	C_9H_{10}	68.0
Diphenyl.	$(C_6H_5)_2$	68.5 (70°C)
Dibenzyl.	$(C_6H_5CH_2)_2$	61.8 (60°C)
Triphenyl methane	$CH(C_6H_5)_3$	49.8 (95°C)
Anisole.	C_7H_8O	60.0
Phenetole	$C_8H_{10}O$	49.6
Styrene.	C_6H_8	56
Chlorobenzene	C_6H_5Cl	57.5
Benzoyl chloride.	C_7H_5OCl	65
Bromobenzene	C_6H_5Br	62
Benzyl chloride.	C_7H_7Cl	56.0
Benzal chloride.	$C_7H_6Cl_2$	52.0
Aniline.	C_6H_7N	60
Methyl aniline	C_7H_9N	65.7
Benzylamine	C_7H_9N	52.7
Nitrobenzene	$C_6H_5NO_2$	68
o-Nitrotoluene	$C_7H_7NO_2$	80
m-Nitrotoluene.	$C_7H_7NO_2$	81
Benzonitrile.	C_6H_5CN	68
Methyl benzoate.	$C_6H_8O_2$	63
Ethyl benzoate	$C_9H_{10}O_2$	60
Naphthalene.	$C_{10}H_8$	70.6
Decaline.	$C_{10}H_{18}$	17.6
α-Chloronaphthalene.	$C_{10}H_7Cl$	78
Indole.	C_8H_7N	46
Quinoline	C_9H_7N	79
Pyridine	C_5H_5N	46

Table VIII. Temperature Dependence of the Depol

Triacetin

t,°C	18	—5	—17	—39	—51	—58	—73
$\Delta_u \cdot 10^2$	32	34	36	36	38	39	50

β- Ether

t,°C	128	115	103	81	67	46	36	13	1	—1
$\Delta_u \cdot 10^2$	49	49.5	50.5	52.5	54.5	57.7	59	60.5	60.5	60.

Salol

t,°C	134	104	91	75	48	42	33	18	—23	—3
$\Delta_u \cdot 10^2$	64	64.8	65.7	66.8	69.8	70.5	71.5	72	72	72

Table IX. Depolarization of Scattered Light in Phenol–Water Mixtures at Different Temperatures*

Wt. % phenol	Mixing temp., °C	Δ_h	$\Delta_v \cdot 10^2$	$\Delta_u \cdot 10^2$	Relative intensity of scattered light
15	60	1	0.5	1.0	6.6
28	68.5	4.55	0.15	0.62	71
34	69	8.35	0.07	0.62	186
50	64.5	2.44	0.27	1.0	15
60	61	1	1.9	3.8	3
70	30	1	6.0	11.0	1

Carbon bisulfide–methyl alcohol (critical mixture)

Mixing temp., °C	Δ_h	$\Delta_v \cdot 10^2$	$\Delta_u \cdot 10^2$
62.2	1	4.1	8.2
57.5	1.07	2.8	5.8
55	1.11	2.1	4.5
52.5	1.20	1.7	3.95
50	11.37	1.3	2.8
47.5	1.47	1.1	2.8
45	1.70	0.93	2.4
42.5	2.2	0.64	2.0
42.5	3.7	0.34	1.5
40	4.0	0.27	1.5

*In the works of Krishnan [262] Δ_h is defined as I_{yz}/I_{yx}, i.e., it is the reciprocal of (19.33). In Table IX, we have given the data for Δ_h of Krishnan in correspondence with our definition of this quantity.

Table X. Relative Intensity of Light Scattered by the Surface and Volume of Various Liquids*

Liquid	Surface tension γ^*, dynes/cm	Surface scattering		Volume scattering
		Computed	Measured	
Water	72	1.0	1.0	1.0
Hexane	16.34	5.8	8.5	5.9
Octane	20.00	5.5	7.8	4.8
Ethylene chloride	30.1	4.9	3.8	8.5
Chloroform	25.3	5.9	6.3	7.4
Carbon tetrachloride	24.6	6.6	12.6	6.0
Silicone	15.8	8.0	7.4	—
Formic acid	23.50	4.0	4.9	5.6
Acetic acid	26.60	3.8	5.8	6.2
Propionic acid	26.60	3.8	5.8	6.2
Butyric acid	26.70	4.1	6.4	6.0
Ethyl ether	15.27	5.3	7.3	5.9
Methyl alcohol	23.00	3.0	4.0	2.7
Ethyl alcohol	22.00	4.0	7.2	3.1
Isopropyl alcohol	21.34	4.6	7.4	3.5
Butyl alcohol	24.40	4.6	7.4	3.5
Allyl alcohol	23.15	5.5	7.8	6.1
Benzyl alcohol	39.70	6.43	10.0	12.4
Ethyl formate	22.00	3.9	5.2	5.0
Propyl formate	22.20	4.4	7.4	4.7
Propyl acetate	22.00	4.6	5.6	4.8
Mercury	547		0.1	—

* Table X is taken from the data of Ramdas [399], and Raman and Ramdas [398].

Table XI. Light Intensity Scattered by the Surface of Liquid CO_2, as a Function of the Temperature [399]*

Temp., °C	γ,* dynes/cm	Intensity
31.10	0.00015	3000
31.05	0.00060	1100
31.00	0.0012	500
30.80	0.0032	270
30.60	0.009	120
30.40	0.027	90
30.20	0.035	70
30.00	0.049	60

*The critical temperature of CO_2, taken as the standard of the various authors, lies in the range from 30.98°C to 31.40°C.

Table XII. Molecular Weight M_n, M_w, and Constants of Interaction B of Polystyrene Solutions, Determined from the Osmotic Pressure and Light Scattering [413]

Fraction	Solvent	M_n, osmotic pressure	M_w, light scattering	B, osmotic pressure	B, light scattering	Literature
1	Methyl ethyl ketone	400 000	445 000			[404]
2		200 000	250 000			
3		198 000	182 000			
4		200 000	190 000			
5		172 000	178 000			
7		91 000	107 000			
8		100 000	95 000			
	Benzene	108 000	109 000	$3.7 \cdot 10^{-4}$	$3.5 \cdot 10^{-4}$	[414]
		249 000	260 000	$4.4 \cdot 10^{-4}$	$4.5 \cdot 10^{-4}$	
C	Toluene	440 000	451 000	$4.4 \cdot 10^{-4}$	$5.29 \cdot 10^{-4}$	[415]
D		910 000	1 065 000	$4.05 \cdot 10^{-4}$	$3.8 \cdot 10^{-4}$	
*F*1	Methyl ethyl ketone	160 000	178 000			[416]
*F*7		42 000	42 000			
*PS*1		29 000	53 000			
*PS*3		50 000	103 000			
*PS*4		63 000	119 000			
*PS*5		100 000	220 000			

Table XIII. Molecular Weight of Certain Proteins [43]

Protein	Molecular weight $M \cdot 10^{-3}$		$\Delta u \cdot 10^2$	Method of calibration*
	λ 5780 Å λ 5461 Å	λ 4358 Å		
Antibody of rabbit serum	158			1
Ovalbumen	45.7		2.4	2
Lysozyme	14.8		3.0	2
β-Lactoglobulin	35.7		2.1	2
Bovine serum albumen		73	2.7	2
		77	0.2	3
	71			4
		68		5
α-Chemotrisinogen		25		2,5
Insulin		12		2,5
Bovine fibrinogen		540		5
		340		2
d-Glyceraldehyde-3-phosphate dihydrogenase	140	140		5

*Method of calibration:

(1) CS_2 was used as the standard: $R_{90} = 47.8 \cdot 10^{-6}$ cm^{-1} for λ = 5461 Å.

(2) Absorption method of determination.

(3) Benzene was used as the standard: $R_{90} = 49 \cdot 10^{-6}$ cm^{-1} for λ = 4358 Å.

(4) Scattering in yellow turnip virus was used for calibration measurements according to the degree of transparency.

(5) Scattering in ludox by the same method as in (4).

Table XIV. Velocity of Hypersound According to Data of Various Authors [53]

Benzene				Toluene				Carbon tetrachloride				Authors
Velocity of ultrasound v_0 = 1324 m/sec				Velocity of ultrasound v_0 = 1341 m/sec				Velocity of ultrasound v_0 = 920 m/sec				
Δλ; Δν*	θ°	λ, Å	v, m/sec	Δλ; Δν	θ°	λ, Å	v, m/sec	Δλ; Δν	θ°	λ, Å	v, m/sec	
0.047 ± 0.0003	90	4358	1504 ± 96									Gross [27] (1930)
0.047 0.056 0.055 0.063	90 135 90 135	4047 4047 5791 4358	1600 1470 1340 1560									Gross and Khvostikov [442] (1932)
0.044	90		1410									Meyer and Ramm [253] (1932)
0.042 ± 0.002	90	4358	1345 ± 65									Khvostikov [443] (1934)
0.062	150	4358	1450 ± 187	0.060 ± 0.008 0.057 ± 0.008	150	4358	1420 ± 185 1245 ± 185	0.042 ± 0.004	150	4358	945 ± 95	Ramm [235] (1934)
0.320 0.320 0.306	180 180 180	4047 4078 4358	1280 1280 1300	0.320 0.320 0.29 0.21	180 180 180 180	4047 4078 4358 5461	1255 1255 1230 1220	0.230 0.25 0.25	180 180 180	4358 4078 4047	1030 1210 1210	Rao [438, 439] (1937-1941)
0.329 0.268	180 180	4078 4722	1290 1231					0.226 0.195	180 180	4078 4722	932 938	Venkateswaran [441] (1942)

*Differences in frequency Δν or wavelength Δλ between the central lines and the MB components are expressed in cm^{-1} and Å, respectively.

Table XV. Some Characteristics of Liquids, Studied in [31-33, 36, 55]

(Excitation of scattered light of λ 4358 Å of Hg, observed at an angle $\theta = 90°$. Temperature of the experiment 20 °C.)

Substance	n	$\Lambda \cdot 10^5$, cm	$v_0, \frac{m}{sec}$	$\frac{\Delta v}{\Delta T}, \frac{m}{sec \cdot deg}$
Benzene	1.524	2.04	1324	—4.6
Carbon bisulfide	1.697	1.85	1153	—3.2
Carbon tetrachloride	1.472	2.10	920	—3.1
Methylene chloride	1.424	2.16	1092	—4.4
Methylene bromide	1.545	2.0	971	—
Chloroform	1.446	2.14	1005	—3.6
Toluene	1.519	2.02	1328	—4.5
Acetone	1.366	2.26	1192	—5.5
Water	1.340	2.30	1485	+2.5
Acetic acid	—	2.26	1144	—

Table XVI. Temperature Dependence of the Scattered Light Intensity in the Transition from Liquid to Glassy State in Various Individual Substances [188, 189]

Triacetin

t, °C	18	3	1	0	—3	—7	—17	—21	—22	—28	—33	—35	—41	—42	—45	—46	—48	—50	—54	—58	—67
I/I_{18} mean	1	0.9	0.89	0.88	0.87	0.85	0.82	0.81	0.81	0.80	0.79	0,79	0.78	0.78	0.78	0.78	0.78	0.78	0.78	0.78	0.78

β-Ether

t, °C	56	39	18	3	—7	—30	—39	—45	—58	—62	—71	—75	—92
I/I_{18} mean	1.11	1.06	1	0.96	0.94	0.89	0.87	0.86	0.84	0.83	0.82	0.81	0.79

Salol

t, °C	18	—1	—3	—7	—19	—22	—26	—35	—39	—47	—48	—53	—60	—68	—72	—73
I/I_{18} mean	1	0,93	0,92	0.90	0.86	0.85	0.84	0.80	0.78	0.76	0.76	0.74	0.73	0.70	0.70	0.70

References

1. J. Tyndall, Phil. Mag. 37:384 (1869); Proc. Roy. Soc. (London) 17:223 (1869).
2. Lord Rayleigh, Scientific Papers, Vol. I (University Press, Cambridge, 1899), pp. 87-110.
3. Lord Rayleigh, Scientific Papers, Vol. I (University Press, Cambridge, 1899), pp. 518-536.
4. Lord Rayleigh, Scientific Papers, Vol. IV (University Press, Cambridge, 1903), pp. 397-405.
5. L. I. Mandel'shtam, Collected Works, Vol. I (Izd. Akad. Nauk SSSR, 1948), p. 109; Ann. Physik 23:626 (1907).
6. G. S. Landsberg, Introduction to the book by K. Kohlrausch, Combination Scattering Spectra (Izd. Akad. Nauk SSSR, 1958), p. 355.
7. A. Schuster, Theory of Optics (Edward Arnold, London, 1909).
8. H. A. Lorentz, Statistical Theories in Thermodynamics (Russian translation, ONTI, 1935).
9. H. Dember, Ann. Physik 49:599 (1916).
10. J. Cabannes, Compt. Rend. 160:62 (1915).
11. R. J. Strutt (Lord Rayleigh, the younger), Proc. Roy. Soc. (London) 94:453; 95:155 (1918).
12. R. Wood, Phil. Mag. 39:423 (1920).
13. M. Smoluchowski, Ann. Physik 25:205 (1908).
14. A. Einstein, Ann. Physik 33:1275 (1910).
15. L. I. Mandel'shtam, Ann. Physik 41:609 (1913); Collected Works, Vol. I (Izd. Akad. Nauk SSSR, 1948).
16. M. A. Leontovich, Statistical Physics (Gostekhizdat, 1944).
17. G. S. Landsberg, Z. Phys. 43:773 (1927); Collected Works (Izd. Akad. Nauk SSSR, 1957).
18. M. A. Leontovich and S. L. Mandel'shtam, Z. Phys. Sowjetunion 1:317 (1932).

19. F. Matossi, Z. Phys. 92:425 (1934); 96:698 (1935).
20. G. P. Motulevich, Fiz. Inst. Akad. Nauk SSSR 5:9 (1950); Izv. Akad. Nauk SSSR, Ser. Fiz. 11:390 (1947).
21. K. S. Vul'fson and M. I. Lombert, Dokl. Akad. Nauk SSSR 21:26 (1938).
22. I. A. Yakovlev, T. S. Velichkina, and L. F. Mikheeva, Kristallog. 1:123 (1956); Dokl. Akad. Nauk SSSR 107:675 (1956); [translation: Soviet Physics – Doklady 1:215 (1956)].
23. V. L. Ginzburg, Dokl. Akad. Nauk SSSR 105:240 (1955).
24. L. I. Mandel'shtam, Zh. Russ. Fiz. Khim. Obshchestva 58:381 (1926); Collected Works, Vol. I (Izd. Akad. Nauk SSSR, 1957).
25. P. Debye, Ann. Physik 39:789 (1912).
26. L. Brillouin, Ann. Physique (Paris) 17:88 (1922).
27. E. Gross, Z. Phys. 63:685 (1930).
28. E. Gross, Nature 126:201, 400, 603 (1930).
29. I. L. Fabelinskii, Usp. Fiz. Nauk 63:355 (1957).
30. I. L. Fabelinskii, Usp. Fiz. Nauk 77:649 (1962); [translation: Soviet Physics – Uspekhi 5:667 (1963)].
31. I. L. Fabelinskii and O. A. Shustin, Dokl. Akad. Nauk SSSR 92:285 (1953).
32. V. A. Molchanov and I. L. Fabelinskii, Dokl. Akad. Nauk SSSR 105:248 (1955).
33. M. S. Pesin and I. L. Fabelinskii, Dokl. Akad. Nauk SSSR 122:575 (1958); [translation: Soviet Physics – Doklady 3:974 (1959)].
34. M. S. Pesin and I. L. Fabelinskii, Dokl. Akad. Nauk SSSR 129:299 (1959); [translation: Soviet Physics – Doklady 4:1264 (1960)].
35. M. S. Pesin and I. L. Fabelinskii, Dokl. Akad. Nauk SSSR 135:1114 (1960); [translation: Soviet Physics – Doklady 5:1290 (1961)].
36. M. S. Pesin, Dissertation, Fiz. Inst. Akad. Nauk SSSR, Moscow (1961); Tr. Fiz. Inst. Akad. Nauk SSSR 30:158 (1964).
37. J. Cabannes and P. Daure, Compt. Rend. 186:1533 (1928).
38. C. V. Raman and K. S. Krishnan, Nature 122:278, 882 (1928).
39. M. Leontovich, J. Phys. USSR 4:499 (1941).
40. S. M. Rytov, Zh. Eks. i Teor. Fiz. 33:514, 669 (1957); [translation: Soviet Physics – JETP 6:401, 513 (1958)].
41. V. S. Starunov, Dokl. Akad. Nauk SSSR 153:1055 (1963); [translation: Soviet Physics – Doklady 8:1205 (1964)].

42. P. Debye, J. Appl. Phys. 15:338 (1944).
43. K. A. Stacey, Light Scattering in Physical Chemistry (Academic Press, New York, 1956).
44. H. van de Hulst, Light Scattering by Small Particles (Wiley, New York, 1957).
45. V. L. Ginzburg, Dokl. Akad. Nauk SSSR 42:172 (1944).
46. V. V. Vladimirskii, Zh. Eks. i Teor. Fiz. 9:1226 (1939).
47. L. D. Landau and E. M. Lifshitz, Electrodynamics of Continuous Media (Gostekhizdat, 1957); [translation published by Addison-Wesley, Reading, Mass., 1960].
48. L. D. Landau and E. M. Lifshitz, The Classical Theory of Fields (Addison-Wesley, Reading, Mass., 1961).
49. J. A. Stratton, Electromagnetic Theory (McGraw-Hill, New York, 1941).
50. G. Placzek, Rayleigh Scattering and the Raman Effect (GNTIU, Kharkov, 1935).
51. L. D. Landau and E. M. Lifshitz, Statistical Physics (Addison-Wesley, Reading, Mass., 1958).
52. I. L. Fabelinskii, Izv. Akad. Nauk SSSR, Ser. Fiz. 17:538 (1953).
53. I. L. Fabelinskii, Tr. Fiz. Inst. Akad. Nauk SSSR 9:181 (1958).
54. Sunanda Bai, Proc. Indian Acad. Sci. A15:338 (1942).
55. G. de Vaucouleurs, Compt. Rend. 229:35 (1949).
56. G. de Vaucouleurs, Ann. Physique (Paris) 6:211 (1951).
57. Y. Rocard, Compt. Rend. 229:1137 (1949).
58. M. A. Leontovich, Introduction to Thermodynamics (Gostekhizdat, 1951).
59. C. V. Raman and C. S. Venkateswaran, Proc. Roy. Soc. (London) A171:137 (1939).
60. L. M. Levin, Izv. Akad. Nauk SSSR, Ser. Fiz. 4:111 (1940).
61. W. G. McMillan, Jr. and J. E. Mayer, J. Chem. Phys. 13:276 (1945).
62. A. I. Akhiezer and V. B. Berestetskii, Quantum Electrodynamics (Interscience, New York, 1965).
63. R. H. Gans, Handbuch Exp. Phys., Vol. 19 (1928).
64. K. R. Ramanathan, Indian J. Phys. 1:413 (1927).
65. C. V. Raman and K. S. Krishnan, Phil. Mag. 5:498 (1928).
66. S. Bhagavantam, Scattering of Light and the Raman Effect (Chemical Publishing Co., New York, 1942).

67. C. Fabry, General Introduction to Photometry (Russian translation, ONTI, 1934).
68. D. S. Volosov and M. V. Tsivkin, Theory and Calculation of Optical Systems (Iskusstvo, 1960).
69. A. I. Ansel'm, Zh. Eks. i Teor. Fiz. 17:489 (1947).
70. I. E. Tamm, Fundamentals of the Theory of Electricity (Gostekhizdat, 1957).
71. B. Trumpy, Z. Phys. 88:226 (1934).
72. A. Rousset, Ann. Physique (Paris) 5:5 (1936).
73. I. L. Fabelinskii, Izv. Akad. Nauk SSSR, Ser. Fiz. 9:186 (1945).
74. J. Cabannes, J. Phys. Radium 8:321 (1927).
75. J. Cabannes, La Diffusion Moleculaire de la Lumière (Presse Univ. France, Paris, 1929).
76. G. S. Landsberg and L. I. Mandel'shtam, Z. Phys. Sowjetunion 8:376 (1935); L. I. Mandel'shtam, Collected Works, Vol. 2 (Izd. Akad. Nauk SSSR, 1947); G. S. Landsberg, Collected Works (Izd. Akad. Nauk SSSR, 1957).
77. P. Debye and H. Sack, Theory of the Electrical Properties of Molecules (Russian translation, ONTI, 1936); P. Debye, Usp. Fiz. Nauk 17:459 (1937).
78. M. Altschul, Z. Phys. Chem. 11:577 (1893).
79. P. A. Bazhulin, Tr. Fiz. Inst. Akad. Nauk 5:261 (1950).
80. L. S. Ornstein and F. Zernike, Amsterdam Proc. 17:793 (1914); 18:1520 (1916); 19:1321 (1917).
81. L. S. Ornstein and F. Zernike, Phys. Z. 19:134 (1918).
82. L. S. Ornstein and F. Zernike, Phys. Z. 19:134 (1918); 27:761 (1926).
83. M. Leontowitsch (Leontovich), Z. Phys. Sowjetunion 3:35 (1933).
84. A. Münster, Theory of Fluctuations, in the collection: Thermodynamics of Irreversible Processes (IL, 1962).
85. Y. Rocard, J. Phys. Radium 4:165 (1933).
86. J. Hirschfelder, C. Curtiss, and R. Bird, Molecular Theory of Gases and Liquids (Wiley, New York, 1954).
87. I. Z. Fisher, Statistical Theory of Liquids (Fizmatgiz, 1961).
88. A. S. Davydov, Zh. Eks. i Teor. Fiz. 10:263 (1940).
89. H. A. Lorentz, Theory of Electrons and Its Application to the Phenomena of Light and Radiant Heat (Russian translation, Gostekhizdat, 1956; 1915 English edition, Dover, New York).

90. M. Fixman, J. Chem. Phys. 23:2074 (1955).
91. K. R. Ramanathan, Proc. Roy. Soc. (London) 102:151 (1922).
92. M. J. Klein and L. Tisza, Phys. Rev. 76:1861 (1949).
93. J. Yvon, Actualités Sci. et Ind. No. 542 (1937); No. 543 (1937).
94. B. H. Zimm, J. Chem. Phys. 13:141 (1945).
95. V. I. Davydov, Scattering of X Rays in Liquids (ONTI,1935).
96. M. Goldstein and E. R. Michelin, J. Appl. Phys. 26:1450 (1955).
97. R. James, Optical Principles of the Diffraction of X Rays (Macmillan, New York, 1933-1953).
98. R. Fürth and C. L. Williams, Proc. Roy. Soc. (London) A224:104 (1954).
99. F. J. Pearson, Proc. Phys. Soc. (London) 75:633 (1960).
100. F. Zernike and T. A. Prins, Z. Phys. 41:184 (1927).
101. P. Debye and H. Menke, Phys. Z. 31:797 (1930).
102. H. A. Stuart and H. G. Trieschmann, Hand-u. Jahrbuch der Chem. Phys., Vol. 8, Parts II-III (Leipzig, 1936).
103. L. I. Komarov and I. Z. Fisher, Zh. Eks. i Teor. Fiz. 43:1928 (1962); [translation: Soviet Physics – JETP 16:1358 (1963)].
104. I. Z. Fisher, Usp. Fiz. Nauk 76:499 (1962); [translation: Soviet Physics – Uspekhi 5:239 (1962)].
105. V. L. Ginzburg, in: 10th Conf. on Spectroscopy, L'vov, Vol. I (1957), p. 115.
106. V. L. Ginzburg and A. P. Levanyuk, in: Collected Papers in Memory of G. S. Landsberg (Izd. Akad. Nauk SSSR,1959).
107. L. D. Landau, Zh. Eks. i Teor. Fiz. 7:19 (1937).
108. L. D. Landau, Zh. Eks. i Teor. Fiz. 7:1232 (1937).
109. V. L. Ginzburg, Usp. Fiz. Nauk 46:348 (1952).
110. G. S. Landsberg, Optics, 4th ed., rev. (Gostekhizdat, 1957).
111. A. A. Andronov and M. A. Leontovich, Z. Phys. 38:485 (1926); A. A. Andronov, Collected Works (Izd. Akad. Nauk SSSR, 1956).
112. R. Gans, Ann. Physik 74:231 (1924).
113. R. Gans, Ann. Physik 79:204 (1926).
114. Lord Rayleigh, Proc. Roy. Soc. (London) A79:399 (1907).
115. H. Stuart, Structure of Molecules (ONTIU, 1937).
116. H. A. Stuart, Hand-u. Jahrbuch der Chem. Phys., Vol. 10, Part III, p. 27.
117. L. Silberstein, Phil. Mag. 33:215, 521 (1927).

118. R. Gans, Ann. Physik 65: 97 (1921).
119. J. Kerr, Phil. Mag. 50: 337 (1875).
120. J. Kerr, Phil. Mag. 8: 85, 229 (1879).
121. D. V. Bims, Usp. Fiz. Nauk 13: 209 (1933); Rev. Mod. Phys. 4: 133 (1932).
122. W. Voigt, Ann. Physik 4: 197 (1901).
123. P. Langevin, Le Radium 7: 249 (1910).
124. M. Born, Ann. Physik 55: 177 (1918).
125. M. Born, Optik (J. Springer, 1933).
126. R. Gans, Ann. Physik 64 :481 (1921).
127. R. de Mallemann, Ann. Physique (Paris) 2: 5 (1924).
128. G. Szivessy, Handbuch der Physik, Vol. 21 (Berlin, 1929), p. 724.
129. L. I. Mandel'shtam, Collected Works, Vol. 5 (Izd. Akad. Nauk SSSR, 1950).
130. G. S. Landsberg, Molecular Light Scattering, in: Jubilee Collection, 30th Anniversary of the Great October Socialist Revolution (Izd. Akad. Nauk SSSR, 1947).
131. G. S. Landsberg, Collected Works (Izd. Akad. Nauk SSSR, 1958), p. 429.
132. G. S. Landsberg and L. I. Mandel'shtam, Zh. Prikl. Fiz. 6: 155 (1929).
133. L. I. Mandel'shtam, G. S. Landsberg, and M. A. Leontovich, Phys. Z. 60: 334 (1930); L. I. Mandel'shtam, Collected Works, Vol. 2 (Izd. Akad. Nauk SSSR, 1950).
134. G. S. Landsberg, Usp. Fiz. Nauk 36: 284 (1948).
135. M. A. Leontovich, Z. Phys. 72 :247 (1931).
136. H. Lamb, Hydrodynamics (Russian translation, Gostekhizdat, 1947 ; English edition, Dover, New York).
137. Lord Rayleigh, Theory of Sound, Vol. 2 (Russian translation, Gostekhizdat, 1944; English edition, Dover, New York).
138. M. A. Leontovich, Izv. Akad. Nauk SSSR, p. 633 (1936).
139. P. A. Bazhulin and M. A. Leontovich, Dokl. Akad. Nauk SSSR 57: 29 (1947).
140. L. D. Landau and G. Placzek, Z. Phys. Sowjetunion 5: 172 (1934).
141. V. L. Ginzburg, Dokl. Akad. Nauk SSSR 30: 397 (1941).
142. V. L. Ginzburg, Izv. Akad. Nauk SSSR, Ser. Fiz. 9: 174 (1945).
143. E. F. Gross, Zh. Eks. i Teor. Fiz. 16: 129 (1946).

144. I. L. Fabelinskii, Izv. Akad. Nauk SSSR, Ser. Fiz. 11:382 (1947).
145. I. L. Fabelinskii, Dokl. Akad. Nauk SSSR 106 : 822 (1956).
146. G. S. Landsberg, Z. Phys. 45:442 (1927).
147. V. L. Ginzburg, Zh. Eks. i Teor. Fiz. 13 : 243 (1943).
148. I. E. Tamm, Z. Phys. 60 : 345 (1930).
149. L. D. Landau, J. Phys. USSR 5: 71 (1941).
150. V. P. Peshkov, Zh. Eks. i Teor. Fiz. 16 : 1000 (1946).
151. J. Frenkel, Kinetic Theory of Liquids (Oxford University Press, 1946).
152. P. Debye, Polar Molecules (Dover, New York).
153. L. Natanson, Phil. Mag. 2 : 442 (1901); M. A. Isakovich, Dokl. Akad. Nauk SSSR 23 :782 (1939).
154. M. A. Leontovich, Izv. Akad. Nauk SSSR, Ser. Fiz. 5: 148 (1941).
155. S. M. Rytov, Correlation Theory of Rayleigh Light Scattering, in Experimental and Theoretical Studies in Physics, in: Collected Papers in Memory of G. S. Landsberg (Izd. Akad. Nauk SSSR, 1959).
156. S. M. Rytov, Zh. Eks. i Teor. Fiz. 33 : 166 (1957); [translation: Soviet Physics – JETP 6 : 130 (1958)].
157. L. D. Landau and E. M. Lifshitz, Mechanics of Continuous Media (Addison-Wesley, Reading, Mass.).
158. D. V. Sivukhin, Zh. Eks. i Teor. Fiz. 9 : 1253 (1939).
159. L. D. Landau and E. M. Lifshitz, Zh. Eks. i Teor. Fiz. 32 : 618 (1957); [translation: Soviet Physics – JETP 5 : 512 (1957)].
160. F. Pockels, Lehrbuch der Kristalloptik (Teubner, Leipzig-Berlin, 1906); J. Nye, Physical Properties of Crystals, Their Representation by Tensors and Matrices (Oxford University Press, 1957).
161. V. L. Ginzburg, Zh. Eks. i Teor. Fiz. 10 : 601 (1940).
162. V. V. Vladimirskii, Dokl. Akad. Nauk SSSR 31 : 866 (1941).
163. S. Bhagavantam, Indian J. Phys. 7 : 107 (1932).
164. J. Weiler, Ann. Physik 23 : 493 (1935).
165. H. Volkmann, Ann. Physik 24 : 457 (1935).
166. R. Ananthakrishnan, Proc. Indian Acad. Sci. A2 : 133 (1935).
167. R. Rao, Indian J. Phys. 2 : 61 (1927).
168. B. A. Brice, M. Halwer, and R. Speiser, J. Opt. Soc. Amer. 40 : 768 (1950); 44 : 340 (1954).

169. R. Wood, Physical Optics (Macmillan, New York).
170. I. L. Fabelinskii, Zh. Eks. i Teor. Fiz. 16 : 728 (1946).
171. W. H. Martin, J. Phys. Chem. 24 : 478 (1920).
172. C. S. Venkateswaran, Proc. Indian Acad. Sci. A15:362 (1942).
173. M. I. Rozengart, Techniques of Laboratory Distillation and Rectification (Goskhimizdat, 1951).
174. I. L. Fabelinskii, Zh. Eks. i Teor. Fiz. 10 : 150 (1940); J. Phys. USSR 2 : 277 (1940).
175. N. O. Chechik, S. M. Fainshtein, and T. M. Lifshitz, Electron Multipliers (Gostekhizdat, 1957).
176. L. A. Tumerman, Dokl. Akad. Nauk SSSR 58 : 1945 (1947).
177. R. Gans, Phys. Z. 28 : 661 (1927).
178. P. Daure, Ann. Physique (Paris) 12 : 375 (1929).
179. P. Peyrot, Ann. Physique (Paris) 9 : 335 (1938).
180. K. S. Krishnan, Phil. Mag. 50: 697 (1925).
181. P. A. Bazhulin, A. F. Plate, O. P. Solovova, and B. A. Kazanskii, Izv. Akad. Nauk SSSR, Ser. Khim. 1941 (1) : 13.
182. H. G. Jerrard, J. Opt. Soc. Amer. 38 : 35 (1948).
183. E. Whittaker and G. Robinson, The Calculus of Observations (Blackie and Son, London, 1940).
184. S. L. Mandel'shtam, Introduction to Spectral Analysis (Gostekhizdat, 1935).
185. V. K. Prokof'ev, Photographic Methods of the Quantitative Analysis of Metals and Alloys, Part II (Gostekhizdat, 1951).
186. P. Kh. Pruss, Izv. Akad. Nauk SSSR, Ser. Fiz. 21 : 1526 (1957); V. A. Burgov, Fundamentals of the Recording and Reproduction of Sound (Izkusstvo, 1954).
187. C. V. Sweitzer, J. Phys. Chem. 31 : 1150 (1927).
188. T. S. Velichkina, Izv. Akad. Nauk SSSR, Ser. Fiz. 17 : 546 (1953).
189. T. S. Velichkina, Fiz. Inst. Akad. Nauk SSSR 9 : 59 (1958).
190. L. S. Ornstein, W. J. Moll, and H. C. Burger, Objektive Spektralphotometrie (Samml. Viewieg 108/109, 1932).
191. J. Strong, Procedures in Experimental Physics (Prentice-Hall, New York, 1938).
192. E. Angerer, Techniques of Physical Experiment (Fizmatgiz, 1962).
193. F. S. Baryshanskaya, Zh. Eks. i Teor. Fiz. 7 : 51 (1937).

194. Chow Quantie, Proc. Roy. Soc. (London) A224: 90 (1954).
195. R. O. Gumprecht and C. M. Sliepcevich, J. Phys. Chem. 57: 90 (1953).
196. J. H. Chin, C. M. Sliepcevich, and M. Tribus, J. Phys. Chem. 59: 841 (1955); 59: 845 (1955).
197. W. Aughey and F. J. Baum, J. Opt. Soc. Amer. 44: 833 (1954).
198. G. I. Carr and B. H. Zimm, J. Chem. Phys. 18:1616 (1950).
199. P. Daure, Compt. Rend. 180: 2032 (1925).
200. R. F. Stamm and P. A. Button, J. Chem. Phys. 21: 1304 (1953).
201. B. H. Zimm, J. Chem. Phys. 16: 1093, 1099 (1948).
202. C. Wippler and C. J. Scheibling, J. Chim. Phys. 51: 201 (1954).
203. G. V. Schulz, H. J. Cantow, and G. Meyerhoff, J. Polymer Sci. 10: 79 (1953).
204. P. Bosworth, C. R. Masson, H. W. Melville, and F. W. Peaker, J. Polymer Sci. 9: 565 (1952).
205. R. F. Stamm and T. Mariner, Rev. Sci. Instr. 22: 61 (1951).
206. A. Rousset and R. Lochet, J. Chim. Phys. 51: 13 (1954).
207. J. Cabannes and P. Daure, Compt. Rend. 184: 520 (1927).
208. A. Rousset and R. Lochet, J. Polymer Sci. 10: 319 (1953).
209. A. A. Shishlovskii, Applied Physical Optics (Fizmatgiz, 1961).
210. S. E. Frish, Techniques of Spectroscopy (Izd. Leningr. Gos. Univ., 1936).
211. A. W. Reitz, Z. Phys. Chem. B33: 368 (1936).
212. Ya. S. Bobovich and M. V. Vol'kenshtein, Izv. Akad. Nauk SSSR, Ser. Fiz. 12:553 (1948).
213. M. M. Sushchinskii, Tr. Fiz. Inst. Akad. Nauk SSSR 5: 185 (1950).
214. M. M. Sushchinskii, Tr. Fiz. Inst. Akad. Nauk SSSR 12: 54 (1960).
215. Ya. S. Bobovich and V. A. Aleksandrov, Zav. Lab. 16: 37 (1950).
216. D. H. Rank, R. J. Pfister, and P. D. Coleman, J. Opt. Soc. Amer. 32: 390 (1942).
217. K. Kohlrausch, Ramanspektren (J. W. Edwards, 1945).
218. P. Gill and D. W. Heddle, J. Opt. Soc. Amer. 53: 847 (1963).

219. G. S. Landsberg and L. I. Mandel'shtam, Zh. Russ. Fiz. Khim. Obshchestva 60 : 335 (1928); L. I. Mandel'shtam, Collected Works, Vol. 1 (Izd. Akad. Nauk SSSR, 1948).
220. M. F. Vuks, Izv. Akad. Nauk SSSR, Ser. Fiz. 5 : 150 (1941); Dokl. Akad. Nauk SSSR 48 : 99 (1945).
221. V. K. Ablekov, Opt. i Spektr. 6 : 562 (1959).
222. R. S. Krishnan and V. Chandrasekharan, Nature 165 : 406 (1950).
223. R. S. Krishnan and V. Chandrasekharan, Proc. Indian Acad. Sci. A31 : 427 (1950).
224. V. Chandrasekharan, Proc. Indian Acad. Sci. A32 : 379 (1950).
225. R. S. Krishnan, Nature 165 : 933 (1950).
226. V. Chandrasekharan, Current Sci. 19 : 371 (1950).
227. P. P. Shorygin, Dokl. Akad. Nauk SSSR 78: 469 (1951).
228. A. Rousset, Diffusion of Light by Rigid Molecules (Gauthier-Villars, Paris, 1947).
229. M. F. Vuks and V. L. Litvinov, Dokl. Akad. Nauk SSSR 105 : 696 (1955).
230. I. L. Fabelinskii, Opt. i Spectr. 2 : 510 (1957).
231. I. L. Fabelinskii, Materials of the 10th Conference on Spectroscopy, L'vov (1957), Vol. I, p. 117.
232. S. C. Sirkar and G. S. Kastha, Nature 166 : 1073 (1950).
233. G. S. Kastha, Indian J. Phys. 23 : 247 (1949); 27 : 67 (1953).
234. O. A. Shustin, Diploma research, Fiz. Inst. Akad. Nauk SSSR (1952).
235. W. Ramm, Phys. Z. 35 : 111, 756 (1934).
236. K. Birus, Phys. Z. 39 : 80 (1938).
237. P. Zhakino, Usp. Fiz. Nauk 72 :799 (1960); 78 : 123 (1962).
238. G. Harrison, R. Lord, and J. Loofbourow, Practical Spectroscopy (Prentice-Hall, Englewood Cliffs, New Jersey, 1948).
239. S. Tolansky, High Resolution Spectroscopy (Methuen and Co., London, 1947).
240. K.W. Meissner, J. Opt. Soc. Amer. 31 : 405 (1941); 32 : 185 (1942).
241. Kh. E. Sterin, Fiz. Inst. Akad. Nauk SSSR 9 : 13 (1958).
242. R. Sawyer, Experimental Spectroscopy (Prentice-Hall, Englewood Cliffs, New Jersey, 1951).
243. D. H. Rank, J.M. McCortney, and J. Szasz, J. Opt. Soc. Amer. 38 : 287 (1948).

244. P. Flubacher, A. J. Leadbetter, J. A. Morrison, and B. P. Stoicheff, Int. J. Phys. Chem. Solids 12:53 (1960).
245. G. S. Landsberg and F. S. Baryshanskaya, Izv. Akad. Nauk SSSR, Ser. Fiz. 10: 509 (1946); G. S. Landsberg, Collected Works, p. 142.
246. D. I. Mash, V. S. Starunov, and I. L. Fabelinskii, Zh. Eks. i Teor. Fiz. 47: 783 (1964); [translation: Soviet Physics – JETP 20: 523 (1964)].
247. G. V. Mikhailov, Tr. Fiz. Inst. Akad. Nauk 27: 150 (1964)
248. H. L. Welsh, C. Cumming, and E. J. Standburg, J. Opt. Soc. Amer. 41: 712 (1951).
249. G. Placzek and E. Teller, Z. Phys. 81: 209 (1933).
250. V. Weisskopf, Z. Phys. 85: 451 (1933).
251. M. L. Sosinskii, Izv. Akad. Nauk SSSR, Ser. Fiz. 17: 621 (1953).
252. H. Schuller and E. O. Tones, Z. Phys. 74: 631 (1932); A. Kleiman and S. P. Davis, J. Opt. Soc. Amer. 53: 822 (1963).
253. E. H. Meyer and W. Ramm, Phys. Z. 33: 270 (1932).
254. A. P. Ivanov, Electric Light Sources. Gas Discharge Tubes (Gosenergoizdat, 1948).
255. V. K. Ablekov, V. P. Zaitsev, and M. S. Pesin, Pribory i Tekh. Eks. 1960(2): 140.
256. A. A. Michelson, Studies in Optics (Univ. Chicago Press, 1927).
257. C. S. Venkateswaran, Proc. Indian Acad. Sci. A15: 323 (1942).
258. Sunanda Bai, Proc. Indian Acad. Sci. A15: 349, 357 (1942).
259. New Wavelength Standards Proposed, J. Opt. Soc. Amer. 53: 401 (1963).
260. P. G. Wilkinson and K. L. Andrew, J. Opt. Soc. Amer. 53: 710 (1963); K. M. Baird, P. S. Smith, and K. H. Heart, J. Opt. Soc. Amer. 53: 717 (1963).
261. A. Javan, E. Balik, and V. Bond, Frequency Characteristics of Continuous Lasers Operating with Mixtures of Helium – Neon, in the collection of translations: Lasers (IL, 1963).
262. R. S. Krishnan, Proc. Indian Acad. Sci. A1: 782 (1934); A7: 21 (1938).
263. V. P. Zaitsev, V. S. Starunov, and I. L. Fabelinskii, Pribory i Tekh. Eks. 1964(1): 221.

264. D. I. Mash, V. S. Starunov, E. Tiganov, and I. L. Fabelinskii, Zh. Eks. i Teor. Fiz. (in press).
265. A. A. Andronov, Radiofizika 3 : 645 (1960).
266. A. Sommerfeld, Optics (Academic Press, New York, 1954).
267. A. Unsöld, Physik der Sternatmosphären (J. Springer, Berlin, 1955).
268. D. J. Coumou, J. Colloid Sci. 15 : 408 (1960).
269. H. C. Burger and P. H. van Cittert, Z. Phys. 79 : 722 (1932); 81 : 428 (1933).
270. M. M. Sushchinskii, Zh. Eks. i Teor. Fiz. 25 : 87 (1953).
271. P. Crosby and B. W. Koerber, J. Opt. Soc. Amer. 53 : 358 (1963).
272. G. P. Motulevich and I. L. Fabelinskii, Zh. Eks. i Teor. Fiz. 25 : 604 (1953).
273. W. C. Röntgen and L. Zehnder, Ann. Physik und Chemie 44 : 24 (1891).
274. J. Eisele, Ann. Physik 76 : 396 (1925).
275. F. Himstedt and J. Wertheimer, Ann. Physik 67 : 395 (1922).
276. C. V. Raman and C. V. Venkateswaran, Nature 143 : 798 (1939).
277. G. P. Motulevich and I. L. Fabelinskii, Izv. Akad. Nauk SSSR, Ser. Fiz. 14 : 542 (1950).
278. G. P. Motulevich and I. L. Fabelinskii, Dokl. Akad. Nauk SSSR 83 : 203 (1952).
279. G. P. Motulevich and I. L. Fabelinskii, Dokl. Akad. Nauk SSSR 106 : 637 (1956); [translation: Soviet Physics – Doklady 1 : 81 (1956)].
280. T. S. Velichkina and I. L. Fabelinskii, Dokl. Akad. Nauk SSSR 75 : 177 (1950).
281. A. R. von Hippel, Dielectrics and Waves (Wiley, New York, 1954).
282. S. M. Rytov, Izv. Akad. Nauk SSSR, Ser. Fiz. 1937(2) :223.
283. A. Schallamach, Proc. Phys. Soc. (B) 62:70 (1949).
284. P. Debye and F. W. Sears, Proc. Nat. Acad. Sci. USA 18 : 409 (1932).
285. R. Lucas and P. Biquard, Compt. Rend. 194 :2132 (1932).
286. G. P. Motulevich, I. L. Fabelinskii, and L. N. Shteingauz, Dokl. Akad. Nauk SSSR 70 :29 (1950).
287. I. P. Golyamina, Akust. Zh. 1 : 40 (1955).

288. G. P. Motulevich and I. L. Fabelinskii, Dokl. Akad. Nauk SSSR 81: 787 (1951).
289. J. E. Piercy and J. Lamb, Proc. Roy. Soc. (London) A226 :43 (1954).
290. D. H. Hall and J. Lamb, Proc. Phys. Soc. (London) 73 : 354 (1959).
291. G. P. Motulevich and I. L. Fabelinskii, Akust. Zh. 3 : 205 (1957).
292. V. P. Zaitsev, G. P. Motulevich, and I. L. Fabelinskii, Akust. Zh. 4 : 137 (1958).
293. S. M. Raiskii, Zh. Eks. i Teor. Fiz. 20 : 378 (1950).
294. J. M. M. Pinkerton, Proc. Phys. Soc. (London) B62 : 286 (1949).
295. J. H. Andreae, R. Bass, E. L. Heasell, and J. Lamb, Acustica 8 : 131 (1958).
296. S. V. Krivokhizha and I. L. Fabelinskii, Zh. Eks. i Teor. Fiz. (in press).
297. P. Bridgman, The Physics of High Pressures (G. Bell, London, 1949).
298. I. V. Obreimov, Application of Fresnel Diffraction to Physical and Technical Measurements (Izd. Akad. Nauk SSSR, 1945).
299. B. A. Brice and M. Halwer, J. Opt. Soc. Amer. 41 : 1033 (1951).
300. S. R. Caplan, J. Scient. Instr. 31 : 295 (1954).
301. A. Kruis, Z. Phys. Chem. 34B : 1, 13, 82 (1936).
302. L. W. Tilton and T. K. Taylor, J. Res. Nat. Bur. Stand. 20 : 419 (1938).
303. J. Cabannes, Ann. Physique (Paris) 15: 5 (1921).
304. I. I. Sobel'man, Dokl. Akad. Nauk SSSR 88: 653 (1963).
305. Tien Kiu, J. Phys. Radium 9 : 247 (1938).
306. A. Vassy and M. E. Vassy, J. Phys. Radium 10 : 401 (1939).
307. R. Penndorf, J. Opt. Soc. Amer. 47 : 176 (1957).
308. D. Deirmendjian, J. Opt. Soc. Amer. 48 : 1018 (1958).
309. L. Wijnberg, J. Opt. Soc. Amer. 51 : 916 (1961).
310. H. Barrell, J. Opt. Soc. Amer. 41 : 295 (1951).
311. B. Challot and G. Morlot, J. Chim. Phys. 55 : 515 (1958).
312. Landolt-Börnstein, Zahlenwerte und Funktionen aus Physik, Chemie, Astronomie, Geophysik, und Technik, 6th ed. (Springer-Verlag, Berlin).

313. V. L. Ginzburg, Zh. Eks. i Teor. Fiz. 34 : 246 (1958); [translation: Soviet Physics – JETP 7:170 (1958)].
314. J. Cabannes and J. Granier, J. Phys. Radium 4 : 429 (1923).
315. J. Cabannes and A. Lepape, Compt. Rend. 179 : 235 (1924).
316. R. J. Strutt (Lord Rayleigh, the younger), Proc. Roy. Soc. (London) A95 : 155 (1918); 97 : 435 (1920); 98 :57 (1920).
317. S. Parthasarathy, Indian J. Phys. 7 : 139 (1932).
318. E. V. Pyaskovskaya-Fesenkova, Investigation of Light Scattering in the Earth's Atmosphere (Izd. Akad. Nauk SSSR, 1957).
319. C. Bhagavantam, Nature 130 : 740 (1932).
320. C. M. Levis and W. V. Houston, Phys. Rev. 44 : 903 (1933).
321. C. Bhagavantam, Nature 129 : 830 (1932).
322. R. G. Dickinson, R. T. Dillon, and F. Rasetti, Phys. Rev. 34 : 582 (1929).
323. C. V. Raman and R. S. Krishnan, Proc. Roy. Soc. (London) A122 : 23 (1929).
324. J. Cabannes and J. Rocard, J. Phys. Radium 10 : 52 (1929).
325. C. Bhagavantam, Indian J. Phys. 6 : 19 (1931).
326. R. H. Dicke, Phys. Rev. 89 : 472 (1953); 103 : 620 (1956).
327. G. S. Landsberg and A. A. Shubin, Zh. Eks. i Teor. Fiz. 6 : 52 (1936).
328. G. Wolfsohn, Z. Phys. 63 : 634 (1930).
329. M. F. Vuks and N. B. Rozhdestvenskaya, Dokl. Akad. Nauk SSSR 147 : 573 (1962); [translation: Soviet Physics – Doklady 7 : 1010 (1963)].
330. C. Venkateswaran, Proc. Indian Acad. Sci. A15 : 316 (1942).
331. R. Lennuier, Compt. Rend. 226 : 708 (1948).
332. J. Cabannes, Compt. Rend. 226 : 710 (1948).
333. I. F. Zartman, J. Acoust. Soc. Amer. 21 : 171 (1949).
334. J. J. Markham, R. T. Beyer, and R. B. Lindsay, Rev. Mod. Phys. 23 : 353 (1951).
335. I. L. Fabelinskii, Experimental and Theoretical Studies in Physics, in: Collection of Papers in Memory of G. S. Landsberg (Izd. Akad. Nauk SSSR, 1959).
336. W. H. Martin and S. Lehrman, J. Phys. Chem. 26 : 75 (1922).
337. F. Picker, Techniques of Polymer Characterization (London, 1959).
338. C. V. Raman and R. Rao, Phil. Mag. 46 : 626 (1923).
339. P. Doty and R. F. Steiner, J. Chem. Phys. 18 : 1211 (1950).

340. A. Oth, J. Oth, and V. Desreux, J. Polymer Sci. 10: 551 (1953).
341. M. Halwer, G. C. Nutting, and B. A. Brice, J. Chem. Phys. 21: 1425 (1953).
342. M. Harrand, Ann. Physique (Paris) 8: 646 (1953).
343. G. Oster, Analyt. Chem. 25: 1165 (1953).
344. H. J. L. Trap and J. J. Hermans, Rec. Trav. Chim. Pays-Bas 73: 167, 354 (1954).
345. B. Sedlacek, Chem. Listy 47: 1113, 1120, 1591, 1865 (1953).
346. S. H. Maron and R. L. Lou, J. Polymer Sci. 24: 273 (1954).
347. L. M. Kushner, J. Opt. Soc. Amer. 44: 155 (1954).
348. H. J. Cantow and G. V. Schulz, Z. Phys. Chem. 1: 365 (1954); H. J. Cantow, Die Makromolekulare Chem. 18-19: 367 (1956).
349. R. F. Stamm and P. A. Button, J. Chem. Phys. 23: 2456 (1955).
350. L. Trossarelli and G. Saini, Atti della Accad. Science di Torino 90: 382 (1956).
351. D. K. Carpenter and W. R. Krigbaum, J. Chem. Phys. 24: 1041 (1956).
352. D. A. I. Goring and P. G. Napier, J. Chem. Phys. 22: 147 (1954).
353. R. W. Fessenden and R. S. Stein, J. Chem. Phys. 22: 1778 (1954).
354. J. Kraut and W. B. Dandliker, J. Chem. Phys. 23: 1544 (1955).
355. R. H. Blaker, R. M. Badger, and T. S. Gilmann, J. Phys. Chem. 53: 794 (1949).
356. J. Cabannes, Ann. Physique (Paris) 1: 857 (1956).
357. J. Rocard, Ann. Physique (Paris) 10: 116 (1928).
358. S. Guinond and J. Tonnelat, J. Chim. Phys. 51: 539 (1954).
359. J. Timmermons and F. Martin, J. Chim. Phys. 23: 746 (1926); 25: 411 (1928); J. Timmermons and Hennout-Roland, J. Chim. Phys. 27: 401 (1930).
360. W. H. Perkin, J. Chem. Soc. 61: 287 (1892).
361. R. Weegmann, Z. Phys. Chem. 2: 218 (1888).
362. D. Tyrer, J. Chem. Soc. 105: 2534 (1914).
363. E. B. Freyer, J. C. Hubbard, and D. H. Andrews, J. Amer. Chem. Soc. 51: 759 (1929).
364. Handbook of Chemistry and Physics, 44th ed. (The Chemical Rubber Publishing Co., 1962).
365. Taschenbuch für Chemiker und Physiker (Springer-Verlag, Berlin, 1949).

366. W. H. Martin, Trans. Roy. Soc. Canada 17 : 151 (1923).
367. W. H. Martin and S. Lehrman, J. Phys. Chem. 26 : 75 (1922); 27 : 558 (1923).
368. R. S. Krishnan, Phil. Mag. 50 : 697 (1925).
369. R. Ananthakrishnan, Proc. Indian Acad. Sci. 1 : 201 (1934).
370. S. Parthasarathy, Indian J. Phys. 8 : 275 (1933).
371. R. Rao, Indian J. Phys. 2 : 7 (1927).
372. L. Landau, J. Phys. USSR 5 : 71 (1941).
373. F. London, Phys. Rev. 54 : 947 (1938).
374. A. D. Galanin, Zh. Eks. i Teor. Fiz. 10 : 1267 (1940).
375. L. I. Schiff, Phys. Rev. 57 : 844 (1940).
376. E. Burton, H. D. Smith, and J. O. Wilhelm, Phenomena at the Temperature of Liquid Helium (New York, 1940).
377. J. McLennan, H. G. Smith, and J. O. Wilhelm, Phil. Mag. 14 : 161 (1932).
378. I. A. Yakovlev, J. Phys. USSR 7 : 307 (1943).
379. A. W. Lawson and L. Meyer, Phys. Rev. 93 : 259 (1954).
380. H. A. Stuart and H. Volkmann, Z. Phys. 83 : 444, 461 (1933).
381. H. Rehaag and H. A. Stuart, Phys. Z. 38 : 1027 (1937).
382. A. Peterlin and H. A. Stuart, Z. Phys. 113 : 663 (1939).
383. H. Müller, Phys. Rev. 50 : 547 (1936).
384. M. V. Vol'kenshtein, Molecular Optics (Gostekhizdat, 1951).
385. L. Onsager, J. Amer. Chem. Soc. 58 : 1486 (1936).
386. K. L. Wolf, G. Briegleb, and H. A. Stuart, Z. Phys. Chem. B6 : 163 (1929).
387. B. Trumpy, Z. Phys. 66 : 790 (1930).
388. A. D. Buckingham and M. J. Stephen, Trans. Faraday Soc. 53 : 884 (1957).
389. S. Kielich, Bull. Acad. Polon. 6 : 215 (1958).
390. W. Heitler, Quantum Theory of Radiation (Oxford University Press, 1954).
391. M. Born and E. Wolf, Principles of Optics (Pergamon Press, 1959).
392. A. D. Buckingham and J. A. Pople, Proc. Phys. Soc. (London) A68 : 905 (1955).
393. A. D. Buckingham, Trans. Faraday Soc. 52 : 611 (1956).
394. H. Benoit and W. H. Stockmayer, J. Phys. Radium 17 : 21 (1956).
395. J. A. Pople, Proc. Roy. Soc. (London) A215 : 67 (1952).
396. A. Eisenstein and N. S. Gingrich, Phys. Rev. 62 : 261 (1942).

397. C. V. Raman and L. A. Ramdas, Proc. Roy. Soc. (London) A108:561 (1925).
398. C. V. Raman and L. A. Ramdas, Proc. Roy. Soc. (London) A109:150, 272 (1925).
399. L. A. Ramdas, Indian J. Phys. 1:199 (1926).
400. L. A. Ramdas, Proc. Indian Assn. Sci. 9, Part II, p. 129 (1925); Part IV, p. 323 (1925).
401. L. A. Ramdas, Indian J. Phys. 1:1 (1926).
402. S. Jagannathan, Proc. Indian Acad. Sci. 1:115 (1934).
403. P. Debye, J. Phys. Chem. 51:18 (1947).
404. P. M. Doty, B. H. Zimm, and H. Mark, J. Chem. Phys. 12:144 (1944); 13:159 (1945).
405. T. Neugebauer, Ann. Physik 42:509 (1943).
406. B. H. Zimm, R. S. Stein, and P. Doty, Polymer Bull. 1:90 (1955).
407. Lord Rayleigh, Phil. Mag. 35:373 (1918).
408. H. J. Benoit, J. Polymer Sci. 11:507 (1953).
409. H. Benoit, A. M. Holtzer, and P. M. Doty, J. Phys. Chem. 58:635 (1954).
410. P. Horn, H. Benoit, and G. Oster, J. Chim. Phys. 48:530 (1951).
411. K. S. Shifrin, Scattering of Light in a Turbid Medium (Gostekhizdat, 1951).
412. R. C. C. Leite, R. S. Moore, P. S. Porto, and J. E. Ripper, Phys. Rev. Letters 14:7 (1965).
413. H. A. Stuart, Die Strukture des Freien Moleküls, in: Der Physik der Hochpolymeren, Vol. 1-2 (Springer-Verlag, Berlin, 1952).
414. E. D. Kunst, Dissertation, Groningen, 1950 (cited in [413]).
415. J. Hengstenberg, Makromol. Chem. 6:127 (1951).
416. J. W. Breitenbach and E. Duch, Monatsheft Chem. 82:628 (1951).
417. M. Halwer, G. C. Nutting, and B. A. Brice, J. Amer. Chem. Soc. 73:2786 (1951).
418. P. Doty and T. Edsall, Advances Protein Chem. 6:34 (1951).
419. H. Z. Blumer, Z. Phys. 32:119 (1925).
420. L. I. Mandel'shtam and M. A. Leontovich, Dokl. Akad. Nauk SSSR 3:111 (1936); L. I. Mandel'shtam, Collected Works, Vol. II (Izd. Akad. Nauk SSSR, 1947), p. 170.

421. L. I. Mandel'shtam and M. A. Leontovich, Zh. Eks. i Teor. Fiz. 7:438 (1937); L. I. Mandel'shtam, Collected Works, Vol. II (Izd. Akad. Nauk SSSR, 1947), p. 176.
422. J. J. Markham, R. T. Beyer, and R. B. Lindsay, Rev. Mod. Phys. 23:353 (1951).
423. I. G. Mikhailov and V. A. Solov'ev, Usp. Fiz. Nauk 50:2 (1953).
424. K. F. Herzfeld and T. A. Litovitz, Absorption and Dispersion of Ultrasonic Waves (Academic Press, New York, 1959).
425. R. Bär, Proc. Indian Acad. Sci. A8:289 (1938).
426. R. Krishnan, Proc. Indian Acad. Sci. A13:281 (1941).
427. G. K. Randall, Proc. Indian Acad. Sci. A16:369 (1942).
428. J. Kremen and J. Shapiro, J. Opt. Soc. Amer. 54:500 (1954).
429. P. Biquard, Compt. Rend. 197:309 (1933).
430. R. Bär, Helv. Phys. Acta 10:332 (1937).
431. R. T. Beyer and M. C. Smith, J. Acoust. Soc. Amer. 18:424 (1946).
432. B. G. Shpakovskii, Dokl. Akad. Nauk SSSR 18:177 (1938).
433. L. Zachoval, J. Phys. Radium 10:350 (1939).
434. J. Lamb and J. M. M. Pinkerton, Proc. Roy. Soc. (London) A199:114 (1949).
435. J. Lamb and D. H. A. Huddart, Trans. Faraday Soc. 46:540 (1950).
436. K. N. Baranskii, Dokl. Akad. Nauk SSSR 114:517 (1957); [translation: Soviet Physics – Doklady 2:237 (1958)]. J. Lamb, M. Redwood, and Z. Shteinshleifer, Phys. Rev. Letters 3:28 (1959).
437. G. R. Ringo, J. W. Fitzgerald, and B. G. Hurdle, Phys. Rev. 72:87 (1947).
438. R. Rao, Nature 139:885 (1937); Proc. Indian Acad. Sci. A8:163 (1935).
439. R. Rao and S. Samaya, Phys. Rev. 60:615 (1941).
440. V. L. Ginzburg, Dokl. Akad. Nauk SSSR 36:9 (1942).
441. C. S. Venkateswaran, Proc. Indian Acad. Sci. A15:371 (1942).
442. E. F. Gross and I. A. Khvostikov, Phys. Rev. 42:579 (1932).
443. I. A. Khvostikov, Z. Phys. Sowjetunion 6:343 (1934).
444. Kh. E. Sterin, Dokl. Akad. Nauk SSSR 62:219 (1948).
445. D. H. Rank, E. L. Shull, and D. W. E. Axford, Nature 164:67 (1949).

446. M. S. Tunin, Candidate's dissertation, Moscow Oblast Pedagogical Institute (MOPI), 1961.
447. M. I. Shakhparonov and M. S. Tunin, Hypersonic Properties of Liquids and the Structure of Molecules (MOPI, Moscow, 1962).
448. J. Lamb and J. Andreae, Nature 167: 898 (1951).
449. J. H. Andreae and J. Lamb, Proc. Phys. Soc. (London) B64: 1021 (1951).
450. R. A. Rapuano, Phys. Rev. 72: 78 (1947).
451. E. L. Heasell and J. Lamb, Proc. Phys. Soc. (London) B69: 869 (1956).
452. J. H. Andreae, Proc. Phys. Soc. (London) B70: 71 (1957).
453. J. H. Andreae, P. L. Ioce, and R. I. Oliver, Proc. Phys. Soc. (London) 75: 82 (1960).
454. L. L. Myasnikov and L. S. Freiman, Contemporary Problems of Physical Acoustics (GONTI, 1935).
455. H. O. Kneser, Ann. Physik 11: 761 (1931); 16: 337 (1933).
456. J. M. M. Pinkerton, Proc. Phys. Soc. (London) B62: 129 (1949).
457. H. O. Kneser, Ann. Physik 32: 277 (1938).
458. H. O. Kneser, Ann. Physik 43: 465 (1943).
459. K. F. Herzfeld, J. Acoust. Soc. Amer. 13: 33 (1941).
460. K. F. Herzfeld, R. U. Schwartz, and Z. I. Slawsky, J. Chem. Phys. 20: 1658 (1952).
461. J. H. Andreae and J. Lamb, Proc. Phys. Soc. (London) B69: 814 (1956).
462. M. S. deGroot and J. Lamb, Proc. Roy. Soc. (London) A242: 36 (1957).
463. J. Lamb, in: Symposium, Ultrasonic Investigations of the Properties of Liquids (Royal Flemish Acad. of Sciences, 1959).
464. O. Nomoto, J. Phys. Soc. Japan 12: 85 (1957).
465. T. Kishimoto and O. Nomoto, Proc. Phys. Soc. Japan 9: 620 (1954).
466. T. Kishimoto and O. Nomoto, Proc. Phys. Soc. Japan 9: 1021 (1954).
467. T. Kishimoto and O. Nomoto, Proc. Phys. Soc. Japan 10: 933 (1955).
468. R. Bass and J. Lamb, Proc. Roy. Soc. (London) A247: 168 (1958).

469. R. Bass and J. Lamb, Proc. Roy. Soc. (London) A243 : 94 (1957).
470. K. Krebs and J. Lamb, Proc. Roy. Soc. (London) A244 : 558 (1958).
471. J. H. Andreae and P. L. Toyce, Brit. J. Appl. Phys. 13 : 462 (1962).
472. R. E. Nettleton, Phys. Fluids 2 : 256 (1959).
473. J. Ziman, Electrons and Phonons (Oxford University Press, 1960).
474. F. A. Korolev, High Resolution Spectroscopy (Gostekhizdat, 1953).
475. E. F. Gross and A. A. Syromyatnikov, Izv. Akad. Nauk SSSR, Ser. Fiz. 5 : 144 (1941).
476. C. V. Venkateswaran, Current Sci. 20 :14 (1951).
477. N. S. Sen, Science and Culture 25 :203 (1959).
478. K. C. Medhi, Indian J. Phys. 33 : 135 (1959).
479. D. H. Rank, J. Opt. Soc. Amer. 47 : 103 (1957).
480. R. J. Strutt (Lord Rayleigh, the younger), Proc. Roy. Soc. A95 : 476 (1919).
481. C. V. Raman, J. Opt. Soc. Amer. 15 : 185 (1927).
482. R. S. Krishnan, Proc. Indian Acad. Sci. A3 : 211 (1936).
483. D. H. Rank and A. E. Douglas, J. Opt. Soc. Amer. 38:966 (1948).
484. D. I. Levin, Rayleigh Scattering in Glass and the Structure of Glass, in the collection: Composition of Glass (Izd. Akad. Nauk SSSR, 1955).
485. R. D. Maurer, J. Chem. Phys. 25 : 1206 (1956).
486. L. Prudhomme, Compt. Rend. 245 : 300 (1957).
487. V. V. Vladimirskii, Dokl. Akad. Nauk SSSR 36 : 251 (1942).
488. V. V. Vladimirskii, Rayleigh Scattering of Light, Candidate's dissertation, Moscow State Univ. (1940).
489. A. I. Kolyadin, Opt. i Spektr. 1 : 907 (1956).
490. N. A. Voishvillo, Opt. i Spektr. 3: 281 (1957).
491. H. Mueller, Proc. Roy. Soc. (London) A166 : 425 (1938).
492. T. S. Velichkina, Molecular Scattering of Light in Viscous Liquids and Amorphous Solids, Candidate's dissertation Fiz. Inst. Akad. Nauk SSSR, Moscow (1953).
493. R. B. Sosman, The Properties of Silica (The Chemical Catalog Co., New York, 1927).
494. R. S. Krishnan, R. V. G. S. Rao, and K. Vedam, Current Sci. 13 : 89 (1950).

495. K. Vedam, Phys. Rev. 78 : 472 (1950).
496. M. V. Vol'kenshtein, Dokl. Akad. Nauk SSSR 116 :207 (1957); [translation: Soviet Physics – Doklady 2 :430 (1958)].
497. M. V. Vol'kenshtein, Experimental and Theoretical Investigations in Physics, in: Collection of Papers in Memory of G. S. Landsberg (Izd. Akad. Nauk SSSR, 1959).
498. M. Kornfel'd, Elasticity and the Transparency of Liquids (Gostekhizdat, 1951).
499. C. V. Raman and B. V. R. Rao, Nature 139 : 58 (1937); 141 : 242 (1938).
500. G. H. Tammann and H. Hesse, Z. Anorg. Allg. Chem. 156 : 245 (1926).
501. M. A. Schulz, Compt. Rend. 238 :1014 (1954).
502. D. Sette and T. A. Litovitz, J. Chem. Phys. 21 : 17 (1953).
503. R. Piccerelli and T. A. Litovitz, J. Acoust. Soc. Amer. 29 : 1009 (1957).
504. Yu. Zhivlyuk, Some Problems on Relaxation Theory in Liquids. Diploma research, Fiz.-Tekhn. Inst. (1959).
505. V. S. Starunov, Opt. i Spektr. 18(2) (1965).
506. V. V. Vladimirskii, Zh. Eks. i Teor. Fiz. 12 : 199 (1942)..
507. H. A. Kramers, Physica 7 : 284 (1940).
508. R. A. Sack, Proc. Phys. Soc. (London) B70 : 414 (1957).
509. S. Bhagavantam and A. V. Rao, Indian J. Phys. 8 : 97 (1933).
510. I. L. Fabelinskii, Dokl. Akad. Nauk SSSR 57:341 (1947).
511. S. Bhagavantam, Proc. Indian Acad. Sci. A2 : 63 (1935).
512. E. Gross and M. Vuks, J. Phys. Radium 6: 457 (1935).
513. E. Gross and M. Vuks, J. Phys. Radium 7 : 113 (1936).
514. E. Gross and M. Vuks, Acta Physicochimica 9 : 209 (1938).
515. E.F. Gross and Sh. Raskin, Izv. Akad. Nauk SSSR, Ser. Fiz. 9 : 184 (1945).
516. C. S. Venkateswaran, Proc. Indian Acad. Sci. A8 : 448 (1938).
517. E. F. Gross, Izv. Akad. Nauk SSSR, Ser. Fiz. 5 : 19 (1941).
518. J. Weiler, Z. Phys. 68 : 782 (1931).
519. K. Bapagya, Proc. Indian Acad. Sci. A9 :404 (1939); A10 : 253 (1940).
520. J. Broudmüller, Z. Phys. 140 : 75 (1955).
521. K. Venkateswarlu and G. Thyagarajan, Z. Phys. 154 : 81 (1959).
522. S. C. Sirkar, S. B. Roy, and D. K. Ghosh, Indian J. Phys. 35 : 377 (1961).

523. M. F. Crawford, H. L. Welsh, and J. H. Harrold, Canad. J. Phys. 30 : 81 (1952).
524. A. Langseth, Z. Phys. 77 : 60 (1932).
525. W. Hanle and O. Mäerks, Z. Phys. 114 : 407 (1939).
526. V. A. Marinin, Zh. Eks. i Teor. Fiz. 22 : 34 (1952).
527. G. S. Simkin, V. P. Naberezhnykh, and N. V. Lukin, Izmeritel. Tekhn. 8 : 41 (1960).
528. B. D. Saxena, Proc. Indian Acad. Sci. A8 : 460 (1938).
529. Sunanda Bai, Proc. Indian Acad. Sci. A13 : 439 (1941).
530. M. Wien, Phys. Z. 37 : 869 (1936).
531. D. I. Mash, L. S. Mayants, and I. L. Fabelinskii, Zh. Tekhn. Fiz. 19 : 1192 (1949).
532. S. I. Vavilov, Izv. Akad. Nauk SSSR, Ser. Fiz. 2 : 395 (1937).
533. C. J. F. Böttcher, Theory of Electric Polarization (Elsevier, Amsterdam, 1952).
534. C. P. Smyth, Dielectric Behavior and Structure (McGraw-Hill, New York, 1955).
535. V. Braun, Dielectrics (Russian translation, IL, 1962).
536. H. Fröhlich, Theory of Dielectrics: Dielectric Constant and Dielectric Loss (Oxford University Press, 1958).
537. I. G. Mikhailov and S. B. Gurevich, Zh. Eks. i Teor. Fiz. 19 : 193 (1949).
538. P. P. Kobeko, Amorphous Matter (Izd. Akad. Nauk SSSR, 1952).
538a. G. Tamman, The Glasslike State (Gostekhizdat, 1935).
539. V. K. Ablekov and I. L. Fabelinskii, Dokl. Akad. Nauk SSSR 125 : 297 (1959).
540. I. L. Fabelinskii, Zh. Eks. i Teor. Fiz. 45 : 822 (1963); [translation: Soviet Physics – JETP 18 : 564 (1963)].
541. S. M. Rytov, Tr. Fiz. Inst. Akad. Nauk SSSR 2 : 41 (1940).
542. D. F. Holshouser, H. von Foester, and G. L. Clark, J. Opt. Soc. Amer. 51 : 1360 (1961).
543. P. Connes, Puong Hong Tuan, and J. Pinard, J. Phys. Radium 23 : 173 (1962).
544. P. G. Tager, The Kerr Cell (Iskusstvo Press, 1937).
545. P. Pershan and N. Bloembergen, in: Advances in Quantum Electronics (Academic Press, New York, 1961).
546. I. P. Kaminow, Appl. Phys. Letters 2 : 41 (1961).
547. I. P. Kaminow, Proc. IEEE 51 : 132, 137 (1963).
548. C. J. Peters, Proc. IEEE 51 : 147 (1963).

549. G. S. Landsberg and K. S. Vul'fson, Z. Phys. 58 : 95 (1929); G. S. Landsberg, Collected Works (Izd. Akad. Nauk SSSR, 1958).
550. G. S. Landsberg and S. L. Mandel'shtam, Z. Phys. 73 : 502 (1931); G. S. Landsberg, Collected Works (Izd. Akad. Nauk SSSR, 1958).
551. G. P. Motulevich, I. L. Fabelinskii, and I. A. Yakovlev, Experimental and Theoretical Investigations in Physics, in: Collection of Papers in Memory of G. S. Landsberg (Izd. Akad. Nauk SSSR, 1959).
552. A. V. Shubnikov, Quartz and Its Uses (Izd. Akad. Nauk SSSR, 1940).
553. M. Born, Proc. Math. Phys. Soc. Egypt 3 : 35 (1947).
554. R. Fürth, Proc. Phys. Soc. (London) B69 :255 (1956).
555. S. P. F. Humphreys-Owen, Proc. Phys. Soc. (London) B69 : 350 (1956).
556. S. P. F. Humphreys-Owen, Proc. Phys. Soc. (London) B68 : 325 (1955).
557. L. Tourel and J. Chapelle, Compt. Rend. 240: 2507 (1955).
558. J. Chapelle and L. Tourel, J. Chim. Phys. 52 : 307 (1955).
559. L. Tourel, Compt. Rend. 243 :1755 (1956).
560. L. Tourel, Compt. Rend. 246 : 932 (1958).
561. L. Tourel and G. Fontaine, Compt. Rend. 252 : 1934 (1961).
562. L. Tourel and B. Baudelt, Compt. Rend. 256 : 914 (1963).
563. L. Tourel and S. P. F. Humphreys-Owen, Proc. Phys. Soc. (London) 75 : 473 (1960).
564. O. Theimer and C. A. Plint, Annals Phys. 3 : 408 (1958).
565. R. S. Krishnan, Proc. Indian Acad. Sci. A26 :339 (1947).
566. K. S. Vul'fson, Dokl. Akad. Nauk SSSR 28 : 794 (1940).
567. I. A. Yakovlev, Study of Second-Order Phase Transitions in Solids, Doctoral Dissertation, Moscow State Univ. (1957).
568. I. A. Yakovlev and T. S. Velichkina, Usp. Fiz. Nauk 63 : 411 (1957).
569. T. S. Velichkina, L. F. Mikheeva, and I. A. Yakovlev, in: 10th Conference on Spectroscopy, Vol. I (Izd. L'vov Univ., 1957).
570. L. I. Mandel'shtam, Dokl. Akad. Nauk SSSR 2 : 219 (1934); Collected Works, Vol. II (Izd. Akad. Nauk SSSR, 1947).
571. G. S. Landsberg and A. A. Shubin, Zh. Eks. i Teor. Fiz. 9 : 1309 (1939); G. S. Landsberg, Collected Works (Izd. Akad. Nauk SSSR, 1958).

572. M. A. Leontovich, Dokl. Akad. Nauk SSSR 1: 93 (1935).
573. M. A. Leontovich, Zh. Eks. i Teor. Fiz. 9: 1314 (1939).
574. V. L. Kirpichev, Complete Works, Vol. I (1917), p. 171.
575. J. Lamb, M. Redwood, and Z. Shteinshleifer, Phys. Rev. Letters 3: 28 (1959).
576. H. E. Bömmel and K. Dransfeld, Phys. Rev. 117: 1245 (1960).
577. W. P. Mason, Physical Acoustics and the Properties of Solids (Van Nostrand, New York, 1958).
578. V. V. Vladimirskii, Dokl. Akad. Nauk SSSR 38: 229 (1942).
579. A. I. Akhiezer, Zh. Eks. i Teor. Fiz. 8: 1318 (1938)
580. I. Ya. Pomeranchuk, Zh. Eks. i Teor. Fiz. 11: 455 (1941).
581. A. Massoulier, Compt. Rend. 255: 1093 (1962).
582. G. V. Rozenberg, Optics of Thin Layers (Fizmatgiz, 1958).
583. L. Holland, Vacuum Deposition of Thin Films (Chapman and Hall, 1958).
584. O. S. Heavens, Optical Properties of Thin Solid Films (Butterworths, London, 1955).
585. S. Metfessel', Thin Films and Their Preparation (Gosenergoizdat, 1963).
586. H. Kuhn and B. A. Wilson, Proc. Phys. Soc. (London) B63: 745 (1950).
587. J. Burridge, H. Kuhn, and A. Pary, Proc. Phys. Soc. (London) B66: 963 (1953).
588. P. Giacomo, Rev. Optique 35: 317 (1956); J. Ring and L. Wilcock, Nature 171: 648 (1933); 173: 994 (1956).
589. S. Penselin and A. Studel, Z. Phys. 142: 21 (1955).
590. R. Chabbal, J. Rech. Cent. Nat. Rech. Sci. Lab. Bellevue (Paris) 1953(24): 138.
591. G. Hansen, Handbuch Phys. Opt., Vol. II, Leipzig (1927).
592. N. J. Bridge and A. D. Buckingham, J. Chem. Phys. 40: 2733 (1964).
593. H. Z. Cummins, N. Knable, and Y. Yeh, Phys. Rev. Letters 12: 150 (1964).
594. R. C. C. Leite, R. S. Moore, and P. S. Porto, J. Chem. Phys. 40: 3741 (1964).
595. R. Y. Chiao, C. H. Townes, and B. P. Stoicheff, Phys. Rev. Letters 12: 592 (1964).
596. J. L. Stewart and E. S. Stewart, J. Acoust. Soc. Amer. 35: 913 (1963).
597. E. S. Stewart and J. L. Stewart, J. Acoust. Soc. Amer. 35: 975 (1963).

598. G. B. Benedek, J. B. Lastovka, K. Fritsch, and T. Greytak, J. Opt. Soc. Amer. 54 : 1284 (1964).
599. E. Garmire and C. H. Townes, Appl. Phys. Letters 5 : 84 (1964).
600. V. E. Eskin, Usp. Fiz. Nauk 82 : 649 (1964).
601. R. G. Brewer and K. E. Rieckhoff, Phys. Rev. Letters 13 : 334 (1964).
602. R. Y. Chiao and B. P. Stoicheff, J. Opt. Soc. Amer. 54 : 1286 (1964).
603. K. D. Mielenz, R. B. Stephens, K. F. Nefflen, and A. Fabry, J. Res. Nat. Bur. Standards 68C : 1 (1964).
604. R. Pecora, J. Chem. Phys. 40 : 1604 (1964).
605. E. S. Stewart and J. L. Stewart, Phys. Rev. Letters 13 : 437 (1964).
606. M. Sh. Giterman and V. M. Kontorovich, Zh. Eks. i Teor. Fiz. 47 : 2134 (1964); [translation: Soviet Physics — JETP 20 : 1433 (1964)].
607. S. A. Akhmanov and R. V. Khokhlov, in: Problems of Nonlinear Optics, 1962-63 (Akad. Nauk SSSR, Inst. Nauch. Inf., Moscow, 1964).
608. N. Bloembergen, Nonlinear Optics (W. A. Benjamin, New York, 1965).
609. R. G. Brewer, Appl. Phys. Letters 5 : 127 (1964).
610. H. Takuma and D. A. Jennings, Appl. Phys. Letters 5 : 239 (1964).
611. D. A. Jennings and H. Takuma, Appl. Phys. Letters 5 : 241 (1964).
612. H. B. Huntington, Solid State Physics 7: 213 (1958).
613. A. Shavlov, Usp. Fiz. Nauk 84: 303 (1964).
614. I. L. Fabelinskii, Usp. Fiz. Nauk 87 : 9 (1965); [translation: Soviet Physics — Uspekhi 8 : 637 (1966)].
615. D. I. Mash, V. V. Morozov, V. S. Starunov, and I. L. Fabelinskii, Zh. Eks. i Teor. Fiz. Pis'ma 2 : 349 (1965).
616. S. A. Akhmanov and Chin Dong-A, Usp. Fiz. Nauk (in press).
617. Chin Dong-A, The Theory of Stimulated MB Scattering, Diploma research, Moscow State Univ. (1964).
618. E. Garmire, F. Pendarese, and C. H. Townes, Phys. Rev. Letters 11 : 160 (1963).
619. N. M. Kroll, Phys. Rev. 127 : 1207 (1962).
620. L. I. Mandel'shtam and N. D. Papaleski, Zh. Eks. i Tekh. Fiz. 4 : 117 (1934); L. I. Mandel'shtam, Collected Works, Vol. 2 (1947), p. 130.

621. A. A. Andronov, A. A. Vitt, and S. E. Khaikin, Theory of Oscillations, 2nd ed. (Fizmatgiz, Moscow, 1959).
622. A. Korpel, R. Adler, and B. Alpiner, Appl. Phys. Letters 5: 86 (1964).
623. E. V. Pogorelova and R. V. Khokhlov, Vestn. Mosk. Gos. Univ., Ser. III, 1962(5) : 62.
624. A. Jurkys, B. Sca, and P. N. Robson, Proc. IEE 107B : 119 (1960).
625. P. Tienkin, J. Appl. Phys. 29 : 95 (1958).
626. J. M. Manley and H. E. Rowe, Proc. IRE 44:904 (1956).
627. R. M. White, J. Appl. Phys. 34 : 2123 (1963).
628. E. F. Carome, N. A. Clark, and C. E. Moller, Appl. Phys. Letters 4 : 95 (1964).
629. C. R. Giuliano, Appl. Phys. Letters 5 : 137 (1964).
630. D. I. Mash, V. V. Morozov, V. S. Starunov, E. V. Tiganov, and I. L. Fabelinskii, Zh. Eks. i Teor. Fiz. Pis'ma 2 : 246 (1965).
631. D. I. Mash, V. V. Morozov, V. S. Starunov, and I.L. Fabelinskii, Zh. Eks. i Teor. Fiz. Pis'ma 2 : 25 (1965).
632. M. A. Isakovich and I. A. Chaban, Dokl. Akad. Nauk SSSR 65(2) (1965).
633. M. A. Isakovich and I. A. Chaban, Zh. Eks. i Teor. Fiz. 50:1343 (1966); [translation, Soviet Phys. JETP 23:893 (1966)].
634. E. V. Tiganov, Pribory i Tekhn. Eksp. (in press).
635. M. A. Bionti, Rev. Sci. Instr. 27: 36 (1956).
636. G. Benedek and T. Greytak, Proc. IEEE 53 : 1623 (1965).
637. H. Z. Cummins and R. W. Gammon, J. Chem. Phys. 44 : 2785 (1966).
638. G. R. Hanes, R. Turner, and J. E. Piercy, J. Acoust. Soc. Amer. 38 : 1057 (1965).
639. S. S. Alpert, R. Novick, L. Seigal, and Y. Yeh (Preprint).
640. N. C. Ford, Jr. and G. B. Benedek, Phys. Rev. Letters 15 : 649 (1965).
641. S. L. Shapiro, M. McClintock, D. A. Jennings, and R. L. Barger, IEEE J. Quantum Electronics QE-2 : 89 (1966).
642. N.A. Clark, C. E. Moeller, J. A. Bucaro, and E. F. Carome, J. Chem. Phys. 44 : 2528 (1966).
643. E. V. Tiganov, Zh. Eks. i Teor. Fiz. Pis'ma 4:385 (1966).
644. V.S. Starunov, E. V. Tiganov, and I. L. Fabelinskii, Zh. Eks. i Teor. Fiz. Pis'ma 4:262 (1966).

645. I. B. Lezhnev, Dissertation, Fiz.-Tekhn. Inst. Akad. Nauk, TurkmenSSR, Ashkhabad (1963).
646. H. Z. Cummins and R. W. Gammon, Appl. Phys. Letters 6 : 171 (1965).
647. D. H. Rank, E. M. Kiess, U. Fink, and T. A. Wiggins, J. Opt. Soc. Amer. 55 : 925 (1965).
648. B. S. Guberman and V. V. Morozov, Opt. i Spektr. (in press).
649. R. D. Mountain, Rev. Mod. Phys. 38 : 205 (1966).
650. D. H. Rank, E. M. Kiess, and U. Fink, J. Opt. Soc. Amer. 56 : 163 (1966).
651. E. E. Hagenlocker and M. G. Rado, Appl. Phys. Letters 7 : 236 (1965).
652. D. H. Rank, T. A. Wiggins, R. V. Wick, and D. P. Eastman, J. Opt. Soc. Amer. 56 : 174 (1966).
653. S. V. Krivokhizha, D. I. Mash, V. V. Morozov, V. S. Starunov, and I. L. Fabelinskii, Zh. Eks. i Teor. Fiz. Pis'ma 3 : 245 (1966).
654. Laser Research. Solid State Research, Lincoln Laboratory, MIT 1965(1) : 29-37.
655. H. Hsu and W. Kovage, Physics Letters 15 : 207 (1965).
656. T. P. Belikova and E. A. Sviridenkov, Zh. Eks. i Teor. Fiz. Pis'ma 1 : 171 (1965).
657. D. W. Harper, Brit. J. Appl. Phys. 16 :751 (1965).
658. R. Nava, R. Arzt, I. Ciccarello, and R. Dransfeld, Phys. Rev. 134A : 581 (1964).
659. E. M. Ganapol'skii and A. N. Chenets, Zh. Eks. i Teor. Fiz. 51 : 383 (1966); [translation: Soviet Physics – JETP 24 : 255 (1967)].
660. V. I. Bespalov and V. I. Talanov, Zh. Eks. i Teor. Fiz. Pis'ma 3 : 47 (1966).
661. Yu. P. Raizer, Zh. Eks. i Teor. Fiz. Pis'ma 4 : 3 (1966).
662. L. M. Belyaev, V. V. Nabatov, Yu. V. Pisarevskii, and Yu. V. Shaldin, Kristallogr. 10 : 767 (1965); [translation: Soviet Physics – Crystallography 10 :647 (1966)].
663. L. K. Zarembo and V. A. Krasil'nikov, Usp. Fiz. Nauk 68 : 687 (1959); [translation: Soviet Physics – Uspekhi 2 : 580 (1960)].
664. A. L. Polyakova, Zh. Eks. i Teor. Fiz. Pis'ma 4 : 132 (1966).
665. A. A. Chaban, Zh. Eks. i Teor. Fiz. Pis'ma 3 : 79 (1966).

666. F. De Martini, Appl. Phys. Letters 9 : 31 (1966).
667. E. K. Zavoisky and S. D. Fanchenko, Appl. Optics 4 : 1155 (1965).
668. R. G. Brewer, Phys. Rev. 140A : 800 (1965).
669. V. S. Starunov, Dissertation, Fiz. Inst. im P. N. Lebedev, Akad. Nauk SSSR, Moscow (1965).
670. T. J. Greytak and G. B. Benedek, Phys. Rev. Letters 1 : 179 (1966).
671. C. R. Giuliano, Appl. Phys. Letters 7 : 279 (1965).
672. T. A. Wiggins, R. V. Wick, D. H. Rank, and A. H. Guenther, Appl. Optics 4 : 1203 (1965).
673. R. Brewer, Appl. Phys. Letters 9 : 51 (1966).
674. R. Brewer, Appl. Phys. Letters 8 : 165 (1965).
675. S. M. Shapiro, R. W. Gammon, and H. Cummins, Preprint (1966).
676. R. W. Mink, E. E. Hagenlocker, and W. G. Rado, Phys. Rev. Letters 17 : 229 (1966).
677. O. A. Shustin, Zh. Eks. i Teor. Fiz. Pis'ma 3 : 491 (1966).
678. R. W. Gammon and H. Z. Cummins, Phys. Rev. Letters 17 : 193 (1966).
679. L. D. Landau and I. M. Khalatnikov, Dokl. Akad. Nauk SSSR 96 : 469 (1954).
680. E. J. O'Brien and T. A. Litovitz, J. Appl. Phys. 35: 180 (1964).
681. Y. R. Shen and N. Bloembergen, Phys. Rev. 137A : 1787 (1965).
682. A. Yariv, IEEE J. Quantum Electronics QE-1 : 28 (1965).
683. G. F. Quate, C. D. Wilkinson, and D. K. Winslow, Proc. IEEE 53 : 1800 (1965).
684. D. A. Sealer and H. Hsu, IEEE J. Quantum Electronics QE-1 : 116 (1965).
685. N. M. Kroll, J. Appl. Phys. 36 : 34 (1965).
686. C. L. Tang, J. Appl. Phys. 37 : 2945 (1966).
687. D. E. Caddes, C. F. Quate, and C. D. W. Wilkinson, Appl. Phys. Letters 8 : 309 (1966).
688. N. Bloembergen and P. Lallemand, Phys. Rev. Letters 16 : 81 (1965).
689. P. Lallemand, Appl. Phys. Letters 8 : 276 (1966).
690. A. G. Askar'yan, Zh. Eks. i Teor. Fiz. 42 : 1567 (1962); [translation: Soviet Physics — JETP 15 : 1088 (1962)].

691. V. I. Talanov, Izv. Vysshikh. Uchebn. Zavedenii Radiofiz. 7: 564 (1964).
692. R. Y. Chiao, E. Garmire, and C. H. Townes, Phys. Rev. Letters 13: 479 (1964).
693. P. L. Kellog, Phys. Rev. Letters 15: 1005 (1965).
694. Y. R. Shen, Physics Letters 20: 378 (1966).
695. Ya. B. Zel'dovich and Yu. P. Raizer, Zh. Eks. i Teor. Fiz. Pis'ma 3: 137 (1966).
696. G. Hauchecorne and G. Mayer, Compt. Rend. 261: 4014 (1965).
697. P. Lallemand and N. Bloembergen, Phys. Rev. Letters 15: 1010 (1965).
698. P. Whiteman and G. W. Wilson, Nature 208: 66 (1965).
699. B. M. Ashkinadze, V. I. Vladimirov, V. A. Likhachev, S.M. Ryvkin, V. M. Salimanov, and I. D. Yaroshetskii, Zh. Eks. i Teor. Fiz. 50: 1187 (1966).
700. E. H. Jacobsen, Experiments with Phonons at Microwave Frequencies, in: Advances in Quantum Electronics, Columbia University Press, New York, p. 469; J. Ilukor and E. H. Jacobsen, Science 153: 1113 (1966).
701. W. Eisenmenger, H. Kinder, and K. Lassmann, Acustica 16: 1 (1965-66).

Index

E

T

U

MIX
Papier aus verantwortungsvollen Quellen
Paper from responsible sources
FSC® C105338

If you have any concerns about our products,
you can contact us on
ProductSafety@springernature.com

In case Publisher is established outside the EU,
the EU authorized representative is:
Springer Nature Customer Service Center GmbH
Europaplatz 3, 69115 Heidelberg, Germany

Printed by Libri Plureos GmbH
in Hamburg, Germany